MOLECULAR LIGHT SCATTERING
AND OPTICAL ACTIVITY

Using classical and quantum methods with a strong emphasis on symmetry principles, this book develops the theory of a variety of optical activity and related phenomena from the perspective of molecular scattering of polarized light. In addition to the traditional topic of optical rotation and circular dichroism in the visible and ultraviolet region associated with electronic transitions, the newer topic of optical activity associated with vibrational transitions, which may be studied using both infrared and Raman techniques, is also treated. Ranging from the physics of elementary particles to the structure of viruses, the subject matter of the book reflects the importance of optical activity and chirality in much of modern science and will be of interest to a wide range of physical and life scientists.

LAURENCE BARRON worked with Professor Peter Atkins for his doctorate in theoretical chemistry from Oxford University, followed by postdoctoral work with Professor David Buckingham at Cambridge University. He was appointed to a faculty position at Glasgow University in 1975, where he is currently the Gardiner Professor of Chemistry. His research interests are in the electric, magnetic and optical properties of molecules, especially chiral phenomena including Raman optical activity which he pioneered and is developing as a novel probe of the structure and behaviour of proteins, nucleic acids and viruses.

MOLECULAR LIGHT SCATTERING AND OPTICAL ACTIVITY

Second edition, revised and enlarged

LAURENCE D. BARRON, F.R.S., F.R.S.E.

Gardiner Professor of Chemistry, University of Glasgow

CAMBRIDGE
UNIVERSITY PRESS

CAMBRIDGE UNIVERSITY PRESS
Cambridge, New York, Melbourne, Madrid, Cape Town, Singapore,
São Paulo, Delhi, Dubai, Tokyo

Cambridge University Press
The Edinburgh Building, Cambridge CB2 8RU, UK

Published in the United States of America by Cambridge University Press, New York

www.cambridge.org
Information on this title: www.cambridge.org/9780521121378

First published 1983
Second edition published 2004
This digitally printed version (with corrections) 2009

A catalogue record for this publication is available from the British Library

Library of Congress Cataloguing in Publication data

Barron, L. D.
Molecular light scattering and optical activity / Laurence D. Barron – 2nd edn., rev. and enl.
p.   cm.
Includes bibliographical references and index.
ISBN 0 521 81341 7
1. Optical rotatory dispersion.   2. Circular dichroism.   I. Title.
QD473.B37 2004
541.7–dc22   2004043552

ISBN 978-0-521-81341-9 Hardback
ISBN 978-0-521-12137-8 Paperback

For Sharon

There are some enterprises in which a careful disorderliness is the true method.

Herman Melville, Moby Dick

Contents

Preface to the first edition

Scientists have been fascinated by optical activity ever since its discovery in the early years of the last century, and have been led to make major discoveries in physics, chemistry and biology while trying to grapple with its subtleties. We can think of Fresnel's work on classical optics, Pasteur's discovery of enantiomeric pairs of optically active molecules which took him into biochemistry and then medicine, and Faraday's conclusive demonstration of the intimate connection between electromagnetism and light through his discovery of magnetic optical activity. And of course the whole subject of stereochemistry, or chemistry in space, has its roots in the realization by Fresnel and Pasteur that the molecules which exhibit optical rotation must have an essentially helical structure, so from early on molecules were being thought about in three dimensions.

A system is called 'optically active' if it has the power to rotate the plane of polarization of a linearly polarized light beam, but in fact optical rotation is just one of a number of optical activity phenomena which can all be reduced to the common origin of a different response to right- and left-circularly polarized light. Substances that are optically active in the absence of external influences are said to exhibit 'natural' optical activity. Otherwise, all substances in magnetic fields are optically active, and electric fields can sometimes induce optical activity in special situations.

It might be thought that a subject originating at the start of the nineteenth century would be virtually exhausted by now, but nothing could be further from the truth. The recent dramatic developments in optical and electronic technology have led to large increase in the sensitivity of conventional optical activity measurements, and have enabled completely new optical activity phenomena to be observed and applied. Traditionally, optical activity has been associated almost exclusively with electronic transitions; but one particularly significant advance over the last decade has been the extension of natural optical activity measurements into the vibrational spectrum using both infrared and Raman techniques. It is now becoming clear

xi

that vibrational optical activity makes possible a whole new world of fundamental studies and practical applications quite undreamt of in the realm of conventional electronic optical activity.

Optical activity measurements are expected to become increasingly important in chemistry and biochemistry. This is because 'conventional' methods have now laid the groundwork for the determination of gross molecular structure, and emphasis is turning more and more towards the determination of the precise three-dimensional structures of molecules in various environments: in biochemistry it is of course the fine detail in three dimensions that is largely responsible for biological function. Whereas X-ray crystallography, for example, provides such information completely, it is restricted to studies of molecules in crystals in which the three dimensional structures are not necessarily the same as in the environment of interest. Natural optical activity measurements are a uniquely sensitive probe of molecular stereochemistry, both conformation and absolute configuration, but unlike X-ray methods can be applied to liquid and solution samples, and even to biological molecules *in vivo*. The significance of magnetic optical activity measurements, on the other hand, can probably be summarized best by saying that they inject additional structure into atomic and molecular spectra, enabling more information to be extracted.

Following the recent triumph of theoretical physics in unifying the weak and electromagnetic forces into a single 'electroweak' force, the world of physics has also started to look at optical activity afresh. Since weak and electromagnetic forces have turned out to be different aspects of the same, more fundamental, unified force, the absolute parity violation associated with the weak force is now thought to infiltrate to a tiny extent into all electromagnetic phenomena, and this can be studied in the realm of atoms and molecules by means of delicate optical activity experiments. So just as optical activity acted as a catalyst in the progress of science in the last century, in our own time it appears set to contribute to further fundamental advances. One could say that optical activity provides a peephole into the fabric of the universe!

In order to deal with the optical properties of optically active substances in a unified fashion, and to understand the relationship between the conventional 'birefringence' phenomena of optical rotation and circular dichroism and the newer 'scattering' phenomena of Rayleigh and Raman optical activity, the theory is developed in this book from the viewpoint of the scattering of polarized light by molecules. In so doing, a general theory of molecular optics is obtained and is applied to the basic phenomena of refraction, birefringence and Rayleigh and Raman scattering. Optical activity experiments are then regarded as applications of these phenomena in ways that probe the asymmetry in the response of the optically active system to right- and left-circularly polarized light. As well as using the results of the general theory to obtain expressions for the observables in each particular optical activity phenomenon, where possible the expressions are also derived separately in as simple a fashion as possible for the benefit of the reader who is interested in one topic in isolation.

There are several important topics within the general area of optical activity that I have either omitted or mentioned only briefly, mainly because they are outwith the theme of molecular scattering of polarized light, and also because of my lack of familiarity with them. These include circular polarization of luminescence, and chiral discrimination. I have also not treated helical polymers: to do justice to this very important topic would divert us too far from the fundamental theory. Where I have discussed specific atomic or molecular systems, this has been to illuminate the theory rather than to give an exhaustive explanation of the optical activity of any particular system. For a much broader view of *natural* optical activity, including experimental aspects and a detailed account of a number of specific systems, the reader is referred to S. F. Mason's new book 'Molecular Optical Activity and the Chiral Discriminations' (Mason, 1982).

So this is not a comprehensive treatise on optical activity. Rather, it is a personal view of the theory of optical activity and related polarized light scattering effects that reflects my own research interests over the last 14 years or so. During the earlier part of this period I was fortunate to work with, and learn from, two outstanding physical chemists: Dr P. W. Atkins in Oxford and Professor A. D. Buckingham in Cambridge; and their influence extends throughout the book.

I wish to thank the many colleagues who have helped to clarify much of the material in this book through discussion and correspondence over the years. I am particularly grateful to Dr J. Vrbancich for working through the entire manuscript and pointing out many errors and obscure passages.

Glasgow
May 1982

Preface to the second edition

Interest in optical activity has burgeoned since the first edition of this book was published in 1982. The book anticipated a number of new developments and helped to fuel this interest, but has become increasingly hard to find since going out of print in 1990. Numerous requests about where a copy might be found, often accompanied by 'our library copy has been stolen' and the suggestion that a second edition would be well-received, have encouraged me to prepare this new edition. The book has been considerably revised and enlarged, but the general plan and style remain as before.

Traditionally, the field of optical activity and chirality has been largely the preserve of synthetic and structural chemistry due to the inherent chirality of many molecules, especially natural products. It has also been important in biomolecular science since proteins, nucleic acids and oligosaccharides are constructed from chiral molecular building blocks, namely the L-amino acids and the D-sugars, and the chemistry of life is exquisitely stereospecific. The field is becoming increasingly important in these traditional areas. For example, chirality and enantioselective chemistry are now central to the pharmaceutical industry since many drugs are chiral and it has been recognized that they should be manufactured as single enantiomers; and chiroptical spectroscopies are used ever more widely for studying the solution structure and behaviour of biomolecules, a subject at the forefront of biomedical science. But in recent years optical activity and chirality have also been embraced enthusiastically by several other disciplines. Physicists, for example, are becoming increasingly interested in the field due to the subtle new optical phenomena, linear and nonlinear, supported by chiral fluids, crystals and surfaces. Furthermore, since homochiral chemistry is the signature of life, and considerable effort is being devoted to searches for evidence of life, or at least of prebiotic chemistry, elsewhere in the cosmos including interstellar dust clouds, cometary material and the surfaces of extrasolar planets, chirality has captured the interest of some astrophysicists and space scientists. It has even caught the attention of applied mathematicians and electrical engineers on account of the novel and potentially useful electromagnetic properties of chiral media. •

Although containing a significant amount of new material the second edition, like the first, is not a comprehensive treatise on optical activity and remains a personal view of the theory of optical activity and related polarized light scattering effects that reflects my own research interests. The material on symmetry and chirality has been expanded to include motion-dependent enantiomorphism and the associated concepts of 'true' and 'false' chirality, and to expose productive analogies between the physics of chiral molecules and that of elementary particles which are further emphasized by considering the violation of parity and time reversal invariance. Another significant addition is a detailed treatment of *magnetochiral* phenomena, which are generated by a subtle interplay of chirality and magnetism and which were unknown at the time of writing the first edition. Since vibrational optical activity has now 'come of age' thanks to new developments in instrumentation and theory in the 1980s and 1990s, the treatment of this topic has been considerably revised and expanded. Of particular importance is a new treatment of vibrational circular dichroism in Chapter 7; serious problems in the quantum chemical theory, now resolved, were unsolved at the time of writing the first edition, which contains an error in the way in which the Born–Oppenheimer approximation was applied. The revised material on natural Raman optical activity now reflects the fact that it has become an incisive chiroptical technique giving information on a vast range of chiral molecular structures, from the smallest such as CHFClBr to the largest such as intact viruses. New developments in magnetic Raman optical activity are also described which illustrate how it may be used as a novel probe of magnetic structure.

Another subject to come of age in recent years is nonlinear optical activity, manifest as a host of different optical phenomena generated by intense laser beams incident on both bulk and surface chiral samples. However the subject has become too large and important, and too specialized with respect to its theoretical development, to do it justice within this volume which is therefore confined to linear optical activity phenomena.

I have benefited greatly from interactions with many colleagues who have helped directly and indirectly with the identification and correction of errors in the first edition, and with the preparation of new material. I am especially grateful in this respect to E. W. Blanch, I. H. McColl, A. D. Buckingham, J. H. Cloete, R. N. Compton, J. D. Dunitz, K.-H. Ernst, R. A. Harris, L. Hecht, W. Hug, T. A. Keiderling, L. A.

Nafie, R. D. Peacock, P. L. Polavarapu, M. Quack, R. E. Raab, G. L. J. A. Rikken, A. Rizzo, P. J. Stephens, G. Wagnière and N. I. Zheludev.

I hope that workers in many different areas of pure and applied science will find something of value in this second edition.

Glasgow
2004

Symbols

The symbols below are grouped according to context. In some cases the same symbol has more than one meaning, but it is usually clear from the context which meaning is to be taken. A tilde above a symbol, for example $\tilde{A}$, denotes a complex quantity, the complex conjugate being denoted by an asterisk, for example $\tilde{A}^*$. A dot over a symbol, for example $\dot{A}$, denotes the time derivative of the corresponding quantity. An asterisk is also used to denote an antiparticle or an antiatom, for example ν^* and Co^*.

Historical review

α	optical rotation angle
$[\alpha]$	specific rotation
ψ	ellipticity
$[\psi]$	specific ellipticity
ϵ	decadic molar extinction coefficient
g	dissymmetry factor
V	Verdet constant
Δ	dimensionless Rayleigh or Raman circular intensity difference
R,S	absolute configuration in the Cahn–Ingold–Prelog notation. (R)-(+) etc. specifies the sense of optical rotation associated with a particular absolute configuration
P,M	helicity designation of the absolute configuration of helical molecules

Electric and magnetic fields and electromagnetic waves

λ	wavelength
c	velocity of light
v	wave velocity
ω	angular frequency, magnitude $2\pi v/\lambda$ ($2\pi c/\lambda$ in free space)
n	refractive index, magnitude c/v
n'	absorption index

$\tilde{n}$	complex refractive index $n + in'$
n	propagation vector, magnitude n
κ	wavevector, magnitude ω/v (may be written $\omega\mathbf{n}/c$)
E	electric field vector in free space
B	magnetic field vector in free space
D	electric field vector within a medium
H	magnetic field vector within a medium
ρ	electric charge density
J	electric current density
N	Poynting vector
I	intensity (time average of $|\mathbf{N}|$)
ϕ	scalar potential
A	vector potential
P	bulk polarization
M	bulk magnetization
Q	bulk quadrupole polarization
ϵ	dielectric constant
μ	magnetic permeability
ϵ_0	permittivity of free space
μ_0	permeability of free space

Polarized light

η	ellipticity of the polarization ellipse
θ	azimuth of the polarization ellipse
S_0, S_1, S_2, S_3	Stokes parameters
P	degree of polarization
$\tilde{\Pi}$	complex polarization vector
$\tilde{\rho}_{\alpha\beta}$	complex polarization tensor

Geometry and symmetry

i,j,k	unit vectors along space-fixed axes x,y,z.
I,J,K	unit vectors along molecule-fixed axes X,Y,Z.
r	position vector
$l_{\lambda'\alpha}$	direction cosine between the λ' and α axes ($\cos^{-1} l_{\lambda'\alpha}$ is the angle between the λ' and α axes)
$\delta_{\alpha\beta}$	Kronecker delta
$\varepsilon_{\alpha\beta\gamma}$	alternating tensor
$T_{\alpha\beta...}$	$\nabla_\alpha \nabla_\beta \ldots R^{-1}$
P	parity operation
T	classical time reversal operation

$\tilde{C}$	charge conjugation operation
p	eigenvalue of P
$2\pi b$	helix pitch
a	helix radius
$[\Gamma^2]$	symmetric part of the direct product of the representation Γ with itself
$\{\Gamma^2\}$	antisymmetric part of the direct product of the representation Γ with itself
$D^{(j)}$	irreducible representation of the proper rotation group R_3^+
T_q^k	irreducible spherical tensor operator

Classical mechanics

v	velocity vector
p	linear momentum vector
L	angular momentum vector
F	Lorentz force vector
W	total energy
T	kinetic energy
V	potential energy
L	Lagrangian function
H	Hamiltonian function
$\mathbf{p}'$	generalized momentum vector
Q_p	normal coordinate for the pth normal mode of vibration
P	momentum conjugate to Q_p, namely $\dot{Q}_p$
s_q	qth internal vibrational coordinate
L	vibrational **L**-matrix

Quantum mechanics

h	Planck constant
$\hbar$	$h/2\pi$
ψ	wavefunction
H	Hamiltonian operator
e_j, v_j, r_j	electronic, vibrational, rotational parts of the jth quantum state
j, m	general angular momentum quantum number, associated magnetic quantum number, of a particle
l, m_l	orbital angular momentum quantum number, associated magnetic quantum number, of a particle
s, m_s	spin angular momentum quantum number, associated magnetic quantum number, of a particle
J, M	total angular momentum quantum number, associated magnetic quantum number, of an atom or molecule

K	quantum number specifying the projection of the total angular momentum onto the principal axis of a symmetric top
g_i	g-value of the ith particle spin
Θ	quantum mechanical time reversal operator
ϵ	eigenvalue of Θ^2
A^T	transpose of operator A
$A^\dagger = A^{T*}$	Hermitian conjugate of operator A
Y_{lm}	spherical harmonic function
2δ	tunnelling splitting
2ϵ	parity-violating energy difference between chiral enantiomers
G	Fermi weak coupling constant
α	fine structure constant
g	weak charge
Q_W	effective weak charge
θ_W	Weinberg electroweak mixing angle
σ	Pauli spin operator
Z	proton number
$[a, b]$	commutator $ab - ba$
$\{a, b\}$	anticommutator $ab + ba$

Molecular properties

e_i	electric charge of the ith particle ($+e$ for the proton, $-e$ for the electron)
q	net charge or electric monopole moment
μ	electric dipole moment vector
$\mathbf{m}$	magnetic dipole moment vector
$\Theta_{\alpha\beta}$	traceless electric quadrupole moment tensor
$\alpha_{\alpha\beta}$	real part of the electric dipole–electric dipole polarizability tensor
$\alpha'_{\alpha\beta}$	imaginary part of the electric dipole–electric dipole polarizability tensor
$G_{\alpha\beta}$	real part of the electric dipole–magnetic dipole optical activity tensor
$G'_{\alpha\beta}$	imaginary part of the electric dipole–magnetic dipole optical activity tensor
$A_{\alpha,\beta\gamma}$	real part of the electric dipole–electric quadrupole optical activity tensor
$A'_{\alpha,\beta\gamma}$	imaginary part of the electric dipole–electric quadrupole optical activity tensor
$\mathscr{G}_{\alpha\beta}$	real part of the magnetic dipole–electric dipole optical activity tensor
$\mathscr{G}'_{\alpha\beta}$	imaginary part of the magnetic dipole–electric dipole optical activity tensor

$\mathscr{A}_{\alpha,\beta\gamma}$	real part of the electric quadrupole–electric dipole optical activity tensor
$\mathscr{A}'_{\alpha,\beta\gamma}$	imaginary part of the electric quadrupole–electric dipole optical activity tensor
$\tilde{\alpha}_{\alpha\beta}$, etc.	complex polarizability $\alpha_{\alpha\beta} - i\alpha'_{\alpha\beta}$, etc. (the minus sign arises from the choice of sign in the exponents of the complex dynamic electric and magnetic fields)
α	isotropic invariant of $\alpha_{\alpha\beta}$
G'	isotropic invariant of $G'_{\alpha\beta}$
$\beta(\alpha)^2$	anisotropic invariant of $\alpha_{\alpha\beta}$
$\beta(G')^2$	anisotropic invariant of $G'_{\alpha\beta}$
$\beta(A)^2$	anisotropic invariant of $A_{\alpha,\beta\gamma}$
κ	dimensionless polarizability anisotropy

Spectroscopy

$[\Delta\theta]$	specific rotation
η	ellipticity
I^R, I^L	Rayleigh or Raman scattered intensity in right (R)- or left (L)-circularly polarized incident light
$D(j \leftarrow n)$	dipole strength for the $j \leftarrow n$ transition
$R(j \leftarrow n)$	rotational strength for the $j \leftarrow n$ transition
$\hbar\delta$	Zeeman splitting
A, B, C	Faraday A-, B- and C-terms

1

A historical review of optical activity phenomena

> Yet each in itself – this was the uncanny, the antiorganic, the life-denying
> character of them all – each of them was absolutely symmetrical, icily
> regular in form. They were too regular, as substance adapted to life never
> was to this degree – the living principle shuddered at this perfect preci-
> sion, found it deathly, the very marrow of death – Hans Castorp felt he
> understood now the reason why the builders of antiquity purposely and
> secretly introduced minute variations from absolute symmetry in their
> columnar structures.
>
> *Thomas Mann* (The Magic Mountain)

1.1 Introduction

In the Preface, an optical activity phenomenon was defined as one whose origin may
be reduced to a different response of a system to right- and left-circularly polarized
light. This first chapter provides a review, from a historical perspective, of the main
features of a range of phenomena that can be classified as manifestations of optical
activity, together with a few effects that are related but are not strictly examples
of optical activity. The reader is referred to the splendid books by Lowry (1935),
Partington (1953) and Mason (1982) for further historical details.

The symbols and units employed in this review are those encountered in the
earlier literature, which uses CGS units almost exclusively; but these are not necessarily the same as those used in the rest of the book in which the theory of many
of the phenomena included in the review are developed in detail from the unified
viewpoint of the molecular scattering of polarized light. In particular, the theoretical
development in subsequent chapters employs SI units since these are currently in
favour internationally.

1

1.2 Natural optical rotation and circular dichroism

Optical activity was first observed by Arago (1811) in the form of colours in sunlight that had passed along the optic axis of a quartz crystal placed between crossed polarizers. Subsequent experiments by Biot (1812) established that the colours were due to two distinct effects: optical rotation, that is the rotation of the plane of polarization of a linearly polarized light beam; and optical rotatory dispersion, that is the unequal rotation of the plane of polarization of light of different wavelengths. Biot also discovered a second form of quartz which rotated the plane of polarization in the opposite direction. Biot (1818) recognized subsequently that the angle of rotation α was inversely proportional to the square of the wavelength λ of the light for a fixed path length through the quartz. The more accurate experimental data available to Drude (1902) enabled him to replace Biot's law of inverse squares by

$$\alpha = \sum_j \frac{A_j}{\lambda^2 - \lambda_j^2}, \tag{1.2.1}$$

where A_j is a constant appropriate to the visible or near ultraviolet absorption wavelength λ_j. Modern molecular theories of optical rotation all provide equations of this form for transparent regions.

Optical rotation was soon discovered in organic liquids such as turpentine (Biot, 1815), as well as in alcoholic solutions of camphor and aqueous solutions of sugar and tartaric acid, the last being reported in 1832 (Lowry, 1935). It was appreciated that the optical activity of fluids must reside in the individual molecules, and may be observed even when the molecules are oriented in random fashion; whereas that of quartz is a property of the crystal structure and not of the individual molecules, since molten quartz is not optically active. As discussed in detail in Section 1.9 below, it was eventually realized that the source of natural optical activity is a *chiral* (handed) molecular or crystal structure which arises when the structure has a sufficiently low symmetry that it is not superposable on its mirror image. The two distinct forms that can exist are said to have opposite *absolute configurations*, and these generate optical rotations of equal magnitude but opposite sense at a given wavelength.

The relationship between absolute configuration and the sense of optical rotation is subtle and has exercised theoreticians for a good many years. The modern system for specifying the absolute configuration of most chiral molecules is based on the R (for *rectus*) and S (for *sinister*) system of Cahn, Ingold and Prelog, supplemented with the P (for plus) and M (for minus) designation for molecules that have a clear helical structure. The sense of optical rotation (usually measured at the sodium D-line wavelength of 589 nm) associated with a particular absolute configuration is given in brackets, for example (R)-$(-)$ or (S)-$(+)$. Eliel and Wilen (1994) may be consulted for further details. The definitive method of determining

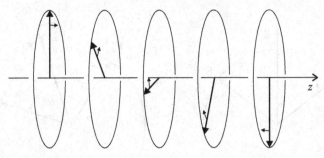

Fig. 1.1 The instantaneous electric field vectors of a right-circularly polarized light beam propagating along z. A vector in a fixed plane rotates clockwise when viewed in the $-z$ direction.

absolute configuration is via anomalous X-ray scattering associated with the presence of a relatively heavy atom substituted into the molecule, first demonstrated by Bijvoet *et al.* (1951) in a study of sodium rubidium tartrate. However, many chiral molecules are not accessible to X-ray crystallography: for these cases optical activity phenomena such as optical rotation, which are intrinsically sensitive to molecular chirality, are being used with increasing success. An optical method that can differentiate between the two enantiomers of a chiral compound is referred to as a *chiroptical* technique.

Fresnel's celebrated theory of optical rotation (Fresnel, 1825) followed from his discovery of circularly polarized light. In a circularly polarized light beam, the tip of the electric field vector in a fixed plane perpendicular to the direction of propagation traces out a circle with time: traditionally, the circular polarization is said to be right handed (positive) or left handed (negative) depending on whether the electric field vector rotates clockwise or anticlockwise, respectively, when viewed in this plane by an observer looking towards the source of the light. At a given instant, the tips of the electric field vectors distributed along the direction of propagation of a circularly polarized light beam constitute a helix, as shown in Fig. 1.1. Since the helix moves along the direction of propagation, but does not rotate, the previous definition of right and left handedness corresponds with the handedness of the helix, for as the helix moves through the fixed plane, the point of intersection of the tip of the electric field vector when viewed towards the light source rotates clockwise for a right-handed helix and anticlockwise for a left-handed helix. A particularly clear account of circularly polarized light and of the pitfalls that may arise in its graphical description may be found in the book by Kliger, Lewis and Randall (1990).

Fresnel realized that linearly polarized light can be regarded as a superposition of coherent left- and right-circularly polarized light beams of equal amplitude, the orientation of the plane of polarization being a function of the relative phases of the two components. This is illustrated in Fig. 1.2*a*. He attributed optical rotation to a

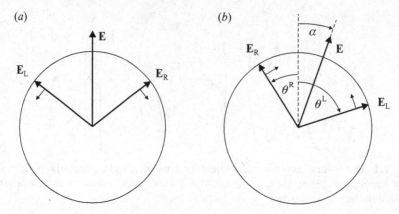

Fig. 1.2 (*a*) The electric field vector of a linearly polarized light beam decomposed into coherent right- and left-circularly polarized components. The propagation direction is out of the plane of the paper. (*b*) The rotated electric field vector at some further point in the optically active medium. Take note of Fig. 1.1 if confused by Fig. 1.2*b*.

difference in the velocity of propagation of the left- and right-circularly polarized components of the linearly polarized beam in the medium, for the introduction of a phase difference between the circularly polarized components would change the orientation of the plane of polarization, as shown in Fig. 1.2*b*. Suppose that a linearly polarized light beam of angular frequency $\omega = 2\pi c/\lambda$ enters a transparent optically active medium at $z = 0$. If, *at a given instant*, the electric field vectors of the right- and left-circularly polarized components at $z = 0$ are parallel to the direction of polarization of the linearly polarized light beam, then *at the same instant* the electric field vectors of the right- and left-circularly polarized components at some point $z = l$ in the optically active medium are inclined at angles $\theta^R = -2\pi cl/\lambda v^R$ and $\theta^L = 2\pi cl/\lambda v^L$, respectively, to this direction, where v^R and v^L are the velocities of the right- and left-circularly polarized components in the medium. The angle of rotation in radians is then

$$\alpha = \tfrac{1}{2}(\theta^R + \theta^L) = \frac{\pi cl}{\lambda}\left(\frac{1}{v^L} - \frac{1}{v^R}\right). \tag{1.2.2}$$

Since the refractive index is $n = c/v$, the angle of rotation in radians per unit length (measured in the same units as λ) can be written

$$\alpha = \frac{\pi}{\lambda}(n^L - n^R), \tag{1.2.3}$$

and is therefore a function of the *circular birefringence* of the medium, that is the difference between the refractive indices n^L and n^R for left- and right-circularly polarized light.

In the chemistry literature, the medium is said to be *dextro rotatory* if the plane of polarization rotates clockwise (positive angle of rotation), and *laevo rotatory* if the plane of polarization rotates anticlockwise (negative angle of rotation), when viewed towards the source of the light. The path of a linearly polarized light beam in a transparent optically active medium is characterized by a helical pattern of electric field vectors, since the orientation of each electric field vector is a function only of its position in the medium, although its amplitude is a function of time.

The form of the Drude equation (1.2.1) follows from (1.2.3) if an expression for the wavelength dependence of the refractive index such as

$$n^2 = 1 + \sum_j \frac{C_j \lambda^2}{\lambda^2 - \lambda_j^2} \tag{1.2.4}$$

is used, where C_j is a constant appropriate to the visible or near ultraviolet absorption wavelength λ_j. This is a version of Sellmeier's equation (1872). Thus if the C_js are slightly different for right- and left-circularly polarized light, an expression for $(n^L)^2 - (n^R)^2$ is found. But $(n^L)^2 - (n^R)^2 = (n^L - n^R)(n^L + n^R)$, and since n^L and n^R are close to n, the refractive index for unpolarized light, the value of $(n^L + n^R)$ may be taken as $2n$, and Drude's equation (1.2.1) is obtained with $A_j = \pi\lambda(C_j^L - C_j^R)/2n$. This simple argument serves to illustrate how optical rotation can be generated if a mechanism exists giving $C_j^L \neq C_j^R$.

Since refraction and absorption are intimately related, an optically active medium should absorb right- and left-circularly polarized light differently. This was first observed by Haidinger (1847) in amethyst quartz crystals, and later by Cotton (1895) in solutions of copper and chromium tartrate. Furthermore, linearly polarized light becomes elliptically polarized in an absorbing optically active medium: since elliptically polarized light can be decomposed into coherent right- and left-circularly polarized components of different amplitude, as illustrated in Fig. 1.3, the traditional theory ascribes the generation of an ellipticity to a difference in the absorption of the two circular components. The *ellipticity* ψ is obtained from the ratio of the minor and major axes of the ellipse, which are simply the difference and sum of the amplitudes of the two circular components:

$$\tan \psi = (E_R - E_L)/(E_R + E_L). \tag{1.2.5}$$

When $E_R > E_L$, ψ is defined to be positive, corresponding to a clockwise rotation of the electric field vector of the elliptically polarized beam in a fixed plane. The attenuation of the amplitude of a light beam by an absorbing medium is related to the absorption index n' and path length l by

$$E_l = E_0 e^{-2\pi n' l/\lambda}. \tag{1.2.6}$$

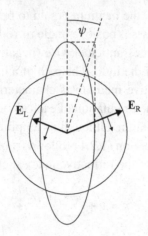

Fig. 1.3 Elliptical polarization, specified by the angle ψ, resolved into coherent right- and left-circular polarizations of different amplitude.

The ellipticity is then

$$
\begin{aligned}
\tan \psi &= \frac{e^{-2\pi n'^{R}l/\lambda} - e^{-2\pi n'^{L}l/\lambda}}{e^{-2\pi n'^{R}l/\lambda} + e^{-2\pi n'^{L}l/\lambda}} \\
&= \frac{e^{\pi(n'^{L}-n'^{R})/\lambda} - e^{-\pi l(n'^{L}-n'^{R})/\lambda}}{e^{\pi l(n'^{L}-n'^{R})/\lambda} + e^{-\pi l(n'^{L}-n'^{R})/\lambda}} \\
&= \tanh\left[\frac{\pi l}{\lambda}(n'^{L} - n'^{R})\right],
\end{aligned}
\tag{1.2.7}
$$

where n'^{L} and n'^{R} are the absorption indices for left- and right-circularly polarized light. For small ellipticities, in radians per unit length (measured in the same units as λ),

$$
\psi \approx \frac{\pi}{\lambda}(n'^{L} - n'^{R}).
\tag{1.2.8}
$$

The ellipticity is therefore a function of $(n'^{L} - n'^{R})$, the *circular dichroism* of the medium.

Apart from the fact that they are signed quantities, circular dichroism and optical rotatory dispersion have wavelength dependence curves in the region of an electronic absorption very similar to those for conventional absorption and refraction, respectively. These are illustrated in Fig. 1.4. Circular dichroism, together with the anomalous optical rotatory dispersion which accompanies it in the absorption region, are known collectively as the *Cotton effect*. The ellipticity maximum coincides with the point of inflection in the curve of optical rotatory dispersion, which ideally coincides with the maximum of an electronic absorption band at λ_j. The ellipticity

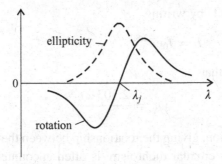

Fig. 1.4 The ellipticity and anomalous optical rotatory dispersion in the region of the electronic absorption wavelength λ_j. The signs shown here correspond to a positive Cotton effect.

and optical rotatory dispersion curves always have the relative signs shown in Fig. 1.4 for an isolated absorption band in a given sample. At wavelengths far removed from any λ_j, the rotatory dispersion is given by the Drude equation (1.2.1), but in the anomalous region the Drude equation must be modified to remove the singularity and to allow for the finite absorption width. If there are several adjacent absorption bands, the net Cotton effect will be a superposition of the individual Cotton effect curves.

Optical rotation measurements are usually presented as the *specific optical rotatory power* (often called simply the *specific rotation*)

$$[\alpha] = \frac{\alpha V}{ml}, \tag{1.2.9}$$

where α is the optical rotation in degrees, V is the volume containing a mass m of the optically active substance, and l is the path length. In much of the chemistry literature, CGS units are used and l is specified in decimetres. Similarly, circular dichroism measurements are usually presented as the *specific ellipticity*

$$[\psi] = \frac{\psi V}{ml}, \tag{1.2.10}$$

where ψ is measured in degrees. Circular dichroism is now usually obtained directly by measuring the difference in the decadic molar extinction coefficients

$$\epsilon = \frac{1}{cl} \log \frac{I_0}{I_l}, \tag{1.2.11}$$

where I is the intensity of the light wave and c is the concentration of absorbing molecules in moles per litre, of separate left- and right-circularly polarized light beams, rather than via the ellipticity induced in an initially linearly polarized light beam. Since the intensity of a wave is proportional to the square of the amplitude, the relationship between extinction coefficient and absorption index is obtained from (1.2.6) and (1.2.11) by writing

$$I_l = I_0 e^{-2.303\epsilon cl} = I_0 e^{-4\pi n'l/\lambda}, \tag{1.2.12}$$

from which it follows that

$$n' = \frac{2.303\lambda c\epsilon}{4\pi}. \tag{1.2.13}$$

The following expression, giving the relationship between the ellipticity in degrees and the decadic molar circular dichroism, is often encountered in the chemistry literature:

$$[\theta] = 3300(\epsilon^L - \epsilon^R) = 3300\Delta\epsilon. \tag{1.2.14}$$

This obtains from (1.2.8), (1.2.10) and (1.2.13) if CGS units are used and it is remembered that the path length is specified in decimetres.

A useful dimensionless quantity is the *dissymmetry factor* (Kuhn, 1930)

$$g = \frac{\epsilon^L - \epsilon^R}{\epsilon} = \frac{\epsilon^L - \epsilon^R}{\frac{1}{2}(\epsilon^L + \epsilon^R)}, \tag{1.2.15}$$

which is the ratio of the circular dichroism to the conventional absorption. The constants that arise in the determination of absolute absorption intensities therefore cancel out, and g often reduces to simple expressions involving just the molecular geometry. Since circular dichroism is of necessity always determined in the presence of absorption, g is also an appropriate criterion of whether or not circular dichroism in a particular absorption band is measurable, given the available instrumental sensitivity.

Although optical rotatory dispersion and circular dichroism have been known for more than 100 years, until the middle of the twentieth century most applications in chemistry utilized just the optical rotation at some transparent wavelength, usually the sodium D line at 589 nm. Then in the early 1950s a revolution in the study of optically active molecules was brought about through the introduction of instruments to measure optical rotatory dispersion routinely: this was possible as a result of developments in electronics, particularly the advent of photomultiplier tubes, so that the recording of visible and ultraviolet spectra no longer depended on the use of photographic plates. Steroid chemistry was one of the first areas to benefit, mainly as a result of the pioneering work of Djerassi (1960). Instruments to measure circular dichroism routinely were developed in the early 1960s when electro-optic modulators, which switch the polarization of the incident light between right and left circular at a suitable frequency, became available, and this technique is now generally preferred over optical rotatory dispersion because it provides better discrimination between overlapping absorption bands (the circular dichroism lineshape function drops to zero much more rapidly than the optical rotatory dispersion lineshape function).

Conventional optical rotation and circular dichroism utilize visible or ultra-violet radiation: since this excites the electronic states of the molecule, these techniques can be regarded as forms of polarized electronic spectroscopy. Thus it is the spatial distribution of the electronic states responsible for a particular circular dichroism band, for example, that is probed. This can often be related to the stereochemistry of the molecular skeleton in ways that are elaborated in later chapters. It is often stated that optical rotatory dispersion and circular dichroism are used to look at the stereochemistry of the molecule through the eyes of the chromophore (the structural group absorbing the visible or near ultraviolet radiation). The first successful application of this anthropomorphic viewpoint was the celebrated *octant rule* of Moffit *et al.* (1961), which relates the sign and magnitude of Cotton effects induced in the inherently optically inactive carbonyl chromophore by the spatial arrangement of perturbing groups in the rest of the molecule. The theoretical basis of the octant rule is discussed in detail in Chapter 5.

There are two topics closely related to circular dichroism that should be mentioned, namely circular polarization of luminescence, and fluorescence detected circular dichroism. The latter is simply an alternative method of measuring circular dichroism in samples, usually biological, with poor transmission, and involves measurement of a difference in the fluorescence intensity excited by right- and left-circularly polarized incident light with wavelength in the vicinity of an electronic absorption band (Turner, Tinoco and Maestre, 1974). The former refers to a circularly polarized component in the light spontaneously emitted from an optically active molecule in an excited state. The well-known relationship between the Einstein coefficients for absorption and spontaneous emission suggests that the circular dichroism and circular polarization of luminescence associated with a particular molecular electronic transition will provide identical structural information. However, differences between these observables will occur when the structure of the molecule in the ground electronic state differs from the structure in the excited luminescent state. Thus circular dichroism is a probe of ground state structure and circular polarization of luminescence is a probe of excited state structure. Under certain conditions, circular polarization of luminescence can be used to study aspects of excited state molecular dynamics such as photoselection and reorientational relaxation. A detailed development of these topics is outside the scope of this book, and the interested reader is referred to reviews by Richardson and Metcalf (2000) and Dekkers (2000).

An interesting variant of fluorescence detected circular dichroism has been mooted: circular differential photoacoustic spectroscopy (Saxe, Faulkner and Richardson, 1979). In conventional photoacoustic spectroscopy, light energy is absorbed by a sample, and that portion of the absorbed energy which is subsequently dissipated into heat is detected in the following manner. If the exciting light is modulated in time, the sample heating and cooling will also be modulated. The resulting temperature fluctuations lead to the transformation of the thermal energy into mechanical energy carried by sound waves in the sample which are detected with a microphone. In circular differential photoacoustic spectroscopy, the polarization of the incident light is modulated between right- and left-circular and the intensity of any sound waves detected at the modulation frequency will be a function of the circular dichroism of the absorbing chiral sample. It could be more widely applicable than fluorescence detected circular dichroism because a fluorescing chromophore is not required, and could be particularly attractive for studying molecules on surfaces.

As well as their general importance in stereochemistry, natural optical activity techniques, especially ultraviolet circular dichroism, have become central physical methods in biochemistry and biophysics since they are sensitive to the delicate stereochemical features that determine biological function (Fasman, 1996; Berova, Nakanishi and Woody, 2000).

1.3 Magnetic optical rotation and circular dichroism

Faraday's conviction of the connection between electromagnetism and light led him to the discovery of the rotation of the plane of polarization of a linearly polarized light beam on traversing a rod of lead borate glass placed between the poles of an electromagnet (Faraday, 1846). A Faraday rotation is found when light is transmitted through any medium, isotropic or oriented, in the direction of a magnetic field. The sense of rotation depends on the relative directions of the light beam and the magnetic field, and is reversed on reversing either the direction of the light beam or the magnetic field. Thus magnetic rotatory power differs from natural rotatory power in that the rotations are added, rather than cancelled, on reflecting the light back through the medium. It was soon discovered that magnetic optical rotation varies inversely with the square of the wavelength, in accordance with Biot's law for natural optical rotation; although it was subsequently found that a better approximation is provided by a formula similar to Drude's equation (1.2.1).

The quantitative investigations of Verdet (1854) are summarized in Verdet's law for the angle of rotation per unit path length in a magnetic field $\mathbf{B}$ making an angle θ with the direction of propagation of the light beam:

$$\alpha = VB\cos\theta, \tag{1.3.1}$$

where V is the Verdet constant for the material for a given wavelength and temperature. For light passing through the medium in the direction of the magnetic field (north pole to south pole) most diamagnetic materials rotate the plane of polarization

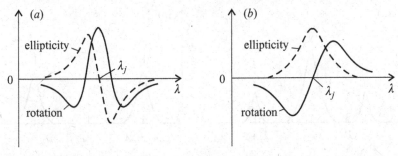

Fig. 1.5 The magnetic ellipticity and anomalous optical rotatory dispersion shown by (*a*) diamagnetic and (*b*) paramagnetic samples in the region of the electronic absorption wavelength λ_j.

in an anticlockwise sense when viewed towards the light source, corresponding to a negative rotation in the chemistry convention. This optical rotation is in the same sense as the circulation of current in a solenoid producing an equivalent magnetic field.

Magnetic optical rotation can be described in terms of different refractive indices for left- and right-circularly polarized light, and (1.2.3) applies equally well to natural and magnetic rotation, although the origin of the circular birefringence is different in the two cases. In regions of absorption there is a difference in the absorption of left- and right-circularly polarized light in the direction of the magnetic field, and linearly polarized light acquires an ellipticity given by the same equation (1.2.8) that describes natural circular dichroism.

Verdet also discovered that iron salts in aqueous solution show a magnetic rotation which is in the opposite sense to that of water, arising from the paramagnetism of iron salts. In general, only the magnetic rotatory dispersions of diamagnetic materials follow the laws of Drude and Verdet; those of paramagnetic materials are more complicated. The influence of temperature on the magnetic rotation of diamagnetic materials is slight, but paramagnetic materials show a pronounced variation with temperature which is related to the temperature dependence of paramagnetism.

The dispersion with wavelength of the magnetic rotation and ellipticity in a region of absorption depends on the relative magnitudes of the diamagnetic and paramagnetic contributions. The two ideal cases are illustrated in Fig. 1.5. The diamagnetic rotation curve shown is actually the resultant of two equal and opposite optical rotatory dispersion curves for two adjacent electronic absorption bands, and is usually symmetric. The paramagnetic rotation curve is like an optical rotatory dispersion curve for a single absorption band, and is usually unsymmetric.

Faraday had looked for the effect of a magnetic field on a source of radiation, but without success because strong fields and spectroscopes of good resolution were not available to him. The first positive results were obtained by Zeeman (1896), and were described as a broadening of the two lines of the first principal doublet from a sodium

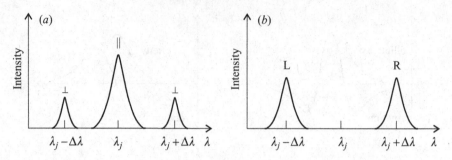

Fig. 1.6 The normal Zeeman effect (*a*) for light emitted perpendicular to the magnetic field and (*b*) for light emitted in the direction of the magnetic field.

flame placed between the poles of a powerful electromagnet. Soon afterwards, Lorentz showed that his electron theory of radiation and matter accommodated this observation: when viewed perpendicular to the magnetic field, the spectral lines should be split into three linearly polarized components with the central (unshifted) line linearly polarized parallel ($\parallel$) to the field and the other two lines linearly polarized perpendicular ($\perp$) to the field; when the magnetic field points towards the observer, there should be two lines on either side of the original line with the high and low wavelength lines showing right- and left-circular polarizations, respectively. This is illustrated in Fig. 1.6. The displacements $\Delta\lambda$ should be proportional to the magnetic field strength. These predictions were verified later by Zeeman, but only for certain spectral lines showing what is now called the *normal* Zeeman effect; other lines (including the components of the first principal sodium doublet) split into a greater number of components and are said to show the *anomalous* Zeeman effect. The normal effect is simply a special case in which the effects of electron spin are absent.

Since the right- and left-circularly polarized components of light emitted by an atom in the presence of a magnetic field are differentiated, the Zeeman effect can be regarded as a manifestation of optical activity. Indeed, it was soon recognized that the main features of the Faraday effect can be explained in terms of the Zeeman effect. Since right- and left-circularly polarized light beams are also absorbed at the slightly different wavelengths $\lambda_j^R = \lambda_j + \Delta\lambda$ and $\lambda_j^L = \lambda_j - \Delta\lambda$ in a magnetic field along the direction of propagation of the incident beam, one could use, for example, equation (1.2.4) for the refractive index with two absorption wavelengths λ_j^R and λ_j^L:

$$(n^L)^2 - (n^R)^2 = C_j\lambda^2 \left[\frac{1}{\lambda^2 - \left(\lambda_j^L\right)^2} - \frac{1}{\lambda^2 - \left(\lambda_j^R\right)^2} \right], \tag{1.3.2}$$

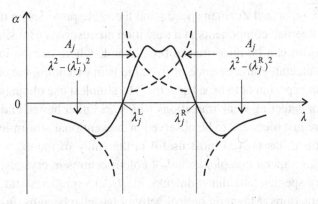

Fig. 1.7 The diamagnetic optical rotatory dispersion curve generated from two equal and opposite Drude-type curves centred on λ_j^L and λ_j^R. The sign shown here obtains when the magnetic field is in the direction of propagation of the light beam.

which is simply the sum of two equal and opposite optical rotatory dispersion curves centred on λ_j^L and λ_j^R, so that

$$\alpha \approx \frac{\pi C_j \lambda}{2n} \left[\frac{1}{\lambda^2 - \left(\lambda_j^L\right)^2} - \frac{1}{\lambda^2 - \left(\lambda_j^R\right)^2} \right]. \tag{1.3.3}$$

If (1.3.3) is modified to remove the singularities and allow for a finite absorption width, the general form of a diamagnetic optical rotation curve is reproduced, as illustrated in Fig. 1.7. Similarly, the general form of a diamagnetic ellipticity curve is reproduced from the sum of two equal and opposite ellipticity curves centred on λ_j^L and λ_j^R.

Notice that in justifying Drude's equation (1.2.1) for the dispersion of natural optical rotation we invoked a slight difference in the constants C_j in Sellmeier's equation (1.2.4) for the refractive indices for right- and left-circularly polarized light, but assumed that the resonance wavelengths were the same, whereas in developing the form (1.3.3) for the diamagnetic rotation curve we assumed that the C_js are the same for the opposite circular polarizations but that the resonance wavelengths are different. This illustrates two distinct mechanisms by which optical rotation (and circular dichroism) can be generated, and we shall see later, when general quantum mechanical theories are developed, that analogues of both mechanisms can contribute to both natural and magnetic optical rotation and circular dichroism.

The main significance of magnetic optical activity in chemistry is that it provides information about ground and excited electronic states of atoms and molecules. As indicated above, magnetic circular dichroism is the difference between left- and right-circularly polarized Zeeman spectra and therefore provides no new information when the Zeeman components of a transition are resolved. But since magnetic circular dichroism can be measured in broad bands where conventional Zeeman effects are undetectable, the essence of its value is in extending the circularly polarized Zeeman experiment to broad spectra. The simplest use of magnetic circular dichroism is for detecting weak transitions which are either buried under a stronger transition or are just too weak to be observed in conventional absorption. Magnetic circular dichroism has proved most useful in the study of the excited electronic states of transition metal complexes and of colour centres in crystals; particularly their symmetry species, angular momenta, electronic splittings and vibrational–electronic interactions. Magnetic optical activity has also been useful in the study of organic and biological systems, especially for cyclic π electron molecules such as porphyrins.

Not surprisingly, there is a magnetic version of the circular polarization of luminescence (outlined in Section 1.2) that is shown by all molecules in a magnetic field parallel to the direction of observation of the luminescence. Again this gives information about excited state molecular properties, and we refer to Richardson and Riehl (1977) for further details. There are also magnetic versions of fluorescence detected circular dichroism, and circular differential photoacoustic spectroscopy.

1.4 Light scattering from optically active molecules

Optical rotation and circular dichroism are concerned with the polarization characteristics of light transmitted through an optically active medium, and are therefore associated with refraction. Refraction is one consequence of the scattering of light by the electrons and nuclei in the constituent molecules of the medium, and can be accompanied by Rayleigh and Raman scattering in all directions. Rayleigh scattered light has the same frequency as the incident light, whereas the frequency of Raman scattered light is shifted from that of the incident light by amounts corresponding to molecular rotational, vibrational and electronic transitions. Specular reflection by polished surfaces of glass and metals, and diffuse reflection by, for example, a sheet of paper, can also be attributed ultimately to molecular scattering.

The scattering description of refraction is subtle, and involves interference between the unscattered component of the light wave and the net plane wavefront in the forward direction from planar arrays of individual scatterers in the medium. This process is discussed in detail in Chapter 3. This interference modifies the polarization properties of the light from individual molecular scatterers, so the light refracted through an optically active medium has different polarization properties from the Rayleigh- and Raman-scattered light (and the reflected light). Thus with linearly polarized light incident on isotropic optically active samples at transparent wavelengths, the refracted light suffers a rotation of the plane of polarization with no ellipticity produced, whereas the scattered light acquires an ellipticity but no rotation of the plane of polarization.

The origin of the ellipticity in Rayleigh and Raman scattered light is easily understood in general terms because optically active molecules respond differently to right- and left-circularly polarized light, which are therefore scattered to different extents. Consequently, the coherent right- and left-circularly polarized components into which a linearly polarized beam can be resolved are scattered differently, and are no longer of equal amplitude in the scattered light, which is therefore elliptically polarized. A dramatic example is provided by cholesteric liquid crystals which have enormous optical rotatory powers so that the light reflected from the surface is almost completely circularly polarized (Giesel, 1910).

Instead of measuring an ellipticity in Rayleigh and Raman scattered light, a difference in the scattered intensities in right- and left-circularly polarized incident light (the *circular intensity difference*) may be measured directly instead. At transparent wavelengths, the ellipticity (or the associated degree of circular polarization) of the scattered light and the circular intensity difference provide equivalent information about optically active molecules, but subtle differences can arise at absorbing wavelengths. Both the degree of circular polarization and the circular intensity difference are manifestations of *Rayleigh and Raman optical activity*.

The first attempts to observe Rayleigh and Raman optical activity concentrated on the circular intensity difference. The chequered history of these attempts is briefly as follows. Gans (1923) considered additional contributions to Rayleigh scattering from optically active molecules, but omitted a crucial interference term that generates the ellipticity and the circular intensity difference; he claimed to have observed optical activity effects in the depolarization ratio, but de Mallemann (1925) pointed out that the depolarization ratio anomalies originated in optical rotation of the incident and scattered beams. Shortly after the discovery of the Raman effect, Bhagavantam and Venkateswaran (1930) found differences in the relative intensities of some of the vibrational Raman lines of two optical isomers in unpolarized incident light, but these were subsequently attributed to impurities. Although he had no explicit theory, Kastler (1930) thought that, since optically active molecules respond differently to right- and left-circularly polarized light, a difference might exist in the vibrational Raman spectra of optically active molecules in right- and left-circularly polarized incident light, but the instrumentation at that time was far too primitive for him to observe the effect. Perrin (1942) alluded to the existence of additional polarization effects in light scattered from optically active molecules; but it was not until the theoretical work of Atkins and Barron (1969) that the interference mechanism (between light waves scattered via the molecular polarizability and optical activity property tensors) responsible for the ellipticity in the scattered light and the circular intensity difference was discovered. Barron and Buckingham (1971) subsequently developed a more definitive version of the theory and introduced the following definition of the dimensionless Rayleigh and Raman circular intensity difference,

$$\Delta = \frac{I^R - I^L}{I^R + I^L},$$ (1.4.1)

where I^R and I^L are the scattered intensities in right- and left-circularly polarized incident light, as an appropriate experimental quantity in Rayleigh and Raman optical activity. The first reported natural Raman circular intensity difference spectra by Bosnich, Moskovits and Ozin (1972) and by Diem, Fry and Burow (1973) originated in instrumental artifacts, but the spectra subsequently reported in the chiral molecules 1-phenylethylamine and 1-phenylethanol, $(C_6H_5)CH(CH_3)(NH_2)$ and $(C_6H_5)CH(CH_3)(OH)$, by Barron, Bogaard and Buckingham (1973) were confirmed by Hug *et al.* (1975) as genuine. On account of experimental difficulties, the natural Rayleigh circular intensity difference has not yet been observed in small chiral molecules, but has been reported in large biological structures (Maestre *et al.*, 1982; Tinoco and Williams, 1984).

Since all molecules can show optical rotation and circular dichroism in a magnetic field, it is not surprising that all molecules in a strong magnetic field should show Rayleigh and Raman optical activity (Barron and Buckingham, 1972). More specifically, the magnetic field must be parallel to the incident light beam to generate a circular intensity difference, and parallel to the scattered light beam to generate an ellipticity. The signs of these observables reverse on reversing the magnetic field direction. The first observation of this effect was in the resonance Raman spectrum of a dilute aqueous solution of ferrocytochrome c, a haem protein (Barron, 1975*a*). It should be mentioned, however, that there is a much older phenomenon that probably falls within the definition of magnetic optical activity in light scattering, namely the Kerr magneto-optic effect (Kerr, 1877). Here, linearly polarized light becomes elliptically polarized when reflected from the polished pole of an electromagnet: the incident light must be linearly polarized either in, or perpendicular to, the plane of incidence, otherwise elliptical polarization results from metallic reflection.

More surprisingly, although there are no simple electrical analogues of magnetic optical rotation and circular dichroism (they would violate parity and reversality, as discussed in Section 1.9), Rayleigh and Raman optical activity should also be shown by any fluid in a static electric field perpendicular to both the incident and scattered directions (Buckingham and Raab, 1975). Electric Rayleigh optical activity was first observed by Buckingham and Shatwell (1980) in gaseous methyl chloride.

There has been interest in the influence on circular dichroism spectra of the differential scattering of right- and left-circularly polarized light by turbid optically active media: light scattered out of the sides of the sample removes an intensity from the transmitted beam additional to that from absorption (Tinoco and Williams, 1984). A dramatic example of the effect is provided by cholesteric liquid crystals (de Gennes and Prost, 1993): an initially linearly polarized beam can become almost completely circularly polarized after passing through a slab on account of the preferential scattering (reflection) of one of the coherent circularly polarized components.

The main significance of Rayleigh optical activity is that, from appropriate measurements in light scattered at 90°, it provides a measure of the anisotropy in the molecular optical activity using an isotropic sample such as a liquid or solution. Such information can only be obtained from optical rotation or circular dichroism measurements using an oriented sample such as a crystal, or a fluid in a static electric field (Tinoco, 1957). The main significance of Raman optical activity is that it provides an alternative method to infrared optical rotation and circular dichroism for measuring vibrational optical activity: this is discussed further in the next section.

1.5 Vibrational optical activity

It had been appreciated for some time that the measurement of optical activity associated with molecular vibrations could provide a wealth of delicate stereochemical information. But only since the early 1970s, thanks mainly to developments in optical and electronic technology, have the formidable technical difficulties been overcome and vibrational optical activity spectra been observed using both infrared and Raman techniques.

The significance of vibrational optical activity becomes apparent when it is compared with conventional electronic optical activity in the form of optical rotation and circular dichroism of visible and near ultraviolet radiation. These conventional techniques have proved most valuable in stereochemical studies, but since the electronic transition frequencies of most structural units in a molecule occur in the far ultraviolet, they are restricted to probing limited regions of molecules, in particular chromophores and their immediate intramolecular environments, and cannot be used at all when a molecule lacks a chromophore (although optical rotation measurements at transparent frequencies can still be of value). But since a vibrational spectrum, infrared or Raman, contains bands from vibrations associated with most parts of a molecule, measurements of some form of vibrational optical activity could provide much more information.

The obvious method of measuring vibrational optical activity is by extending optical rotatory dispersion and circular dichroism into the infrared. But in addition to the technical difficulties in manipulating polarized infrared radiation, there is a fundamental physical difficulty: optical activity is a function of the frequency of the exciting light and infrared frequencies are several orders of magnitude smaller than visible and near ultraviolet frequencies. On the other hand, the Raman effect provides vibrational spectra using visible exciting light, the molecular vibrational frequencies being measured as small displacements from the frequency of the incident light in the visible spectrum of the scattered light. Consequently, the fundamental frequency problem does not arise if vibrational optical activity is measured by means of the Raman circular intensity difference (or degree of circular polarization), outlined in the previous section.

Natural infrared optical rotation was first observed as long ago as 1836 by Biot and Melloni, who passed linearly polarized infrared radiation along the optic axis of a column of quartz, but this probably originated mainly in near infrared electronic transitions. Further progress was slow, and Lowry (1935) concluded a review of infrared optical activity with the unenthusiastic statement: 'Very few measurements of rotatory dispersion have been made in the infrared, since this phenomenon shows no points of outstanding interest, the rotatory power decreasing steadily with increasing wavelength, even when passing through an infrared absorption band'. Anomalous infrared optical rotatory dispersion in quartz was reported by Gutowsky (1951), but this work was challenged by West (1954). Katzin (1964) reanalyzed the early near infrared optical rotatory dispersion data of Lowry and Snow (1930) and concluded that, while electronic transitions were mainly responsible, contributions from infrared vibrational transitions were certainly present. Hediger and Günthard (1954) reported the observation of anomalous optical rotatory dispersion associated with an overtone in the vibrational spectrum of 2-butanol, but Wyss and Günthard (1966) subsequently questioned the results and in further experiments failed to observe any effects.

Schrader and Korte (1972) reported anomalous optical rotatory dispersion in the vibrational spectrum of N-(p-methoxybenzylidene) butylaniline perturbed into the cholesteric mesophase by the addition of an optically active solute. Soon afterwards, Dudley, Mason and Peacock (1972) reported vibrational circular dichroism in a similar sample. The reason that vibrational optical activity is so readily accessible in cholesteric liquid crystals is that the helix pitch length is of the order of the wavelength of the infrared radiation.

The first ray of hope for practical chemical applications of infrared vibrational optical activity came in 1973 when Hsu and Holzwarth reported well defined circular dichroism bands arising from vibrations of water molecules in optically active crystals such as nickel sulphate, α $NiSO_4 \cdot 6H_2O$. This ray intensified when Holzwarth *et al.* (1974) reported circular dichroism in the $2920\,cm^{-1}$ band of 2,2,2-trifluoro 1-phenylethanol, $(C_6H_5)C^*H(CF_3)(OH)$, due to the C^*–H stretching mode. The publication by Nafie, Keiderling and Stephens (1976) of vibrational circular dichroism spectra down to about $2000 \, cm^{-1}$ in a number of typical optically active molecules served notice that infrared vibrational circular dichroism had become a routine technique.

While this frontal attack on vibrational optical activity through infrared optical rotation and circular dichroism was under way, the outflanking manoeuvre involving Raman optical activity, described in the previous section, was passing relatively unnoticed. In fact the first observations of Raman optical activity reported by Barron, Bogaard and Buckingham (1973), mentioned previously, constituted the first observations of genuine natural vibrational optical activity of small chiral molecules in the liquid phase. High quality infrared circular dichroism and Raman optical activity spectra of chiral molecules may now be measured routinely and are proving increasingly valuable for solving a wide range of stereochemical problems. The Raman optical activity spectrum of α-pinene is shown in Fig. 1.8 as a typical example of a vibrational optical activity spectrum. The fact that there is almost perfect mirror symmetry in the spectra of the two enantiomers, which were studied in microgram quantities in throw-away capillary tubes, emphasizes the ease and reliability of such measurements using the latest generation of instrument (Hug, 2003). Typical infrared circular dichroism spectra have a similar general appearance, except that they do not penetrate much below $800 \, cm^{-1}$ due to both technical problems and the fundamental frequency problem mentioned above. Also the signs and magnitudes of infrared circular dichroism bands associated with particular vibrations generally bear no relation to the corresponding Raman optical activity bands due to the completely different mechanisms responsible for the two phenomena (see Chapter 7).

Vibrational optical activity techniques, both infrared and Raman, have become especially valuable in biochemistry and biophysics, enormous progress having been made since the publication of the first edition of this book. Important milestones were the first reports of the vibrational optical activity spectra of proteins using infrared circular dichroism by Keiderling (1986) and Raman optical activity by Barron, Gargaro and Wen (1990). Raman optical activity spectra may even be recorded routinely on intact live viruses in aqueous solution to provide information on the structures and mutual interactions of the protein coat and the nucleic acid core (see Section 7.6).

Vibrational optical activity induced by a magnetic field using infrared circular dichroism was first observed by Keiderling (1981) but, as mentioned in Section 1.4, it had been observed previously as a circular intensity difference in resonance Raman scattering. Just as conventional magnetic optical activity injects additional structure into an electronic spectrum, so magnetic infrared and Raman optical activity inject additional structure into a vibrational spectrum, thereby facilitating the assignment of bands, for example. Magnetic vibrational circular dichroism

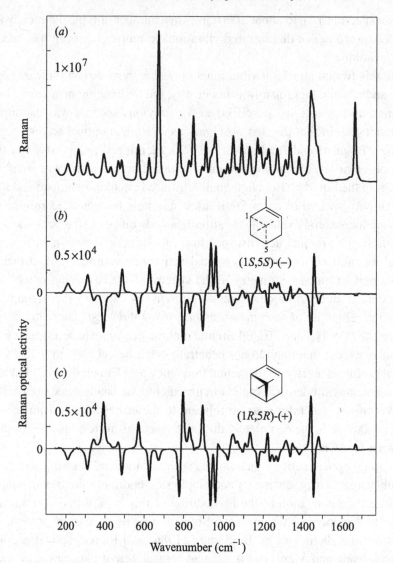

Fig. 1.8 The Raman (a) and Raman optical activity (b, c) spectra of the two enantiomers of α-pinene measured as the degree of circular polarization in backscattered light. Adapted from Hug (2003). Spectrum (b) is a little less intense than (c) because the ($1S$, $5S$)-($-$) sample had a slightly lower enantiomeric excess. The absolute intensities are not defined but the relative Raman and Raman optical activity intensities are significant.

is valuable for studies of small molecules in the gas phase, where it can yield vibrational g-values from rotationally resolved bands (Bour, Tam and Keiderling, 1996). On the other hand, systems in degenerate ground states, most commonly encountered as Kramers degeneracy in molecules with an odd number of electrons, add another dimension to magnetic Raman optical activity studies, since transitions between the magnetically split components of the degenerate ground electronic level superimposed upon the vibrational transition may be observed. This 'Raman electron paramagnetic resonance' effect was first observed by Barron and Meehan (1979) in resonance scattering from dilute solutions of transition metal complexes such as iridium (IV) hexachloride, $IrCl_6^{2-}$. Raman electron paramagnetic resonance provides information about the magnetic structure of ground and low-lying excited electronic states, including the sign of the g-factor and how the magnetic structure changes when the molecule is in an excited vibrational state.

There has also been discussion of optical activity associated with pure rotational transitions of chiral molecules in the gas phase, including optical rotation and circular dichroism in the microwave region and Raman optical activity (Salzman, 1977; Barron and Johnston, 1985; Polavarapu, 1987), but to date no experimental observations have been reported.

1.6 X-ray optical activity

Since the appearance of the first edition of this book, optical activity measurements have been extended to X-ray wavelengths, thanks to developments in the X-ray synchrotron beams that are essential for such measurements. This was first achieved for magnetic field induced circular dichroism by Schutz *et al.* (1987), who studied magnetized iron. The first observation of natural circular dichroism in chiral molecules was made a decade later by Alagna *et al.* (1998) in crystals of a chiral neodymium complex. Magnetic X-ray circular dichroism was observed first because the X-ray magnetic dissymmetry factors can be several orders of magnitude larger than the X-ray natural dissymmetry factors.

Both magnetic and natural X-ray circular dichroism originate in near-edge atomic absorptions and their associated structure. The magnetic effect is now widely used to explore the magnetic properties of magnetically ordered materials. The natural effect, studies of which are still in their infancy (Peacock and Stewart, 2001; Goulon *et al.*, 2003), is sensitive to absolute chirality in the molecular environment around the absorbing atom. An interesting aspect of natural X-ray circular dichroism is that it relies mainly on an unusual electric dipole–electric quadrupole mechanism, discussed in detail in later chapters, that survives only in oriented samples such as crystals. The electric dipole–magnetic dipole mechanism that dominates infrared, visible and ultraviolet circular dichroism and which survives in isotropic media such as liquids and solutions is small in the X-ray region. In this respect magnetochiral dichroism, described in the next section, could be favourable for the study of chiral samples in the X-ray region because an electric dipole–electric quadrupole contribution survives in isotropic media, and linearly polarized synchrotron radiation, which is easier to generate than circularly polarized, could be employed with the measurements effected by reversing the magnetic field direction.

1.7 Magnetochiral phenomena

Shortly after the appearance of the first edition of this book, a remarkable new class of optical phenomena that depend on the interplay of chirality and magnetism came to prominence. Wagnière and Meier (1982) predicted that a static magnetic field parallel to the propagation direction of an incident light beam would induce a small shift in the absorption coefficient of a medium composed of chiral molecules. This shift is independent of the polarization characteristics of the light beam and so appears even in unpolarized light. The shift changes sign either on replacing the chiral molecule by its mirror image enantiomer or on reversing the relative directions of the magnetic field and the propagation direction of the light beam. Portigal and Burstein (1971) had earlier shown, on the basis of symmetry arguments, that an extra term exists in the dielectric constant of a chiral medium which is proportional to $\mathbf{k} \cdot \mathbf{B}$, where $\mathbf{k}$ is the unit propagation vector of the light beam and $\mathbf{B}$ is the external static magnetic field; and Baranova and Zeldovich (1979a) had predicted a shift in the refractive index of a fluid composed of chiral molecules in a static magnetic field applied parallel to the direction of propagation of a light beam.

The associated difference in absorption of a light beam parallel ($\uparrow\uparrow$) and antiparallel ($\uparrow\downarrow$) to the magnetic field was subsequently christened *magnetochiral dichroism* by Barron and Vrbancich (1984), with the corresponding difference in refractive index called *magnetochiral birefringence*. The magnetochiral dichroism experiment is illustrated in Fig. 1.9. The corresponding magnetochiral birefringence and dichroism observables, $n^{\uparrow\uparrow} - n^{\uparrow\downarrow}$ and $n'^{\uparrow\uparrow} - n'^{\uparrow\downarrow}$, are linear in the

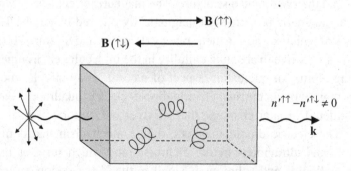

Fig. 1.9 The magnetochiral dichroism experiment. The absorption index n' of a medium composed of chiral molecules is slightly different for *unpolarized* light when a static magnetic field is applied parallel ($\uparrow\uparrow$) and antiparallel ($\uparrow\downarrow$) to the direction of propagation of the beam.

magnetic field strength just like the Faraday effect. Magnetochiral dichroism was first observed by Rikken and Raupach (1997) in a chiral europium(III) complex in dimethylsulphoxide solution, and magnetochiral birefringence by Kleindienst and Wagnière (1998) in chiral organic fluids such as a camphor derivative and carvone. At the time of writing there are some unresolved problems with measurements of magnetochiral birefringence, since different experimental strategies appear to give quite different results (Vallet *et al.*, 2001).

It might appear at first sight that magnetochiral dichroism is simply the result of cascade mechanisms involving successive natural circular dichroism and magnetic circular dichroism steps, and *vice versa*. Thus natural circular dichroism of the incoherent right- and left-circularly polarized components of equal amplitude into which unpolarized light may be decomposed leads to the initially unpolarized light beam acquiring a circular component as it progresses through the medium, which will subsequently be absorbed differently depending on whether the applied magnetic field is parallel or antiparallel to the propagation direction. Equivalently, magnetic circular dichroism will induce a circular component in the initially unpolarized light beam, followed by natural circular dichroism. Although these cascade mechanisms may provide an initial insight into the physical origin of the phenomenon, and will indeed provide higher-order contributions (Rikken and Raupach, 1998), as elaborated in Chapter 6 magnetochiral dichroism originates primarily in a single-step scattering process in which the chiral and magnetic interactions interfere.

Although the magnetochiral effects observed to date are very weak, they are of fundamental interest. For example, they provide a new source of absolute enantioselection *via* photochemical reactions in unpolarized light in a static magnetic field that may be significant for the origin of biological homochirality (Rikken and Raupach, 2000). Also, they might be exploited in new phenomena of technological significance in chiral magnetic media such as an anisotropy in electrical resistance through a chiral conductor in directions parallel and antiparallel to a static magnetic field (Rikken, Fölling and Wyder, 2001).

1.8 The Kerr and Cotton–Mouton effects

The Kerr and Cotton–Mouton effects refer to the linear birefringence induced in a fluid or an isotropic solid by a static electric or magnetic field, respectively, applied perpendicular to the propagation direction of a light beam (Kerr, 1875; Cotton and Mouton, 1907). The effects originate mainly in a partial orientation of the molecules in the medium. The sample behaves, in fact, like a uniaxial crystal with the optic axis parallel to the direction of the field. Although these phenomena are not manifestations of optical activity (they do not originate in a difference in response to right- and left-circularly polarized light) we describe them briefly since equations for the associated polarization changes emerge automatically from the birefringent scattering treatment presented in Chapter 3.

If the light beam is linearly polarized at 45° to the direction of the applied field, elliptical polarization is produced on account of a phase difference induced in the two coherent resolved components of the light beam linearly polarized parallel and perpendicular to the static field direction. Since the phase difference is

$$\delta = \frac{2\pi l}{\lambda}(n_\| - n_\perp),\tag{1.8.1}$$

where $n_\|$ and $n_\perp$ are the refractive indices for light linearly polarized parallel and perpendicular to the static field direction, the resulting ellipticity is simply $\delta/2$ so that, in radians per unit path length,

$$\psi = \frac{\pi}{\lambda}(n_\| - n_\perp).\tag{1.8.2}$$

At absorbing wavelengths, the two different refractive indices for light linearly polarized parallel and perpendicular to the static field direction are accompanied by different absorption coefficients. This results in a rotation of the major axis of the polarization ellipse because a difference in amplitude develops between the two orthogonal resolved components for which no phase difference exists. Again, this optical rotation due to linear dichroism is not a manifestation of optical activity. The lineshapes for the dispersion of linear birefringence and linear dichroism are the same as for ordinary refraction and absorption. Further information on linear dichroism and its applications in chemistry may be found in the books by Michl and Thulstrup (1986) and Rodger and Nordén (1997).

1.9 Symmetry and optical activity

The subject of symmetry and optical activity reviewed in this section impacts on many different areas of science, ranging from classical crystal optics to elementary particle physics, cosmology and the origin of life. Some of the topics mentioned here are revisited in detail in Chapter 4, but for others a more detailed account is beyond the scope of this book.

1.9.1 Spatial symmetry and optical activity

Fresnel's analysis of optical rotation in terms of different refractive indices for left- and right-circularly polarized light immediately provided a physical insight into the symmetry requirements for the structure of an optically active medium. In the words of Fresnel (1824):

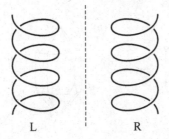

L | R

Fig. 1.10 A right-handed helix and its left-handed mirror image.

There are certain refracting media, such as quartz in the direction of its axis, turpentine, essence of lemon, etc., which have the property of not transmitting with the same velocity circular vibrations from right to left and those from left to right. This may result from a peculiar constitution of the refracting medium or of its molecules, which produces a difference between the directions right to left and left to right; such, for instance, would be a helicoidal arrangement of the molecules of the medium, which would present inverse properties according as these helices were dextrogyrate or laevogyrate.

A finite cylindrical helix is the archetype for all figures exhibiting what Pasteur (1848) called *dissymmetry* to describe objects 'which differ only as an image in a mirror differs from the object which produces it.' Thus a helix and its mirror image cannot be superposed since reflection reverses the screw sense, as illustrated in Fig. 1.10. Systems which exist in two nonsuperposable mirror image forms are said to exhibit *enantiomorphism*. Dissymmetric figures are not necessarily *asymmetric*, that is devoid of all symmetry elements, since they may possess one or more proper rotation axes (the finite cylindrical helix has a twofold rotation axis C_2 through the mid point of the coil, perpendicular to the long helix axis). However, dissymmetry excludes improper rotation axes, that is centres of inversion, reflection planes and rotation–reflection axes. In recent years the word dissymmetry has been replaced by *chirality*, meaning handedness (from the Greek *chir* = hand), in the more modern literature of stereochemistry and other branches of science. 'Chirality' was first used in this context by Lord Kelvin, Professor of Natural Philosophy at the University of Glasgow. His complete definition is as follows (Lord Kelvin, 1904):

I call any geometrical figure, or group of points, *chiral*, and say that it has chirality if its image in a plane mirror, ideally realized, cannot be brought to coincide with itself. Two equal and similar right hands are homochirally similar. Equal and similar right and left hands are heterochirally similar or 'allochirally' similar (but heterochirally is better). These are also called 'enantiomorphs', after a usage introduced, I believe, by German writers. Any chiral object and its image in a plane mirror are heterochirally similar.

The first sentence is essentially the definition used today. Strictly speaking, the term 'enantiomorph' is usually reserved for a macroscopic object such as a crystal, and 'enantiomer' for a molecule, but because of the ambiguity of scale in the case

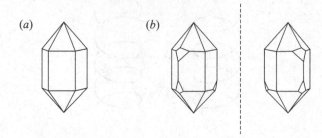

Fig. 1.11 (*a*) A holohedral hexagonal crystal. (*b*) A hemihedral hexagonal crystal and its mirror image.

of general physical systems these two terms are used as synonyms in this book. The group theoretical criterion for an object to be chiral is that it must not possess improper rotation symmetry elements such as a centre of inversion, reflection planes or rotation–reflection axes and so must belong to one of the point groups C_n, D_n, O, T or I.

Direct evidence that the structure of optically active materials is in some way chiral followed from the observation by Hauy in 1801 that the apparent hexagonal symmetry of quartz crystals was in fact reduced by the presence of small facets on alternate corners of the crystal. These hemihedral facets destroy the centre and planes of symmetry of the basic holohedral hexagonal crystal, and reduce the sixfold principal rotation axis with six perpendicular twofold rotation axes to a threefold principal axis with three perpendicular twofold axes, giving rise to two mirror image forms of quartz, as in Fig. 1.11. The two forms of quartz which Biot had found to provide opposite senses of optical rotation were subsequently identified by Herschel (1822) as the two hemihedral forms of quartz. This early example is very instructive since it illustrates a feature common to the generation of natural optical activity in many systems; namely a small chiral perturbation of a basic structure that is inherently symmetric.

Pasteur extended the concept of chirality from the realm of the structures of optically active crystals to that of the individual molecules which provide optically active fluids or solutions. He worked with tartaric acid, which Biot had shown to be optically active, and with paratartaric acid, which was chemically identical but optically inactive. The crystal forms of tartaric acid and most of its salts are hemihedral, whereas those of paratartaric acid and most of its salts are holohedral. But an anomaly in the case of sodium ammonium tartrate was discovered by Mitscherlich: the crystals of both active and inactive forms are hemihedral (in fact this was fortuitous since sodium ammonium paratartrate only gives hemihedral crystals when crystallized below 26°C). In 1848 Pasteur, in following up this discovery, observed that although both were indeed hemihedral, in the tartrate the hemihedral facets

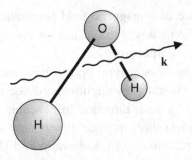

Fig. 1.12 A water molecule oriented in such a way that it appears as part of a left-handed helix to a light beam travelling parallel or antiparallel to the unit vector **k**.

were all turned the same way, whereas in the paratartrate there were equal amounts of crystals with opposite hemihedral facets. Pasteur reports (quoted by Lowry, 1935):

I carefully separated the crystals which were hemihedral to the right from those hemihedral to the left, and examined their solutions separately in the polarizing apparatus. I then saw with no less surprise than pleasure that the crystals hemihedral to the right deviated the plane of polarization to the right, and that those hemihedral to the left deviated to the left.

Paratartaric acid was therefore identified as a mixture, now known as a racemic mixture, of equal parts of mirror image forms of tartaric acid which neutralized the optical activity. This work, together with Fresnel's earlier statements, was instrumental in establishing the tetrahedral valencies of the carbon atom because a molecule must be assigned a three dimensional structure in order to be chiral.

While the absence of a centre of inversion, reflection planes and rotation–reflection axes in individual molecules is mandatory if an isotropic ensemble is to show optical activity, some crystals and oriented molecules which lack a centre of inversion but possess reflection planes or a rotation–reflection axis (so that they are superposable on their mirror images) can show optical activity for certain directions of propagation of the light beam. For example, an oriented water molecule (point group symmetry C_{2v}) appears as part of a helix to a light beam in any direction not contained in either of the two reflection planes of the molecule, as illustrated in Fig. 1.12. For every direction of the light beam for which the molecule appears as part of a left-handed helix, there is a direction for which the molecule appears as part of a corresponding right-handed helix. The optical rotations in the two directions are equal and opposite, so an isotropic ensemble of water molecules does not show optical activity. Although optical activity in oriented water has never been observed, optical rotation has been observed in certain directions in two other non-enantiomorphous systems: crystals of silver gallium sulphide, $AgGaS_2$, which have D_{2d} ($\bar{4}2\,m$) symmetry (Hobden, 1967); and the planar molecules of para-azoxyanisole, which form nematic liquid crystals, by orienting the molecules on a glass plate in the presence of a magnetic field perpendicular to the direction of propagation of the light (Williams, 1968). Thus natural optical activity is not exclusively related to enantiomorphism.

Although the measurement of natural optical rotation in chiral cubic crystals such as those of sodium chlorate is straightforward due to their spatial isotropy, for light propagation in a general direction in noncubic crystals natural optical activity is obscured by linear birefringence. It was only with the introduction of high accuracy universal polarimetry by Kobayashi and Uesu (1983) that optical rotation in crystals of any symmetry could be measured reliably and accurately. This enabled the optical rotation of tartaric acid crystals to be measured for the first time (Brożek *et al.*, 1995), 163 years after Biot's observation of the optical activity of tartaric acid in solution and 147 years after Pasteur's manual resolution of the enantiomorphous crystals. Kaminsky (2000) has provided a comprehensive review of the subtle topic of crystal optical activity and its measurement.

1.9.2 Inversion symmetry and physical laws

The discussion so far has been concerned with the intrinsic spatial symmetry, known as point group symmetry, of optically active molecules and crystals. The objects of the physical world display many kinds of spatial symmetry. For example stars, planets, water droplets and atoms have the high degree of symmetry associated with a sphere; and even the plant and animal world exhibit some degree of symmetry, although the symmetry of a butterfly is not as fundamental as that of a crystal or molecule. An object is said to have spatial symmetry if, after subjecting it to a symmetry operation such as inversion, reflection or rotation with respect to a symmetry element within the object, it looks the same as it did before. But more remarkable than these spatial symmetries is the fact that symmetries exist in the laws which determine the operation of the physical world. One consequence is that if a complete experiment is subjected to space inversion or time reversal, the resulting experiment should, in principle, be realizable (Wigner, 1927).

The symmetry operation of space inversion, represented by the parity operator P, inverts the coordinates used to specify the system through the coordinate origin, which may be located arbitrarily. This is equivalent to a reflection of the actual physical system in any plane containing the coordinate origin, followed by a rotation R_π through 180° about an axis perpendicular to the reflection plane, as illustrated in Fig. 1.13. Most physical laws, in particular those of electromagnetism (but not those responsible for β-decay), are unchanged by space inversion; in other words the equations representing the physical laws are unchanged if the coordinates (x, y, z) are replaced everywhere by $(-x, -y, -z)$, and the physical processes described by these laws are said to conserve parity.

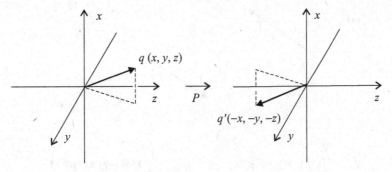

Fig. 1.13 The operation of space inversion P. An object at the point $q(x, y, z)$ is moved to the point $q'(-x, -y, -z)$.

The symmetry operation of time reversal, represented by the operator T, reverses the motions of all the physical entities in the system. If replacing the time coordinate (t) by ($-t$) everywhere in equations describing physical laws leaves those equations unchanged, the physical processes represented by those laws are said to be time reversal invariant, or have reversality. The reversality of a process referred to here must not be confused with the thermodynamic notion of reversability: a process will have reversality as long as the process with all motions reversed is in principle a possible process, however improbable it may be; thermodynamics is concerned with calculating the probability. The mechanical shuffling of a pack of cards is, in principle, a reversible process, although thermodynamics would classify it as an irreversible process. As Sachs (1987) has emphasized, the time coordinate has little to do with the thermodynamic concept of the 'arrow of time'. Time reversal is best thought of as motion reversal. It does not mean going backwards in time! A remarkable book by the philosopher–physicist Costa de Beauregard (1987) provides a comprehensive critical review of time as a measurable entity and the relation between its intrinsic reversibility and the asymmetry between past and future.

A *scalar* physical quantity such as temperature has magnitude but no directional properties; a *vector* quantity such as velocity has magnitude and an associated direction; and a *tensor* quantity such as electric polarizability has magnitudes associated with two or more directions. Scalars, vectors and tensors are classified according to their behaviour under the operations P and T. A vector whose sign is changed by P is called a *polar* or *true* vector; for example a position vector $\mathbf{r}$, as shown in Fig. 1.14a. A vector whose sign is not changed by P is called an *axial* or *pseudo* vector; for example the angular momentum is $\mathbf{L} = \mathbf{r} \times \mathbf{p}$, the vector product of the position vector $\mathbf{r}$ and the momentum $\mathbf{p}$, and since the polar vectors $\mathbf{r}$ and $\mathbf{p}$ change sign under P, the axial vector $\mathbf{L}$ does not. In other words $\mathbf{L}$ is defined relative to the sense of rotation by a 'right hand rule', and P does not change the sense of rotation, as illustrated in Fig. 1.14b. A vector whose sign is not changed by T is

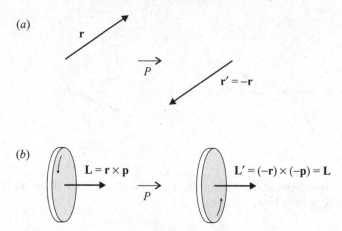

Fig. 1.14 The space inversion operator P changes the sign of the polar position vector $\mathbf{r}$ in (a) but does not change the sign of the axial angular momentum vector $\mathbf{L}$ in (b).

called *time-even*; for example the position vector, which is not a function of time. A vector whose sign is changed by T is called *time-odd*; for example, velocity and angular momentum, which are linear functions of time. Figure 1.15 illustrates the effect of T on $\mathbf{r}$, $\mathbf{v}$ and $\mathbf{L}$.

Pseudoscalar quantities are, in accordance with the classification outlined in the previous paragraph, numbers with no directional properties but which change sign under space inversion. Pseudoscalars are of central importance in natural optical activity phenomena because the quantities that are measured, such as optical rotation angle or circular intensity difference, are pseudoscalars. Since a helix is the archetype for all chiral objects, it is instructive to identify the pseudoscalar helix parameter. A circular helix can be defined by the radius vector from the origin $\mathbf{O}$ of a coordinate system to a point on the curve (see Fig. 1.16):

$$\mathbf{r} = \mathbf{i}a\cos\theta + \mathbf{j}a\sin\theta + \mathbf{k}b\theta, \qquad (1.9.1)$$

where $\mathbf{i}$, $\mathbf{j}$, $\mathbf{k}$ are unit vectors along the x, y, z axes. The helix pitch is $2\pi b$, this being the distance between successive turns. A right-handed helical screw sense is characterized by a positive value of b since a positive change in θ (taking x into y) is associated with a positive translation through $b\theta$ along z. This assumes that a right-handed system of axes, as in Fig. 1.16, is used. Similarly, a left-handed screw sense is characterized by a negative value of b since a positive change in θ is now associated with a negative translation along z. Since P reverses the screw sense of the helix, it changes b to $-b$ so the helix pitch is therefore a pseudoscalar. Since, as discussed in Section 1.9.3 below, the pattern of electric field vectors of a linearly polarized light beam established in an optical rotatory medium constitutes a circular

(a)

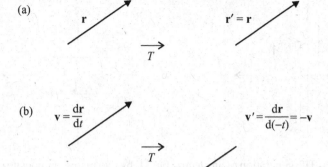

(b)

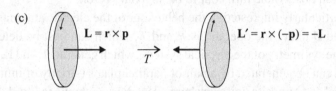

(c)

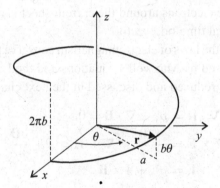

Fig. 1.15 The time reversal operator T does not change the sign of the time-even position vector $\mathbf{r}$ in (a) but changes the sign of the time-odd velocity vector $\mathbf{v}$ in (b) and angular momentum vector $\mathbf{L}$ in (c).

Fig. 1.16 A circular helix. a is the helix radius and $2\pi b$ is the pitch.

helix, this analysis shows that the optical rotation angle α is a pseudoscalar because, for a path length l, $\alpha = -l/b$. The minus sign arises from the fact that b is defined above to be positive for a right-handed helical screw sense, whereas the chemical convention for a positive angle of optical rotation is that it be associated with a left-handed helical light path (see Fig. 1.18a). In fact it is shown later (Section 4.3.3)

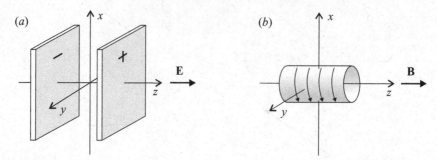

Fig. 1.17 (*a*) The generation of an electric field **E** by two plates of opposite charge. (*b*) The generation of a magnetic field **B** by a cylindrical current sheet.

that only the *natural* optical rotation observable is a pseudoscalar: the *magnetic* optical rotation observable turns out to be an axial vector.

We are particularly interested in the behaviour of the electric and magnetic field vectors **E** and **B** under the operations P and T, which can best be determined by examining the symmetry of the physical systems which generate **E** and **B**. A uniform electric field can be generated by a pair of parallel plates (strictly of infinite extent) carrying equal and opposite uniform charge densities, as shown in Fig. 1.17*a*. Under P, the two plates exchange positions, while retaining their respective charges, so **E** changes sign. Since the charges are stationary, T does not affect the system. Thus **E** is a polar time-even vector. A uniform magnetic field can be generated by a cylindrical current sheet (strictly of infinite length), as shown in Fig. 1.17*b*. The sense of rotation of the electrons around the current sheet is reversed by T but not by P. Thus **B** is an axial time-odd vector.

We can now see that the laws of electromagnetism conserve parity and reversality. The laws are summarized by Maxwell's equations and the Lorentz force equation (these equations are introduced and discussed in the next chapter):

$$\nabla \cdot \mathbf{D} = \rho, \quad \nabla \cdot \mathbf{B} = 0,$$
$$\nabla \times \mathbf{E} = -\frac{\partial \mathbf{B}}{\partial t}, \quad \nabla \times \mathbf{H} = \mathbf{J} + \frac{\partial \mathbf{D}}{\partial t},$$
$$\mathbf{F} = \rho \mathbf{E} + \mathbf{J} \times \mathbf{B}.$$

Thus the third equation, for example, which summarizes Faraday's and Lenz's law of electromagnetic induction, is easily seen to be invariant under P and T:

$$\nabla \times \mathbf{E} = -\frac{\partial \mathbf{B}}{\partial t} \begin{cases} P \atop \rightarrow & (-\nabla) \times (-\mathbf{E}) = -\dfrac{\partial(+\mathbf{B})}{\partial(+t)} \\ T \atop \rightarrow & (+\nabla) \times (+\mathbf{E}) = -\dfrac{\partial(-\mathbf{B})}{\partial(-t)}. \end{cases}$$

The remaining equations are easily shown to be similarly invariant. Consequently, any physical process involving only the electromagnetic interaction, for example the interaction of light with a molecule, must conserve parity and reversality.

For completeness, a third fundamental symmetry operation, that of charge conjugation, should be mentioned. Charge conjugation, represented by the operator C, arises in relativistic quantum field theory and interconverts particles and antiparticles (Berestetskii, Lifshitz and Pitaevskii, 1982). For charged particles this implies a reversal of charge. Although it has no classical counterpart, it nonetheless has conceptual value in certain contexts and is useful for checking the consistency of equations. For example, by interpreting C simply as reversing the signs of all the charges in a system, it is easily seen that Maxwell's equations given above are invariant under this operation.

1.9.3 Inversion symmetry and optical rotation

It is now demonstrated that the natural and magnetic optical rotation experiments conserve parity and reversality. Similar arguments can be applied to all other optical activity phenomena and, as illustrated below, can be used to discount or predict possible new effects without recourse to mathematical theories. This section is based on articles by Rinard and Calvert (1971) and Barron (1972).

The natural optical rotation experiment consists of a chiral medium, such as a quartz crystal or a fluid containing inherently chiral molecules, in which the electric field vectors of a linearly polarized light beam are established in a helical pattern. A convenient representation of this helical pattern is a twisted ribbon extending through the medium, with the electric field vectors vibrating in the plane of the ribbon, as illustrated in Fig. 1.18a. The helical pattern of electric field vectors is a physical object with well-defined symmetry properties. Since only electromagnetic interactions are involved, the physical processes giving rise to optical rotation must conserve parity and reversality. In other words, if P and T are applied to the entire experiment, the result must also be a possible experiment. Since P is not a point group symmetry operation for chiral molecules, it is sometimes implied in the literature that processes involving such molecules do not conserve parity (Ulbricht, 1959): this incorrect notion presumably arises because the experiment is not considered in its entirety. Thus under P, the screw sense of the helical pattern of electric field vectors in the medium is inverted (the optical rotation angle being a pseudoscalar) and the direction of propagation of the light beam is reversed (Fig. 1.18a); at the same time the chiral medium is converted into its nonsuperposable mirror image (Fig. 1.18b). This result is itself a possible experiment since replacing the chiral medium by its enantiomorph results in an opposite sense of optical rotation, and reversing the direction of propagation of the light beam does not affect the optical rotation sense. Thus natural optical rotation conserves parity. Under T,

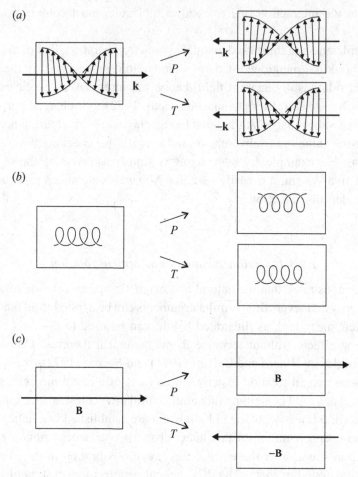

Fig. 1.18 The effect of P and T on (a) the helical pattern established by the electric field vectors of a linearly polarized light beam propagating along the direction of a unit vector **k** in a rotatory medium, on (b) a chiral medium and on (c) an achiral medium in the presence of a static magnetic field. Notice that the negative optical rotation angle in the initial state on the left in (a) is associated with a right-handed screw sense in the helical pattern of electric field vectors.

the direction of propagation is reversed, but the screw sense of the helical pattern of electric field vectors is preserved (Fig. 1.18a). Since T does not affect the chiral medium (if nonmagnetic), the time-reversed experiment is physically realizable, and corresponds simply to reversing the direction of propagation of the light beam, which does not change the sense of the optical rotation. Thus natural optical rotation has reversality.

The Faraday rotation experiment consists of an achiral medium in a static externally applied magnetic field parallel to the direction of propagation of a linearly polarized light beam whose electric field vectors are established in a helical pattern in the medium. Again P reverses the direction of propagation of the light beam and inverts the screw sense of the helical pattern of electric field vectors (Fig. 1.18a), but the achiral medium and the magnetic field are unchanged (Fig. 1.18c). This corresponds with what is found experimentally, namely that reversing the direction of propagation of the light beam relative to the magnetic field direction reverses the sense of the Faraday rotation. Thus Faraday rotation conserves parity. Under T, the direction of propagation of the light beam is reversed, but its sense of optical rotation is preserved. Since T reverses the direction of the magnetic field but does not affect the medium (if nonmagnetic in the absence of the applied magnetic field) (Fig. 1.18c), the time-reversed experiment is physically realizable, for reversing the directions of both the magnetic field and the light beam preserves the sense of the Faraday rotation. Thus Faraday rotation has reversality. It can also be seen that Faraday rotation must depend on odd powers of **B** since these change sign under T, whereas even powers do not.

These symmetry arguments can also be used to demonstrate that there is no simple electrical analogue of the Faraday effect; in other words, that optical rotation cannot be induced in a linearly polarized light beam traversing an isotropic achiral medium by a static electric field in the direction of propagation. Thus P does not affect the medium, although the direction of the electric field and the direction of propagation and optical rotation sense of the light beam are reversed: as all directions in the unperturbed medium are equivalent, any optical rotation induced by odd (or even) powers of **E** would violate parity. Similarly, T does not affect the electric field, the medium, or the sense of optical rotation, but reverses the direction of propagation of the light beam relative to the electric field direction. Consequently, any optical rotation induced by odd (but *not* even) powers of E would also violate reversality. It might be thought that this effect could be induced in a fluid of *chiral* molecules: certainly, pictorial arguments show that parity would not be violated, but they also show that reversality would be violated. The extension of these pictorial arguments to more exotic media is cumbersome, so we refer to the group theoretical discussions of Buckingham, Graham and Raab (1971) and Gunning and Raab (1997) for demonstrations that an electric analogue of the Faraday effect is possible in certain crystals, and to Kaminsky (2000) for an account of theoretical and experimental aspects of this phenomenon, which is called electrogyration in crystal optics.

Although rotation of the plane of polarization of linearly polarized light in an isotropic achiral medium in the absence of magnetic fields would violate parity, rotation of the major axis of the polarization ellipse of an elliptically polarized light beam in the same medium would not. Elliptically polarized light is a coherent superposition of linearly and circularly polarized components. The tip of the electric field vector in a fixed plane perpendicular to the direction of propagation of a circularly polarized light beam traces out a circle with time: thus P reverses the handedness because, although the rotation sense of the electric field vector is maintained, the direction of propagation is reversed; whereas T preserves the handedness because both the rotation sense of the electric field vector and the direction of propagation are reversed. Thus under P, the medium is not affected but the direction of propagation, the sense of the optical rotation and the handedness of an elliptically polarized light beam are reversed; that is, reversing the handedness of the ellipticity reverses the sense of the optical rotation. Under T, the medium is not affected, and the sense of the optical rotation and the handedness of the ellipticity are maintained (with the direction of propagation of the light beam reversed). These conclusions agree with the observations of the effect known as the auto rotation of the polarization ellipse in which the major axis of the polarization ellipse of an intense elliptically polarized laser beam rotates on passing through an isotropic achiral fluid (Maker, Terhune and Savage, 1964). Reversing the handedness of the ellipticity reverses the sense of optical rotation. The effect is due to the intensity dependence of the refractive index: an elliptically polarized light beam may be considered as a superposition of two coherent circularly polarized beams of different intensity, so the two components will propagate through the medium with different velocities thereby causing the major axis of the ellipse to rotate. An interesting speculation is that a mechanism could exist for producing optical rotation of an elliptically polarized light beam in a racemic mixture which would be a function of the optical activity of one of the enantiomers, for such an effect would not violate parity or reversality.

It is easy to see that optical rotation induced by a rapid rotation of a complete isotropic sample about an axis parallel to the propagation direction of the light beam would not violate parity or reversality and is therefore a possible phenomenon. This effect, called the *rotatory ether drag*, was observed by Jones (1976) in a rapidly rotating rod made of Pockels glass. The symmetry aspects lead it to be classified along with the Faraday effect as optical activity induced by a time-odd external influence.

1.9.4 Inversion symmetry and optical activity in light scattering

Similar pictorial arguments can be applied to Rayleigh and Raman optical activity. These are illustrated most simply for ellipticity in Rayleigh scattered light in linearly polarized incident light: the method applies equally well to the circular intensity difference, but the exposition is more cumbersome.

Fig. 1.19a shows an experiment in which a small right ellipticity is detected in a light beam scattered at 90° from an isotropic chiral medium in an incident light beam linearly polarized perpendicular to the scattering plane. Under P, the directions of

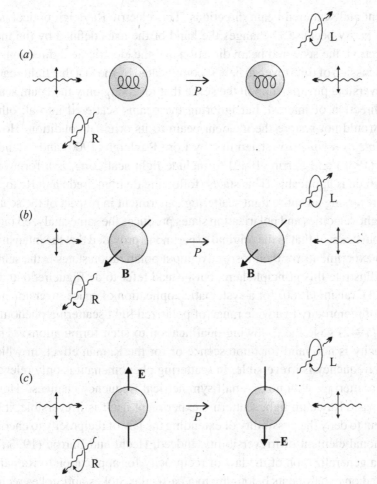

Fig. 1.19 The effect of P on (a) the natural, (b) the magnetic and (c) the electric Rayleigh optical activity experiments.

the incident and scattered beams are reversed, with the scattered beam now carrying a left ellipticity and the chiral medium replaced by its nonsuperposable mirror image. Assuming that space is isotropic, this is a realizable experiment because replacing the medium by its enantiomorph results in an opposite sense of ellipticity in the scattered beam. Thus natural Rayleigh optical activity conserves parity.

If the ellipticity in the scattered beam were generated in an achiral medium by a static magnetic field parallel to the scattered beam, application of P would reverse the direction of the scattered beam relative to the magnetic field (Fig. 1.19b) and so magnetic Rayleigh optical activity also conserves parity.

Fig. 1.19c illustrates the more subtle phenomenon of an ellipticity in the scattered beam generated in an achiral medium by a static electric field perpendicular to both the incident and scattered beam directions. This electric Rayleigh optical activity conserves parity because P changes the hand of the axes defined by the incident beam direction, the scattered beam direction and the electric field direction.

The discussion of time reversal here is complicated by the fact that light scattering is not a reversible phenomenon in the sense that reversing only the beam scattered into the direction of interest, but ignoring the beams scattered into all other directions, would not restore the incident beam to its original condition. However, the *principle of reciprocity*, stated first by Lord Rayleigh (1900) and extended by Krishnan (1938) and Perrin (1942) to include light scattering, is a form of time reversal which is applicable. This states (following de Figueiredo and Raab, 1980) that time-reversing an entire light scattering experiment in respect of the scattering system, light velocities and polarization states produces the same analyzed intensity in the output beam as that in the original experiment, provided that the intensity used in the two experiments for their respective input polarization states is the same. We shall not illustrate this principle here, but instead refer to de Figueiredo and Raab (1980) and Graham (1980) for a systematic application of space inversion and the principle of reciprocity to a wide range of polarized light scattering phenomena.

Perrin (1942) gave the following qualification to such formulations: 'The law of reciprocity is not valid for fluorescence or for the Raman effect, in which the change in frequency is irreversible. In scattering phenomena it is only relevant for Rayleigh scattering, with no or small symmetrical frequency changes.' However, since the basic Rayleigh light scattering experiment itself is irreversible, it seems inconsistent to deny the possibility of extending the law of reciprocity to encompass this additional element of irreversibility. Indeed, Hecht and Barron (1993*a*) have provided a generalization of the law of reciprocity for application to Raman scattering based on experiments belonging to a particular Stokes/antiStokes reciprocal pair.

1.9.5 *Motion-dependent enantiomorphism: true and false chirality*

Optical activity is not necessarily the hallmark of chirality. The failure to distinguish properly between natural and magnetic optical rotation, for example, has been a source of confusion in the literature of both chemistry and physics. Lord Kelvin (1904) was fully aware of the fundamental distinction, for his *Baltimore Lectures* contain the statement:

The magnetic rotation has neither left-handed nor right-handed quality (that is to say, no chirality). This was perfectly understood by Faraday, and made clear in his writings, yet even to the present day we frequently find the chiral rotation and the magnetic rotation of the plane of polarized light classed together in a manner against which Faraday's original description of his discovery contains ample warning.

He may have had Pasteur in mind. For example, because a magnetic field induces optical rotation, Pasteur thought that by growing crystals, normally holohedral, in a magnetic field a magnetically induced dissymmetry would be manifest in hemihedral crystal forms. The resulting crystals, however, retained their usual holohedral forms (Mason, 1982). Lord Kelvin's viewpoint was reinforced much later by Zocher and Török (1953), who discussed the space–time symmetry aspects of natural and magnetic optical activity from a general classical viewpoint and recognized that quite different asymmetries are involved. Similarly Post (1962) emphasized the fundamental distinction between natural and magnetic optical activity in terms of the reciprocal and nonreciprocal characteristics, respectively, of the two phenomena (reciprocal and nonreciprocal refer here to the fact that the natural optical rotation sense is the same on reversing the direction of propagation of the light beam whereas the magnetic optical rotation sense reverses).

It is already clear from Section 1.9.3 above that natural and magnetic optical rotation have different symmetry characteristics. Further considerations (see Section 4.3.3) show that the natural optical rotation observable is a time-even pseudoscalar, whereas the magnetic optical rotation observable is a time-odd axial vector. These and other arguments suggest that the hallmark of a chiral system is that it can support time-even pseudoscalar observables. This leads to the following definition which enables chirality to be distinguished from other types of dissymmetry (Barron, 1986*a*,*b*):

True chirality is exhibited by systems that exist in two distinct enantiomeric (enantiomorphic) states that are interconverted by space inversion, but not by time reversal combined with any proper spatial rotation.

This means that the spatial enantiomorphism shown by truly chiral systems is time invariant. Spatial enantiomorphism that is time noninvariant has different characteristics that this author has called false chirality to emphasize the distinction. Originally, it was not intended that the terminology 'true' and 'false' chirality should become standard nomenclature, but these terms have gradually crept into the literature of stereochemistry. Notice that a magnetic field on its own is not even falsely chiral because there is no associated spatial enantiomorphism. Essentially, for a truly chiral system, parity P is not a symmetry operation (since it generates a different system, namely the enantiomer) but time reversal T is a symmetry operation; whereas for a falsely chiral system neither P nor T are symmetry operations on their own but the combination PT is a symmetry operation.

A stationary object such as a finite helix that is chiral according to Lord Kelvin's original definition is accommodated by the first part of this definition: space inversion is a more fundamental operation than mirror reflection, but provides an equivalent result. Time reversal is irrelevant for a stationary object, but the full

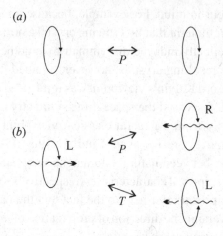

Fig. 1.20 The effect of parity P and time reversal T on the motions of (a) a stationary spinning particle and (b) a translating spinning particle. The designations L and R for left and right handed follow the convention used in elementary particle physics, which is opposite to the classical optics convention used elsewhere in this book.

definition is required to identify more subtle sources of chirality in which motion is an essential ingredient. A few examples will make this clear.

Consider an electron, which has a spin quantum number $s = \frac{1}{2}$ with $m_s = \pm\frac{1}{2}$ corresponding to the two opposite projections of the spin angular momentum onto a space-fixed axis. A stationary spinning electron is not a chiral object because space inversion P does not generate a distinguishable P-enantiomer (Fig. 1.20a). However, an electron translating with its spin projection parallel or antiparallel to the direction of propagation has true chirality because P interconverts distinguishable left and right spin-polarized versions propagating in opposite directions, whereas time reversal T does not (Fig. 1.20b). In elementary particle physics, the projection of the spin angular momentum $\mathbf{s}$ of a particle along its direction of motion is called the helicity $\lambda = \mathbf{s} \cdot \mathbf{p}/|\mathbf{p}|$ (Gibson and Pollard, 1976). Spin-$\frac{1}{2}$ particles can have $\lambda = \pm\hbar/2$, the positive and negative states being called right and left handed. This, however, corresponds to the opposite sense of handedness to that used in the usual definition of right- and left-circularly polarized light in classical optics as employed in this book.

The photons in a circularly polarized light beam propagating as a plane wave are in spin angular momentum eigenstates characterized by $s = 1$ with $m_s = \pm1$ corresponding to projections of the spin angular momentum vector parallel or antiparallel, respectively, to the propagation direction. The absence of states with $m_s = 0$ is connected with the fact that photons, being massless, have no rest frame and so always move with the velocity of light (Berestetskii, Lifshitz and Pitaevskii,

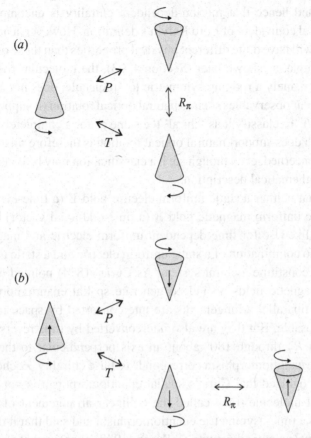

Fig. 1.21 The effect of P, T and R_π on (a) a stationary spinning cone, which has false chirality, and on (b) a translating spinning cone, which has true chirality. The systems generated by P and T may be interconverted by R_π in (a) but not in (b).

1982). Considerations the same as those in Fig. 1.20b show that a circularly polarized photon has true chirality.

Now consider a cone spinning about its symmetry axis. Because P generates a version that is not superposable on the original (Fig. 1.21a), it might be thought that this is a chiral system. The chirality, however, is false because T followed by a rotation R_π through 180° about an axis perpendicular to the symmetry axis generates the same system as space inversion (Fig. 1.21a). If, however, the spinning cone is also translating along the axis of spin, T followed by R_π now generates a system different from that generated by P alone (Fig. 1.21b). Hence a *translating spinning cone* has true chirality.

Mislow (1999) has argued that a nontranslating spinning cone belongs to the spatial point group C_∞ and so is chiral. More generally, he has suggested that objects that exhibit enantiomorphism, whether T-invariant or not, belong to chiral point groups and hence that motion-dependent chirality is encompassed in the group theoretical equivalent of Lord Kelvin's definition. However, a nontranslating spinning cone will have quite different physical properties than those of, say, a finite helix. For example, as shown later (Section 4.3.3), the molecular realization of a spinning cone, namely a rotating symmetric top molecule, does not support time-even pseudoscalar observables such as natural optical rotation (it supports magnetic optical rotation). To classify it as 'chiral' the same as for a completely asymmetric molecule which does support natural optical rotation is therefore misleading as far as physics is concerned, even though such a classification may be consistent within a particular mathematical description.

It is clear that neither a static uniform electric field $\mathbf{E}$ (a time-even polar vector) nor a static uniform magnetic field $\mathbf{B}$ (a time-odd axial vector) constitutes a chiral system; likewise for time dependent uniform electric and magnetic fields. Furthermore, no combination of a static uniform electric and a static uniform magnetic field can constitute a chiral system. As Curie (1894) pointed out, collinear electric and magnetic fields do indeed generate spatial enantiomorphism. Thus parallel and antiparallel arrangements are interconverted by space inversion and are not superposable. But they are also interconverted by time reversal combined with a rotation R_π through 180° about an axis perpendicular to the field directions and so the enantiomorphism corresponds to false chirality. Zocher and Török (1953) also recognized that Curie's spatial enantiomorphism is not the same as that of a chiral molecule: they called the collinear arrangement of electric and magnetic fields a time-asymmetric enantiomorphism and said that it does not support time-symmetric optical activity. Tellegen (1948) conceived of a medium with novel electromagnetic properties comprising microscopic electric and magnetic dipoles tied together with their moments either parallel or antiparallel. Such media clearly exhibit enantiomorphism corresponding to false chirality. Although much discussed (Post, 1962; Lindell *et al.*, 1994; Raab and Sihvola, 1997; Weiglhofer and Lakhtakia, 1998), Tellegen media have never been observed in nature and do not appear to have been fabricated.

In fact the basic requirement for two collinear vectorial influences to generate chirality is that one transforms as a polar vector and the other as an axial vector, with both either time even or time odd. The second case is exemplified by the magnetochiral phenomena described in Section 1.7 above, where a birefringence and a dichroism may be induced in an isotropic chiral sample by a uniform magnetic field $\mathbf{B}$ collinear with the propagation vector $\mathbf{k}$ of a light beam of arbitrary polarization. Thus parallel and antiparallel arrangements of $\mathbf{B}$ and $\mathbf{k}$, which are interconverted by space inversion, are true chiral enantiomers because they cannot be interconverted by time reversal since $\mathbf{k}$ and $\mathbf{B}$ are both time odd.

The new definition of chirality described here has proved useful in areas as diverse as the scattering of spin-polarized electrons from molecules (Blum and Thompson, 1997) and absolute enantioselection (Avalos *et al.*, 1998). A new dimension has been added to the physical background of these areas by the suggestion that processes involving chiral molecules may exhibit a breakdown of conventional microscopic reversibility, but preserve a new and deeper principle of *enantiomeric* microscopic reversibility, in the presence of a falsely chiral influence such as collinear electric and magnetic fields (Barron, 1987). The conventional microscopic reversibility of a process is based on the invariance of the quantum mechanical scattering amplitude under time reversal so that the amplitudes for the forward and reverse processes are identical. The enantiomeric microscopic reversibility of a process, which is only relevant when chiral particles are involved, is based on the invariance of the scattering amplitude under both time reversal and parity so that the amplitude for the forward process equals that for the reverse process involving the mirror-image chiral particles. In other words, the process is not invariant under P and T separately but is invariant under the combined PT operation. This exposes an analogy with CP violation in elementary particle physics, mentioned in the next section, in which the concept of false chirality also arises but with respect to CP-enantiomorphism rather than P-enantiomorphism and CPT-invariance rather than PT-invariance.

1.9.6 Symmetry violation: the fall of parity and time reversal invariance

Prior to 1957 it had been accepted as self evident that handedness is not built into the laws of nature. If two objects exist as nonsuperposable mirror images of each other, such as the two enantiomers of a chiral molecule, it did not seem reasonable that nature should prefer one over the other. Any difference between enantiomeric systems was thought to be confined to the sign of pseudoscalar observables: the mirror image of any complete experiment involving one enantiomer should be realizable, with any pseudoscalar observable (such as optical rotation angle) changing sign but retaining *exactly* the same magnitude. Then Lee and Yang (1956) pointed out that, unlike the strong and electromagnetic interactions, there was no evidence for parity conservation in processes involving the weak interaction. Of the experiments they suggested, that performed by Wu *et al.* (1957) is the most famous.

The Wu experiment studied the β-decay process

$$^{60}\text{Co} \rightarrow {}^{60}\text{Ni} + e^- + v_e^*$$

in which, essentially, a neutron decays *via* the weak interaction into a proton, an electron e^- and an electron antineutrino v_e^*. The nuclear spin magnetic moment $\mathbf{I}$ of

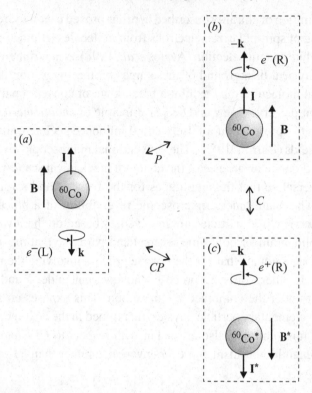

Fig. 1.22 Parity violation in β-decay. Only experiment (*a*) is observed; the space-inverted version (*b*) cannot be realized. Symmetry is recovered in experiment (*c*), obtained from (*a*) by invoking the combined *CP* operation. Anti-Co is represented by Co*, and **B*** and **I*** are reversed relative to **B** and **I** because the charges on the moving source particles change sign under *C*.

each ^{60}Co nucleus was aligned with an external magnetic field **B**, and the angular distribution of the emitted electrons measured. It was found that the electrons were emitted preferentially in the direction *antiparallel* to that of **B** (Fig. 1.22a). As discussed in Section 1.9.2, **B** and **I** are axial vectors and so do not change sign under space inversion, whereas the electron propagation vector **k** does because it is a polar vector. Hence in the corresponding space-inverted experiment the electrons should be emitted *parallel* to the magnetic field (Fig. 1.22b). It is only possible to reconcile the opposite electron propagation directions in Figs. 1.22a and 1.22b with parity conservation if there is no preferred direction for electron emission (an isotropic distribution), or if the electrons are emitted in the plane perpendicular to **B**. The observation depicted in Fig. 1.22a provides unequivocal evidence for parity violation. Another important aspect of parity violation in β-decay is that the emitted electrons have a left-handed longitudinal spin polarization, being accompanied by right-handed antineutrinos.

In fact symmetry is recovered by invoking invariance under the combined *CP* operation in which charge conjugation and space inversion are applied together. This would mean that the missing experiment is to be found in the antiworld! In other words, nature has no preference between the original experiment depicted in Fig. 1.22*a* and that depicted in Fig. 1.22*c*, generated from the original by *CP*, in which anti-^{60}Co decays into a right-handed spin-polarized positron moving antiparallel to the antimagnetic field. This has not been tested directly in the decay of nuclei due to the unavailability of antinuclei, but *CP* invariance has been established experimentally for the decay of certain elementary particles (however, as outlined below, it has been shown to be violated in the neutral *K* meson system). This result implies that *P* violation is accompanied here by *C* violation: absolute charge is distinguished since the charge that we call negative is carried by electrons, which are emitted with a left-handed spin polarization.

Following the Wu experiment described above, the original Fermi theory of the weak interaction was upgraded in order to take account of parity violation. This was achieved by reformulating the theory in such a way that the interaction takes the form of a left-handed pseudoscalar. However, a number of technical problems remained. These were finally overcome in the 1960s in the celebrated work of S. Weinberg, A. Salam and S. L. Glashow, which unified the theory of the weak and electromagnetic interactions into a single electroweak interaction theory. The conceptual basis of the theory rests on two pillars; gauge invariance and spontaneous symmetry breaking (Gottfried and Weisskopf, 1984; Weinberg, 1996). In addition to accommodating the massless photon and the two massive charged W^+ and W^- particles, which mediate the charge-changing weak interactions, a new massive particle, the neutral intermediate vector boson Z^0, was predicted which can generate a new range of neutral current phenomena including parity-violating effects in atoms and molecules. In one of the most important experiments of all time, these three particles were detected in 1983 at CERN in proton–antiproton scattering experiments.

It is clear from Section 1.9.3 that optical rotation in vapours of free atoms would violate parity (but not time-reversal invariance). In fact tiny optical rotations and related observables are now measured routinely in atomic vapours of heavy atoms such as bismuth and thallium (Khriplovich, 1991; Bouchiat and Bouchiat, 1997). One source of such effects is a weak neutral current interaction between the nucleus and the orbital electrons. Hegstrom *et al.* (1988) have provided an appealing pictorial representation of the associated atomic chirality in terms of a helical electron probability current density. Such experiments are remarkable in that they address issues in particle physics from 'bench top' experiments. For example, they are uniquely sensitive to a variety of 'new physics' (beyond the standard model) because they measure a set of model-independent electron–quark electroweak coupling constants that are different from those that are probed by high energy experiments requiring accelerators (Wood *et al.*, 1997).

Chiral molecules support a unique manifestation of parity violation in the form of a slight lifting of the exact degeneracy of the energy levels of mirror-image enantiomers (Rein, 1974; Khriplovich, 1991). Being a time-even pseudoscalar, the weak neutral current interaction largely responsible for this parity-violating energy difference is the quintessential truly chiral influence in molecular physics. It lifts only the degeneracy of the space-inverted (P) enantiomers of a truly chiral system; the P-enantiomers of a falsely chiral system such as a nontranslating spinning cone remain strictly degenerate (Barron, 1986a). Although not yet observed experimentally, this tiny parity-violating energy difference between enantiomers may be calculated (Hegstrom, Rein and Sandars, 1980) and has attracted considerable discussion as a possible source of biological homochirality (see, for example, MacDermott, 2002 and Quack, 2002). Initial results appeared to support the idea, but these are contradicted by the most recent and sophisticated studies (Wesendrup *et al.*, 2003; Sullivan *et al.*, 2003). Much more theoretical and experimental work is needed to find out whether or not there is any connection between parity violation and biological homochirality.

Since, on account of parity violation, the P-enantiomers of a truly chiral object are not exactly degenerate, they are not strict enantiomers (because the concept of enantiomers implies the *exact* opposites). So where is the strict enantiomer of a chiral object to be found? In the antiworld, of course! Just as symmetry is recovered in the Wu experiment above by invoking CP rather than P alone, one might expect strict enantiomers to be interconverted by CP; in other words, the molecule with the opposite absolute configuration but composed of antiparticles should have exactly the same energy as the original (Barron, 1981a,b; Jungwirth, Skála and Zahradník, 1989), which means that a chiral molecule is associated with two distinct pairs of strict enantiomers (Fig. 1.23). Since P violation automatically implies C violation here, it also follows that there is a small energy difference between a chiral molecule in the real world and the corresponding chiral molecule with the same absolute configuration in the antiworld. Furthermore, the P- and C-violating energy differences must be identical. This general definition of the strict enantiomer of a chiral system is consistent with the chirality that free atoms display on account of parity violation. The weak neutral current generates only one type of chiral atom in the real world: the conventional enantiomer of a chiral atom obtained by space inversion alone does not exist. Clearly the enantiomer of a chiral atom is generated by the combined CP operation. Thus the corresponding atom composed of antiparticles will of necessity have the opposite 'absolute configuration' and will show an opposite sense of parity-violating optical rotation (Barron, 1981a).

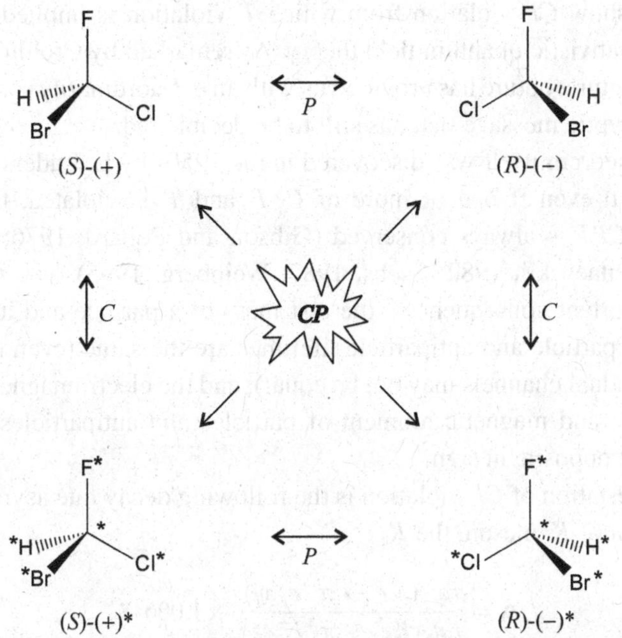

Fig. 1.23 The two pairs of strict enantiomers (exactly degenerate) of a chiral molecule that are interconverted by *CP*. The structures with atoms marked by asterisks are antimolecules built from the antiparticle versions of the constituents of the original molecules. The strict degeneracy remains even if *CP* is violated provided *CPT* is conserved. The absolute configurations shown for CHFClBr were determined by Raman optical activity and specific rotation (Costante *et al.*, 1997; Polavarapu, 2002*a*).

The *P*-enantiomers of objects such as translating spinning electrons or cones that only exhibit chirality on account of their motion also show parity-violating energy differences. One manifestation is that left-handed and right-handed particles (or antiparticles) have different weak interactions (Gibson and Pollard, 1976; Gottfried and Weisskopf, 1984). Again, strict enantiomers are interconverted by *CP*: for example, a left-handed electron and a right-handed positron. Notice that, since a photon is its own antiparticle (Berestetskii, Lifshitz and Pitaevskii, 1982; Weinberg, 1995), right- and left-handed circularly polarized photons are automatically strict enantiomers.

Violation of time reversal was first observed in the famous experiment of Christenson *et al.* (1964) involving measurements of rates for different decay modes of the neutral *K*-meson, the K^0 (Gottfried and Weisskopf, 1984; Sachs, 1987; Branco, Lavoura and Silva, 1999). Although unequivocal, the effects are very small; certainly nothing like the parity-violating effects in weak processes which can sometimes be absolute. In fact *T* violation itself is not observed directly: rather, the observations show *CP* violation from which *T* violation is implied *via* the *CPT* theorem of relativistic quantum field theory. As remarked by Cronin (1981) in his Nobel Prize lecture, nature has provided us with an extraordinarily sensitive system to convey a cryptic message that has still to be deciphered.

The *CPT* theorem itself was discovered in the 1950s by L. Lüders and W. Pauli and states that, even if one or more of *C*, *P* and *T* is violated, the combined operation of *CPT* is always conserved (Gibson and Pollard, 1976; Berestetskii, Lifshitz and Pitaevskii, 1982; Sachs, 1987; Weinberg, 1995). The *CPT* theorem has three important consequences: the rest mass of a particle and its antiparticle are equal; the particle and antiparticle lifetimes are the same (even though decay rates for individual channels may not be equal); and the electromagnetic properties such as charge and magnetic moment of particles and antiparticles are equal in magnitude but opposite in sign.

One manifestation of *CP* violation is the following decay rate asymmetry of the long-lived neutral *K*-meson, the K_L:

$$\Delta = \frac{rate(K_L \rightarrow \pi^- e_r^+ \nu_l)}{rate(K_L \rightarrow \pi^+ e_l^- \nu_r^*)} \approx 1.00648.$$

As the formula indicates, K_L can decay into either positive pions π^+ plus left-helical electrons e_l^- plus right-helical antineutrinos ν_r^*; or into negative antipions π^- plus right-helical positrons e_r^+ plus left-helical neutrinos ν_l. Because the two sets of decay products are interconverted by *CP*, this decay rate asymmetry is a manifestation of *CP* violation. Since a particle and its antiparticle have the same rest mass if *CPT* invariance holds, the *CP*-violating interaction responsible for the decay rate asymmetry of the K_L does not lift the degeneracy of the two sets of *CP*-enantiomeric products. This type of *CP* violation therefore falls within the conceptual framework of chemical catalysis because only the kinetics, but not the thermodynamics, of the decay process are affected (Barron, 1994).

The original proof of the exact degeneracy of the strict (*CP*) enantiomers of a chiral molecule which appear in Fig. 1.23 was based on the *CPT* theorem with the assumption that *CP* is not violated. However it was subsequently shown, using an extension of the proof that a particle and its antiparticle have identical rest mass even if *CP* is violated provided *CPT* is conserved, that the *CP*-enantiomers of a chiral molecule remain strictly degenerate even in the presence of *CP* violation provided *CPT* invariance holds (Barron, 1994). This suggests that forces responsible for *CP* violation exhibit false chirality with respect to *CP*-enantiomorphism: the two distinct enantiomeric forces that are interconverted by *CP* (only one of which exists in our world, hence *CP* violation) are also interconverted by *T* due to *CPT* invariance. This perception is reinforced by a remark by Okun (1985) that *CP*-violating interaction terms used in quantum chromodynamics transform with respect to *CP* and *T* in the same way as **E.B**. (In fact **E.B** transforms in the same way under *CP* and *T* as it does under *P* and *T* because **E** and **B** are both *C* odd.) Hence if *P*-violating forces are regarded as quintessential truly chiral influences, *CP*-violating forces may be regarded as quintessential falsely chiral influences!

Another consequence of the *CPT* theorem, that particles and antiparticles have electromagnetic properties equal in magnitude but opposite in sign, immediately reveals a fatal flaw in the suggestion that a circularly polarized photon supports a static magnetic field parallel or antiparallel to the propagation direction depending on the sense of circular polarization, thereby introducing the concept of a 'light magnet' which can generate a new range of magneto-optical phenomena (Evans, 1993). This is because, since a photon is its own antiparticle, any such magnetic field must be zero. The nonexistence of the photon's static magnetic field may also be proved from pictorial arguments based on conservation of charge conjugation symmetry (Barron, 1993).

Despite being the cornerstone of elementary particle physics, the possibility that even *CPT* symmetry might be violated to a very small extent should nonetheless be contemplated (Sachs, 1987). The simplest tests focus on the measurement of the rest mass of a particle and its associated antiparticle, because any difference would reveal a violation of *CPT*. Also, the photon's static magnetic field might be sought experimentally as a signature of *CPT* violation. However, the world of atoms and molecules might ultimately prove the best testing ground (Quack, 2002). For example, if antihydrogen were to be manufactured in sufficient quantities, ultrahigh resolution spectroscopy could be used to compare energy intervals in atomic hydrogen and antihydrogen as a test of *CPT* invariance to much higher precision than any previous measurements (Eades, 1993; Walz *et al.*, 2003). Cold antihydrogen atoms suitable for precision spectroscopy experiments were first produced in 2002 at CERN. Looking even further ahead to a time when chiral molecules made of antimatter might be available, detection of energy differences between *CP*-enantiomers might be attempted since, as mentioned above, *CPT* violation would lift their degeneracy.

1.9.7 Chirality and relativity

It was demonstrated in Section 1.9.5 that a spinning sphere or cone translating along the axis of spin possesses true chirality. This is an interesting concept because it exposes a link between chirality and special relativity. Consider a particle with a right-handed helicity moving away from an observer. If the observer accelerates to a sufficiently high velocity that she starts to catch up with the particle, it will then appear to be moving towards the observer and so takes on a left-handed helicity. In its rest frame, the helicity of the particle is undefined so its chirality vanishes.

Only for massless particles such as photons and neutrinos is the chirality conserved since they always move at the velocity of light in any reference frame.

This relativistic aspect of chirality is in fact a central feature of modern elementary particle theory, especially in connection with the weak interaction where the parity violating aspects are velocity dependent. A good example is provided by the interaction of electrons with neutrinos. Neutrinos are quintessential chiral objects since only CP enantiomers corresponding to left-helical neutrinos and right-helical antineutrinos exist. Consider first the extreme case of electrons moving close to the velocity of light. Only left-helical relativistic electrons interact with left-helical neutrinos *via* the weak force; right-helical relativistic electrons do not interact at all with neutrinos. But right-helical relativistic positrons interact with right-helical antineutrinos. For nonrelativistic electron velocities, the weak interaction still violates parity but the amplitude of the violation is reduced to order v/c (Gottfried and Weisskopf, 1984). This is used to explain the interesting fact that the $\pi^- \to e^- v_e^*$ decay is a factor of 10^4 smaller than the $\pi^- \to \mu^- v_e^*$ decay, even though the available energy is much larger in the first decay. Thus in the rest frame of the pion, the lepton (electron or muon) and the antineutrino are emitted in opposite directions so that their linear momenta cancel. Also, since the pion is spinless, the lepton must have a right-handed helicity in order to cancel the right-handed helicity of the antineutrino. Thus both decays would be forbidden if e and μ had the velocity c because the associated maximal parity violation dictates that both be pure left-handed. However, on account of its much greater mass, the muon is emitted much more slowly than the electron so there is a much greater amplitude for it to be emitted with a right-handed helicity. This discussion applies only to charge-changing weak processes, mediated by W^+ or W^- particles. Weak neutral current processes, mediated by Z^0 particles, are rather different since, even in the relativistic limit, both left- and right-handed electrons participate but with slightly different amplitudes.

1.9.8 Chirality in two dimensions

Chirality in two dimensions arises when there are two distinct enantiomers, confined to a plane or surface, that are interconverted by parity but not by any rotation within the plane about an axis perpendicular to the plane (symmetry operations out of the plane require an inaccessible third dimension). In two dimensions, however, the parity operation is no longer equivalent to an inversion through the coordinate origin as in three dimensions because this would not change the handedness of the two coordinate axes. Instead, an inversion of just one of the two axes is required (Halperin, March-Russel and Wilczek, 1989). For example, if the axes x, y are in the plane with z being perpendicular, then the parity operation could be taken as producing either $-x, y$ or $x, -y$, which are equivalent to mirror reflections across the lines defined by the y or x axes, respectively. Hence an object such as a scalene triangle (one with three sides of different length), which is achiral in three dimensions, becomes chiral in the two dimensions defined by the plane of the triangle because mirror reflection across any line within the plane generates a triangle which cannot be superposed on the original by any rotation about the z axis. Notice that a subsequent reflection across a second line, perpendicular to the first, generates a triangle superposable on the original, which demonstrates why an inversion of both axes, so that $x, y \rightarrow -x, -y$ is not acceptable as the parity operation in two dimensions.

Consider a surface covered with an isotropic layer (meaning no preferred orientations in the plane) of molecules. If the molecules are achiral, there will be an infinite number of mirror reflection symmetry operations possible across lines within the plane, which generate an indistinguishable isotropic layer. But if the surface molecules are chiral, such mirror reflections would generate the distinct isotropic surface composed of the enantiomeric molecules and so are not symmetry operations.

Such considerations are not purely academic. For example, chiral molecules on an isotropic surface were observed by Hicks, Petralli-Mallow and Byers (1994) to generate huge circular intensity differences in pre-resonance second harmonic scattering via pure electric dipole interactions. This is a genuine chiroptical phenomenon since it distinguishes between chiral enantiomers, and a plethora of related polarization effects can be envisaged (Hecht and Barron, 1996). The equivalent time-even pseudoscalar observables in light scattered from chiral molecules in bulk three-dimensional samples are approximately three orders of magnitude smaller because, as discussed in later chapters, electric dipole–magnetic dipole and electric dipole–electric quadrupole processes are required. Similar effects should exist in linear Rayleigh and Raman scattering from chiral surfaces and interfaces (Hecht and Barron, 1994) but have not been observed at the time of writing. In another manifestation of two-dimensional chirality, a rotation of the plane of polarization and an induced ellipticity have been observed in light diffracted from the surface of artificial chiral planar gratings based on chiral surface nanostructures (Papakostas *et al.*, 2003), with intriguing polarized colour images also observed (Schwanecke *et al.*, 2003).

Natural optical activity in reflection from the surface of a chiral medium has been an elusive and controversial phenomenon, but it has now been observed for a chiral liquid, namely a solution of camphorquinone in methanol, by Silverman, Badoz and Briat (1992), and for a chiral crystal, namely α-HgS (cinnabar) which belongs to the D_3 (32) point group, by Bungay, Svirko and Zheludev (1993). The phenomenon was subsequently observed in certain nonchiral crystals such as zinc blende semiconductors belonging to the T_d ($\bar{4}3$ m) point group, GaAs being an example (Svirko and Zheludev, 1994; 1998).

Arnaut (1997) has provided a generalization of the geometrical aspects of chirality to spaces of any dimensions. Essentially, an N-dimensional object is chiral in an N-dimensional space if it cannot be brought into congruence with its enantiomorph through a combination of translation and rotation within the N-dimensional space. As a consequence, an N-dimensional object which is N-dimensionally chiral loses its chirality in an M-dimensional space where $M > N$ because it can be rotated in the $(M-N)$-subspace onto its enantiomorph. Arnaut (1997) refers to chirality in one, two and three dimensions as axi-chirality, plano-chirality and chirality, respectively, and provides a detailed analysis of plano-chirality with examples such as a swastika, a logarithmic spiral and a jagged ring.

The concept of false chirality arises in two dimensions as well as in three. For example, the sense of a spinning electron on a surface with its axis of spin perpendicular to the surface is reversed under the two-dimensional parity operation (unlike in three dimensions). Because electrons with opposite spin sense are non-superposable in the plane, a spinning electron on a surface would seem to be chiral. However, the apparent chirality is false because the sense of spin is also reversed by time reversal (as in three dimensions). The enantiomorphism is therefore time noninvariant, the system being invariant under the combined PT operation but not under P and T separately.

2

Molecules in electric and magnetic fields

Are not gross bodies and light convertible into one another; and may not bodies receive much of their activity from the particles of light which enter into their composition? The changing of bodies into light, and light into bodies, is very comformable to the course of Nature, which seems delighted with transmutations.

Isaac Newton (*Opticks*)

2.1 Introduction

The theory of optical activity developed in this book is based on a semi-classical description of the interaction of light with molecules; that is, the molecules are treated as quantum mechanical objects perturbed by classical electromagnetic fields. Although quantum electrodynamics, in which the radiation field is also quantized, provides the most complete account to date of the radiation field and its interactions with molecules (Craig and Thirunamachandran, 1984), it is not used in this book since the required results can be obtained more directly with semiclassical methods.

The present chapter reviews those aspects of classical electrodynamics and quantum mechanical perturbation theory required for the semiclassical description of the scattering of polarized light by molecules. The methods are based on theories developed in the 1920s and 1930s when the new Schrödinger–Heisenberg formulation of quantum mechanics was applied to the interaction of light with atoms and molecules. Thus close parallels will be found with works such as Born's *Optik* (1933), Born and Huang's *Dynamical Theory of Crystal Lattices* (1954), Placzek's article on the theory of the Raman effect (1934) and also parts of the *Course of Theoretical Physics* by Landau and Lifshitz (1960, 1975, 1977), Lifshitz and Pitaevskii (1980) and Berestetskii, Lifshitz and Pitaevskii (1982). The extension in this chapter of this classic work to the higher-order molecular property tensors responsible, among other things, for optical activity, follows a treatment due to

53

Buckingham (1967, 1978). Like all of these works, this book makes considerable use of a cartesian tensor notation, which is elaborated in Chapter 4: this is essential if the delicate couplings between electromagnetic field components and components of the molecular property tensors responsible for optical activity are to be manipulated succinctly. For further details of the many complexities of Raman scattering, the recently published comprehensive treatise on the theory of the Raman effect by Long (2002) should be consulted.

2.2 Electromagnetic waves

2.2.1 Maxwell's equations

A charge density ρ and current density $\mathbf{J} = \rho\mathbf{v}$ generate electromagnetic fields. The sources and fields are related by *Maxwell's equations*

$$\nabla \cdot \mathbf{D} = \rho, \tag{2.2.1a}$$

$$\nabla \cdot \mathbf{B} = 0, \tag{2.2.1b}$$

$$\nabla \times \mathbf{E} = -\frac{\partial \mathbf{B}}{\partial t}, \tag{2.2.1c}$$

$$\nabla \times \mathbf{H} = \mathbf{J} + \frac{\partial \mathbf{D}}{\partial t}. \tag{2.2.1d}$$

$\mathbf{E}$ and $\mathbf{B}$ are the electric and magnetic fields in free space and $\mathbf{D}$ and $\mathbf{H}$ are the corresponding modified fields in material media. If the medium is isotropic the fields are related by

$$\mathbf{D} = \epsilon\epsilon_0\mathbf{E}, \tag{2.2.2a}$$

$$\mathbf{H} = \frac{1}{\mu\mu_0}\mathbf{B}, \tag{2.2.2b}$$

where ϵ and μ are the *dielectric constant* and *magnetic permeability* of the medium, and ϵ_0 and μ_0 are the *permittivity* and *permeability* of free space.

Maxwell's equations summarize the following laws of electromagnetism: (2.2.1a) is the differential form of Gauss's theorem applied to electrostatics; (2.2.1b) is the corresponding result for magnetostatics since magnetic charges do not exist; (2.2.1c) is Faraday's and Lenz's law of electromagnetic induction; and (2.2.1d) is Ampere's law for magnetomotive force with the important modification that the *displacement current*, which arises when the electric displacement $\mathbf{D}$ changes with time, is added to the *conduction current* $\mathbf{J}$ which is simply the current flow due to the motion of electric charges.

In an infinite homogeneous medium (including free space) containing no free charges and having zero conductivity, Maxwell's equations reduce to

$$\nabla \cdot \mathbf{D} = 0, \tag{2.2.3a}$$

$$\nabla \cdot \mathbf{B} = 0, \tag{2.2.3b}$$

$$\nabla \times \mathbf{E} = -\frac{\partial \mathbf{B}}{\partial t}, \tag{2.2.3c}$$

$$\nabla \times \mathbf{H} = \frac{\partial \mathbf{D}}{\partial t}. \tag{2.2.3d}$$

Using the vector identity

$$\nabla \times (\nabla \times \mathbf{F}) = \nabla(\nabla \cdot \mathbf{F}) - \nabla^2 \mathbf{F},$$

these four equations reduce to two equivalent wave equations:

$$\nabla^2 \mathbf{E} = \mu \mu_0 \epsilon \epsilon_0 \frac{\partial^2 \mathbf{E}}{\partial t^2}, \tag{2.2.4a}$$

$$\nabla^2 \mathbf{B} = \mu \mu_0 \epsilon \epsilon_0 \frac{\partial^2 \mathbf{B}}{\partial t^2}. \tag{2.2.4b}$$

The wave velocity is

$$v = (\mu \mu_0 \epsilon \epsilon_0)^{-\frac{1}{2}} \tag{2.2.5a}$$

with a free space value of

$$c = (\mu_0 \epsilon_0)^{-\frac{1}{2}}. \tag{2.2.5b}$$

In fact $v = c/n$, where

$$n = (\mu \epsilon)^{\frac{1}{2}} \tag{2.2.6}$$

is the refractive index of the medium.

2.2.2 Plane monochromatic waves

Of particular importance is the special case of electromagnetic waves in which the fields depend on only one space coordinate. Such waves are said to be plane, and if propagating in the z direction the fields have the same value over any plane, $z = $ a constant, normal to the direction of propagation. This means that all partial derivatives of the fields with respect to x and y are zero so that, from (2.2.3a) and (2.2.3b),

$$\frac{\partial E_z}{\partial z} = \frac{\partial B_z}{\partial z} = 0,$$

and from (2.2.3*c*) and (2.2.3*d*),

$$\frac{\partial E_z}{\partial t} = \frac{\partial B_z}{\partial t} = 0.$$

Thus the waves are completely transverse, with no field components in the direction of propagation. The wave equations (2.2.4) now take the form

$$\frac{\partial^2 \mathbf{E}}{\partial z^2} - \frac{1}{v^2}\frac{\partial^2 \mathbf{E}}{\partial t^2} = 0. \tag{2.2.7}$$

If the plane wave is associated with a single frequency, it is said to be monochromatic, and a solution of (2.2.7) is

$$\mathbf{E} = \mathbf{E}^{(0)} \cos(\omega t - 2\pi z/\lambda), \tag{2.2.8}$$

which is conveniently written as the real part of the complex expression

$$\tilde{\mathbf{E}} = \tilde{\mathbf{E}}^{(0)} e^{-\mathrm{i}(\omega t - 2\pi z/\lambda)}, \tag{2.2.9}$$

where ω, the angular frequency of the wave, is related to the wavelength by $\omega = 2\pi v/\lambda$. The sign of the exponent in (2.2.9) is a matter of convention since it does not affect the real part. Although most works on classical optics choose a plus sign, the choice of a minus sign is universal in quantum mechanics (it leads to a positive photon momentum) and is therefore advantageous in a work on molecular optics such as this. We also use a tilde throughout the book to denote a complex quantity.

We now introduce a wavevector κ, with magnitude ω/v, in the direction of propagation. It is convenient to write κ in terms of a propagation vector $\mathbf{n}$ with magnitude equal to the refractive index in the direction of propagation:

$$\kappa = \frac{\omega}{c}\mathbf{n}. \tag{2.2.10}$$

$\mathbf{n}$ becomes a unit propagation vector in free space. Equation (2.2.9) can now be written

$$\tilde{\mathbf{E}} = \tilde{\mathbf{E}}^{(0)} e^{\mathrm{i}(\kappa \cdot \mathbf{r} - \omega t)}. \tag{2.2.11}$$

Since the momentum of individual photons in a plane wave is $\hbar\kappa$, the reason for the choice of the minus sign in the exponent of (2.2.9) is now clear, for it gives a positive photon momentum.

From (2.2.3*c*), (2.2.3*d*) and (2.2.11) we obtain the following important relationships between the electric and magnetic field vectors in a plane wave:

$$\mathbf{B} = \frac{1}{c}\mathbf{n} \times \mathbf{E}, \tag{2.2.12a}$$

$$\mathbf{E} = -\frac{c}{n^2}\mathbf{n} \times \mathbf{B}. \tag{2.2.12b}$$

2.2.3 Force and energy

The *Lorentz force* density acting on a region of charge and current density in an electromagnetic field is

$$\mathbf{F} = \rho\mathbf{E} + \mathbf{J} \times \mathbf{B}. \tag{2.2.13}$$

The rate at which the Lorentz forces within a finite volume V do work is

$$\int \mathbf{F} \cdot \mathbf{v}\,dV = \int \rho[\mathbf{v} \cdot \mathbf{E} + \mathbf{v} \cdot (\mathbf{v} \times \mathbf{B})]dV = \int \mathbf{J} \cdot \mathbf{E}\,dV.$$

This power represents a conversion of electromagnetic energy into mechanical or thermal energy, and must be balanced by a corresponding rate of decrease of electromagnetic energy within V. Use (2.2.1c) and (2.2.1d) and the vector identity

$$\nabla \cdot (\mathbf{E} \times \mathbf{H}) = \mathbf{H} \cdot (\nabla \times \mathbf{E}) - \mathbf{E} \cdot (\nabla \times \mathbf{H})$$

to write

$$\int \mathbf{J} \cdot \mathbf{E}\,dV = -\int \left[\mathbf{E} \cdot \frac{\partial \mathbf{D}}{\partial t} + \mathbf{H} \cdot \frac{\partial \mathbf{B}}{\partial t} + \nabla \cdot (\mathbf{E} \times \mathbf{H}) \right] dV$$

$$= -\frac{1}{2}\frac{\partial}{\partial t} \int (\mathbf{D} \cdot \mathbf{E} + \mathbf{B} \cdot \mathbf{H})dV - \int (\mathbf{E} \times \mathbf{H}) \cdot d\mathbf{S}.$$

The last term has been transformed into an integral over the surface $\mathbf{S}$ bounding V. The rate at which the fields do work can now be equated with the rate at which the energy stored in the field diminishes plus the rate at which energy flows into V. Thus we take

$$U = \tfrac{1}{2}(\mathbf{D} \cdot \mathbf{E} + \mathbf{B} \cdot \mathbf{H}) \tag{2.2.14}$$

to be the electromagnetic energy density, and

$$\mathbf{N} = \mathbf{E} \times \mathbf{H} \tag{2.2.15}$$

to be the rate at which electromagnetic energy flows across unit area at the boundary. $\mathbf{N}$ is called the *Poynting vector*, and gives the instantaneous rate of energy flow in the direction of propagation of an electromagnetic wave.

The intensity I is the mean rate of energy flow, which is the average of the magnitude of $\mathbf{N}$ over a complete period of the wave. For a plane wave, the magnitude of $\mathbf{N}$ is

$$|\mathbf{N}| = \frac{1}{\mu\mu_0}|\mathbf{E} \times \mathbf{B}| = \frac{1}{\mu\mu_0 c}|\mathbf{E} \times (\mathbf{n} \times \mathbf{E})| = \left(\frac{\epsilon\epsilon_0}{\mu\mu_0}\right)^{\frac{1}{2}} E^{(0)2}.$$

Since a plane wave is sinusoidal, the intensity, which is the time average of $|\mathbf{N}|$, is simply

$$I = \frac{1}{2}\left(\frac{\epsilon\epsilon_0}{\mu\mu_0}\right)^{\frac{1}{2}} E^{(0)^2}. \tag{2.2.16}$$

2.2.4 The scalar and vector potentials

The four Maxwell equations (2.2.1) can be reduced to two equations involving a scalar potential ϕ and a vector potential $\mathbf{A}$. Thus since $\nabla \cdot \mathbf{B} = 0$ and the divergence of a curl is always zero, we can write $\mathbf{B}$ in terms of $\mathbf{A}$:

$$\mathbf{B} = \nabla \times \mathbf{A}. \tag{2.2.17}$$

Equation (2.2.1c) now becomes

$$\nabla \times \left(\mathbf{E} + \frac{\partial \mathbf{A}}{\partial t}\right) = 0$$

so the electric field vector can be written

$$\mathbf{E} = -\frac{\partial \mathbf{A}}{\partial t} + \mathbf{a},$$

where $\mathbf{a}$ is a vector whose curl is zero. But since the curl of the gradient of a scalar function is zero, we can write

$$\mathbf{E} = -\nabla\phi - \frac{\partial \mathbf{A}}{\partial t}. \tag{2.2.18}$$

The four Maxwell equations now reduce to

$$\nabla^2\mathbf{A} - \nabla(\nabla \cdot \mathbf{A}) - \frac{1}{v^2}\left(\nabla\frac{\partial\phi}{\partial t} + \frac{\partial^2\mathbf{A}}{\partial t^2}\right) = -\mu\mu_0\,\mathbf{J}, \tag{2.2.19a}$$

$$\nabla^2\phi + \frac{\partial}{\partial t}(\nabla \cdot \mathbf{A}) = -\frac{\rho}{\epsilon\epsilon_0}. \tag{2.2.19b}$$

These two equations are uncoupled by exploiting the arbitrariness in the definition of the potentials.

Two electromagnetic fields are physically identical if they are characterized by the same $\mathbf{B}$ and $\mathbf{E}$, even though $\mathbf{A}$ and ϕ are different for the two fields. Consider the potentials $\mathbf{A}$ and ϕ determined from $\mathbf{A}_0$ and ϕ_0 by the *gauge transformation*

$$\mathbf{A} = \mathbf{A}_0 - \nabla\Lambda, \tag{2.2.20a}$$

$$\phi = \phi_0 + \frac{\partial\Lambda}{\partial t}, \tag{2.2.20b}$$

where Λ is an arbitrary function of the coordinates and time. The **B** and **E** calculated from (2.2.17) and (2.2.18) using **A** and ϕ are the same as the **B** and **E** calculated using $\mathbf{A}_0$ and ϕ_0. This enables restrictions to be placed on **A** and ϕ which simplify the Maxwell equations (2.2.19).

If we choose Λ so that

$$\nabla^2 \Lambda = \nabla \cdot \mathbf{A}_0, \qquad (2.2.21)$$

we obtain

$$\nabla \cdot \mathbf{A} = 0 \qquad (2.2.22)$$

and (2.2.19) become

$$\nabla^2 \mathbf{A} - \frac{1}{v^2}\left(\nabla \frac{\partial \phi}{\partial t} + \frac{\partial^2 \mathbf{A}}{\partial t^2}\right) = -\mu\mu_0 \mathbf{J}, \qquad (2.2.23a)$$

$$\nabla^2 \phi = -\frac{\rho}{\epsilon\epsilon_0}. \qquad (2.2.23b)$$

Any choice of gauge which has $\nabla \cdot \mathbf{A} = 0$ is called a *Coulomb* gauge since ϕ is then determined from Poisson's equation (2.2.23b) by the charges alone, as if they were at rest.

If we choose Λ so that

$$\nabla^2 \Lambda - \frac{1}{v^2}\frac{\partial^2 \Lambda}{\partial t^2} = \nabla \cdot \mathbf{A}_0 + \frac{1}{v^2}\frac{\partial \phi_0}{\partial t}, \qquad (2.2.24)$$

we obtain

$$\nabla \cdot \mathbf{A} + \frac{1}{v^2}\frac{\partial \phi}{\partial t} = 0, \qquad (2.2.25)$$

and (2.2.19) are now uncoupled completely:

$$\nabla^2 \mathbf{A} - \frac{1}{v^2}\frac{\partial^2 \mathbf{A}}{\partial t^2} = -\mu\mu_0 \mathbf{J}, \qquad (2.2.26a)$$

$$\nabla^2 \phi - \frac{1}{v^2}\frac{\partial^2 \phi}{\partial t^2} = -\frac{\rho}{\epsilon\epsilon_0}. \qquad (2.2.26b)$$

In most books equation (2.2.25) is called the *Lorentz condition* and any choice of gauge which satisfies it is called a *Lorentz* gauge. However, it was pointed out recently that this is a case of mistaken paternity. This condition and the associated gauge should really be attributed to the Danish physicist L. Lorenz rather than the Dutch physicist H. A. Lorentz (van Bladel, 1991).

In free space, or in a medium without sources, ρ and **J** are zero. ϕ is then automatically zero in the Coulomb gauge, and can be made to vanish in the Lorentz gauge by a further specialization of Λ in (2.2.20). The field is then determined by

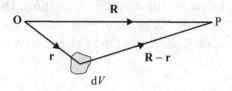

Fig. 2.1 The system of vectors used to specify the position of a point P at which are detected the electromagnetic fields generated by a volume element dV containing charge and current density.

A alone, and is entirely transverse:

$$\nabla \cdot \mathbf{A} = 0, \tag{2.2.27a}$$

$$\nabla^2 \mathbf{A} - \frac{1}{v^2} \frac{\partial^2 \mathbf{A}}{\partial t^2} = 0. \tag{2.2.27b}$$

The corresponding electric and magnetic field vectors are simply

$$\mathbf{E} = -\frac{\partial \mathbf{A}}{\partial t}, \tag{2.2.28a}$$

$$\mathbf{B} = \nabla \times \mathbf{A}. \tag{2.2.28b}$$

General solutions of the uncoupled equations (2.2.26) are now required in order to find the electromagnetic fields generated by the charge and current density ρ and $\mathbf{J}$. If these sources are static, $\mathbf{A}$ and ϕ are independent of time and the general solutions have the form

$$\mathbf{A}(\mathbf{R}) = \frac{\mu\mu_0}{4\pi} \int \frac{\mathbf{J}\,dV}{|\mathbf{R} - \mathbf{r}|}, \tag{2.2.29a}$$

$$\phi(\mathbf{R}) = \frac{1}{4\pi\epsilon\epsilon_0} \int \frac{\rho\,dV}{|\mathbf{R} - \mathbf{r}|}, \tag{2.2.29b}$$

where $\mathbf{R}$ is the position vector of the point P at which the fields are determined, $\mathbf{r}$ is the position vector of the volume element containing ρ and $\mathbf{J}$, and $|\mathbf{R} - \mathbf{r}|$ is the distance from the volume element to P, as illustrated in Fig. 2.1. If the charge and current densities are functions of time, the solutions are

$$\mathbf{A}(\mathbf{R}, t) = \frac{\mu\mu_0}{4\pi} \int \frac{[\mathbf{J}]\,dV}{|\mathbf{R} - \mathbf{r}|}, \tag{2.2.30a}$$

$$\phi(\mathbf{R}, t) = \frac{1}{4\pi\epsilon\epsilon_0} \int \frac{[\rho]\,dV}{|\mathbf{R} - \mathbf{r}|}, \tag{2.2.30b}$$

where the square brackets mean that ρ and $\mathbf{J}$ are to be taken at the time $t - |\mathbf{R} - \mathbf{r}|/v$. This is because the disturbances set up by ρ and $\mathbf{J}$ propagate with velocity v and take a time $|\mathbf{R} - \mathbf{r}|/v$ to travel the distance $|\mathbf{R} - \mathbf{r}|$. Thus the potentials at $\mathbf{R}$ at a time t are related to what happened at the element of charge and current density at an earlier time $t - |\mathbf{R} - \mathbf{r}|/v$, and are known as *retarded potentials*.

2.3 Polarized light

2.3.1 Pure polarization

A plane monochromatic wave travelling in the z direction can be written as a sum of two coherent waves linearly polarized in the x and y directions,

$$\tilde{\mathbf{E}} = \tilde{E}_x \mathbf{i} + \tilde{E}_y \mathbf{j}, \qquad (2.3.1)$$

where $\mathbf{i}, \mathbf{j}, \mathbf{k}$, the unit vectors along x, y, z, form a right-handed system such that $\mathbf{i} \times \mathbf{j} = \mathbf{k}$. The polarization of the wave is determined by the relative phases and magnitudes of the complex amplitudes $\tilde{E}_x$ and $\tilde{E}_y$. For example, if $\tilde{E}_x$ and $\tilde{E}_y$ have the same phase the polarization is linear and, if they are equal in magnitude and $\pi/2$ out of phase, the polarization is circular. Using the traditional convention that right- or left-circular polarization is a clockwise or anticlockwise rotation of the electric field vector in a plane when viewed by an observer receiving the wave, we can write

$$\tilde{\mathbf{E}}_{\substack{R \\ L}} = \frac{1}{\sqrt{2}} E^{(0)} (\mathbf{i} + e^{\mp i\pi/2} \mathbf{j}) e^{i(\kappa z - \omega t)}$$

$$= \frac{1}{\sqrt{2}} E^{(0)} (\mathbf{i} \mp i\mathbf{j}) e^{i(\kappa z - \omega t)}. \qquad (2.3.2)$$

Notice that the sign of $i\mathbf{j}$ in (2.3.2) is determined by the choice of sign in the exponent of (2.2.9).

The most general pure polarization state is represented by an ellipse, illustrated in Fig. 2.2. The *ellipticity* η is determined by the ratio of the minor and major axes of the ellipse, b and a, through

$$\tan \eta = \frac{b}{a}. \qquad (2.3.3)$$

The orientation of the ellipse is specified by the angle θ, called the *azimuth*, between the a and the x axes. Since a and b are the relative amplitudes of two waves that are $\pi/2$ out of phase, a phase factor $\exp(i\pi/2) = i$ is associated with the b axis: with this choice of sign, a positive η corresponds to a right-handed ellipticity (remembering that the wave function is $\exp(-i\omega t)$). If $a^2 + b^2 = 1$, a and b can be regarded as the real and imaginary parts of a complex unit polarization vector $\tilde{\mathbf{\Pi}}$ such that

$$\tilde{\mathbf{E}} = E^{(0)} (\tilde{\Pi}_x \mathbf{i} + \tilde{\Pi}_y \mathbf{j}) e^{i(\kappa z - \omega t)}. \qquad (2.3.4)$$

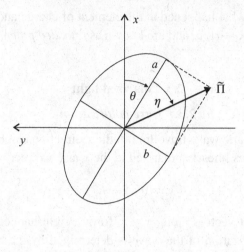

Fig. 2.2 The polarization ellipse referred to space-fixed axes x and y. The propagation direction z is out of the plane of the paper. η is the ellipticity, θ is the azimuth and $\tilde{\Pi}$ is the complex unit polarization vector. The ellipticity and azimuth so defined are in accord with the conventions for a positive ellipticity and angle of optical rotation used in Chapter 1.

Projecting $\tilde{\Pi}$ onto the real a and imaginary b axes, and then onto the x and y axes, gives the following expressions for its complex components:

$$\tilde{\Pi}_x = \cos\eta\cos\theta - i\sin\eta\sin\theta, \tag{2.3.5a}$$

$$\tilde{\Pi}_y = -\cos\eta\sin\theta - i\sin\eta\cos\theta. \tag{2.3.5b}$$

On substituting $\theta = 0$ and $\eta = \pm\pi/4$ into (2.3.5) and (2.3.4), (2.3.2) is recovered.

Three quantities are required to specify the state of a monochromatic plane wave light beam: the intensity I, the azimuth θ, and the ellipticity η. These can be extracted from the complex representation (2.3.4) of the wave by taking suitable real products of components such as the four *Stokes parameters* (Stokes, 1852)

$$S_0 = \tilde{E}_x\tilde{E}_x^* + \tilde{E}_y\tilde{E}_y^*, \tag{2.3.6a}$$

$$S_1 = \tilde{E}_x\tilde{E}_x^* - \tilde{E}_y\tilde{E}_y^*, \tag{2.3.6b}$$

$$S_2 = -(\tilde{E}_x\tilde{E}_y^* + \tilde{E}_y\tilde{E}_x^*), \tag{2.3.6c}$$

$$S_3 = -i(\tilde{E}_x\tilde{E}_y^* - \tilde{E}_y\tilde{E}_x^*). \tag{2.3.6d}$$

Our notation and sign convention follows that of Born and Wolf (1980), except for our definition of a positive azimuth, which leads to a sign difference in S_2. For a completely polarized beam only three of the Stokes parameters are independent since

$$S_0^2 = S_1^2 + S_2^2 + S_3^2. \tag{2.3.7}$$

From (2.3.4) and (2.3.5), the Stokes parameters can be written

$$S_0 = E^{(0)^2}(\tilde{\Pi}_x \tilde{\Pi}_x^* + \tilde{\Pi}_y \tilde{\Pi}_y^*) = E^{(0)^2}, \tag{2.3.8a}$$

$$S_1 = E^{(0)^2}(\tilde{\Pi}_x \tilde{\Pi}_x^* - \tilde{\Pi}_y \tilde{\Pi}_y^*) = E^{(0)^2} \cos 2\eta \cos 2\theta, \tag{2.3.8b}$$

$$S_2 = -E^{(0)^2}(\tilde{\Pi}_x \tilde{\Pi}_y^* + \tilde{\Pi}_y \tilde{\Pi}_x^*) = E^{(0)^2} \cos 2\eta \sin 2\theta, \tag{2.3.8c}$$

$$S_3 = -iE^{(0)^2}(\tilde{\Pi}_x \tilde{\Pi}_y^* - \tilde{\Pi}_y \tilde{\Pi}_x^*) = E^{(0)^2} \sin 2\eta, \tag{2.3.8d}$$

from which the intensity, azimuth and ellipticity can be extracted:

$$I = \frac{1}{2}\left(\frac{\epsilon\epsilon_0}{\mu\mu_0}\right)^{\frac{1}{2}} S_0, \tag{2.3.9a}$$

$$\theta = \frac{1}{2}\tan^{-1}\left(\frac{S_2}{S_1}\right), \tag{2.3.9b}$$

$$\eta = \frac{1}{2}\tan^{-1}\left[\frac{S_3}{(S_1^2 + S_2^2)^{\frac{1}{2}}}\right]. \tag{2.3.9c}$$

It might be thought that η is given more directly by

$$\eta = \frac{1}{2}\sin^{-1}\left(\frac{S_3}{S_0}\right).$$

This is certainly true if the wave is completely polarized, but it is shown below that when the wave is only partially polarized this quantity is no longer the ellipticity.

The Stokes parameters correspond to the set of four intensity measurements required to determine completely the state of a light beam. Two optical elements are required: an analyzer, such as a Nicol prism, for which the emergent beam is linearly polarized along the transmission axis of the analyzer; and a retarder, such as a quarter-wave plate, which alters the phase relationship between coherent orthogonal polarization components of the beam. If $I(\sigma, \tau)$ denotes the intensity of the light transmitted through a retarder which subjects the y component of the light to a retardation τ with respect to the x component, followed by an analyzer with its transmission axis oriented at an angle σ to the x axis, the Stokes parameters are given by the following measurements:

$$S_0 \propto I(0, 0) + I(\pi/2, 0),$$
$$S_1 \propto I(0, 0) - I(\pi/2, 0),$$
$$S_2 \propto I(\pi/4, 0) - I(3\pi/4, 0),$$
$$S_3 \propto I(3\pi/4, \pi/2) - I(\pi/4, \pi/2).$$

Thus S_0 gives the total intensity, S_1 gives the excess in intensity transmitted by an analyzer which accepts linear polarization with an azimuth $\theta = 0$ over that transmitted by an analyzer which accepts linear polarization with an azimuth $\theta = \pi/2$. S_2 has a similar interpretation with respect to the azimuths $\theta = \pi/4$ and $\theta = 3\pi/4$. S_3 is the excess in intensity transmitted by a device which accepts right-circularly polarized light over that transmitted by a device which accepts left-circularly polarized light.

An alternative method of specifying the polarization, entirely equivalent to the Stokes parameters, involves a Hermitian polarization density matrix, called a *coherency matrix* by Born and Wolf (1980) and a *polarization tensor* by Landau and Lifshitz (1975), with elements

$$\tilde{\rho}_{\alpha\beta} = \frac{\tilde{E}_\alpha \tilde{E}_\beta^*}{E^{(0)2}} = \tilde{\Pi}_\alpha \tilde{\Pi}_\beta^*. \tag{2.3.10}$$

Using (2.3.5), these elements can be written in terms of the Stokes parameters or in terms of the azimuth and ellipticity:

$$\begin{aligned}
\tilde{\rho}_{\alpha\beta} &= \frac{1}{2S_0} \begin{pmatrix} S_0 + S_1 & -S_2 + iS_3 \\ -S_2 - iS_3 & S_0 - S_1 \end{pmatrix} \\
&= \frac{1}{2} \begin{pmatrix} 1 + \cos 2\eta \cos 2\theta & -\cos 2\eta \sin 2\theta + i \sin 2\eta \\ -\cos 2\eta \sin 2\theta - i \sin 2\eta & 1 - \cos 2\eta \cos 2\theta \end{pmatrix}.
\end{aligned} \tag{2.3.11}$$

2.3.2 Partial polarization

Strictly monochromatic light is always completely polarized, with the tip of the electric field vector at each point in space moving around an ellipse, which may in particular cases reduce to a circle or a straight line. In practice, we usually have to deal with waves which are only approximately monochromatic, containing frequencies in a small interval $\Delta\omega$ centred on an apparent monochromatic frequency ω. Such waves are called *quasi-monochromatic*, and can be represented as a superposition, such as a Fourier sum, of strictly monochromatic waves with various frequencies. Quasi-monochromatic light has an extra 'dimension' in its range of possible polarizations, because the component monochromatic waves can have different polarizations and phases. At one extreme, the net electric field vector of quasi-monochromatic light can have the polarization properties of a completely monochromatic wave, and the light is said to be completely polarized. The opposite extreme is unpolarized or natural light, where the tip of the net electric field vector moves irregularly and shows no preferred directional properties. In general, the variation of the electric field vectors is neither completely regular nor completely irregular, and the light is said to be *partially* polarized. This is usually the condition of scattered light.

If ω is the average frequency of a quasi-monochromatic wave, its electric field vector at a fixed point in space can be written

$$\tilde{\mathbf{E}} = \tilde{\mathbf{E}}^{(0)}(t)e^{i(\boldsymbol{\kappa}\cdot\mathbf{r}-\omega t)}, \tag{2.3.12}$$

where the complex vector amplitude $\tilde{\mathbf{E}}^{(0)}(t)$ is a slowly varying function of the time ($\tilde{\mathbf{E}}^{(0)}$ would be a constant if the wave were strictly monochromatic). In fact both the polarization vector and the scalar amplitude can vary with time:

$$\tilde{\mathbf{E}}^{(0)}(t) = \tilde{\boldsymbol{\Pi}}(t)\tilde{E}^{(0)}(t). \tag{2.3.13}$$

Complete polarization results when only the amplitude of the polarization ellipse varies over a long time, that is $\tilde{E}^{(0)}(t)$ varies and $\tilde{\boldsymbol{\Pi}}(t)$ is constant. The wave is unpolarized if $\tilde{\boldsymbol{\Pi}}(t)$ shows no preferred azimuth or ellipticity over a long time. These distinctions apply only when the duration of the observation is large compared with the reciprocal of the frequency width $\Delta\omega$ of the quasi-monochromatic wave, which is usually the case.

Measured intensities are time averages of real quadratic functions of the fields. Thus the Stokes parameters and the polarization tensor of a quasi-monochromatic beam are defined in terms of time averaged products of electric field vectors. If the light is completely polarized, the time averages of the products of components of $\tilde{\boldsymbol{\Pi}}$ are

$$\overline{\tilde{\Pi}_\alpha \tilde{\Pi}_\beta^*} = \tilde{\Pi}_\alpha \tilde{\Pi}_\beta^*, \tag{2.3.14}$$

so the time-averaged Stokes parameters of a completely polarized quasi-monochromatic wave are still related by

$$S_0^2 = S_1^2 + S_2^2 + S_3^2,$$

the same as (2.3.7) for a monochromatic beam. If the light is completely unpolarized, all orientations of $\tilde{\boldsymbol{\Pi}}$ in the xy plane are equally probable, so the time average is effectively an average over all orientations in two dimensions:

$$\overline{\tilde{\Pi}_\alpha \tilde{\Pi}_\beta^*} = \tfrac{1}{2}\delta_{\alpha\beta}. \tag{2.3.15}$$

(Averages of tensor components are developed in Section 4.2.5.) The time-averaged Stokes parameters of a completely unpolarized beam are therefore

$$S_0^2 = E^{(0)^2}, \tag{2.3.16a}$$
$$S_1^2 = S_2^2 = S_3^2 = 0. \tag{2.3.16b}$$

Consequently, partial polarization must be characterized by

$$S_0^2 > S_1^2 + S_2^2 + S_3^2, \tag{2.3.17}$$

the inequality arising from the presence of an unpolarized component.

It is convenient to introduce a *degree of polarization P*, which takes values between 0 for unpolarized light and 1 for polarized light, such that

$$S_0 = E^{(0)^2}, \tag{2.3.18a}$$

$$S_1 = PE^{(0)^2} \cos 2\eta \cos 2\theta, \tag{2.3.18b}$$

$$S_2 = PE^{(0)^2} \cos 2\eta \sin 2\theta, \tag{2.3.18c}$$

$$S_3 = PE^{(0)^2} \sin 2\eta, \tag{2.3.18d}$$

$$P = \left(S_1^2 + S_2^2 + S_3^2\right)^{\frac{1}{2}}/S_0. \tag{2.3.18e}$$

Evidently P is the ratio of the intensity of the polarized part of the beam to the total intensity. Thus a partially polarized beam can be decomposed into a polarized and an unpolarized part, and its Stokes parameters are simply the sums of the Stokes parameters of the polarized and unpolarized components. The azimuth and ellipticity of the polarized part are given by (2.3.9b and c), but the ratio S_3/S_0 now gives the *degree of circularity*, which is the ratio of the intensity of the circularly polarized component to the total intensity, rather than the ellipticity.

The polarization tensor can also be used to specify partial polarization. Equation (2.3.11) is now generalized to

$$\rho_{\alpha\beta} = \frac{1}{2} \begin{pmatrix} 1 + P \cos 2\eta \cos 2\theta & P(-\cos 2\eta \sin 2\theta + i \sin 2\eta) \\ P(-\cos 2\eta \sin 2\theta - i \sin 2\eta) & 1 - P \cos 2\eta \cos 2\theta \end{pmatrix}. \tag{2.3.19}$$

The determinant is

$$|\rho_{\alpha\beta}| = \rho_{xx}\rho_{yy} - \rho_{xy}\rho_{yx} = \tfrac{1}{4}(1 - P^2).$$

Thus for completely polarized light $|\rho_{\alpha\beta}| = 0$, and for completely unpolarized light $|\rho_{\alpha\beta}| = \frac{1}{4}$. A convenient representation of the polarization tensor is

$$\rho_{\alpha\beta} = \tfrac{1}{2}[i_\alpha i_\beta + j_\alpha j_\beta + (i_\alpha i_\beta - j_\alpha j_\beta)P \cos 2\eta \cos 2\theta$$

$$- (i_\alpha j_\beta + i_\beta j_\alpha)P \cos 2\eta \sin 2\theta + i(i_\alpha j_\beta - i_\beta j_\alpha)P \sin 2\eta], \tag{2.3.20}$$

where i_α and j_α are the α components of the unit vectors **i** and **j**.

The representation

$$\tilde{\mathbf{E}} = \tilde{\mathbf{\Pi}}E^{(0)}e^{i(\boldsymbol{\kappa}\cdot\mathbf{r}-\omega t)} \tag{2.3.21}$$

is known as the *Jones vector* description of polarized light (Jones, 1941). Since the Jones vector carries the absolute phase of the wave, the state of a combination of coherent light beams is obtained by first summing the individual Jones vectors, then extracting I, θ, η and P from the Stokes parameters formed from the net Jones vector. This procedure is the basis of the calculation in Chapter 3 of birefringent polarization changes, which originate in interference between the transmitted and the forward-scattered waves. In contrast, the state of a combination of incoherent beams is obtained by summing immediately the Stokes parameters of the individual components and then extracting I, θ, η and P.

The Stokes parameters constitute a vector of length PS_0 in a three-dimensional real space; the locus of the tip of the vector is a sphere, called the *Poincaré sphere*, and all possible polarization conditions are encompassed by its surface. Alternatively, the three components of the Stokes vector in Poincaré space, together with PS_0, can be regarded as a vector in a four-dimensional real space. The latter viewpoint exposes the mathematical connection between the Stokes parameters and the polarization tensor, for $\rho_{\alpha\beta}$ has the form of a second-rank spinor, and therefore represents a real four-dimensional vector in a two-dimensional complex space. The Jones vector, on the other hand, has the mathematical form of a first-rank spinor.

The Jones vector is analogous to a wave function description, and the Stokes vector or polarization tensor is analogous to a density matrix description, of the state of a system in quantum mechanics. Thus the Jones vector can only specify a pure polarized light beam, and a quantum mechanical wave function can only specify a pure state. A partially polarized light beam is an incoherent superposition of pure polarized beams and must be specified by a Stokes vector or polarization tensor, and a mixed quantum mechanical state is an incoherent superposition of pure states and must be specified by a density matrix. Light is usually generated as a result of a transition between two quantum states of an atom or molecule. Complete polarization results if the quantum states of the emitter are precisely defined both before and after the transition; if either is incompletely defined, the emitted light is incompletely polarized. Fano (1957) has discussed this question in detail.

2.4 Electric and magnetic multipole moments

The structures of charge and current distributions giving rise to scalar and vector potentials are now investigated. Charge distributions are developed in terms of electric multipole moments, and current distributions in terms of magnetic multipole moments.

2.4.1 Electric multipole moments

Our treatment of electric multipole moments follows Landau and Lifshitz (1975) and Buckingham (1970). The zeroth moment of a collection of point charges e_i is the *net charge* or *electric monopole moment*

$$q = \sum_i e_i, \tag{2.4.1}$$

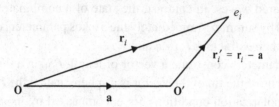

Fig. 2.3 The effect of shifting the coordinate origin from $\mathbf{O}$ to $\mathbf{O}'$ on the position vector of a point charge e_i.

where $e_i = +e$ for the proton and $-e$ for the electron (in SI, $e = 1.603 \times 10^{-19}$C).

The first moment of a collection of charges is the *electric dipole moment* vector

$$\boldsymbol{\mu} = \sum_i e_i \mathbf{r}_i, \qquad (2.4.2)$$

where $\mathbf{r}_i$ is the position vector of the ith charge. Notice that if the net charge is zero, the electric dipole moment is independent of the choice of the origin. Thus if the origin is moved from $\mathbf{O}$ to a point $\mathbf{O}' = \mathbf{O} + \mathbf{a}$, where $\mathbf{a}$ is some constant vector, the position vector $\mathbf{r}_i$ in the old coordinate system becomes $\mathbf{r}'_i = \mathbf{r}_i - \mathbf{a}$ in the new (Fig. 2.3), and the electric dipole moment becomes

$$\boldsymbol{\mu}' = \sum_i e_i \mathbf{r}'_i = \boldsymbol{\mu} - q\mathbf{a}. \qquad (2.4.3)$$

If q is not zero there is a unique point, called the centre of charge, relative to which $\boldsymbol{\mu} = 0$.

The second moment of a collection of charges is the *electric quadrupole moment* tensor

$$\boldsymbol{\Theta} = \frac{1}{2} \sum_i e_i \left(3\mathbf{r}_i\mathbf{r}_i - r_i^2 \mathbf{1} \right), \qquad (2.4.4)$$

where r_i^2 is the scalar product $\mathbf{r}_i \cdot \mathbf{r}_i = x_i^2 + y_i^2 + z_i^2$, and $\mathbf{1}$ is the symmetric unit tensor $\mathbf{ii} + \mathbf{jj} + \mathbf{kk}$. In cartesian tensor notation, (2.4.4) is written

$$\Theta_{\alpha\beta} = \frac{1}{2} \sum_i e_i \left(3r_{i\alpha}r_{i\beta} - r_i^2 \delta_{\alpha\beta} \right). \qquad (2.4.5)$$

It is clear that $\Theta_{\alpha\beta}$ is a symmetric second-rank tensor,

$$\Theta_{\alpha\beta} = \Theta_{\beta\alpha}, \qquad (2.4.6)$$

with zero trace,

$$\Theta_{\alpha\alpha} = \Theta_{xx} + \Theta_{yy} + \Theta_{zz} = 0, \qquad (2.4.7)$$

so that it has five independent components:

$$\Theta = \begin{pmatrix} \Theta_{xx} & \Theta_{xy} & \Theta_{xz} \\ \Theta_{yx} = \Theta_{xy} & \Theta_{yy} & \Theta_{yz} \\ \Theta_{zx} = \Theta_{xz} & \Theta_{zy} = \Theta_{yz} & \Theta_{zz} = -\Theta_{xx} - \Theta_{yy} \end{pmatrix}. \qquad (2.4.8)$$

The electric quadrupole moment is only independent of the choice of origin if the net charge and the dipole moment are both zero. Thus, on moving the origin from O to $O' = O + \mathbf{a}$,

$$\Theta'_{\alpha\beta} = \frac{1}{2} \sum_i e_i \left(3 r'_{i\alpha} r'_{i\beta} - r_i'^2 \delta_{\alpha\beta} \right)$$

$$= \Theta_{\alpha\beta} - \tfrac{3}{2}\mu_\alpha a_\beta - \tfrac{3}{2}\mu_\beta a_\alpha + \mu_\gamma a_\gamma \delta_{\alpha\beta} + \tfrac{1}{2}q(3a_\alpha a_\beta - a^2 \delta_{\alpha\beta}). \qquad (2.4.9)$$

If q is zero but μ is not zero, a centre of dipole exists relative to which $\Theta_{\alpha\beta} = 0$.

The electric quadrupole moment is sometimes defined as $\frac{1}{2}\sum_i e_i r_{i\alpha} r_{i\beta}$. However, the traceless definition (2.4.5) is preferred here since it automatically emerges as the source of a well-defined part of the scalar potential generated by a static charge distribution (see Section 2.4.3). Another related reason for preferring the traceless definition is that it vanishes for a spherical charge distribution.

The general nth order electric multipole moment is defined as

$$\xi^{(n)}_{\alpha\beta\gamma\cdots\nu} = \frac{(-1)^n}{n!} \sum_i e_i r_i^{2n+1} \nabla_{i\alpha} \nabla_{i\beta} \nabla_{i\gamma} \cdots \nabla_{i\nu} \left(\frac{1}{r_i} \right), \qquad (2.4.10)$$

where $\nabla_{i\alpha} = \partial/\partial r_{i\alpha}$, and is symmetric in all n suffixes. It is instructive to evaluate explicitly

$$\nabla_\alpha \nabla_\beta r^{-1} = \nabla_\alpha \nabla_\beta (r_\gamma r_\gamma)^{-\frac{1}{2}} = \nabla_\alpha \left[-r_\beta (r_\gamma r_\gamma)^{-\frac{3}{2}} \right]$$

$$= 3 r_\alpha r_\beta (r_\gamma r_\gamma)^{-\frac{5}{2}} - \delta_{\alpha\beta} (r_\gamma r_\gamma)^{-\frac{3}{2}} = (3 r_\alpha r_\beta - r^2 \delta_{\alpha\beta}) r^{-5}, \qquad (2.4.11)$$

and also the Laplace equation

$$\nabla_\alpha \nabla_\alpha r^{-1} = \nabla^2 r^{-1} = 0. \qquad (2.4.12)$$

This enables us to verify that $\xi^{(2)}_{\alpha\beta} = \Theta_{\alpha\beta}$, and to see that if any tensor suffix in (2.4.10) is repeated (with summation over repeated suffixes implied), the corresponding multipole moment vanishes. In general, the maximum number of independent components of $\xi^{(n)}_{\alpha\beta\gamma\cdots\nu}$ is $2n + 1$, though symmetry may reduce this number. Such a tensor is said to be *irreducible* because the vanishing on contraction (summing over repeated suffixes) means that no tensor of lower rank can be constructed from the components.

Instead of real multipole moments expressed as cartesian tensors, it is possible to define complex multipole moments expressed as spherical harmonics. But molecules have a natural cartesian frame rather than a natural polar frame (except linear molecules), so for our purposes the real form is preferable.

2.4.2 Magnetic multipole moments

Magnetic monopoles have not been observed. The first moment of a circulating current distribution is the *magnetic dipole moment*; in the absence of external magnetic fields this is

$$\mathbf{m} = \sum_i \frac{e_i}{2m_i} \mathbf{r}_i \times \mathbf{p}_i, \qquad (2.4.13)$$

where m_i and $\mathbf{p}_i$ are the mass and linear momentum of the ith charge. On moving the origin from $\mathbf{O}$ to $\mathbf{O}' = \mathbf{O} + \mathbf{a}$, the magnetic dipole moment changes to

$$\mathbf{m}' = \sum_i \frac{e_i}{2m_i} \mathbf{r}'_i \times \mathbf{p}'_i = \mathbf{m} - \frac{1}{2} \mathbf{a} \times \dot{\boldsymbol{\mu}}, \qquad (2.4.14)$$

where $\dot{\boldsymbol{\mu}} = \partial \boldsymbol{\mu} / \partial t$, so the magnetic dipole moment is independent of the choice of the origin only in the absence of a time-dependent electric dipole moment.

The vector product $\mathbf{r} \times \mathbf{p}$ is the orbital angular momentum $\mathbf{l}$ of the particle. Spin angular momentum $\mathbf{s}$ also contributes to the magnetic moment, and in general

$$\mathbf{m} = \sum_i \frac{e_i}{2m_i} (\mathbf{l}_i + g_i \mathbf{s}_i), \qquad (2.4.15)$$

where g_i is the *g-value* of the ith particle spin ($g = 2.0023$ for a free electron). When the particles are electrons, the factor $e/2m$ is often replaced by $-\mu_B/\hbar$, where $\mu_B = e\hbar/2m = 9.274 \times 10^{-24} \, \text{JT}^{-1}$ is the *Bohr magneton*.

The definition of the *magnetic quadrupole moment* is not clear-cut, and several different versions have been proposed. The following definition in the absence of external magnetic fields has been given by Buckingham and Stiles (1972):

$$M_{\alpha\beta} = \tfrac{1}{2}(3m_{\alpha\beta} - m_{\gamma\gamma}\delta_{\alpha\beta}), \qquad (2.4.16)$$

where

$$m_{\alpha\beta} = \sum_i \frac{e_i}{2m_i} \left[r_{i\alpha} \left(\tfrac{2}{3} l_{i\beta} + g_i s_{i\beta} \right) + r_{i\beta} \left(\tfrac{2}{3} l_{i\alpha} + g_i s_{i\alpha} \right) \right]. \qquad (2.4.17)$$

While this definition is satisfactory for 'static' classical current distributions, Raab (1975) has shown that the symmetry with respect to exchange of the suffixes leads to an uncharacteristic form for the electromagnetic fields radiated by an oscillating magnetic quadrupole moment. This point is not pursued here since we do not invoke magnetic quadrupole moments.

2.4.3 Static electric multipole fields

We now consider the electric field generated by a stationary charge distribution. If the charge density is written, with the aid of a δ function, in terms of point charges as

$$\rho = \sum_i e_i \delta(\mathbf{r} - \mathbf{r}_i), \tag{2.4.18}$$

the static scalar potential (2.2.29b) becomes

$$\phi(\mathbf{R}) = \frac{1}{4\pi\epsilon_0} \sum_i \frac{e_i}{|\mathbf{R} - \mathbf{r}_i|}, \tag{2.4.19}$$

where $|\mathbf{R} - \mathbf{r}_i|$ is the distance from the ith charge to the point P where the potential is required. We are interested in values of $|\mathbf{R}|$ sufficiently large that $|\mathbf{R}| \gg |\mathbf{r}_i|$. We can then use the following expansion:

$$\frac{1}{|\mathbf{R} - \mathbf{r}_i|} = (R_\alpha R_\alpha - 2R_\alpha r_{i\alpha} + r_{i\alpha} r_{i\alpha})^{-\frac{1}{2}}$$

$$= \frac{1}{R} + \frac{R_\alpha r_{i\alpha}}{R^3} + \frac{1}{2} \left(\frac{3 R_\alpha R_\beta r_{i\alpha} r_{i\beta}}{R^5} - \frac{r_i^2}{R^3} \right) + \cdots. \tag{2.4.20}$$

The scalar potential (2.4.19) can now be written in terms of the electric multipole moments of the collection of charges:

$$\phi(\mathbf{R}) = \frac{1}{4\pi\epsilon_0} \left(\frac{q}{R} + \frac{R_\alpha \mu_\alpha}{R^3} + \frac{R_\alpha R_\beta \Theta_{\alpha\beta}}{R^5} + \cdots \right). \tag{2.4.21}$$

Using $E_\alpha = -\nabla_\alpha \phi$, the associated static electric field is

$$E_\alpha(\mathbf{R}) = \frac{1}{4\pi\epsilon_0} \left(\frac{R_\alpha q}{R^3} + \frac{3 R_\alpha R_\beta \mu_\beta - R^2 \mu_\alpha}{R^5} \right.$$

$$\left. + \frac{5 R_\alpha R_\beta R_\gamma \Theta_{\beta\gamma} - 2R^2 R_\beta \Theta_{\alpha\beta}}{R^7} + \cdots \right). \tag{2.4.22}$$

The physical interpretation of (2.4.21) is as follows. At very large distances, the collection of charges looks like a point charge and we can set $|\mathbf{R} - \mathbf{r}_i| = R$ so that the potential is given by the first term, a point charge potential. But if the collection of charges is neutral, the first term vanishes. Since all the charges are not at one point, there should still be a residual potential, and the successive terms correspond to using an increasingly accurate expression for $|\mathbf{R} - \mathbf{r}_i|$.

This is an appropriate point at which to introduce the tensors

$$T = R^{-1}, \tag{2.4.23a}$$

$$T_\alpha = \nabla_\alpha R^{-1} = -R_\alpha R^{-3}, \tag{2.4.23b}$$

$$T_{\alpha\beta} = T_{\beta\alpha} = \nabla_\alpha \nabla_\beta R^{-1} = (3R_\alpha R_\beta - R^2 \delta_{\alpha\beta})R^{-5}, \tag{2.4.23c}$$

$$T_{\alpha\beta\gamma} = \nabla_\alpha \nabla_\beta \nabla_\gamma R^{-1} = -3[5R_\alpha R_\beta R_\gamma - R^2(R_\alpha \delta_{\beta\gamma}$$
$$+ R_\beta \delta_{\alpha\gamma} + R_\gamma \delta_{\alpha\beta})]R^{-7}, \tag{2.4.23d}$$

$$T_{\alpha\beta\gamma\cdots\nu} = \nabla_\alpha \nabla_\beta \nabla_\gamma \cdots \nabla_\nu R^{-1}. \tag{2.4.23e}$$

These tensors are symmetric in all suffixes, and (2.4.12) shows that a repeated suffix reduces a tensor to zero.

The scalar potential (2.4.21) and electric field (2.4.22) due to a charge distribution can now be given succinct forms:

$$\phi(\mathbf{R}) = \frac{1}{4\pi\epsilon\epsilon_0}(Tq - T_\alpha \mu_\alpha + \tfrac{1}{3}T_{\alpha\beta}\Theta_{\alpha\beta} + \cdots), \tag{2.4.24}$$

$$E_\alpha(\mathbf{R}) = \frac{1}{4\pi\epsilon\epsilon_0}(-T_\alpha q + T_{\alpha\beta}\mu_\beta - \tfrac{1}{3}T_{\alpha\beta\gamma}\Theta_{\beta\gamma} + \cdots). \tag{2.4.25}$$

2.4.4 Static magnetic multipole fields

We now turn to the corresponding static magnetic field generated by a system of charges in 'stationary' motion. If the current density is written in terms of point charges moving with velocity $\dot{\mathbf{r}}$,

$$\mathbf{J} = \sum_i e_i \dot{\mathbf{r}}_i \delta(\mathbf{r} - \mathbf{r}_i), \tag{2.4.26}$$

the static vector potential (2.2.29a) becomes

$$\mathbf{A}(\mathbf{R}) = \frac{\mu\mu_0}{4\pi} \sum_i \frac{e_i \dot{\mathbf{r}}_i}{|\mathbf{R} - \mathbf{r}_i|}. \tag{2.4.27}$$

Here we are interested in a constant current which generates, through some circulatory character, a static vector potential and a corresponding static magnetic field. The field will be a function only of the coordinates, not of the time, so a time average is required:

$$\overline{\mathbf{A}(\mathbf{R})} = \frac{\mu\mu_0}{4\pi} \sum_i \frac{\overline{e_i \dot{\mathbf{r}}_i}}{|\mathbf{R} - \mathbf{r}_i|}, \tag{2.4.28}$$

where the average includes the position vectors $|\mathbf{R} - \mathbf{r}_i|$, which change during the motion of the charges. Using the expansion (2.4.20), this becomes

$$\overline{\mathbf{A}(\mathbf{R})} = \frac{\mu\mu_0}{4\pi} \sum_i \left[\overline{\frac{e_i \dot{\mathbf{r}}_i}{R}} + \overline{\frac{e_i \dot{\mathbf{r}}_i (\mathbf{r}_i \cdot \mathbf{R})}{R^3}} + \cdots \right]. \tag{2.4.29}$$

The first term of (2.4.29) vanishes since the time average of the linear velocity of a particle constrained to move within a small volume is zero. Since $\mathbf{R}$ is constant with time,

$$\dot{\mathbf{r}}(\mathbf{r} \cdot \mathbf{R}) = \tfrac{1}{2} \frac{d}{dt} \mathbf{r}(\mathbf{r} \cdot \mathbf{R}) + \tfrac{1}{2} [\dot{\mathbf{r}}(\mathbf{r} \cdot \mathbf{R}) - \mathbf{r}(\dot{\mathbf{r}} \cdot \mathbf{R})]$$

$$= \tfrac{1}{2} \frac{d}{dt} \mathbf{r}(\mathbf{r} \cdot \mathbf{R}) - \tfrac{1}{2} \mathbf{R} \times (\mathbf{r} \times \dot{\mathbf{r}}). \tag{2.4.30}$$

The time average of the first term of (2.4.30) vanishes for the same reason as the first term of (2.4.29). However, the time average of the angular velocity of a particle constrained to move within a small volume is not zero, so (2.4.29) can now be written in terms of the magnetic dipole moment (2.4.13):

$$\overline{\mathbf{A}(\mathbf{R})} = \frac{\mu\mu_0}{4\pi} \left(\frac{\mathbf{m} \times \mathbf{R}}{R^3} + \cdots \right). \tag{2.4.31}$$

Using $\mathbf{B} = \nabla \times \mathbf{A}$, the associated magnetic field is

$$B_\alpha(\mathbf{R}) = \frac{\mu\mu_0}{4\pi} \left(\frac{3 R_\alpha R_\beta m_\beta - R^2 m_\alpha}{R^5} + \cdots \right). \tag{2.4.32}$$

For simplicity, we have derived only the magnetic dipole contribution to the magnetic field. The general expression is analogous to (2.4.25) for the electric field, except that there is no magnetic analogue of the net electric charge:

$$B_\alpha(\mathbf{R}) = \frac{\mu\mu_0}{4\pi} \left(T_{\alpha\beta} m_\beta - \tfrac{1}{3} T_{\alpha\beta\gamma} M_{\beta\gamma} + \cdots \right). \tag{2.4.33}$$

2.4.5 Dynamic electromagnetic multipole fields

Of particular importance are the electromagnetic fields generated by a system of time-varying charges and currents; we formulate this radiation field in terms of specific contributions from particular time-varying electric and magnetic multipole moments. Our basic assumption is that the charge and current densities vary harmonically with time. Such is the case if a monochromatic light wave is incident on the system, with the fields radiated by the induced oscillating charges and currents constituting scattered light. Hence we write

$$\rho(t) = \rho^{(0)} e^{-i\omega t}, \tag{2.4.34a}$$

$$\mathbf{J}(t) = \mathbf{J}^{(0)} e^{-i\omega t}. \tag{2.4.34b}$$

For simplicity, we omit the tildes over complex quantities in this section.

The radiated fields are determined by the retarded potentials (2.2.30), which require $\rho(t)$ and $\mathbf{J}(t)$ to be evaluated at the retarded time $t' = t - |\mathbf{R} - \mathbf{r}|/v$:

$$\phi(\mathbf{R}, t) = \frac{1}{4\pi \epsilon \epsilon_0} \int \frac{\rho^{(0)} e^{i(\kappa|\mathbf{R}-\mathbf{r}|-\omega t)} dV}{|\mathbf{R} - \mathbf{r}|}, \tag{2.4.35a}$$

$$\mathbf{A}(\mathbf{R}, t) = \frac{\mu \mu_0}{4\pi} \int \frac{\mathbf{J}^{(0)} e^{i(\kappa|\mathbf{R}-\mathbf{r}|-\omega t)} dV}{|\mathbf{R} - \mathbf{r}|}. \tag{2.4.35b}$$

If the dimensions of the charge and current system are small compared with the wavelength, these retarded potentials can be developed in powers of κ. We use the expansions

$$|\mathbf{R} - \mathbf{r}| = R \left[1 - \frac{R_\alpha r_\alpha}{R^2} - \frac{1}{2} \left(\frac{R_\alpha R_\beta r_\alpha r_\beta}{R^4} - \frac{r^2}{R^2} \right) + \cdots \right],$$

$$\frac{1}{|\mathbf{R} - \mathbf{r}|} = \frac{1}{R} \left[1 + \frac{R_\alpha r_\alpha}{R^2} + \frac{1}{2} \left(\frac{3 R_\alpha R_\beta r_\alpha r_\beta}{R^4} - \frac{r^2}{R^2} \right) + \cdots \right],$$

to write

$$\frac{e^{i\kappa|\mathbf{R}-\mathbf{r}|}}{|\mathbf{R} - \mathbf{r}|} = \frac{e^{i\kappa R}}{R} \left[1 + \frac{R_\alpha r_\alpha}{R^2} + \frac{1}{2} \left(\frac{3 R_\alpha R_\beta r_\alpha r_\beta}{R^4} - \frac{r^2}{R^2} \right) - \frac{i\kappa R_\alpha r_\alpha}{R} \right.$$

$$\left. - \frac{i\kappa}{2} \left(\frac{3 R_\alpha R_\beta r_\alpha r_\beta}{R^3} - \frac{r^2}{R} \right) - \frac{\kappa^2 R_\alpha R_\beta r_\alpha r_\beta}{2R^2} + \cdots \right]. \tag{2.4.36}$$

Writing the charge density in terms of point charges, and using (2.4.36) in (2.4.35a), gives the following multipole expansion for the dynamic scalar potential part of the radiation:

$$\phi(\mathbf{R}, t) = \frac{e^{i(\kappa R - \omega t)}}{4\pi \epsilon \epsilon_0 R} \left(\frac{R_\alpha \mu_\alpha^{(0)}}{R^2} + \frac{R_\alpha R_\beta \Theta_{\alpha\beta}^{(0)}}{R^4} - \frac{i\kappa R_\alpha \mu_\alpha^{(0)}}{R} \right.$$

$$\left. - \frac{i\kappa R_\alpha R_\beta \Theta_{\alpha\beta}^{(0)}}{R^3} - \frac{\kappa^2 R_\alpha R_\beta \sum_i e_i r_{i\alpha}^{(0)} r_{i\beta}^{(0)}}{2R^2} + \cdots \right). \tag{2.4.37}$$

The development of the dynamic vector potential is more delicate since it is necessary to relate the current to the moments of the charge distribution. Our starting point is the equation of continuity, which expresses the conservation of charge within a body. The rate at which charge leaves a volume V bounded by a surface $\mathbf{S}$ is $\int_S \mathbf{J} \cdot d\mathbf{S}$. Since charge is conserved,

$$\int_S \mathbf{J} \cdot d\mathbf{S} = -\frac{dq}{dt} = -\int_V \frac{\partial \rho}{\partial t} dV,$$

where q is the net charge contained in V. From Gauss' theorem,

$$\int_S \mathbf{J} \cdot d\mathbf{S} = \int_V \nabla \cdot \mathbf{J} dV,$$

from which the equation of continuity follows:

$$\nabla \cdot \mathbf{J} + \frac{\partial \rho}{\partial t} = 0. \tag{2.4.38}$$

Invoking the harmonic time dependence of $\mathbf{J}$ and ρ, this becomes

$$\nabla \cdot \mathbf{J} = i\omega \rho.$$

Now multiply by an arbitrary scalar or tensor function f of position:

$$\int f(\nabla_\alpha J_\alpha) dV = \int \nabla_\alpha (f J_\alpha) dV - \int J_\alpha (\nabla_\alpha f) dV = i\omega \int \rho f dV.$$

From Gauss' theorem,

$$\int_V \nabla_\alpha (f J_\alpha) dV = \int_S f J_\alpha dS_\alpha,$$

which is zero if the surface of integration is taken beyond the boundary of the charge and current distribution. Therefore

$$\int J_\alpha (\nabla_\alpha f) dV = -i\omega \int \rho f dV. \tag{2.4.39}$$

Putting $f = 1$, we find from (2.4.39) that $\int \rho dV = 0$, so we must assert that the system is neutral overall for this treatment to be consistent. Putting $f = r_\beta$ and $f = r_\beta r_\gamma$ in (2.4.39) we find

$$\int J_\alpha dV = -i\omega \mu_\alpha, \tag{2.4.40a}$$

$$\int (J_\alpha r_\beta + J_\beta r_\alpha) dV = -i\omega \sum_i e_i r_{i\alpha} r_{i\beta}, \tag{2.4.40b}$$

$$\int (J_\alpha r_\beta - J_\beta r_\alpha) dV = -\varepsilon_{\alpha\beta\gamma} \varepsilon_{\gamma\delta\epsilon} \int r_\delta J_\epsilon dV = -2\varepsilon_{\alpha\beta\gamma} m_\gamma. \tag{2.4.40c}$$

Using (2.4.40) together with

$$\int J_\alpha r_\beta \mathrm{d}V = \frac{1}{2}\int (J_\alpha r_\beta + J_\beta r_\alpha)\mathrm{d}V + \frac{1}{2}\int (J_\alpha r_\beta - J_\beta r_\alpha)\mathrm{d}V$$

and (2.4.36) in (2.4.35b) gives the following multipole expansion for the dynamic vector potential:

$$
A_\alpha(\mathbf{R}, t) = -\frac{\mu\mu_0}{4\pi R}e^{i(\kappa R - \omega t)}\left(\frac{\varepsilon_{\alpha\beta\gamma} R_\beta m_\gamma^{(0)}}{R^2} + \frac{i c\kappa\mu_\alpha^{(0)}}{n}\right.
$$
$$
-\frac{i\kappa\varepsilon_{\alpha\beta\gamma} R_\beta m_\gamma^{(0)}}{R} + \frac{i c\kappa R_\beta \sum_i e_i r_{i\alpha}^{(0)} r_{i\beta}^{(0)}}{2nR^2}
$$
$$
\left.+\frac{c\kappa^2 R_\beta \sum_i e_i r_{i\alpha}^{(0)} r_{i\beta}^{(0)}}{2nR} + \cdots\right). \tag{2.4.41}
$$

The radiated electric field can be calculated using (2.4.37) and (2.4.41) in $\mathbf{E} = -\partial\mathbf{A}/\partial t - \nabla\phi$. Since the direction of propagation is along $\mathbf{R}$, it is convenient to write $\mathbf{R}$ in terms of the propagation vector $\mathbf{n}$, $R_\alpha = Rn_\alpha/n$, and we find

$$
E_\alpha(\mathbf{R}, t) = \frac{\mu\mu_0}{4\pi}e^{i(\kappa R - \omega t)}\left[\mu_\alpha^{(0)}\left(\frac{\omega^2}{R} + \frac{i\omega c}{nR^2} - \frac{c^2}{n^2 R^3}\right)\right.
$$
$$
- n_\alpha n_\beta \mu_\beta^{(0)}\left(\frac{\omega^2}{n^2 R} + \frac{3i\omega c}{n^3 R^2} - \frac{3c^2}{n^4 R^3}\right) - \varepsilon_{\alpha\beta\gamma} n_\beta m_\gamma^{(0)}\left(\frac{\omega^2}{cR} + \frac{i\omega}{nR^2}\right)
$$
$$
- n_\beta \sum_i e_i r_{i\alpha}^{(0)} r_{i\beta}^{(0)}\left(\frac{i\omega^3}{2cR} - \frac{3\omega^2}{2nR^2}\right) + n_\alpha n_\beta n_\gamma \sum_i e_i r_{i\beta}^{(0)} r_{i\gamma}^{(0)}\left(\frac{i\omega^3}{2n^2 cR} - \frac{3\omega^2}{2n^3 R^2}\right)
$$
$$
\left.+ n_\beta\Theta_{\alpha\beta}^{(0)}\left(\frac{2i\omega c}{n^2 R^3} - \frac{2c^2}{n^3 R^4}\right) - n_\alpha n_\beta n_\gamma\Theta_{\beta\gamma}^{(0)}\left(\frac{\omega^2}{n^3 R^2} + \frac{5i\omega c}{n^4 R^3} - \frac{5c^2}{n^5 R^4}\right) + \cdots\right].
$$
$$\tag{2.4.42a}$$

Similarly, using $\mathbf{B} = \nabla \times \mathbf{A}$, the radiated magnetic field is found to be

$$
B_\alpha(\mathbf{R}, t) = \frac{\mu\mu_0}{4\pi}e^{i(\kappa R - \omega t)}\varepsilon_{\alpha\beta\gamma}\left\{n_\beta\mu_\gamma^{(0)}\left(\frac{\omega^2}{cR} + \frac{i\omega}{nR^2}\right)\right.
$$
$$
- \varepsilon_{\gamma\delta\epsilon} m_\epsilon^{(0)}\left[\frac{\omega^2 n_\beta n_\delta}{c^2 R} + \frac{i\omega(3n_\beta n_\delta - n^2\delta_{\beta\delta})}{ncR^2} - \frac{(3n_\beta n_\delta - n^2\delta_{\beta\delta})}{n^2 R^3}\right] \tag{2.4.42b}
$$
$$
\left.- \sum_i e_i r_{i\gamma}^{(0)} r_{i\delta}^{(0)}\left(\frac{i\omega^3 n_\beta n_\delta}{2c^2 R} - \frac{3\omega^2 n_\beta n_\delta}{2nc R^2} - \frac{3i\omega n_\beta n_\delta}{2n^2 R^3}\right) + \cdots\right\}.
$$

We now consider two important limits of (2.4.42).

At distances large compared with the wavelength ($\kappa R \gg 1$), we retain only terms in $1/R$ in (2.4.42):

$$
E_\alpha(\mathbf{R}, t) = \frac{\omega^2 \mu \mu_0}{4\pi R} e^{i(\kappa R - \omega t)} \left[\left(\mu_\alpha^{(0)} - \frac{n_\alpha n_\beta}{n^2} \mu_\beta^{(0)} \right) \right.
$$
$$
\left. - \frac{1}{c}\varepsilon_{\alpha\beta\gamma} n_\beta m_\gamma^{(0)} - \frac{i\omega}{3c}\left(n_\beta \Theta_{\alpha\beta}^{(0)} - \frac{n_\alpha n_\beta n_\gamma}{n^2}\Theta_{\beta\gamma}^{(0)} \right) + \cdots \right], \quad (2.4.43a)
$$

$$
B_\alpha(\mathbf{R}, t) = \frac{\omega^2 \mu \mu_0}{4\pi R c} e^{i(\kappa R - \omega t)} \varepsilon_{\alpha\beta\gamma} n_\beta
$$
$$
\times \left(\mu_\gamma^{(0)} - \frac{1}{c}\varepsilon_{\gamma\delta\epsilon} n_\delta m_\epsilon^{(0)} - \frac{i\omega}{3c} n_\delta \Theta_{\gamma\delta}^{(0)} + \cdots \right). \quad (2.4.43b)
$$

The meaning of the terms

$$
\left(\mu_\alpha^{(0)} - \frac{n_\alpha n_\beta}{n^2}\mu_\beta^{(0)} \right) \quad \text{and} \quad \left(n_\beta \Theta_{\alpha\beta}^{(0)} - \frac{n_\alpha n_\beta n_\gamma}{n^2}\Theta_{\beta\gamma}^{(0)} \right)
$$

is that the components of the vectors $\mu_\alpha^{(0)}$ and $n_\beta \Theta_{\alpha\beta}^{(0)}$ parallel to the direction of propagation of the radiated field are subtracted out, so that only the perpendicular components remain. Consequently, $\mathbf{E}(\mathbf{R}, t)$ given by (2.4.43a) is entirely transverse; $\mathbf{B}(\mathbf{R}, t)$ is always transverse since $\nabla \cdot \mathbf{B} = 0$. This is consistent with the fact that the electric and magnetic field vectors in (2.4.43a and b) are related by

$$
B_\alpha = \frac{1}{c}\varepsilon_{\alpha\beta\gamma} n_\beta E_\gamma,
$$

which was shown in Section 2.2.2 to be a property of a plane wave. This transversality has also enabled $\frac{3}{2}\sum_i e_i r_{i\alpha}^{(0)} r_{i\beta}^{(0)}$ to be replaced by the traceless electric quadrupole moment $\Theta_{\alpha\beta}^{(0)}$. The region of space at sufficiently large distances for the radiated wave to be considered a plane wave over small regions of space is called the *wave zone*.

At distances small compared with the wavelength ($\kappa R \ll 1$), we can neglect terms in $1/R$ and $1/R^2$ in (2.4.42) and set $\exp(i\kappa R) \approx 1$:

$$
E_\alpha(\mathbf{R}, t) = \frac{e^{-i\omega t}}{4\pi \epsilon \epsilon_0}\left(\frac{3R_\alpha R_\beta \mu_\beta^{(0)} - R^2 \mu_\alpha^{(0)}}{R^5} + \frac{5R_\alpha R_\beta R_\gamma \Theta_{\beta\gamma}^{(0)} - 2R^2 R_\beta \Theta_{\alpha\beta}^{(0)}}{R^7} + \cdots \right),
$$
$$
(2.4.44a)
$$

$$
B_\alpha(\mathbf{R}, t) = \frac{\mu \mu_0 e^{-i\omega t}}{4\pi}\left(\frac{3R_\alpha R_\beta m_\beta^{(0)} - R^2 m_\alpha^{(0)}}{R^5} + \frac{i\omega \varepsilon_{\alpha\beta\gamma} R_\beta R_\delta \Theta_{\gamma\delta}^{(0)}}{R^5} + \cdots \right).
$$
$$
(2.4.44b)
$$

The electric field (2.4.44a) is analogous to the static field (2.4.22) of stationary electric dipole and electric quadrupole moments. In this approximation there is no contribution to the radiated electric field from the magnetic dipole moment. Notice that an oscillating electric quadrupole moment contributes to the radiated magnetic field (2.4.44*b*) in this approximation, whereas there is no analogous contribution to the static magnetic field (2.4.32) from a stationary electric quadrupole moment.

It can be useful to write the wave zone fields (2.4.43) in terms of time derivatives of the oscillating electric and magnetic multipole moments as follows:

$$E_\alpha(\mathbf{R}, t) = -\frac{\mu\mu_0}{4\pi R} e^{i\kappa R} \left[\left(\ddot{\mu}_\alpha - \frac{n_\alpha n_\beta}{n^2} \ddot{\mu}_\beta \right) - \frac{1}{c} \varepsilon_{\alpha\beta\gamma} n_\beta \ddot{m}_\gamma \right.$$
$$\left. + \frac{1}{3c} \left(n_\beta \dddot{\Theta}_{\alpha\beta} - \frac{n_\alpha n_\beta n_\gamma}{n^2} \dddot{\Theta}_{\beta\gamma} \right) + \cdots \right], \qquad (2.4.45a)$$

$$B_\alpha(\mathbf{R}, t) = -\frac{\mu\mu_0}{4\pi R} e^{i\kappa R} \varepsilon_{\alpha\beta\gamma} n_\beta$$
$$\times \left(\ddot{\mu}_\gamma - \frac{1}{c} \varepsilon_{\gamma\delta\epsilon} n_\delta \ddot{m}_\epsilon + \frac{1}{3c} n_\delta \dddot{\Theta}_{\gamma\delta} + \cdots \right). \qquad (2.4.45b)$$

These are the same as the expressions used in Buckingham and Raab (1975), for example, and are equivalent to those derived in Landau and Lifshitz (1975). They are useful, among other things, for checking that each term has the correct behaviour under space inversion and time reversal (see Chapter 4). This is because the propagation vector **n** has well-defined transformation properties (it is *P*-odd and *T*-odd). Similarly for other equations such as (2.4.37) for the scalar potential and (2.4.41) for the vector potential.

2.5 The energy of charges and currents in electric and magnetic fields

We now consider the energy of a system of charges and currents bathed in both static and dynamic external electric and magnetic fields and develop expressions which, in operator form, constitute convenient Hamiltonians for subsequent quantum-mechanical calculations.

The equation of motion of a charged particle in an electromagnetic field is actually the Lorentz force equation

$$\mathbf{F} = e\mathbf{E} + e\mathbf{v} \times \mathbf{B}. \qquad (2.5.1)$$

An equation of motion is generated from a *Lagrangian function L = T − V*, where *T* and *V* are the kinetic and potential energies, through the *Euler–Lagrange equation*,

$$\nabla L - \frac{d}{dt} \frac{\partial L}{\partial \mathbf{v}} = 0. \qquad (2.5.2)$$

It is easily verified that the Lagrangian

$$L = \tfrac{1}{2}mv^2 + e\mathbf{v} \cdot \mathbf{A} - e\phi \qquad (2.5.3)$$

generates the required equation of motion (2.5.1).

From the Lagrangian (2.5.3) we can find the *Hamiltonian H* for a charged particle in an electromagnetic field using

$$H = \mathbf{v} \cdot \frac{\partial L}{\partial \mathbf{v}} - L, \qquad (2.5.4)$$

where $\partial L / \partial \mathbf{v}$ is the *generalized* momentum $\mathbf{p}'$ of the particle. The generalized momentum will only equal the *Newtonian* momentum $\mathbf{p} = m\mathbf{v}$ when V is independent of velocity, which is not the case for a charged particle moving in an electromagnetic field. The generalized momentum is therefore

$$\mathbf{p}' = \frac{\partial L}{\partial \mathbf{v}} = m\mathbf{v} + e\mathbf{A}, \qquad (2.5.5)$$

and the Hamiltonian, expressed in terms of the generalized rather than the Newtonian momentum, is

$$H = \frac{1}{2m}(\mathbf{p}' - e\mathbf{A})^2 + e\phi. \qquad (2.5.6)$$

(Writing the square of a vector expression implies a scalar product.) In applying this result to the interaction of a quantum system with an electromagnetic field, it must be remembered that the operator $-i\hbar\nabla$ replaces $\mathbf{p}'$, not $\mathbf{p}$; also $\mathbf{p}'$ and $\mathbf{A}$ do not necessarily commute.

2.5.1 Electric and magnetic multipole moments in static fields

The Hamiltonian (2.5.6) is now developed to obtain expressions containing explicit multipole terms for the interaction energy between a system of charges and currents and static electric and magnetic fields produced by external sources.

From (2.5.6) the potential energy of the ith charge at $\mathbf{r}_i$ in a static electric field characterized by a scalar potential is $e_i\phi(\mathbf{r}_i)$. We expand $\phi(\mathbf{r})$ in a Taylor series about an origin $\mathbf{O}$ within the system of charges:

$$\phi(\mathbf{r}) = (\phi)_0 + r_\alpha(\nabla_\alpha \phi)_0 + \tfrac{1}{2}r_\alpha r_\beta(\nabla_\alpha\nabla_\beta\phi)_0 + \cdots$$
$$= (\phi)_0 - r_\alpha(E_\alpha)_0 - \tfrac{1}{2}r_\alpha r_\beta(E_{\alpha\beta})_0 + \cdots, \qquad (2.5.7)$$

where $E_{\alpha\beta}$ is used to denote the field gradient $\nabla_\alpha E_\beta$ and a subscript 0 indicates that a field or field gradient is taken at the origin. The potential energy of a system of charges in a static electric field is now obtained in multipole form:

$$V = \sum_i e_i \phi(\mathbf{r}_i) = q(\phi)_0 - \mu_\alpha(E_\alpha)_0 - \tfrac{1}{3}\Theta_{\alpha\beta}(E_{\alpha\beta})_0 + \cdots. \tag{2.5.8}$$

The introduction of the traceless electric quadrupole moment $\Theta_{\alpha\beta}$ is permissible here since the origin is far removed from the external charge distribution producing ϕ so that

$$\delta_{\alpha\beta}(\nabla_\alpha \nabla_\beta \phi)_0 = (\nabla^2 \phi)_0 = -\frac{\rho(\mathbf{0})}{\epsilon\epsilon_0} = 0. \tag{2.5.9}$$

The vector potential describing a static uniform magnetic field may be written

$$\mathbf{A} = \tfrac{1}{2}(\mathbf{B} \times \mathbf{r}) \tag{2.5.10}$$

since this satisfies $\mathbf{B} = \nabla \times \mathbf{A}$ (because $\mathbf{B}$ is independent of $\mathbf{r}$ in a uniform field). In expanding $(\mathbf{p}' - e\mathbf{A})^2$ in the Hamiltonian (2.5.6), notice that, if $\mathbf{p}'$ is the quantum-mechanical operator $-i\hbar\nabla$, $\mathbf{p}'$ and $\mathbf{A}$ do not commute unless $\nabla \cdot \mathbf{A} = 0$, which holds for the vector potential (2.5.10). From (2.5.6), the potential energy of a system of currents is a static uniform magnetic field is therefore

$$\begin{aligned} V &= -\sum_i \frac{e_i}{m_i}\mathbf{p}'_i \cdot \mathbf{A}(\mathbf{r}_i) + \sum_i \frac{e_i^2}{2m_i} A(\mathbf{r}_i)^2 \\ &= -\sum_i \frac{e_i}{2m_i}\mathbf{p}'_i \cdot [\mathbf{B}(\mathbf{r}_i) \times \mathbf{r}_i] + \sum_i \frac{e_i^2}{8m_i}[\mathbf{B}(\mathbf{r}_i) \times \mathbf{r}_i]^2 \\ &= -m_\alpha B_\alpha - \tfrac{1}{2}\chi_{\alpha\beta}^{(d)} B_\alpha B_\beta. \end{aligned} \tag{2.5.11}$$

In developing the second term, we have used the tensor relation

$$\varepsilon_{\alpha\beta\gamma}\varepsilon_{\alpha\delta\lambda} = \delta_{\beta\delta}\delta_{\gamma\lambda} - \delta_{\beta\lambda}\delta_{\gamma\delta},$$

introduced later (Section 4.2.4), to expose the *diamagnetic susceptibility* tensor

$$\chi_{\alpha\beta}^{(d)} = \sum_i \frac{e_i^2}{4m_i}\left(r_{i_\alpha} r_{i_\beta} - r_i^2 \delta_{\alpha\beta}\right). \tag{2.5.12}$$

This can be thought of as generating a magnetic field induced magnetic moment $\tfrac{1}{2}\chi_{\alpha\beta}^{(d)} B_\alpha$ that opposes the inducing field. The magnetic potential energy (2.5.11) contains only dipole interactions because it was derived for a uniform magnetic field. If the field is not uniform and higher multipole interaction terms are required, a general expansion about the origin must be used in place of (2.5.10). For example, although

$$A_\alpha(\mathbf{r}) = \tfrac{1}{2}\varepsilon_{\alpha\beta\gamma}(B_\beta)_0 r_\gamma + \tfrac{1}{3}\varepsilon_{\alpha\gamma\delta} r_\beta(\nabla_\beta B_\gamma)_0 r_\delta + \cdots \tag{2.5.13}$$

is not itself a Taylor expansion, it leads to the correct Taylor expansion for $\mathbf{B}(r)$ using $\mathbf{B} = \nabla \times \mathbf{A}$.

The interaction energy between a system of charges and currents and static electric and magnetic fields is therefore obtained quite naturally in multipole form. But we shall see below that when the fields are dynamic, as in a radiation field, the development of the interaction energy in multipole form is more difficult.

We now consider the interaction energy of two widely separated charge distributions 1 and 2. This is given by an expression similar to (2.5.8) obtained by developing the Coulomb interaction energy between the constituent charges:

$$V = \frac{1}{4\pi\epsilon\epsilon_0} \sum_{i_1,i_2} \frac{e_{i_1}e_{i_2}}{R_{i_1i_2}} = q_2(\phi)_2 - \mu_{2\alpha}(E_\alpha)_2 - \tfrac{1}{3}\Theta_{2\alpha\beta}(E_{\alpha\beta})_2 + \cdots, \quad (2.5.14)$$

where $R_{i_1i_2}$ is the distance between the charge elements e_{i_1} in distribution 1 and e_{i_2} in distribution 2; q_2, $\mu_{2\alpha}$, $\Theta_{2\alpha\beta}$, etc. are the electric multipole moments of distribution 2; and $(\phi)_2$, $(E_\alpha)_2$, $(E_{\alpha\beta})_2$, etc. are the fields and field gradients at the coordinate origin of 2 due to the instantaneous charge distribution 1. Reversing the roles of 1 and 2 gives the same interaction energy. Using (2.4.24) and (2.4.25), the interaction energy becomes

$$V = \frac{1}{4\pi\epsilon\epsilon_0}\big[T_{21}q_1q_2 + T_{21\alpha}(q_1\mu_{2\alpha} - q_2\mu_{1\alpha})$$

$$+ T_{21\alpha\beta}\big(\tfrac{1}{3}q_1\Theta_{2\alpha\beta} + \tfrac{1}{3}q_2\Theta_{1\alpha\beta} - \mu_{1\alpha}\mu_{2\beta}\big) + \cdots\big], \quad (2.5.15)$$

where the subscript 21 on the $\mathbf{T}$ tensors indicates they are functions of the vector $\mathbf{R}_{21} = \mathbf{R}_2 - \mathbf{R}_1$ from the origin on 1 to that on 2. Clearly $\mathbf{T}_{21} = (-1)^n\mathbf{T}_{12}$, where n is the order of the tensor.

The interaction energy of two current distributions is similarly given by

$$V = -m_{2\alpha}(B_\alpha)_2 + \cdots. \quad (2.5.16)$$

Using (2.4.33), this becomes

$$V = \frac{\mu\mu_0}{4\pi}(-T_{21\alpha\beta}m_{1\alpha}m_{2\beta} + \cdots). \quad (2.5.17)$$

Magnetic analogues of the lower order terms in (2.5.15) do not arise since magnetic monopoles do not exist.

2.5.2 *Electric and magnetic multipole moments in dynamic fields*

We now turn to the development of the Hamiltonian (2.5.6) for the important case of charges and currents in dynamic electric and magnetic fields, particularly a radiation field. There are several methods of exposing the multipole interaction terms. The most widely used method in molecular optics is to expand the operator equivalent of (2.5.6) and invoke the quantum mechanical commutation relations between the coordinates and the Hamiltonian of the charges and currents. It was shown in Section 2.2.4 that if the sources of the radiation field are far removed, the condition $\nabla \cdot \mathbf{A} = 0$ and $\phi = 0$ hold in both the Coulomb gauge and the Lorentz gauge. In this case, therefore, the potential energy part of (2.5.6) can be written

$$V = -\sum_i \frac{e_i}{m_i} \mathbf{p}'_i \cdot \mathbf{A}(\mathbf{r}_i) + \sum_i \frac{e_i^2}{2m_i} \mathbf{A}(\mathbf{r}_i)^2. \tag{2.5.18}$$

If the wavelength of the radiation field is large compared with the dimensions of the system of charges and currents, $\mathbf{A}(\mathbf{r})$ can be expanded in a Taylor series about an origin $\mathbf{O}$ within the system of charges and currents. The first term of (2.5.18) then becomes

$$-\sum_i \frac{e_i}{m_i} A_\alpha(\mathbf{r}_i) p'_{i\alpha} = -\sum_i \frac{e_i}{m_i} [(A_\alpha)_0 p'_{i\alpha} + (A_{\beta\alpha})_0 r_{i\beta} p'_{i\alpha} + \cdots]$$

$$= -\sum_i \frac{e_i}{m_i} \left\{ (A_\alpha)_0 p'_{i\alpha} + \tfrac{1}{2}(A_{\beta\alpha})_0 [(r_{i\beta} p'_{i\alpha} + r_{i\alpha} p'_{i\beta}) \right.$$

$$\left. + (r_{i\beta} p'_{i\alpha} - r_{i\alpha} p'_{i\beta})] + \cdots \right\}. \tag{2.5.19}$$

The electric dipole nature of the first term of (2.5.19) is exposed using the commutation relation

$$r_\alpha H - H r_\alpha = \frac{i\hbar}{m} p'_\alpha, \tag{2.5.20}$$

where

$$H = -\frac{\hbar^2}{2m} \nabla^2 + V(\mathbf{r}) \tag{2.5.21}$$

is the Hamiltonian for a particle bound in the molecule. The electric quadrupole nature of the symmetric part of the second term of (2.5.19) is exposed using the commutation relation

$$r_\alpha r_\beta H - H r_\alpha r_\beta = \frac{i\hbar}{m} (r_\beta p'_\alpha + r_\alpha p'_\beta - i\hbar \delta_{\alpha\beta}). \tag{2.5.22}$$

Actually, the commutation relations (2.5.20) and (2.5.22) are only valid if the potential energy $V(\mathbf{r})$ in the Hamiltonian (2.5.21) commutes with the coordinate; as shown later, this is not always true, particularly when spin–orbit coupling is significant. The antisymmetric part of the second term of (2.5.19) already has the form of a magnetic dipole interaction:

$$-\sum_i \frac{e_i}{2m_i}(A_{\beta\alpha})_0(r_{i\beta}p'_{i\alpha} - r_{i\alpha}p'_{i\beta}) = -\sum_i \frac{e_i}{2m_i}(A_{\beta\alpha})_0 r_{i\gamma}p'_{i\delta}(\delta_{\gamma\beta}\delta_{\delta\alpha} - \delta_{\gamma\alpha}\delta_{\delta\beta}).$$

$$= -\sum_i \frac{e_i}{2m_i}\varepsilon_{\epsilon\gamma\delta}r_{i\gamma}p'_{i\delta}\varepsilon_{\epsilon\beta\alpha}(A_{\beta\alpha})_0 = -m_\alpha(B_\alpha)_0. \tag{2.5.23}$$

The interaction Hamiltonian (2.5.18) can therefore be written in the following multipole operator form:

$$V = -\frac{i}{\hbar}(H\mu_\alpha - \mu_\alpha H)(A_\alpha)_0 - \frac{i}{3\hbar}(H\Theta_{\alpha\beta} - \Theta_{\alpha\beta}H)(A_{\beta\alpha})_0$$

$$- m_\alpha(B_\alpha)_0 + \cdots + \sum_i \frac{e_i^2}{2m_i}A(\mathbf{r}_i)^2, \tag{2.5.24}$$

where we have introduced the traceless electric quadrupole moment operator since $A_{\alpha\alpha} \equiv \nabla \cdot \mathbf{A} = 0$.

If the real vector potential is written explicitly as

$$A_\alpha(\mathbf{r}) = \tfrac{1}{2}A^{(0)}\big[\tilde{\Pi}_\alpha e^{i(\kappa_\beta r_\beta - \omega t)} + \tilde{\Pi}_\alpha^* e^{-i(\kappa_\beta r_\beta - \omega t)}\big], \tag{2.5.25}$$

the interaction Hamiltonian (2.5.24) becomes

$$V = \tfrac{1}{2}A^{(0)}\left[-\frac{i}{\hbar}(H\mu_\alpha - \mu_\alpha H)(\tilde{\Pi}_\alpha e^{-i\omega t} + \tilde{\Pi}_\alpha^* e^{i\omega t})\right.$$

$$+ \frac{i\omega}{c}\varepsilon_{\alpha\beta\gamma}n_\beta m_\gamma(\tilde{\Pi}_\alpha e^{-i\omega t} - \tilde{\Pi}_\alpha^* e^{i\omega t})$$

$$\left. + \frac{\omega}{3\hbar c}n_\beta(H\Theta_{\alpha\beta} - \Theta_{\alpha\beta}H)(\tilde{\Pi}_\alpha e^{-i\omega t} - \tilde{\Pi}_\alpha^* e^{i\omega t}) + \cdots\right]$$

$$+ A^{(0)2}\sum_i \frac{e_i}{8m_i^2}(\tilde{\Pi}_\alpha\tilde{\Pi}_\alpha e^{-2i\omega t} + \tilde{\Pi}_\alpha^*\tilde{\Pi}_\alpha^* e^{2i\omega t} + 2\tilde{\Pi}_\alpha\tilde{\Pi}_\alpha^* + \cdots), \tag{2.5.26}$$

which is a convenient form for subsequent applications.

Although the dynamic interaction Hamiltonian (2.5.24) is effectively in multipole form, it is not as 'clean' as the static multipole interaction Hamiltonians (2.5.8) and (2.5.11); also the dynamic diamagnetic interaction has not emerged explicitly. It is possible, however, to transform the fundamental Hamiltonian (2.5.6) into an exact dynamic analogue of the static multipole interaction Hamiltonian that is applicable to both classical and (with an operator interpretation) quantum formulations, and we refer to Woolley (1975a) for a review of the various transformation methods that have been proposed. Here we give a particularly simple method due to Barron and Gray (1973) which shows, by means of a judicious choice of gauge, that the fundamental interaction Hamiltonian (2.5.6) is simply equal to the multipole Hamiltonian. As discussed in Section 2.2.4, provided that a scalar and vector potential generate the correct electric and magnetic fields through

$$\mathbf{E} = -\nabla\phi - \frac{\partial \mathbf{A}}{\partial t}, \tag{2.5.27a}$$

$$\mathbf{B} = \nabla \times \mathbf{A}, \tag{2.5.27b}$$

there is 'gauge freedom' in the choice of ϕ and $\mathbf{A}$. We make an explicit choice with the expansions

$$\phi(\mathbf{r}) = (\phi)_0 - r_\alpha(E_\alpha)_0 - \tfrac{1}{2}r_\alpha r_\beta(E_{\alpha\beta})_0 + \cdots, \tag{2.5.28a}$$

$$A_\alpha(\mathbf{r}) = \tfrac{1}{2}\varepsilon_{\alpha\beta\gamma}(B_\beta)_0 r_\gamma + \tfrac{1}{3}\varepsilon_{\alpha\gamma\delta}r_\beta(B_{\beta\gamma})_0 r_\delta + \cdots, \tag{2.5.28b}$$

which satisfy (2.5.27) if $\mathbf{E}(\mathbf{r})$ and $\mathbf{B}(\mathbf{r})$ can be Taylor expanded:

$$E_\alpha(\mathbf{r}) = (E_\alpha)_0 + r_\beta(E_{\beta\alpha})_0 + \cdots, \tag{2.5.29a}$$

$$B_\alpha(\mathbf{r}) = (B_\alpha)_0 + r_\beta(B_{\beta\alpha})_0 + \cdots. \tag{2.5.29b}$$

It is easy to see how the constant terms $(E_\alpha)_0$ and $(B_\alpha)_0$ arise, but to see how the term $r_\beta(E_{\beta\alpha})_0$ in (2.5.29a) arises requires further explanation: in fact we use the relations

$$\tfrac{1}{2}\nabla_\alpha[r_\beta r_\gamma(E_{\beta\gamma})_0] = \tfrac{1}{2}r_\beta[(E_{\alpha\beta})_0 + (E_{\beta\alpha})_0],$$

$$\tfrac{1}{2}\frac{\partial}{\partial t}\varepsilon_{\alpha\beta\gamma}(B_\beta)_0 r_\gamma = -\tfrac{1}{2}\varepsilon_{\alpha\beta\gamma}[\varepsilon_{\beta\delta\epsilon}(E_{\delta\epsilon})_0]r_\gamma = \tfrac{1}{2}r_\beta[(E_{\alpha\beta})_0 - (E_{\beta\alpha})_0],$$

the second of which makes use of the Maxwell equation $\nabla \times \mathbf{E} = -\partial\mathbf{B}/\partial t$. Substituting (2.5.28) into (2.5.6) now gives the *dynamic multipole interaction Hamiltonian*

$$V = q(\phi)_0 - \mu_\alpha(E_\alpha)_0 - \tfrac{1}{3}\Theta_{\alpha\beta}(E_{\alpha\beta})_0$$
$$- m_\alpha(B_\alpha)_0 - \tfrac{1}{2}\chi_{\alpha\beta}^{(\mathrm{d})}(B_\alpha)_0(B_\beta)_0 + \cdots, \tag{2.5.30}$$

which parallels exactly the static one.

There has been much discussion as to the relative merits of the two dynamic interaction Hamiltonians (2.5.26) and (2.5.30), particularly with regard to the contribution of the term in A^2. However, if applied consistently, the two Hamiltonians should give identical results. An early example of this equivalence was given indirectly by Dirac (1958) in a derivation of the Kramers–Heisenberg dispersion formula for the scattering coefficient of a photon by an atom or molecule. The same dispersion formula is obtained from the interference of two '$\mu \cdot \mathbf{E}$' interactions, describing the separate absorption and emission processes, as is obtained from the interference of two '$\mathbf{p} \cdot \mathbf{A}$' interactions, again describing separate photon absorption and emission processes, added to a single 'A^2' interaction describing simultaneous photon absorption and emission. However, this feature appears to arise only in formulations using a quantized radiation field. In the semi-classical theory of molecular light scattering developed below, the Hamiltonians (2.5.26) and (2.5.30) give identical results even though the A^2 term makes no contribution. This is because we describe light scattering in terms of the radiation emitted by electric and magnetic multipole moments induced in the molecule by the incident light wave and oscillating at the incident frequency; and the A^2 term has no components at the frequency of the incident light wave.

It should be mentioned that, if the Hamiltonian contains the spin–orbit interaction, the transformation to a multipole form is more delicate and new terms arise. Barron and Buckingham (1973) have discussed this matter in detail.

2.6 Molecules in electric and magnetic fields

In this section, perturbation theory is used to derive quantum mechanical expressions for the molecular property tensors that characterize the response of a molecule to a particular electric or magnetic field component. These property tensors appear later in the expressions for the observables, such as the angle of optical rotation, in optical activity experiments.

2.6.1 A molecule in static fields

The electric and magnetic multipole moments appearing in the expressions for the interaction energy of a system of charges and currents with external electric and magnetic fields can be permanent attributes of the system or can be induced by the fields themselves. If the interaction is weak, the situation can be analyzed by expanding the energy W of the system in a Taylor series about the energy in the absence of the field.

Thus for an electrically neutral molecule in a static uniform electric field,

$$W[(\mathbf{E})_0] = (W)_0 + (E_\alpha)_0 \left[\frac{\partial W}{\partial (E_\alpha)_0} \right]_0 + \tfrac{1}{2}(E_\alpha)_0(E_\beta)_0 \left[\frac{\partial^2 W}{\partial (E_\alpha)_0 \partial (E_\beta)_0} \right]_0$$
$$+ \tfrac{1}{6}(E_\alpha)_0(E_\beta)_0(E_\gamma)_0 \left[\frac{\partial^3 W}{\partial (E_\alpha)_0 \partial (E_\beta)_0 \partial (E_\gamma)_0} \right]_0 + \cdots . \quad (2.6.1)$$

The field itself, $(\mathbf{E})_0$, is taken at the molecular origin, and $(W)_0$, $[\partial W / \partial (\mathbf{E})_0]_0$, etc., indicate the energy, its derivative with respect to the field, etc., evaluated for zero field strength at the molecular origin. From (2.5.8) we also have

$$W = (W)_0 - \mu_\alpha (E_\alpha)_0 - \tfrac{1}{3}\Theta_{\alpha\beta}(E_{\alpha\beta})_0 + \cdots , \quad (2.6.2)$$

from which the electric dipole moment is given by

$$\mu_\alpha = -\frac{\partial W}{\partial (E_\alpha)_0}. \tag{2.6.3}$$

Thus from (2.6.1) and (2.6.3) we can write the molecular electric dipole moment in the presence of a static uniform electric field as

$$\mu_\alpha = \mu_{0\alpha} + \alpha_{\alpha\beta}(E_\beta)_0 + \tfrac{1}{2}\beta_{\alpha\beta\gamma}(E_\beta)_0(E_\gamma)_0 + \cdots, \tag{2.6.4a}$$

where

$$\mu_{0\alpha} = -\left[\frac{\partial W}{\partial (E_\alpha)_0}\right]_0, \tag{2.6.4b}$$

$$\alpha_{\alpha\beta} = -\left[\frac{\partial^2 W}{\partial (E_\alpha)_0 \partial (E_\beta)_0}\right]_0, \tag{2.6.4c}$$

$$\beta_{\alpha\beta\gamma} = -\left[\frac{\partial^3 W}{\partial (E_\alpha)_0 \partial (E_\beta)_0 \partial (E_\gamma)_0}\right]_0, \tag{2.6.4d}$$

are, respectively, the *permanent electric dipole moment*, the *electric polarizability* and the *first electric hyperpolarizability*. Thus the tensors $\alpha_{\alpha\beta}$, $\beta_{\alpha\beta\gamma}$ etc., describe the distortion of the molecular charge distribution by successive powers of the electric field.

Similarly, for a molecule in a static electric field gradient,

$$W[(E_{\alpha\beta})_0]_0 = (W)_0 + (E_{\alpha\beta})_0 \left[\frac{\partial W}{\partial (E_{\alpha\beta})_0}\right]_0$$
$$+ \tfrac{1}{2}(E_{\alpha\beta})_0(E_{\gamma\delta})_0 \left[\frac{\partial^2 W}{\partial (E_{\alpha\beta})_0 \partial (E_{\gamma\delta})_0}\right]_0 + \cdots. \tag{2.6.5}$$

From (2.6.2), the electric quadrupole moment is given by

$$\Theta_{\alpha\beta} = -3\frac{\partial W}{\partial (E_{\alpha\beta})_0} \tag{2.6.6}$$

and this, together with (2.6.5), gives the molecular electric quadrupole moment in the presence of a static electric field gradient as

$$\Theta_{\alpha\beta} = \Theta_{0\alpha\beta} + C_{\alpha\beta,\gamma\delta}(E_{\gamma\delta})_0 + \cdots, \tag{2.6.7a}$$

where

$$\Theta_{0\alpha\beta} = -3\left[\frac{\partial W}{\partial (E_{\alpha\beta})_0}\right]_0, \tag{2.6.7b}$$

$$C_{\alpha\beta,\gamma\delta} = -3 \left[\frac{\partial^2 W}{\partial(E_{\alpha\beta})_0 \partial(E_{\gamma\delta})_0} \right]_0, \tag{2.6.7c}$$

are the *permanent electric quadrupole moment* and the *electric quadrupole polarizability*. Thus $C_{\alpha\beta,\gamma\delta}$ describes the distortion of the molecular charge distribution by an electric field gradient.

For a molecule in a static uniform magnetic field,

$$W[(B)_0] = (W)_0 + (B_\alpha)_0 \left[\frac{\partial W}{\partial(B_\alpha)_0} \right]_0$$

$$+ \tfrac{1}{2}(B_\alpha)_0(B_\beta)_0 \left[\frac{\partial^2 W}{\partial(B_\alpha)_0 \partial(B_\beta)_0} \right]_0 + \cdots. \tag{2.6.8}$$

From (2.5.11) we can write

$$W = (W)_0 - m_\alpha(B_\alpha)_0 - \tfrac{1}{2}\chi_{\alpha\beta}^{(d)}(B_\alpha)_0(B_\beta)_0 + \cdots, \tag{2.6.9}$$

which gives the magnetic dipole moment, including the diamagnetic contribution, as

$$m'_\alpha = m_\alpha + \chi_{\alpha\beta}^{(d)}(B_\beta)_0 = -\frac{\partial W}{\partial(B_\alpha)_0}. \tag{2.6.10}$$

From this, together with (2.6.8), we can write the molecular magnetic dipole moment in the presence of a static uniform magnetic field as

$$m'_\alpha = m_{0\alpha} + \chi_{\alpha\beta}(B_\beta)_0 + \cdots, \tag{2.6.11a}$$

where

$$m_{0\alpha} = -\left[\frac{\partial W}{\partial(B_\alpha)_0} \right]_0, \tag{2.6.11b}$$

$$\chi_{\alpha\beta} = \chi_{\alpha\beta}^{(p)} + \chi_{\alpha\beta}^{(d)} = -\left[\frac{\partial^2 W}{\partial(B_\alpha)_0 \partial(B_\beta)_0} \right]_0, \tag{2.6.11c}$$

are the *permanent magnetic dipole moment* and the *magnetic susceptibility*. $\chi_{\alpha\beta}^{(p)}$ is the *temperature-independent paramagnetic susceptibility* and is the magnetic analogue of the electric polarizability $\alpha_{\alpha\beta}$, whereas the diamagnetic contribution $\chi_{\alpha\beta}^{(d)}$ has no electrical analogue.

Time-independent perturbation theory is now introduced to give the static molecular property tensors a quantum mechanical form. We require approximate solutions of the time-independent Schrödinger equation

$$H'\psi' = (H + V)\psi' = W'\psi', \tag{2.6.12}$$

where H is the unperturbed molecular Hamiltonian (2.5.21), V is the operator equivalent of a static interaction Hamiltonian such as (2.5.8) or (2.5.11) whose effect is small compared with that of H, and ψ' and W' are the perturbed molecular wavefunction and energy. Perturbation theory provides approximate expressions for the eigenfunctions ψ'_j and eigenvalues W'_j of the perturbed operator H' in terms of the unperturbed eigenfunctions ψ_j and eigenvalues W_j of the unperturbed operator H. We refer to standard works such as Davydov (1976) for the development of these approximate expressions.

The perturbed energy eigenvalue corresponding to the nondegenerate eigenfunction ψ_n is, to second order in the perturbation,

$$W'_n = W_n + \langle n|V|n \rangle + \sum_{j \neq n} \frac{\langle n|V|j \rangle \langle j|V|n \rangle}{W_n - W_j}, \qquad (2.6.13)$$

where the sum extends over the complete set of eigenfunctions with the exception of the initial state ψ_n. Since the energy of a system correct to the $(2m + 1)$th order in the perturbation is given by wave functions correct to the mth order, we need only take the corresponding perturbed eigenfunction to first order in the perturbation:

$$\psi'_n = \psi_n + \sum_{j \neq n} \frac{\langle j|V|n \rangle}{W_n - W_j} \psi_j. \qquad (2.6.14)$$

If the perturbation is due to a static uniform electric field, $V = -\mu_\alpha (E_\alpha)_0$. Applying (2.6.3) to (2.6.13) and comparing the result with (2.6.4), we find the following expressions for the permanent electric dipole moment and the polarizability of a molecule in the state ψ_n:

$$\mu_{0\alpha} = \langle n|\mu_\alpha|n \rangle, \qquad (2.6.15a)$$

$$\alpha_{\alpha\beta} = -2 \sum_{j \neq n} \frac{\langle n|\mu_\alpha|j \rangle \langle j|\mu_\beta|n \rangle}{W_n - W_j} = \alpha_{\beta\alpha}. \qquad (2.6.15b)$$

These results can also be obtained by taking the expectation value of the electric dipole moment operator with the perturbed eigenfunction (2.6.14), and comparing the result with (2.6.4):

$$\mu_\alpha = \langle n'|\mu_\alpha|n' \rangle$$
$$= \langle n|\mu_\alpha|n \rangle - 2 \sum_{j \neq n} \frac{\langle n|\mu_\alpha|j \rangle \langle j|\mu_\beta|n \rangle}{W_n - W_j} (E_\beta)_0. \qquad (2.6.16)$$

Similar expressions can be found for the other static molecular property tensors, but they are not reproduced here since only the dynamic versions are required in what follows, and these are derived below. Buckingham (1967, 1978) has given a full account of the static electric molecular property tensors to high order.

2.6.2 A molecule in a radiation field

A radiation field induces oscillating electric and magnetic multipole moments in a molecule. These moments are related to the electric and magnetic field components of the radiation field through molecular property tensors which are now functions of the frequency. The first procedure (involving energy eigenvalues) used for obtaining the static induced moments, and hence the static polarizability (2.6.15b), is not applicable here since eigenvalues are not defined in a dynamic field (Born and Huang, 1954). But expectation values are still defined, so to obtain the oscillating induced moments, and hence the dynamic molecular property tensors, we adopt the second procedure of taking expectation values of the multipole moment operators using molecular wave functions perturbed by the radiation field, and identifying the dynamic molecular property tensors in the resulting series.

The periodically perturbed molecular wave functions are obtained by solving the time-dependent Schrödinger equation

$$\left(i\hbar \frac{\partial}{\partial t} - H \right) \psi = V \psi, \tag{2.6.17}$$

where H is the unperturbed molecular Hamiltonian (2.5.21) and V is a dynamic interaction Hamiltonian such as (2.5.26) or (2.5.30). In the absence of V, the general solution of (2.6.17) is the stationary state

$$\psi = \sum_j c_j \psi_j^{(0)} e^{-i\omega_j t}, \tag{2.6.18}$$

where the c_j are time-independent expansion coefficients and ψ_j and $\hbar\omega_j = W_j$ are the eigenfunctions and eigenvalues of H. In the presence of the time-dependent perturbation V, the general solution of (2.6.18) is no longer a stationary state since the expansion coefficients can now be functions of time.

The details of the subsequent development depend on which of the two interaction Hamiltonians (2.5.26) or (2.5.30) is used, although the final results should be identical. Here we employ the multipole Hamiltonian (2.5.30) since it involves less work.

A simple method of solution is to assume that, when the stationary non-degenerate eigenfunction

$$\psi_n = \psi_n^{(0)} e^{-i\omega_n t} \tag{2.6.19}$$

of the unperturbed system is subjected to a small harmonic perturbation of angular frequency ω from a plane-wave radiation field, the corresponding perturbed eigenfunction can be written in the form (Placzek, 1934; Born and Huang, 1954;

Davydov, 1976)

$$\psi'_n = \left\{ \psi_n^{(0)} + \sum_{j \neq n} [\tilde{a}_{jn_\beta}(\tilde{E}_\beta)_0 + \tilde{b}_{jn_\beta}(\tilde{E}_\beta^*)_0 + \tilde{c}_{jn_\beta}(\tilde{B}_\beta)_0 \right.$$

$$\left. + \tilde{d}_{jn_\beta}(\tilde{B}_\beta^*)_0 + \tilde{e}_{jn_{\beta\gamma}}(\tilde{E}_{\beta\gamma})_0 + \tilde{f}_{jn_{\beta\gamma}}(\tilde{E}_{\beta\gamma}^*)_0 + \cdots]\psi_j^{(0)} \right\} e^{-i\omega_n t}. \quad (2.6.20)$$

The first term satisfies (2.6.17) in the absence of V and the other terms are first order in the harmonic perturbation. The coefficients $\tilde{a}_{jn_\beta}$, etc., are now found by using the perturbed eigenfunction (2.6.20) and the multipole interaction Hamiltonian (2.5.30) in the time-dependent Schrödinger equation (2.6.17):

$$-\hbar \sum_{j \neq n} [(\omega_{jn} - \omega)\tilde{a}_{jn_\beta}(\tilde{E}_\beta)_0 + (\omega_{jn} + \omega)\tilde{b}_{jn_\beta}(\tilde{E}_\beta^*)_0$$

$$+ (\omega_{jn} - \omega)\tilde{c}_{jn_\beta}(\tilde{B}_\beta)_0 + (\omega_{jn} + \omega)\tilde{d}_{jn_\beta}(\tilde{B}_\beta^*)_0$$

$$+ (\omega_{jn} - \omega)\tilde{e}_{jn_{\beta\gamma}}(\tilde{E}_{\beta\gamma})_0 + (\omega_{jn} + \omega)\tilde{f}_{jn_{\beta\gamma}}(\tilde{E}_{\beta\gamma}^*)_0]\psi_j^{(0)} e^{-i\omega_n t}$$

$$= -\frac{1}{2}\left\{ \mu_\beta[(\tilde{E}_\beta)_0 + (\tilde{E}_\beta^*)_0] + m_\beta[(\tilde{B}_\beta)_0 + (\tilde{B}_\beta^*)_0] \right.$$

$$\left. + \frac{1}{3}\Theta_{\beta\gamma}[(\tilde{E}_{\beta\gamma})_0 + (\tilde{E}_{\beta\gamma}^*)_0] + \cdots \right\} \psi_n^{(0)} e^{-i\omega_n t}, \quad (2.6.21)$$

where $\omega_{jn} = \omega_j - \omega_n$. Multiplying both sides of (2.6.21) by $\psi_j^{(0)*}$ and integrating over all configuration space, it is found by equating coefficients of identical exponential time factors that

$$\tilde{a}_{jn_\beta} = \langle j|\mu_\beta|n\rangle / 2\hbar(\omega_{jn} - \omega), \quad (2.6.22a)$$

$$\tilde{b}_{jn_\beta} = \langle j|\mu_\beta|n\rangle / 2\hbar(\omega_{jn} + \omega), \quad (2.6.22b)$$

$$\tilde{c}_{jn_\beta} = \langle j|m_\beta|n\rangle / 2\hbar(\omega_{jn} - \omega), \quad (2.6.22c)$$

$$\tilde{d}_{jn_\beta} = \langle j|m_\beta|n\rangle / 2\hbar(\omega_{jn} + \omega), \quad (2.6.22d)$$

$$\tilde{e}_{jn_{\beta\gamma}} = \langle j|\Theta_{\beta\gamma}|n\rangle / 6\hbar(\omega_{jn} - \omega), \quad (2.6.22e)$$

$$\tilde{f}_{jn_{\beta\gamma}} = \langle j|\Theta_{\beta\gamma}|n\rangle / 6\hbar(\omega_{jn} + \omega). \quad (2.6.22f)$$

The oscillating induced electric and magnetic multipole moments of the molecule in the nth eigenstate are now obtained from the expectation values of the corresponding operators using the periodically perturbed eigenfunction (2.6.20). For example, the first few contributions to the induced electric dipole moment are

$$\mu_\alpha = \langle n'|\mu_\alpha|n'\rangle$$

$$= \langle n|\mu_\alpha|n\rangle + \frac{2}{\hbar}\sum_{j \neq n}\frac{\omega_{jn}}{\omega_{jn}^2 - \omega^2}\text{Re}(\langle n|\mu_\alpha|j\rangle\langle j|\mu_\beta|n\rangle)(E_\beta)_0$$

$$- \frac{2}{\hbar}\sum_{j \neq n}\frac{\omega}{\omega_{jn}^2 - \omega^2}\text{Im}(\langle n|\mu_\alpha|j\rangle\langle j|\mu_\beta|n\rangle)\frac{1}{\omega}(\dot{E}_\beta)_0 + \cdots. \quad (2.6.23)$$

In obtaining this result we have written

$$\langle n|\mu_\alpha|j\rangle\langle j|\mu_\beta|n\rangle = \langle n|\mu_\beta|j\rangle^*\langle j|\mu_\alpha|n\rangle^*, \tag{2.6.24}$$

which follows from the Hermiticity of the electric dipole moment operator, and have used the following relationships between real and complex radiation field components:

$$(E_\beta)_0 = \tfrac{1}{2}[(\tilde{E}_\beta)_0 + (\tilde{E}_\beta^*)_0] = \tfrac{1}{2}[\tilde{E}_\beta^{(0)}e^{-i\omega t} + \tilde{E}_\beta^{(0)*}e^{i\omega t}], \tag{2.6.25a}$$

$$(\dot{E}_\beta)_0 = -\frac{i\omega}{2}[\tilde{E}_\beta^{(0)}e^{-i\omega t} - \tilde{E}_\beta^{(0)*}e^{i\omega t}]. \tag{2.6.25b}$$

Extending this procedure, the following expressions for the real induced oscillating electric and magnetic multipole moments are found (Buckingham, 1967, 1978):

$$\mu_\alpha = \alpha_{\alpha\beta}(E_\beta)_0 + \frac{1}{\omega}\alpha'_{\alpha\beta}(\dot{E}_\beta)_0 + \frac{1}{3}A_{\alpha,\beta\gamma}(E_{\beta\gamma})_0$$

$$+ \frac{1}{3\omega}A'_{\alpha,\beta\gamma}(\dot{E}_{\beta\gamma})_0 + G_{\alpha\beta}(B_\beta)_0 + \frac{1}{\omega}G'_{\alpha\beta}(\dot{B}_\beta)_0 + \cdots, \tag{2.6.26a}$$

$$\Theta_{\alpha\beta} = A_{\gamma,\alpha\beta}(E_\gamma)_0 - \frac{1}{\omega}A'_{\gamma,\alpha\beta}(\dot{E}_\gamma)_0 + C_{\alpha\beta,\gamma\delta}(E_{\gamma\delta})_0$$

$$+ \frac{1}{\omega}C'_{\alpha\beta,\gamma\delta}(\dot{E}_{\gamma\delta})_0 + D_{\gamma,\alpha\beta}(B_\gamma)_0 - \frac{1}{\omega}D'_{\gamma,\alpha\beta}(\dot{B}_\gamma)_0 + \cdots, \tag{2.6.26b}$$

$$m'_\alpha = \chi_{\alpha\beta}(B_\beta)_0 + \frac{1}{\omega}\chi'_{\alpha\beta}(\dot{B}_\beta)_0 + \frac{1}{3}D_{\alpha,\beta\gamma}(E_{\beta\gamma})_0$$

$$+ \frac{1}{3\omega}D'_{\alpha,\beta\gamma}(\dot{E}_{\beta\gamma})_0 + G_{\beta\alpha}(E_\beta)_0 - \frac{1}{\omega}G'_{\beta\alpha}(\dot{E}_\beta)_0 + \cdots, \tag{2.6.26c}$$

where the real *dynamic molecular property tensors* that multiply the real radiation field components are

$$\alpha_{\alpha\beta} = \frac{2}{\hbar}\sum_{j\neq n}\frac{\omega_{jn}}{\omega_{jn}^2 - \omega^2}\mathrm{Re}(\langle n|\mu_\alpha|j\rangle\langle j|\mu_\beta|n\rangle) = \alpha_{\beta\alpha}, \tag{2.6.27a}$$

$$\alpha'_{\alpha\beta} = -\frac{2}{\hbar}\sum_{j\neq n}\frac{\omega}{\omega_{jn}^2 - \omega^2}\mathrm{Im}(\langle n|\mu_\alpha|j\rangle\langle j|\mu_\beta|n\rangle) = -\alpha'_{\beta\alpha}, \tag{2.6.27b}$$

$$A_{\alpha,\beta\gamma} = \frac{2}{\hbar}\sum_{j\neq n}\frac{\omega_{jn}}{\omega_{jn}^2 - \omega^2}\mathrm{Re}(\langle n|\mu_\alpha|j\rangle\langle j|\Theta_{\beta\gamma}|n\rangle) = A_{\alpha,\gamma\beta}, \tag{2.6.27c}$$

$$A'_{\alpha,\beta\gamma} = -\frac{2}{\hbar}\sum_{j\neq n}\frac{\omega}{\omega_{jn}^2 - \omega^2}\mathrm{Im}(\langle n|\mu_\alpha|j\rangle\langle j|\Theta_{\beta\gamma}|n\rangle) = A'_{\alpha,\gamma\beta}, \tag{2.6.27d}$$

$$G_{\alpha\beta} = \frac{2}{\hbar}\sum_{j\neq n}\frac{\omega_{jn}}{\omega_{jn}^2 - \omega^2}\mathrm{Re}(\langle n|\mu_\alpha|j\rangle\langle j|m_\beta|n\rangle), \tag{2.6.27e}$$

$$G'_{\alpha\beta} = -\frac{2}{\hbar}\sum_{j\neq n}\frac{\omega}{\omega_{jn}^2 - \omega^2}\mathrm{Im}(\langle n|\mu_\alpha|j\rangle\langle j|m_\beta|n\rangle), \tag{2.6.27f}$$

$$C_{\alpha\beta,\gamma\delta} = \frac{2}{3\hbar} \sum_{j \neq n} \frac{\omega_{jn}}{\omega_{jn}^2 - \omega^2} \text{Re}(\langle n|\Theta_{\alpha\beta}|j\rangle\langle j|\Theta_{\gamma\delta}|n\rangle) = C_{\gamma\delta,\alpha\beta}, \qquad (2.6.27g)$$

$$C'_{\alpha\beta,\gamma\delta} = -\frac{2}{3\hbar} \sum_{j \neq n} \frac{\omega}{\omega_{jn}^2 - \omega^2} \text{Im}(\langle n|\Theta_{\alpha\beta}|j\rangle\langle j|\Theta_{\gamma\delta}|n\rangle) = -C'_{\gamma\delta,\alpha\beta}, \qquad (2.6.27h)$$

$$D_{\alpha,\beta\gamma} = \frac{2}{\hbar} \sum_{j \neq n} \frac{\omega_{jn}}{\omega_{jn}^2 - \omega^2} \text{Re}(\langle n|m_\alpha|j\rangle\langle j|\Theta_{\beta\gamma}|n\rangle) = D_{\alpha,\gamma\beta}, \qquad (2.6.27i)$$

$$D'_{\alpha,\beta\gamma} = -\frac{2}{\hbar} \sum_{j \neq n} \frac{\omega}{\omega_{jn}^2 - \omega^2} \text{Im}(\langle n|m_\alpha|j\rangle\langle j|\Theta_{\beta\gamma}|n\rangle) = D'_{\alpha,\gamma\beta}, \qquad (2.6.27j)$$

$$\chi_{\alpha\beta} = \frac{2}{\hbar} \sum_{j \neq n} \frac{\omega_{jn}}{\omega_{jn}^2 - \omega^2} \text{Re}(\langle n|m_\alpha|j\rangle\langle j|m_\beta|n\rangle)$$

$$+ \sum_i \frac{e_i^2}{4m_i} \langle n|r_{i_\alpha}r_{i_\beta} - r_i^2\delta_{\alpha\beta}|n\rangle = \chi_{\beta\alpha}, \qquad (2.6.27k)$$

$$\chi'_{\alpha\beta} = -\frac{2}{\hbar} \sum_{j \neq n} \frac{\omega}{\omega_{jn}^2 - \omega^2} \text{Im}(\langle n|m_\alpha|j\rangle\langle j|m_\beta|n\rangle) = -\chi'_{\beta\alpha}. \qquad (2.6.27l)$$

Notice that $\alpha_{\alpha\beta}$ is *symmetric* with respect to interchange of the tensor subscripts, whereas $\alpha'_{\alpha\beta}$ is *antisymmetric*. This follows from (2.6.24), which enables us to write

$$\text{Re}(\langle n|\mu_\alpha|j\rangle\langle j|\mu_\beta|n\rangle) = \text{Re}(\langle n|\mu_\beta|j\rangle\langle j|\mu_\alpha|n\rangle), \qquad (2.6.28a)$$

$$\text{Im}(\langle n|\mu_\alpha|j\rangle\langle j|\mu_\beta|n\rangle) = -\text{Im}(\langle n|\mu_\beta|j\rangle\langle j|\mu_\alpha|n\rangle). \qquad (2.6.28b)$$

Similarly for the other molecular property tensors involving products of the same multipole transition moments. No analogous separation into symmetric and antisymmetric parts exists for the property tensors involving products of different multipole transition moments.

This is an appropriate point at which to introduce the dimensionless quantity

$$f_{jn} = \frac{2m\omega_{jn}}{3\hbar} |\langle j|r|n\rangle|^2, \qquad (2.6.29)$$

called the *oscillator strength* of the $j \leftarrow n$ transition between quantum states of a single electron bound in an atom or molecule. The oscillator strength obeys the following *Kuhn–Thomas sum rule*,

$$\sum_j f_{jn} = 1, \qquad (2.6.30)$$

which can be derived as follows. Using the commutation relation (2.5.20) between coordinates and momenta, it is found that coordinate and momentum matrix elements are related by the *velocity–dipole transformation*

$$\langle j|p'_\alpha|n\rangle = im\omega_{jn}\langle j|r_\alpha|n\rangle. \tag{2.6.31a}$$

Using this result, (2.6.29) can be written

$$f_{jn} = \frac{m}{3\hbar}(\omega_{jn}\langle n|r_\alpha|j\rangle\langle j|r_\alpha|n\rangle - \omega_{nj}\langle n|r_\alpha|j\rangle\langle j|r_\alpha|n\rangle)$$

$$= -\frac{i}{3\hbar}(\langle n|r_\alpha|j\rangle\langle j|p'_\alpha|n\rangle - \langle n|p'_\alpha|j\rangle\langle j|r_\alpha|n\rangle).$$

Invoking the closure theorem $(\sum_j |j\rangle\langle j| = 1)$, the required sum is

$$\sum_j f_{jn} = -\frac{i}{3\hbar}\langle n|r_\alpha p'_\alpha - p'_\alpha r_\alpha|n\rangle = -\left(\frac{i}{3\hbar}\right)(3i\hbar) = 1.$$

This applies to a single bound electron, but since every electron in the system will contribute, the Kuhn–Thomas sum rule for an atom or molecule containing Z electrons becomes

$$\sum_j f_{jn} = Z. \tag{2.6.32}$$

The oscillator strength and its sum rule can be useful for writing the real polarizability (2.6.27a) in other forms.

It is convenient to present the real oscillating induced electric and magnetic multipole moments (2.6.26) in a complex form. This facilitates the application of expressions such as (2.4.43) for the fields radiated by oscillating complex multipole moments (the tildes over complex quantities were omitted in Section 2.4.5 in the interests of economy). Introducing the complex dynamic molecular property tensors

$$\tilde{\alpha}_{\alpha\beta} = \alpha_{\alpha\beta} - i\alpha'_{\alpha\beta} = \tilde{\alpha}^*_{\beta\alpha}, \tag{2.6.33a}$$

$$\tilde{A}_{\alpha,\beta\gamma} = A_{\alpha,\beta\gamma} - iA'_{\alpha,\beta\gamma} = \tilde{A}_{\alpha,\gamma\beta}, \tag{2.6.33b}$$

$$\tilde{G}_{\alpha\beta} = G_{\alpha\beta} - iG'_{\alpha\beta}, \tag{2.6.33c}$$

$$\tilde{C}_{\alpha\beta,\gamma\delta} = C_{\alpha\beta,\gamma\delta} - iC'_{\alpha\beta,\gamma\delta} = \tilde{C}^*_{\gamma\delta,\alpha\beta}, \tag{2.6.33d}$$

$$\tilde{D}_{\alpha,\beta\gamma} = D_{\alpha,\beta\gamma} - iD'_{\alpha,\beta\gamma} = \tilde{D}_{\alpha,\gamma\beta}, \tag{2.6.33e}$$

$$\tilde{\chi}_{\alpha\beta} = \chi_{\alpha\beta} - i\chi'_{\alpha\beta} = \tilde{\chi}^*_{\beta\alpha}, \tag{2.6.33f}$$

we obtain the following complex induced oscillating electric and magnetic multipole moments:

$$\tilde{\mu}_\alpha = \tilde{\alpha}_{\alpha\beta}(\tilde{E}_\beta)_0 + \tfrac{1}{3}\tilde{A}_{\alpha,\beta\gamma}(\tilde{E}_{\beta\gamma})_0 + \tilde{G}_{\alpha\beta}(\tilde{B}_\beta)_0 + \cdots$$

$$= \left(\tilde{\alpha}_{\alpha\beta} + \frac{i\omega}{3c}n_\gamma\tilde{A}_{\alpha,\gamma\beta} + \frac{1}{c}\varepsilon_{\delta\gamma\beta}n_\gamma\tilde{G}_{\alpha\delta} + \cdots\right)(\tilde{E}_\beta)_0, \tag{2.6.34a}$$

$$\tilde{\Theta}_{\alpha\beta} = \tilde{A}^*_{\gamma,\alpha\beta}(\tilde{E}_\gamma)_0 + \tilde{D}^*_{\gamma,\alpha\beta}(\tilde{B}_\gamma)_0 + \tilde{C}_{\alpha\beta,\gamma\delta}(\tilde{E}_{\gamma\delta})_0 + \cdots, \tag{2.6.34b}$$

$$\tilde{m}'_\alpha = \tilde{\chi}_{\alpha\beta}(\tilde{B}_\beta)_0 + \tilde{G}^*_{\beta\alpha}(\tilde{E}_\beta)_0 + \tfrac{1}{3}\tilde{D}_{\alpha,\beta\gamma}(\tilde{E}_{\beta\gamma})_0 + \cdots, \tag{2.6.34c}$$

where the complex fields of the plane wave light beam are

$$\tilde{E}_\alpha = \tilde{E}_\alpha^{(0)} e^{i(\kappa_\beta r_\beta - \omega t)},$$

$$\tilde{B}_\alpha = \tilde{B}_\alpha^{(0)} e^{i(\kappa_\beta r_\beta - \omega t)} = \frac{1}{c}\varepsilon_{\alpha\beta\gamma} n_\beta \tilde{E}_\gamma.$$

The minus signs in the complex tensors (2.6.33) arise from our choice of sign in the exponents of these complex field vectors.

It is important to know how the dynamic molecular property tensors change on moving the origin from O to $O + a$. In a neutral system, the changes in the electric dipole, electric quadrupole and magnetic dipole moments were shown in Section 2.4. to be

$$\mu_\alpha \to \mu_\alpha, \tag{2.4.3}$$

$$\Theta_{\alpha\beta} \to \Theta_{\alpha\beta} - \tfrac{3}{2}\mu_\alpha a_\beta - \tfrac{3}{2}\mu_\beta a_\alpha + \mu_\gamma a_\gamma \delta_{\alpha\beta}, \tag{2.4.9}$$

$$m_\alpha \to m_\alpha - \tfrac{1}{2}\varepsilon_{\alpha\beta\gamma} a_\beta \dot{\mu}_\gamma. \tag{2.4.14}$$

If the operator equivalents of these multipole moment changes are used in the property tensors (2.6.27) it is found, using

$$\langle j|\dot{\mu}_\alpha|n\rangle = i\omega_{jn}\langle j|\mu_\alpha|n\rangle, \tag{2.6.31b}$$

which is another version of the velocity–dipole transformation (2.6.31a), that (Buckingham and Longuet–Higgins, 1968)

$$\tilde{\alpha}_{\alpha\beta} \to \tilde{\alpha}_{\alpha\beta}, \tag{2.6.35a}$$

$$\tilde{A}_{\alpha,\beta\gamma} \to \tilde{A}_{\alpha,\beta\gamma} - \tfrac{3}{2}a_\beta \tilde{\alpha}_{\alpha\gamma} - \tfrac{3}{2}a_\gamma \tilde{\alpha}_{\alpha\beta} + a_\delta \tilde{\alpha}_{\alpha\delta}\delta_{\beta\gamma}, \tag{2.6.35b}$$

$$\tilde{G}_{\alpha\beta} \to \tilde{G}_{\alpha\beta} - \tfrac{1}{2}i\omega\varepsilon_{\beta\gamma\delta} a_\gamma \tilde{\alpha}_{\alpha\delta}. \tag{2.6.35c}$$

The contribution of a number of these dynamic molecular property tensors to particular light scattering phenomena are discussed in detail in subsequent chapters. We shall see that, for example, the symmetric polarizability $\alpha_{\alpha\beta}$ provides the major contribution to light scattering and refraction; the antisymmetric polarizability $\alpha'_{\alpha\beta}$, when 'activated' by a magnetic field, generates Faraday optical rotation and circular dichroism; $G'_{\alpha\beta}$ and $A_{\alpha,\beta\gamma}$ generate natural optical rotation and circular dichroism, the latter contributing only in oriented media; and $G_{\alpha\beta}$ and $A'_{\alpha,\beta\gamma}$ generate magnetochiral birefringence and dichroism when activated by a magnetic field.

2.6.3 A molecule in a radiation field at absorbing frequencies

So far, the electronic energy levels of the molecule have been regarded as strictly discrete, which, according to the uncertainty principle, implies that they have an infinite lifetime. One consequence is that the dynamic molecular property tensors derived previously do not apply to the resonance situation, that is when the frequency ω of the plane-wave light beam coincides with one of the natural transition frequencies ω_{jn} of the molecule. Near resonance, the polarizabilities can have greatly enhanced values, and there is the possibility of absorption of radiation by the molecule.

To take account of resonance phenomena, it is necessary to incorporate the finite energy width of the excited states of the molecule, thereby allowing for a finite lifetime. The finite lifetime leads to the spontaneous emission of radiation by molecules in excited states. If the total probability for transitions to all lower states is small, the excited state is called *quasi-discrete* and its amplitude decays exponentially with time:

$$c(t) = c(0)e^{-\frac{1}{2}\Gamma t}, \tag{2.6.36}$$

where Γ is called the damping factor. $1/\Gamma$ has the dimensions of time and is called the lifetime of the excited state. The stationary state

$$\psi_j = \psi_j^{(0)} e^{-i\omega_j t}$$

now becomes the quasi-stationary state

$$\psi_j = \psi_j^{(0)} e^{-i(\omega_j - \frac{1}{2}i\hbar\Gamma_j)t/\hbar}, \tag{2.6.37}$$

so the lifetime of excited states can be incorporated into our formalism simply by changing to complex energies:

$$W_j \to W_j - \tfrac{1}{2}i\hbar\Gamma_j. \tag{2.6.38}$$

For the purposes of this book, it is not necessary to have an explicit quantum mechanical expression for Γ_j since we are only interested in the general form of dispersion and absorption lineshape functions. We refer to Davydov (1976), who follows Weisskopf and Wigner (1930), for further quantum-mechanical discussion of the lifetimes of excited states and the widths of energy levels.

We are usually concerned with molecules whose initial state is the ground state ψ_n. Since the ground state is strictly discrete, its lifetime is infinite and $\Gamma_n = 0$. In the property tensors (2.6.27) we therefore make the replacement

$$\omega_{jn} \to \tilde{\omega}_{jn} = \omega_{jn} - \tfrac{1}{2}i\Gamma_j. \tag{2.6.39}$$

Furthermore, at frequencies ω close to a resonance frequency ω_{jn} we need only use this replacement in the difference term $(\omega_{jn}^2 - \omega^2)$, so the property tensors (2.6.27)

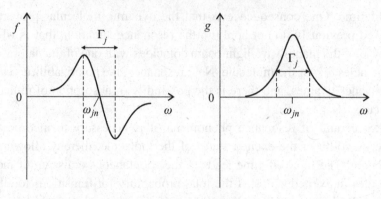

Fig. 2.4 The dispersion and absorption lineshape functions f and g in the region of a resonance frequency ω_{jn}. Γ_j is approximately the width of the band g at half the maximum height.

become valid in regions of absorption through the replacement

$$\frac{1}{\omega_{jn}^2 - \omega^2} \rightarrow \frac{1}{(\tilde{\omega}_{jn} - \omega)(\tilde{\omega}_{jn}^* + \omega)} = \frac{1}{(\omega_{jn}^2 - \omega^2) - i\omega\Gamma_j - \frac{1}{4}\Gamma_j^2}$$

$$\approx \frac{(\omega_{jn}^2 - \omega^2) + i\omega\Gamma_j}{(\omega_{jn}^2 - \omega^2)^2 + \omega^2\Gamma_j^2}. \tag{2.6.40}$$

It is convenient to introduce dispersion and absorption lineshape functions f and g, where

$$\frac{(\omega_{jn}^2 - \omega^2) + i\omega\Gamma_j}{(\omega_{jn}^2 - \omega^2)^2 + \omega^2\Gamma_j^2} = f + ig; \tag{2.6.41a}$$

$$f = \frac{\omega_{jn}^2 - \omega^2}{(\omega_{jn}^2 - \omega^2)^2 + \omega^2\Gamma_j^2}, \tag{2.6.41b}$$

$$g = \frac{\omega\Gamma_j}{(\omega_{jn}^2 - \omega^2)^2 + \omega^2\Gamma_j^2}. \tag{2.6.41c}$$

These functions are drawn in Fig. 2.4. By substituting $\omega_{jn} \pm \frac{1}{2}\Gamma_j$ into the expressions for f and g, and neglecting powers of Γ_j higher than the first, it can be seen that Γ_j is approximately the width of the band g at half the maximum height, and is approximately the separation of the maxima and minima of the band f. In the region of an isolated absorption band from a particular transition $j \leftarrow n$, the dynamic molecular property tensors (2.6.27) are now replaced by

$$\alpha_{\alpha\beta} \rightarrow \alpha_{\alpha\beta}(f) + i\alpha_{\alpha\beta}(g), \tag{2.6.42a}$$

$$\alpha_{\alpha\beta}(f) = \frac{2}{\hbar}f\omega_{jn}\mathrm{Re}(\langle n|\mu_\alpha|j\rangle\langle j|\mu_\beta|n\rangle), \tag{2.6.42b}$$

$$\alpha_{\alpha\beta}(g) = \frac{2}{\hbar}g\omega_{jn}\text{Re}(\langle n|\mu_\alpha|j\rangle\langle j|\mu_\beta|n\rangle); \tag{2.6.42c}$$

$$\alpha'_{\alpha\beta} \to \alpha'_{\alpha\beta}(f) + i\alpha'_{\alpha\beta}(g), \tag{2.6.42d}$$

$$\alpha'_{\alpha\beta}(f) = -\frac{2}{\hbar}f\omega\text{Im}(\langle n|\mu_\alpha|j\rangle\langle j|\mu_\beta|n\rangle), \tag{2.6.42e}$$

$$\alpha'_{\alpha\beta}(g) = -\frac{2}{\hbar}g\omega\text{Im}(\langle n|\mu_\alpha|j\rangle\langle j|\mu_\beta|n\rangle); \tag{2.6.42f}$$

etc.

The expressions for the complex induced oscillating electric and magnetic multipole moments (2.6.34) now need to be modified slightly since in those property tensors where the complex conjugate is specified we do not want the complex conjugate taken of $f + ig$. Therefore we replace (2.6.34) by (Buckingham and Raab, 1975)

$$\tilde{\mu}_\alpha = \tilde{\alpha}_{\alpha\beta}(\tilde{E}_\beta)_0 + \tfrac{1}{3}\tilde{A}_{\alpha,\beta\gamma}(\tilde{E}_{\beta\gamma})_0 + \tilde{G}_{\alpha\beta}(\tilde{B}_\beta)_0 + \cdots, \tag{2.6.43a}$$

$$\tilde{\Theta}_{\alpha\beta} = \tilde{\mathscr{A}}_{\gamma,\alpha\beta}(\tilde{E}_\gamma)_0 + \tilde{\mathscr{D}}_{\gamma,\alpha\beta}(\tilde{B}_\gamma)_0 + \tilde{C}_{\alpha\beta,\gamma\delta}(\tilde{E}_{\gamma\delta})_0 + \cdots, \tag{2.6.43b}$$

$$\tilde{m}'_\alpha = \tilde{\chi}_{\alpha\beta}(\tilde{B}_\beta)_0 + \tilde{\mathscr{G}}_{\alpha\beta}(\tilde{E}_\beta)_0 + \tfrac{1}{3}D_{\alpha,\beta\gamma}(\tilde{E}_{\beta\gamma})_0 + \cdots, \tag{2.6.43c}$$

where

$$\tilde{\mathscr{G}}_{\alpha\beta} = \frac{2}{\hbar}\sum_{j\neq n}\frac{1}{\omega_{jn}^2 - \omega^2}[\omega_{jn}\text{Re}(\langle n|m_\alpha|j\rangle\langle j|\mu_\beta|n\rangle)$$
$$+ i\omega\text{Im}(\langle n|m_\alpha|j\rangle\langle j|\mu_\beta|n\rangle)]$$
$$= G_{\beta\alpha} + iG'_{\beta\alpha}, \tag{2.6.44a}$$

$$\tilde{\mathscr{A}}_{\gamma,\alpha\beta} = \frac{2}{\hbar}\sum_{j\neq n}\frac{1}{\omega_{jn}^2 - \omega^2}[\omega_{jn}\text{Re}(\langle n|\Theta_{\alpha\beta}|j\rangle\langle j|\mu_\gamma|n\rangle)$$
$$+ i\omega\text{Im}(\langle n|\Theta_{\alpha\beta}|j\rangle\langle j|\mu_\gamma|n\rangle)]$$
$$= A_{\gamma,\alpha\beta} + iA'_{\gamma,\alpha\beta}, \tag{2.6.44b}$$

$$\tilde{\mathscr{D}}_{\gamma,\alpha\beta} = \frac{2}{\hbar}\sum_{j\neq n}\frac{1}{\omega_{jn}^2 - \omega^2}[\omega_{jn}\text{Re}(\langle n|\Theta_{\alpha\beta}|j\rangle\langle j|m_\gamma|n\rangle)$$
$$+ i\omega\text{Im}(\langle n|\Theta_{\alpha\beta}|j\rangle\langle j|m_\gamma|n\rangle)]$$
$$= D_{\gamma,\alpha\beta} + iD'_{\gamma,\alpha\beta}. \tag{2.6.44c}$$

It is now shown that the dynamic molecular property tensors that are functions of g are responsible for the absorption of radiation. For simplicity, we consider just $\alpha_{\alpha\beta}(g)$, since this makes the largest contribution to absorption. It is shown later that the much smaller contributions from $\alpha'_{\alpha\beta}(g)$, $G'_{\alpha\beta}(g)$ and $A_{\alpha,\beta\gamma}(g)$ depend on the degree of circularity of the polarization state of the incident light beam and are consequently responsible for circular dichroism. Since the force exerted by an electric field $\mathbf{E}$ on a system of charges is $\sum_i e_i \mathbf{E}_i$, in a time δt the field does work

$$\delta W = \sum_i e_i \mathbf{v}_i \cdot \mathbf{E}\delta t = \dot{\boldsymbol{\mu}} \cdot \mathbf{E}\delta t \tag{2.6.45}$$

on the system of charges. It is the real parts of $\dot{\boldsymbol{\mu}}$ and $\mathbf{E}$ that must be used in (2.6.45), so writing the real electric field in terms of the complex field (2.2.11),

$$E_\alpha = \tfrac{1}{2}(\tilde{E}_\alpha + \tilde{E}_\alpha^*),$$

and the real electric dipole moment in terms of the complex moment (2.6.43a),

$$\mu_\alpha = \tfrac{1}{2}(\tilde{\mu}_\alpha + \tilde{\mu}_\alpha^*) = \tfrac{1}{2}[\tilde{\alpha}_{\alpha\beta}(\tilde{E}_\beta)_0 + \tilde{\alpha}_{\alpha\beta}^*(\tilde{E}_\beta^*)_0],$$

the work becomes

$$\delta W = -\frac{i\omega}{4}[\tilde{\alpha}_{\alpha\beta}(\tilde{E}_\alpha)_0(\tilde{E}_\beta)_0 - \tilde{\alpha}_{\alpha\beta}^*(\tilde{E}_\alpha^*)_0(\tilde{E}_\beta^*)_0$$
$$+ \tilde{\alpha}_{\alpha\beta}(\tilde{E}_\alpha^*)_0(\tilde{E}_\beta)_0 - \tilde{\alpha}_{\alpha\beta}^*(\tilde{E}_\alpha)_0(\tilde{E}_\beta^*)_0]\delta t. \tag{2.6.46}$$

When averaged over the oscillation period the first two terms, which contain $\exp(\pm 2i\omega t)$, vanish and the mean energy absorbed in one second is simply the corresponding mean work:

$$\overline{\Delta W} = \tfrac{1}{2}\omega E^{(0)^2} \text{Im}(\tilde{\alpha}_{\alpha\beta}\tilde{\Pi}_\alpha^* \tilde{\Pi}_\beta). \tag{2.6.47}$$

If the incident light is linearly polarized and the medium is isotropic, containing N molecules per unit volume, this becomes

$$\overline{\Delta W} = \tfrac{1}{6}N\omega E^{(0)^2}\alpha_{\alpha\alpha}(g). \tag{2.6.48}$$

Thus $\alpha_{\alpha\beta}(g)$ is responsible for absorption. Since $\alpha_{\alpha\beta}(f) = \alpha_{\beta\alpha}(f)^*$ and $\alpha_{\alpha\beta}(g) = \alpha_{\beta\alpha}(g)^*$, we can say that absorption arises from the antiHermitian part $i\alpha_{\alpha\beta}(g)$ of the general complex symmetric polarizability tensor $\alpha_{\alpha\beta}(f) + i\alpha_{\alpha\beta}(g)$.

2.6.4 Kramers–Kronig relations

The molecular property tensors, both static and dynamic, that are developed above belong to a class of functions known as response functions. Such functions have some general properties which are independent of any particular theoretical model (such as the semiclassical perturbation model used in this book) of the system which they describe. We illustrate these properties initially for the case of the symmetric polarizability $\alpha_{\alpha\beta}$.

It is necessary to express $\alpha_{\alpha\beta}$ as a sum of dispersive and absorptive parts, as in (2.6.42a), since its behaviour over the complete frequency range is required. Since the dispersion and absorption lineshape functions f and g are functions of ω, we shall now write

$$\tilde{\alpha}_{\alpha\beta}(\omega) = \alpha_{\alpha\beta}(f_\omega) + i\alpha_{\alpha\beta}(g_\omega). \tag{2.6.49}$$

By regarding ω as a complex variable and using the theory of functions of a complex variable, it is possible to derive the following *Kramers–Kronig relations* between the *dispersive* and *absorptive* parts of any response function, here exemplified for $\tilde{\alpha}_{\alpha\beta}(\omega)$:

$$\alpha_{\alpha\beta}(f_\omega) = \frac{1}{\pi}\mathscr{P}\int_{-\infty}^{\infty}\frac{\alpha_{\alpha\beta}(g_\xi)}{\xi - \omega}d\xi, \tag{2.6.50a}$$

$$\alpha_{\alpha\beta}(g_\omega) = -\frac{1}{\pi}\mathscr{P}\int_{-\infty}^{\infty}\frac{\alpha_{\alpha\beta}(f_\xi)}{\xi - \omega}d\xi, \tag{2.6.50b}$$

where $\mathscr{P}$ denotes the Cauchy principal value integral. We refer to works such as Lifshitz and Pitaevski (1980) or Loudon (1983) for a detailed derivation.

The range of integration can be restricted to positive frequencies, which are more meaningful experimentally, by using the following *crossing relations*:

$$\tilde{\alpha}_{\alpha\beta}(-\omega) = \tilde{\alpha}_{\alpha\beta}^*(\omega), \tag{2.6.51a}$$

$$\alpha_{\alpha\beta}(f_{-\omega}) = \alpha_{\alpha\beta}(f_\omega), \tag{2.6.51b}$$

$$\alpha_{\alpha\beta}(g_{-\omega}) = -\alpha_{\alpha\beta}(g_\omega). \tag{2.6.51c}$$

These originate in the necessity for a real field in

$$\mu_\alpha(\omega) = \tilde{\alpha}_{\alpha\beta}(\omega)[E_\beta(\omega)]_0 \tag{2.6.52}$$

to induce a real moment, because $E(-\omega) = E^*(\omega)$, but they also follow directly from the explicit form of the lineshape functions in (2.6.41). Thus, since $\alpha_{\alpha\beta}(g_\xi)$ is an odd function, we can write (2.6.50a) as

$$\alpha_{\alpha\beta}(f_\omega) = \frac{1}{\pi}\mathscr{P}\int_0^\infty\frac{\alpha_{\alpha\beta}(g_\xi)}{\xi + \omega}d\xi + \frac{1}{\pi}\mathscr{P}\int_0^\infty\frac{\alpha_{\alpha\beta}(g_\xi)}{\xi - \omega}d\xi$$

$$= \frac{2}{\pi}\mathscr{P}\int_0^\infty\frac{\xi\alpha_{\alpha\beta}(g_\xi)}{\xi^2 - \omega^2}d\xi. \tag{2.6.53a}$$

Similarly, since $\alpha_{\alpha\beta}(f_\xi)$ is an even function, (2.6.50b) becomes

$$\alpha_{\alpha\beta}(g_\omega) = -\frac{2\omega}{\pi}\mathscr{P}\int_0^\infty\frac{\alpha_{\alpha\beta}(f_\xi)}{\xi^2 - \omega^2}d\xi. \tag{2.6.53b}$$

The Kramers–Kronig relations show that the dispersive and absorptive parts of a response function are intimately connected. A knowledge of one part at all positive frequencies provides, by evaluation of the integral in (2.6.53a) or (2.6.53b), a complete knowledge of the other part at all frequencies. Furthermore, since

$$\mathscr{P}\int_0^\infty \frac{1}{\xi^2 - \omega^2} d\xi = 0, \tag{2.6.54}$$

the absorptive part of a response function is zero if the dispersive part is constant. This means that there can be no absorption of energy from a static applied field.

An important application of Kramers–Kronig relations is to the derivation of sum rules. For any molecule there exists some high frequency ω_{max}, above which the molecule does not absorb. The dispersive part of the symmetric polarizability then has a simple form of frequency dependence which can be taken from the quantum mechanical expression (2.6.27a):

$$\alpha_{\alpha\beta}(f_\omega) = -\frac{2}{\hbar\omega^2} \sum_{j\neq n} \omega_{jn} \mathrm{Re}(\langle n|\mu_\alpha|j\rangle\langle j|\mu_\beta|n\rangle), (\omega > \omega_{max}). \tag{2.6.55}$$

Generalizing the development leading to the Kuhn–Thomas sum rule (2.6.30), we have

$$\sum_{j\neq n} \omega_{jn} \mathrm{Re}(\langle n|r_\alpha|j\rangle\langle j|r_\beta|n\rangle)$$

$$= -\frac{i}{2m} \sum_{j\neq n} (\langle n|r_\alpha|j\rangle\langle j|p_\beta|n\rangle - \langle n|p_\beta|j\rangle\langle j|r_\beta|n\rangle)$$

$$= -\frac{i}{2m}(\langle n|r_\alpha p_\beta - p_\beta r_\alpha|n\rangle - \langle n|r_\alpha|n\rangle\langle n|p_\beta|n\rangle + \langle n|p_\beta|n\rangle\langle n|r_\alpha|n\rangle)$$

$$= \frac{\hbar}{2m}\delta_{\alpha\beta}$$

so that, for a molecule containing Z electrons,

$$\alpha_{\alpha\beta}(f_\omega) = -\frac{Ze^2}{m\omega^2}\delta_{\alpha\beta}, (\omega > \omega_{max}). \tag{2.6.56}$$

Also, we can approximate (2.6.53a) to

$$\alpha_{\alpha\beta}(f_\omega) = -\frac{2}{\pi\omega^2}\int_0^\infty \xi\alpha_{\alpha\beta}(g_\xi)d\xi, (\omega > \omega_{max}), \tag{2.6.57}$$

and comparison with (2.6.56) gives

$$\int_0^\infty \omega\alpha_{\alpha\beta}(g_\omega)d\omega = \frac{\pi Ze^2}{2m}. \tag{2.6.58}$$

Notice that, although in the derivation ω was taken to be some fixed value greater than ω_{max}, the result (2.6.58) is quite general and refers to an integral over the entire absorption spectrum. This can be regarded as an alternative statement of the Kuhn–Thomas sum rule.

Other treatments refer to Kramers–Kronig relations between the real and imaginary parts of a complex response tensor. But here we have carefully refrained from using this terminology, referring instead to the dispersive and absorptive parts. This is to avoid confusion with complex dynamic molecular property tensors such as $\tilde{\alpha}_{\alpha\beta} = \alpha_{\alpha\beta} - i\alpha'_{\alpha\beta}$, introduced earlier, which can contain both real and imaginary parts even at transparent frequencies. Thus in general

$$\tilde{\alpha}_{\alpha\beta}(\omega) = \alpha_{\alpha\beta}(f_\omega) + i\alpha_{\alpha\beta}(g_\omega) - i\alpha'_{\alpha\beta}(f_\omega) + \alpha'_{\alpha\beta}(g_\omega). \tag{2.6.59}$$

However, complex response tensors in this general form are to be used with complex fields, whereas the Kramers–Kronig relations apply to the real and imaginary parts of a response tensor defined for real fields. Thus, just as Kramers–Kronig relations for the symmetric polarizability $\alpha_{\alpha\beta}$ are developed using (2.6.52), the Kramers–Kronig relations for the antisymmetric polarizability $\alpha'_{\alpha\beta}$ are developed using

$$\mu_\alpha(\omega) = \frac{1}{\omega}\tilde{\alpha}'_{\alpha\beta}(\omega)[\dot{E}_\beta(\omega)]_0, \tag{2.6.60}$$

which is taken from the expression (2.6.26a) for the real electric dipole moment induced by a real field. We must now take $\tilde{R}_{\alpha\beta}(\omega) = \alpha'_{\alpha\beta}(\omega)/\omega$ to be the response tensor; in which case, since $\dot{E}(-\omega) = \dot{E}^*(\omega)$, crossing relations of the form (2.6.51) obtain for $\tilde{R}_{\alpha\beta}(\omega)$ (but not for $\tilde{\alpha}'_{\alpha\beta}(\omega)$). This leads to the following relations between the dispersive and absorptive parts of the antisymmetric polarizability:

$$\alpha'_{\alpha\beta}(f_\omega) = \frac{2\omega}{\pi}\,\mathscr{P}\int_0^\infty \frac{\alpha'_{\alpha\beta}(g_\xi)}{(\xi^2 - \omega^2)}\,\mathrm{d}\xi, \tag{2.6.61a}$$

$$\alpha'_{\alpha\beta}(g_\omega) = -\frac{2\omega^2}{\pi}\,\mathscr{P}\int_0^\infty \frac{\alpha'_{\alpha\beta}(f_\xi)}{\xi(\xi^2 - \omega^2)}\,\mathrm{d}\xi. \tag{2.6.61b}$$

We also require the sum rule for the antisymmetric polarizability, analogous to (2.6.58) for the symmetric polarizability. We see from (2.6.27b) that, for frequencies greater than ω_{max}, the dispersive part of $\alpha'_{\alpha\beta}$ becomes

$$\alpha'_{\alpha\beta}(f_\omega) = \frac{2}{\hbar\omega}\sum_{j\neq n}\mathrm{Im}(\langle n|\mu_\alpha|j\rangle\langle j|\mu_\beta|n\rangle), \quad (\omega > \omega_{max}). \tag{2.6.62}$$

Since μ_α and μ_β are commuting Hermitian operators, we can invoke the closure theorem ($\sum_j |j\rangle\langle j| = 1$) and write

$$\sum_{j\neq n}\mathrm{Im}(\langle n|\mu_\alpha|j\rangle\langle j|\mu_\beta|n\rangle)$$
$$= \mathrm{Im}\langle n|\mu_\alpha\mu_\beta|n\rangle - \mathrm{Im}(\langle n|\mu_\alpha|n\rangle\langle n|\mu_\beta|n\rangle) = 0, \tag{2.6.63}$$

which follows from the fact that the product of any two commuting Hermitian operators is pure Hermitian, and that the expectation values of Hermitian operators are real. Thus

$$\alpha'_{\alpha\beta}(f_\omega) = 0, \quad (\omega > \omega_{\max}). \tag{2.6.64}$$

Also, we can approximate (2.6.61a) to

$$\alpha'_{\alpha\beta}(f_\omega) = \frac{2}{\pi\omega} \int_0^\infty \alpha'_{\alpha\beta}(g_\xi) d\xi, \quad (\omega > \omega_{\max}), \tag{2.6.65}$$

and comparison with (2.6.64) gives

$$\int_0^\infty \alpha'_{\alpha\beta}(g_\xi) d\xi = 0. \tag{2.6.66}$$

Care is needed in the extension of these sum rules to other molecular property tensors in the series (2.6.27) because some of the operators specified in the transition moment products do not commute. For example, for a single electron,

$$[\mu_\alpha, m_\beta] = \frac{ie\hbar}{2m} \varepsilon_{\alpha\beta\gamma} \mu_\gamma. \tag{2.6.67}$$

2.6.5 The dynamic molecular property tensors in a static approximation

Direct evaluation of the sum over all excited states in the dynamic molecular property tensors (2.6.27) is often difficult. It can be avoided by invoking a static approximation that is useful in some situations. The tensors $\alpha_{\alpha\beta}$, $G'_{\alpha\beta}$ and $A_{\alpha,\beta\gamma}$, for example, are written

$$\alpha_{\alpha\beta} = 2 \sum_{j \neq n} \frac{1}{W_{jn}} \mathrm{Re}(\langle n|\mu_\alpha|j\rangle \langle j|\mu_\beta|n\rangle), \tag{2.6.68a}$$

$$G'_{\alpha\beta} = -\frac{2\omega}{\hbar} \sum_{j \neq n} \frac{1}{W_{jn}^2} \mathrm{Im}(\langle n|\mu_\alpha|j\rangle \langle j|m_\beta|n\rangle), \tag{2.6.68b}$$

$$A_{\alpha,\beta\gamma} = 2 \sum_{j \neq n} \frac{1}{W_{jn}} \mathrm{Re}(\langle n|\mu_\alpha|j\rangle \langle j|\Theta_{\beta\gamma}|n\rangle), \tag{2.6.68c}$$

where $W_{jn} = W_j - W_n$. This is a reasonable approximation for light scattering at transparent frequencies with the exciting frequency ω much smaller than the molecular absorption frequencies ω_{jn}.

Consider first the real polarizability tensor $\alpha_{\alpha\beta}$. Following Amos (1982), the wavefunction in the presence of a "fake" static electric field is written

$$\psi_n(E_\beta) = \psi_n^{(0)} + E_\beta \psi_n^{(1)}(E_\beta) + \cdots \tag{2.6.69}$$

where, from the perturbation theory result (2.6.14),

$$\psi_n^{(1)}(E_\beta) = \sum_{j \neq n} \frac{1}{W_{jn}} \langle j|\mu_\beta|n\rangle|j\rangle. \tag{2.6.70}$$

The approximate polarizability (2.6.68a) can then be written

$$\alpha_{\alpha\beta} = 2\langle\psi_n^{(0)}|\mu_\alpha|\psi_n^{(1)}(E_\beta)\rangle. \tag{2.6.71}$$

The final computational version is obtained by expressing the wavefunctions in terms of molecular orbitals ϕ_k similarly perturbed by the static electric field:

$$\alpha_{\alpha\beta} = 4 \sum_{k,occ.} \langle\phi_k^{(0)}|\mu_\alpha|\phi_k^{(1)}(E_\beta)\rangle \tag{2.6.72}$$

where the summation is over all occupied molecular orbitals. The real electric dipole–electric quadrupole optical activity tensor $A_{\alpha,\beta\gamma}$ is treated in the same way, giving

$$A_{\alpha,\beta\gamma} = 4 \sum_{k,occ.} \langle\phi_k^{(0)}|\Theta_{\beta\gamma}|\phi_k^{(1)}(E_\alpha)\rangle. \tag{2.6.73}$$

The imaginary electric dipole–magnetic dipole optical activity tensor $G'_{\alpha\beta}$ needs to be treated with more circumspection because it vanishes as $\omega \to 0$ and so does not have a static limit ($G'_{\alpha\beta}$ is purely dynamic, whereas $\alpha_{\alpha\beta}$ and $A_{\alpha,\beta\gamma}$ have both static and dynamic counterparts). However, as pointed out by Amos (1982), $(1/\omega)G'_{\alpha\beta}$ does have a static limit, which can be written in the form

$$\left(\frac{1}{\omega}G'_{\alpha\beta}\right)_{\omega=0} = -2\hbar \, \mathrm{Im}(\langle\psi_n^{(1)}(E_\alpha) \mid \psi_n^{(1)}(B_\beta)\rangle), \tag{2.6.74}$$

where $\psi_n^{(1)}(B_\beta)$ is the corresponding wavefunction perturbed by a static magnetic field. In terms of perturbed molecular orbitals,

$$\left(\frac{1}{\omega}G'_{\alpha\beta}\right)_{\omega=0} = -4\hbar \sum_{k,occ.} \mathrm{Im}(\langle\phi_k^{(1)}(E_\alpha) \mid \phi_k^{(1)}(B_\beta)\rangle). \tag{2.6.75}$$

These results enable the polarizability and optical activity tensors to be obtained from calculations of the molecular orbitals perturbed by a static electric field and a static magnetic field. As mentioned in later chapters, they are especially useful for ab initio calculations of optical rotation and Raman optical activity.

2.7 A molecule in a radiation field in the presence of other perturbations

To discuss field-induced optical activity phenomena such as the Faraday effect, and also the generation of optical activity within molecules through intramolecular interactions between inactive groups, we need to formulate the effects of other perturbations on the dynamic molecular property tensors. Although we exemplify the perturbed tensors for the case of an external perturbation such as a static electric or magnetic field, similar expressions are obtained for internal perturbations such as spin–orbit coupling or vibronic coupling.

The dynamic molecular property tensors are first written as power series in the perturbation; for example, in a static electric field the dynamic polarizability becomes

$$\tilde{\alpha}_{\alpha\beta}(\mathbf{E}) = \tilde{\alpha}_{\alpha\beta} + \tilde{\alpha}^{(\mu)}_{\alpha\beta,\gamma} E_\gamma + \tfrac{1}{2}\tilde{\alpha}^{(\mu\mu)}_{\alpha\beta,\gamma\delta} E_\gamma E_\delta + \cdots. \tag{2.7.1}$$

Quantum-mechanical expressions for the perturbed dynamic polarizability are found using perturbed wavefunctions and energies in (2.6.27a) and (2.6.27b). The eigenfunction ψ'_j and energy eigenvalue W'_j perturbed to first order in the electrostatic interaction $-\mu_\gamma E_\gamma$ are

$$\psi'_j = \psi_j - \frac{E_\gamma}{\hbar} \sum_{k\neq j} \frac{1}{\omega_{jk}} \langle k|\mu_\gamma|j\rangle \psi_k, \tag{2.7.2}$$

$$W'_j = W_j - \langle j|\mu_\gamma|j\rangle E_\gamma. \tag{2.7.3}$$

Such expressions are valid even if the unperturbed eigenfunction ψ_j belongs to a degenerate set, provided the degenerate eigenfunctions are chosen to be diagonal in the perturbation, for then the eigenfunctions ψ_k mixed in cannot belong to the degenerate set containing ψ_j. For example, if the degenerate set is the set of eigenfunctions ψ_{nlm} of the $n=2$ level of the hydrogen atom, and the perturbation is an electric field along z, the functions $(1/\sqrt{2})(\psi_{200}+\psi_{210})$, $(1/\sqrt{2})(\psi_{200}-\psi_{210})$, ψ_{211}, ψ_{21-1} are diagonal in the operator μ_z. In a magnetic field along z, the functions must be diagonal in m_z and are now $\psi_{200}, \psi_{210}, \psi_{211}, \psi_{21-1}$. From (2.7.3), the frequency separation of the perturbed levels is

$$\omega'_{jn} = \omega_{jn} - (\mu_{j\gamma} - \mu_{n\gamma})\frac{E_\gamma}{\hbar}, \tag{2.7.4}$$

where $\mu_{j\gamma} = \langle j|\mu_\gamma|j\rangle$ is the electric dipole moment of the molecule in the unperturbed state ψ_j. Using

$$\frac{1}{\omega'^2_{jn} - \omega^2} \approx \frac{1}{\omega^2_{jn} - \omega^2}\left[1 + \frac{2\omega_{jn}(\mu_{j\gamma} - \mu_{n\gamma})E_\gamma}{\hbar(\omega^2_{jn} - \omega^2)}\right] \tag{2.7.5}$$

and (2.7.2) in (2.6.27a) and (2.6.27b), the perturbed dynamic polarizabilities
are

$$
\alpha_{\alpha\beta,\gamma}^{(\mu)} = \frac{2}{\hbar^2} \sum_{j\neq n} \left\{ \frac{\omega_{jn}^2 + \omega^2}{(\omega_{jn}^2 - \omega^2)^2} (\mu_{j\gamma} - \mu_{n\gamma}) \mathrm{Re}(\langle n|\mu_\alpha|j\rangle\langle j|\mu_\beta|n\rangle) \right.
$$

$$
+ \sum_{k\neq n} \frac{\omega_{jn}}{\omega_{kn}(\omega_{jn}^2 - \omega^2)} \mathrm{Re}[\langle k|\mu_\gamma|n\rangle(\langle n|\mu_\alpha|j\rangle\langle j|\mu_\beta|k\rangle
$$

$$
+ \langle n|\mu_\beta|j\rangle\langle j|\mu_\alpha|k\rangle)]
$$

$$
+ \sum_{k\neq j} \frac{\omega_{jn}}{\omega_{kj}(\omega_{jn}^2 - \omega^2)} \mathrm{Re}[\langle j|\mu_\gamma|k\rangle(\langle n|\mu_\alpha|j\rangle\langle k|\mu_\beta|n\rangle
$$

$$
\left. + \langle n|\mu_\beta|j\rangle\langle k|\mu_\alpha|n\rangle)] \right\}, \tag{2.7.6a}
$$

$$
\alpha_{\alpha\beta,\gamma}^{\prime(\mu)} = -\frac{2}{\hbar^2} \sum_{j\neq n} \left\{ \frac{2\omega\omega_{jn}}{(\omega_{jn}^2 - \omega^2)^2} (\mu_{j\gamma} - \mu_{n\gamma}) \mathrm{Im}(\langle n|\mu_\alpha|j\rangle\langle j|\mu_\beta|n\rangle) \right.
$$

$$
+ \sum_{k\neq n} \frac{\omega}{\omega_{kn}(\omega_{jn}^2 - \omega^2)} \mathrm{Im}[\langle k|\mu_\gamma|n\rangle(\langle n|\mu_\alpha|j\rangle\langle j|\mu_\beta|k\rangle
$$

$$
- \langle n|\mu_\beta|j\rangle\langle j|\mu_\alpha|k\rangle)]
$$

$$
+ \sum_{k\neq j} \frac{\omega}{\omega_{kj}(\omega_{jn}^2 - \omega^2)} \mathrm{Im}[\langle j|\mu_\gamma|k\rangle(\langle n|\mu_\alpha|j\rangle\langle k|\mu_\beta|n\rangle
$$

$$
\left. - \langle n|\mu_\beta|j\rangle\langle k|\mu_\alpha|n\rangle)] \right\}. \tag{2.7.6b}
$$

Expressions for the perturbed optical activity tensors are analogous to the above:
for example, $G_{\alpha\beta,\gamma}^{\prime(\mu)}$ is given by (2.7.6b) with μ_β replaced by m_β; and $A_{\alpha\beta\gamma,\delta}^{(\mu)}$ is given
by (2.7.6a) with μ_β replaced by $\Theta_{\beta\gamma}$.

The frequency dependence of these perturbed dynamic molecular property tensors in the region of an isolated absorption band is easily deduced. In accordance
with the discussion in Section 2.6.3, we use the replacements

$$
\frac{1}{\omega_{jn}^2 - \omega^2} \to f + \mathrm{i}g, \tag{2.7.7a}
$$

$$
\frac{1}{(\omega_{jn}^2 - \omega^2)^2} \to (f^2 - g^2) + 2\mathrm{i}fg, \tag{2.7.7b}
$$

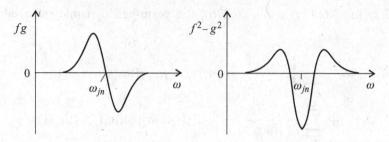

Fig. 2.5 The lineshape functions fg and f^2-g^2. These are valid when the frequency shifts induced by the peturbation are much smaller than the linewidth.

where f and g are given by (2.6.41) and

$$f^2 - g^2 = \frac{\left(\omega_{jn}^2 - \omega^2\right)^2 - \omega^2\Gamma_j^2}{\left[\left(\omega_{jn}^2 - \omega^2\right)^2 + \omega^2\Gamma_j^2\right]^2}, \tag{2.7.7c}$$

$$fg = \frac{\left(\omega_{jn}^2 - \omega^2\right)\omega\Gamma_j}{\left[\left(\omega_{jn}^2 - \omega^2\right)^2 + \omega^2\Gamma_j^2\right]^2}. \tag{2.7.7d}$$

These functions are sketched in Fig. 2.5. The perturbed polarizabilities (2.7.6) are therefore replaced by

$$\alpha_{\alpha\beta,\gamma}^{(\mu)} \to \alpha_{\alpha\beta,\gamma}^{(\mu)}(f) + i\alpha_{\alpha\beta,\gamma}^{(\mu)}(g); \tag{2.7.8a}$$

$$\begin{aligned}
\alpha_{\alpha\beta,\gamma}^{(\mu)}(f) = \frac{2}{\hbar^2} \Big\{ &(f^2 - g^2)(\omega_{jn}^2 + \omega^2)(\mu_{j\gamma} - \mu_{n\gamma})\mathrm{Re}(\langle n|\mu_\alpha|j\rangle\langle j|\mu_\beta|n\rangle) \\
&+ \sum_{k\neq n} \frac{f\omega_{jn}}{\omega_{kn}}\mathrm{Re}[\langle k|\mu_\gamma|n\rangle(\langle n|\mu_\alpha|j\rangle\langle j|\mu_\beta|k\rangle + \langle n|\mu_\beta|j\rangle\langle j|\mu_\alpha|k\rangle)] \\
&+ \sum_{k\neq j} \frac{f\omega_{jn}}{\omega_{kj}}\mathrm{Re}[\langle j|\mu_\gamma|k\rangle(\langle n|\mu_\alpha|j\rangle\langle k|\mu_\beta|n\rangle + \langle n|\mu_\beta|j\rangle\langle k|\mu_\alpha|n\rangle)] \Big\},
\end{aligned} \tag{2.7.8b}$$

$$\begin{aligned}
\alpha_{\alpha\beta,\gamma}^{(\mu)}(g) = \frac{2}{\hbar^2} \Big\{ &2fg(\omega_{jn}^2 + \omega^2)(\mu_{j\gamma} - \mu_{n\gamma})\,\mathrm{Re}(\langle n|\mu_\alpha|j\rangle\langle j|\mu_\beta|n\rangle) \\
&+ \sum_{k\neq n} \frac{g\omega_{jn}}{\omega_{kn}}\mathrm{Re}[\langle k|\mu_\gamma|n\rangle(\langle n|\mu_\alpha|j\rangle\langle j|\mu_\beta|k\rangle + \langle n|\mu_\beta|j\rangle\langle j|\mu_\alpha|k\rangle)] \\
&+ \sum_{k\neq j} \frac{g\omega_{jn}}{\omega_{kj}}\mathrm{Re}[\langle j|\mu_\gamma|k\rangle(\langle n|\mu_\alpha|j\rangle\langle k|\mu_\beta|n\rangle + \langle n|\mu_\beta|j\rangle\langle k|\mu_\alpha|n\rangle)] \Big\},
\end{aligned}$$

$$\tag{2.7.8c}$$

and

$$\alpha_{\alpha\beta,\gamma}^{\prime(\mu)} \rightarrow \alpha_{\alpha\beta,\gamma}^{\prime(\mu)}(f) + i\alpha_{\alpha\beta,\gamma}^{\prime(\mu)}(g); \qquad (2.7.8d)$$

$$\alpha_{\alpha\beta,\gamma}^{\prime(\mu)}(f) = -\frac{2}{\hbar^2} \left\{ 2(f^2 - g^2)\omega\omega_{jn}(\mu_{j\gamma} - \mu_{n\gamma}) \, \mathrm{Im}(\langle n|\mu_\alpha|j\rangle\langle j|\mu_\beta|n\rangle) \right.$$

$$+ \sum_{k\neq n} \frac{f\omega}{\omega_{kn}} \mathrm{Im}[\langle k|\mu_\gamma|n\rangle(\langle n|\mu_\alpha|j\rangle\langle j|\mu_\beta|k\rangle - \langle n|\mu_\beta|j\rangle\langle j|\mu_\alpha|k\rangle)]$$

$$\left. + \sum_{k\neq j} \frac{f\omega}{\omega_{kj}} \mathrm{Im}[\langle j|\mu_\gamma|k\rangle(\langle n|\mu_\alpha|j\rangle\langle k|\mu_\beta|n\rangle - \langle n|\mu_\beta|j\rangle\langle k|\mu_\alpha|n\rangle)] \right\},$$

$$(2.7.8e)$$

$$\alpha_{\alpha\beta,\gamma}^{\prime(\mu)}(g) = -\frac{2}{\hbar^2} \left\{ 4fg\omega\omega_{jn}(\mu_{j\gamma} - \mu_{n\gamma}) \, \mathrm{Im}(\langle n|\mu_\alpha|j\rangle\langle j|\mu_\beta|n\rangle) \right.$$

$$+ \sum_{k\neq n} \frac{g\omega}{\omega_{kn}} \mathrm{Im}[\langle k|\mu_\gamma|n\rangle(\langle n|\mu_\alpha|j\rangle\langle j|\mu_\beta|k\rangle - \langle n|\mu_\beta|j\rangle\langle j|\mu_\alpha|k\rangle)]$$

$$\left. + \sum_{k\neq j} \frac{g\omega}{\omega_{kj}} \mathrm{Im}[\langle j|\mu_\gamma|k\rangle(\langle n|\mu_\alpha|j\rangle\langle k|\mu_\beta|n\rangle - \langle n|\mu_\beta|j\rangle\langle k|\mu_\alpha|n\rangle)] \right\}.$$

$$(2.7.8f)$$

These results apply when the frequency shifts induced by the perturbation are much smaller than the width of the absorption band; in a magnetic field, for example, this corresponds to the Zeeman components being unresolved. When the frequency shifts are much larger than the width of the absorption band (for example, well resolved Zeeman components), the overall lineshape follows simply from the f or g lineshape of each resolved component band. We shall encounter an important example of the latter situation in the generation of characteristic rotatory dispersion and circular dichroism lineshapes through large exciton splittings in chiral dimers (Section 5.3.3).

2.8 Molecular transition tensors

The exposition so far has derived the multipole moments that are oscillating with the same frequency as, and with a definite phase relation to, the inducing light wave. Radiation from such moments is responsible for Rayleigh scattering. On the other hand, the Raman components of the scattered light have frequencies different from, and are usually unrelated in phase to, the incident light wave. Such inelastic light scattering processes can be incorporated into the present semiclassical formalism by introducing dynamic molecular transition tensors which take account of the different initial and final molecular states. These are developed by using, in place of expectation values of multipole moment operators such as (2.6.23), transition moments between initial and final molecular states ψ'_n and ψ'_m perturbed by the light wave.

2.8.1 The Raman transition polarizability

Consistent Raman transition polarizabilities are obtained by developing the following real transition electric dipole moment (Placzek, 1934; Born and Huang, 1954):

$$(\mu_\alpha)_{mn} = \langle m'|\mu_\alpha|n'\rangle + \langle m'|\mu_\alpha|n'\rangle^*. \tag{2.8.1}$$

We use perturbed wave functions of the form (2.6.20) to write the transition electric dipole moment (2.8.1) as follows:

$$(\mu_\alpha)_{mn} = \langle m|\mu_\alpha|n\rangle e^{i\omega_{mn}t}$$

$$+ \frac{1}{2\hbar}\sum_{j\neq n}\left[\frac{\langle m|\mu_\alpha|j\rangle\langle j|\mu_\beta|n\rangle}{\omega_{jn}-\omega}\tilde{E}_\beta^{(0)}e^{-i(\omega-\omega_{mn})t}\right.$$

$$+ \frac{\langle m|\mu_\alpha|j\rangle\langle j|\mu_\beta|n\rangle}{\omega_{jn}+\omega}\tilde{E}_\beta^{(0)*}e^{i(\omega+\omega_{mn})t}\right]$$

$$+ \frac{1}{2\hbar}\sum_{j\neq m}\left[\frac{\langle m|\mu_\beta|j\rangle\langle j|\mu_\alpha|n\rangle}{\omega_{jm}-\omega}\tilde{E}_\beta^{(0)*}e^{i(\omega+\omega_{mn})t}\right.$$

$$+ \frac{\langle m|\mu_\beta|j\rangle\langle j|\mu_\alpha|n\rangle}{\omega_{jm}+\omega}\tilde{E}_\beta^{(0)}e^{-i(\omega-\omega_{mn})t}\right] + \text{complex conjugate}. \tag{2.8.2}$$

The first term,

$$\langle m|\mu_\alpha|n\rangle\, e^{-i\omega_{nm}t} + \langle m|\mu_\alpha|n\rangle^*\, e^{i\omega_{nm}t}, \tag{2.8.3}$$

is a transition moment that describes the generation of spontaneous radiation of frequency ω_{nm} when the molecule is initially in an excited state with $\omega_n > \omega_m$. The remaining terms fall into two types with a frequency dependence $(\omega - \omega_{mn})$ or $(\omega + \omega_{mn})$; these only describe the generation of scattered radiation when $(\omega - \omega_{mn}) > 0$ or $(\omega + \omega_{mn}) > 0$. Since $\omega_{mn} = \omega_m - \omega_n$, the condition on the first type of term can be written $\omega + \omega_n > \omega_m$ and that on the second type can be written $\omega_n - \omega < \omega_m$. Remembering that ω_n is the frequency of the initial molecular state, this means that only terms of the first type describe conventional Raman scattering: thus *Stokes* scattering, which is from a lower to a higher molecular energy level so that $\omega - \omega_{mn} < \omega$, obtains if $\omega_m > \omega_n$; and *antiStokes* scattering, which is from a higher to a lower molecular energy level so that $\omega - \omega_{mn} > \omega$, obtains if $\omega_n > \omega_m$.

According to Placzek (1934), terms of the second type describe an induced emission of two quanta $\omega + \omega_{mn}$ and ω from an excited level of frequency ω_n to a lower level of frequency ω_m; these are not discussed further here.

The Stokes and antiStokes Raman part of (2.8.2) is therefore

$$(\mu_\alpha)_{mn} = \frac{1}{2\hbar} \sum_{j \neq n,m} \left[\frac{\langle m|\mu_\alpha|j\rangle\langle j|\mu_\beta|n\rangle}{\omega_{jn} - \omega} \right.$$
$$\left. + \frac{\langle m|\mu_\beta|j\rangle\langle j|\mu_\alpha|n\rangle}{\omega_{jm} + \omega} \right] \tilde{E}_\beta^{(0)} e^{-i(\omega - \omega_{mn})t}$$
$$+ \text{complex conjugate.} \tag{2.8.4}$$

For simplicity, we have specified $j \neq n, m$ for both terms: this is a good approximation for vibrational Raman scattering since the term that is lost involves $\langle m|\mu_\alpha|m\rangle - \langle n|\mu_\alpha|n\rangle$, the difference in the permanent electric dipole moments of the initial and final states. Using the relation

$$\omega_{jm}\langle m|\mu_\alpha|j\rangle\langle j|\mu_\beta|n\rangle + \omega_{jn}\langle m|\mu_\beta|j\rangle\langle j|\mu_\alpha|n\rangle$$
$$= \tfrac{1}{2}(\omega_{jm} + \omega_{jn})(\langle m|\mu_\alpha|j\rangle\langle j|\mu_\beta|n\rangle + \langle m|\mu_\beta|j\rangle\langle j|\mu_\alpha|n\rangle)$$
$$+ \tfrac{1}{2}\omega_{nm}(\langle m|\mu_\alpha|j\rangle\langle j|\mu_\beta|n\rangle - \langle m|\mu_\beta|j\rangle\langle j|\mu_\alpha|n\rangle), \tag{2.8.5}$$

we can write (2.8.4) in the form

$$(\mu_\alpha)_{mn} = (\alpha_{\alpha\beta})_{mn} E_\beta(\omega - \omega_{mn}) + \frac{1}{\omega - \omega_{mn}}(\alpha'_{\alpha\beta})_{mn} \dot{E}_\beta(\omega - \omega_{mn}), \tag{2.8.6a}$$

where the E_β are functions of the frequency $(\omega - \omega_{mn})$ of the Raman wave, and the transition polarizabilities are

$$(\alpha_{\alpha\beta})_{mn} = \frac{1}{2\hbar} \sum_{j \neq n,m} \frac{1}{(\omega_{jn} - \omega)(\omega_{jm} + \omega)}$$
$$\times [(\omega_{jn} + \omega_{jm}) \operatorname{Re}(\langle m|\mu_\alpha|j\rangle\langle j|\mu_\beta|n\rangle + \langle m|\mu_\beta|j\rangle\langle j|\mu_\alpha|n\rangle)$$
$$+ (2\omega + \omega_{nm}) \operatorname{Re}(\langle m|\mu_\alpha|j\rangle\langle j|\mu_\beta|n\rangle - \langle m|\mu_\beta|j\rangle\langle j|\mu_\alpha|n\rangle)], \tag{2.8.6b}$$

$$(\alpha'_{\alpha\beta})_{mn} = -\frac{1}{2\hbar} \sum_{j \neq n,m} \frac{1}{(\omega_{jn} - \omega)(\omega_{jm} + \omega)}$$
$$\times [(\omega_{jn} + \omega_{jm}) \operatorname{Im}(\langle m|\mu_\alpha|j\rangle\langle j|\mu_\beta|n\rangle + \langle m|\mu_\beta|j\rangle\langle j|\mu_\alpha|n\rangle)$$
$$+ (2\omega + \omega_{nm}) \operatorname{Im}(\langle m|\mu_\alpha|j\rangle\langle j|\mu_\beta|n\rangle - \langle m|\mu_\beta|j\rangle\langle j|\mu_\alpha|n\rangle)]. \tag{2.8.6c}$$

Similar expressions, but without the decomposition into real and imaginary products of transition moments, have been derived by Placzek (1934). The real and imaginary products are displayed explicitly here since this facilitates the application to Raman optical activity phenomena.

Notice that the first terms of both $(\alpha_{\alpha\beta})_{mn}$ and $(\alpha'_{\alpha\beta})_{mn}$ are symmetric with respect to the interchange of the tensor subscripts α and β, and the second terms are antisymmetric. These should be compared with the ordinary dynamic polarizabilities (2.6.27a) and (2.6.27b), the first of which is pure symmetric and the second pure antisymmetric.

According to Placzek (1934), the static transition polarizability that obtains when the frequency ω of the oscillating electric vector of the incident light beam tends to zero describes the effect of an external static electric field on the spontaneous transition amplitude when the molecule is initially in an excited state with $\omega_n > \omega_m$.

The transition polarizabilities (2.8.6) can be written in complex form by introducing a complex transition electric dipole moment:

$$(\tilde{\mu}_\alpha)_{mn} = (\tilde{\alpha}_{\alpha\beta})_{mn}\tilde{E}_\beta, \tag{2.8.7a}$$

$$(\tilde{\alpha}_{\alpha\beta})_{mn} = (\alpha_{\alpha\beta})_{mn} - i(\alpha'_{\alpha\beta})_{mn}. \tag{2.8.7b}$$

Similar expressions are obtained for the transition optical activity tensors $(\tilde{G}_{\alpha\beta})_{mn}$ and $(\tilde{A}_{\alpha,\beta\gamma})_{mn}$, with μ_β replaced by m_β and $\Theta_{\beta\gamma}$, respectively, but now there is no meaningful separation into symmetric and antisymmetric parts when $n = m$.

To facilitate the discussion in subsequent chapters of polarization effects in Rayleigh and Raman scattering, we introduce superscripts 's' and 'a' to denote symmetric and antisymmetric parts of a transition polarizability:

$$(\tilde{\alpha}_{\alpha\beta})_{mn} = (\tilde{\alpha}_{\alpha\beta})^s_{mn} + (\tilde{\alpha}_{\alpha\beta})^a_{mn} = (\alpha_{\alpha\beta})^s_{mn} + (\alpha_{\alpha\beta})^a_{mn} - i(\alpha'_{\alpha\beta})^s_{mn} - i(\alpha'_{\alpha\beta})^a_{mn};$$

$$\tag{2.8.8a}$$

$$(\alpha_{\alpha\beta})^s_{mn} = \frac{1}{2\hbar}\sum_{j\neq n,m}\frac{(\omega_{jn}+\omega_{jm})}{(\omega_{jn}-\omega)(\omega_{jm}+\omega)}$$
$$\times \mathrm{Re}(\langle m|\mu_\alpha|j\rangle\langle j|\mu_\beta|n\rangle + \langle m|\mu_\beta|j\rangle\langle j|\mu_\alpha|n\rangle), \tag{2.8.8b}$$

$$(\alpha_{\alpha\beta})^a_{mn} = \frac{1}{2\hbar}\sum_{j\neq n,m}\frac{(2\omega+\omega_{nm})}{(\omega_{jn}-\omega)(\omega_{jm}+\omega)}$$
$$\times \mathrm{Re}(\langle m|\mu_\alpha|j\rangle\langle j|\mu_\beta|n\rangle - \langle m|\mu_\beta|j\rangle\langle j|\mu_\alpha|n\rangle), \tag{2.8.8c}$$

$$(\alpha'_{\alpha\beta})^s_{mn} = -\frac{1}{2\hbar}\sum_{j\neq n,m}\frac{(\omega_{jn}+\omega_{jm})}{(\omega_{jn}-\omega)(\omega_{jm}+\omega)}$$
$$\times \mathrm{Im}(\langle m|\mu_\alpha|j\rangle\langle j|\mu_\beta|n\rangle + \langle m|\mu_\beta|j\rangle\langle j|\mu_\alpha|n\rangle), \tag{2.8.8d}$$

$$(\alpha'_{\alpha\beta})^a_{mn} = -\frac{1}{2\hbar}\sum_{j\neq n,m}\frac{(2\omega+\omega_{nm})}{(\omega_{jn}-\omega)(\omega_{jm}+\omega)}$$
$$\times \mathrm{Im}(\langle m|\mu_\alpha|j\rangle\langle j|\mu_\beta|n\rangle - \langle m|\mu_\beta|j\rangle\langle j|\mu_\alpha|n\rangle). \tag{2.8.8e}$$

The case $n = m$ in these transition polarizabilities corresponds to Rayleigh scattering, and (2.8.6b) and (2.8.6c) reduce to (2.6.27a) and (2.6.27b) as required, with the real part pure symmetric and the imaginary part pure antisymmetric. The case $n \neq m$ usually corresponds to Raman scattering, and both real and imaginary parts of the complex transition polarizability can contain symmetric and antisymmetric parts. However, $n \neq m$ can also describe Rayleigh scattering between different component states of a degenerate level; interesting possibilities for antisymmetric Rayleigh scattering then arise, as discussed in later chapters.

We can now appreciate that antisymmetric *Rayleigh* scattering is only possible from systems in degenerate states. As discussed later in Chapter 4, time reversal has the effect of replacing the time-independent part of a wavefunction by its complex conjugate. Since atoms and molecules in the absence of external magnetic fields are invariant under time reversal, ψ and ψ^* describe states of the same energy, so if the level is not degenerate, $\psi = \psi^*$ and is pure real. The polarizability (2.6.27b), which is the usual source of antisymmetric Rayleigh scattering, therefore vanishes because it is pure imaginary. We also lose the possibility of antisymmetric Rayleigh scattering from the real transition polarizability (2.8.8c) because n and m must be the same. However, there appears to be no fundamental reason why degeneracy is required for antisymmetric *Raman* scattering.

It is convenient for some applications to introduce the following complex version of the transition electric dipole moment (2.8.4):

$$(\tilde{\mu}_\alpha)_{mn} = (\tilde{\alpha}_{\alpha\beta})_{mn} \tilde{E}_\beta^{(0)} e^{-i(\omega - \omega_{mn})t}, \tag{2.8.9a}$$

where

$$(\tilde{\alpha}_{\alpha\beta})_{mn} = \frac{1}{\hbar} \sum_{j \neq n, m} \left[\frac{\langle m|\mu_\alpha|j\rangle\langle j|\mu_\beta|n\rangle}{\omega_{jn} - \omega} + \frac{\langle m|\mu_\beta|j\rangle\langle j|\mu_\alpha|n\rangle}{\omega_{jm} + \omega} \right] \tag{2.8.9b}$$

is a complex transition polarizability. This can be generated as the matrix element of the following complex scattering operator (Berestetskii, Lifshitz and Pitaevskii, 1982):

$$\tilde{C}_{\alpha\beta} = \frac{1}{\hbar}(b_\beta \mu_\alpha - \mu_\alpha b_\beta), \tag{2.8.10}$$

where b_α is a polar vector operator satisfying

$$\left(i\frac{d}{dt} + \omega\right) b_\alpha = \mu_\alpha. \tag{2.8.11}$$

Taking matrix elements of (2.8.10) and invoking the velocity–dipole transformation (2.6.31b), we find

$$\langle k|\mu_\alpha|j\rangle = (\omega - \omega_{kj})\langle k|b_\alpha|j\rangle. \tag{2.8.12}$$

So by writing

$$\langle j|b_\alpha|n\rangle = \langle j|\mu_\alpha|n\rangle/(\omega - \omega_{jn}),$$
$$\langle m|b_\alpha|j\rangle = \langle m|\mu_\alpha|j\rangle/(\omega + \omega_{jm}),$$

the complex transition polarizability (2.8.9) is generated as follows:

$$\langle m|\tilde{C}_{\alpha\beta}|n\rangle = \frac{1}{\hbar}\langle m|b_\beta\mu_\alpha - \mu_\alpha b_\beta|n\rangle$$

$$= \frac{1}{\hbar}\sum_j (\langle m|b_\beta|j\rangle\langle j|\mu_\alpha|n\rangle - \langle m|\mu_\alpha|j\rangle\langle j|b_\beta|n\rangle)$$

$$= (\tilde{\alpha}_{\alpha\beta})_{mn}. \tag{2.8.13}$$

The scattering operator (2.8.10) is exact; but we shall find more useful instead an approximate operator which breaks down into parts with better defined Hermiticity and time reversal characteristics. Following Child and Longuet–Higgins (1961) we introduce the *effective polarizability operator*

$$\hat{\alpha}_{\alpha\beta} = \hat{\alpha}^s_{\alpha\beta} + \hat{\alpha}^a_{\alpha\beta}, \tag{2.8.14a}$$

$$\hat{\alpha}^s_{\alpha\beta} = \tfrac{1}{2}(\mu_\alpha O^s\mu_\beta + \mu_\beta O^s\mu_\alpha), \tag{2.8.14b}$$

$$\hat{\alpha}^a_{\alpha\beta} = -\tfrac{1}{2}(\mu_\alpha O^a\mu_\beta - \mu_\beta O^a\mu_\alpha), \tag{2.8.14c}$$

where

$$O^s = \left(\frac{1}{H - \bar{W} + \hbar\omega} + \frac{1}{H - \bar{W} - \hbar\omega}\right), \tag{2.8.14d}$$

$$O^a = \left(\frac{1}{H - \bar{W} + \hbar\omega} - \frac{1}{H - \bar{W} - \hbar\omega}\right). \tag{2.8.14e}$$

$\bar{W}$ is the average of the energies W_n and W_m of the initial and final states. By summing over a complete set of states $|j\rangle\langle j|$ inserted after O and using the approximation $\omega_{jn} \approx \omega_{jm}$, it is easily verified that $\langle m|\hat{\alpha}_{\alpha\beta}|n\rangle$ generates the complex transition polarizability (2.8.9). The real transition polarizabilities (2.8.8) are now given by

$$(\alpha_{\alpha\beta})^s_{mn} = \tfrac{1}{2}(\langle m|\hat{\alpha}^s_{\alpha\beta}|n\rangle + \langle m|\hat{\alpha}^s_{\alpha\beta}|n\rangle^*),$$

$$(\alpha_{\alpha\beta})^a_{mn} = \tfrac{1}{2}(\langle m|\hat{\alpha}^a_{\alpha\beta}|n\rangle + \langle m|\hat{\alpha}^a_{\alpha\beta}|n\rangle^*),$$

$$(\alpha'_{\alpha\beta})^s_{mn} = \tfrac{1}{2}i(\langle m|\hat{\alpha}^s_{\alpha\beta}|n\rangle - \langle m|\hat{\alpha}^s_{\alpha\beta}|n\rangle^*),$$

$$(\alpha'_{\alpha\beta})^a_{mn} = \tfrac{1}{2}i(\langle m|\hat{\alpha}^a_{\alpha\beta}|n\rangle - \langle m|\hat{\alpha}^a_{\alpha\beta}|n\rangle^*). \tag{2.8.15}$$

It is shown later (Section 4.3.3) that $\hat{\alpha}^s_{\alpha\beta}$ is a Hermitian time-even operator, whereas $\hat{\alpha}^a_{\alpha\beta}$ is antiHermitian and time odd.

Notice that O^a in (2.8.14e) vanishes when $\omega = 0$, which explains immediately why there is no static antisymmetric polarizability.

We also introduce the *effective optical activity operators*

$$\hat{G}_{\alpha\beta} = \hat{G}^s_{\alpha\beta} + \hat{G}^a_{\alpha\beta}, \tag{2.8.14f}$$

$$\hat{G}^s_{\alpha\beta} = \tfrac{1}{2}(\mu_\alpha O^s m_\beta + m_\beta O^s \mu_\alpha), \tag{2.8.14g}$$

$$\hat{G}^a_{\alpha\beta} = -\tfrac{1}{2}(\mu_\alpha O^a m_\beta - m_\beta O^a \mu_\alpha); \tag{2.8.14h}$$

$$\hat{A}_{\alpha,\beta\gamma} = \hat{A}^s_{\alpha,\beta\gamma} + \hat{A}^a_{\alpha,\beta\gamma}, \tag{2.8.14i}$$

$$\hat{A}^s_{\alpha,\beta\gamma} = \tfrac{1}{2}(\mu_\alpha O^s \Theta_{\beta\gamma} + \Theta_{\beta\gamma} O^s \mu_\alpha), \tag{2.8.14j}$$

$$\hat{A}^a_{\alpha,\beta\gamma} = -\tfrac{1}{2}(\mu_\alpha O^a \Theta_{\beta\gamma} - \Theta_{\beta\gamma} O^a \mu_\alpha). \tag{2.8.14k}$$

The superscripts 's' and 'a' are retained to conform with the corresponding parts of the effective polarizability operator even though there is no longer well defined permutation symmetry. It is shown later that $\hat{G}^s_{\alpha\beta}$ is Hermitian and time odd, $\hat{G}^a_{\alpha\beta}$ is antiHermitian and time even; $\hat{A}^s_{\alpha,\beta\gamma}$ is Hermitian and time even, and $\hat{A}^a_{\alpha,\beta\gamma}$ is antiHermitian and time odd.

It is worth recording that the effective polarizability and optical activity operators can be derived from linear response theory. For example, (2.8.14h) follows from a consideration of temporal correlations between the electric and magnetic dipole moment operators in the absence of the light beam (Harris, 1966).

In order to accommodate resonance phenomena, the transition frequencies ω_{jn} and ω_{jm} in the energy denominators of the complex transition polarizability (2.8.9b) may be replaced by complex transition frequencies of the form (2.6.39) to allow for the finite energy width of the excited states. According to Buckingham and Fischer (2000), both the complex transition frequency and its complex conjugate should appear in the following manner:

$$(\tilde{\alpha}_{\alpha\beta})_{mn} = \frac{1}{\hbar} \sum_{j\neq n,m} \left[\frac{\langle m|\mu_\alpha|j\rangle\langle j|\mu_\beta|n\rangle}{\tilde{\omega}_{jn} - \omega} + \frac{\langle m|\mu_\beta|j\rangle\langle j|\mu_\alpha|n\rangle}{\tilde{\omega}^*_{jm} + \omega} \right]. \tag{2.8.16a}$$

This leads to opposite signs for the damping factors in the two terms,

$$(\tilde{\alpha}_{\alpha\beta})_{mn} = \frac{1}{\hbar} \sum_{j\neq n,m} \left[\frac{\langle m|\mu_\alpha|j\rangle\langle j|\mu_\beta|n\rangle}{\omega_{jn} - \omega - \tfrac{1}{2}i\Gamma_j} + \frac{\langle m|\mu_\beta|j\rangle\langle j|\mu_\alpha|n\rangle}{\omega_{jm} - \omega + \tfrac{1}{2}i\Gamma_j} \right], \tag{2.8.16b}$$

and yields results consistent with those used widely in nonlinear optics (Bloembergen, 1996). That this choice of signs for the damping factors is correct may be confirmed analytically using the causality principle (Hassing and Nørby Svendsen, 2004).

The effective polarizability and optical activity operators (2.8.14) exhibit singularities at resonance frequencies. However, nonsingular versions of these operators may be defined which in addition do not rely on the average energy approximation and are therefore valid for all Raman processes, transparent and resonant. We refer to Hecht and Barron (1993b,c) for further details.

2.8.2 The adiabatic approximation

Most studies of molecular quantum states start by invoking the *adiabatic approximation* (Born and Oppenheimer, 1927) which leads to a separation of the electronic and nuclear motions. We first write the complete molecular Hamiltonian as

$$H = T(r) + T(R) + V(r, R) + V(R), \tag{2.8.17}$$

where r and R denote the sets of electronic and nuclear coordinates, $T(r)$ and $T(R)$ are the electronic and nuclear kinetic energy operators, $V(r, R)$ is the mutual potential energy of the electrons together with the potential energy of the electrons with respect to the nuclei, and $V(R)$ is the potential energy of the nuclei. An approximate solution of the Schrödinger equation for the complete molecule, namely

$$H\Psi_{en}(r, R) = W_{en}\Psi_{en}(r, R), \tag{2.8.18}$$

is sought by writing the true molecular wavefunction $\Psi_{en}(r, R)$ in the approximate form

$$\Psi_{en}(r, R) = \psi_e(r, R)\psi_{en}(R), \tag{2.8.19}$$

where e and n specify the electronic and nuclear quantum states.

The electronic eigenfunction $\psi_e(r, R)$ is a solution of the Schrödinger equation

$$[T(r) + V(r, R)]\psi_e(r, R) = w_e(R)\psi_e(r, R) \tag{2.8.20}$$

which describes the motion of the electrons constrained by a potential energy $V(r, R)$ in which the electron–nuclear part arises from nuclei fixed in a particular configuration R. The electronic energy eigenvalue $w_e(R)$ therefore depends on the nuclear coordinates as parameters. Consequently $\psi_e(r, R)$ characterizes a particular electronic quantum state for infinitely slow changes in the internuclear separations. The electrons are said to follow the nuclear motions adiabatically. In an adiabatic motion, an electron does not make transitions from one state to others; instead, an electronic state itself is deformed progressively by the nuclear displacements. Thus the molecule remains in the same electronic quantum state with energy $w_e(R)$ during the course of a molecular vibration or rotation.

The nuclear eigenfunction $\psi_{en}(R)$ is a solution of the Schrödinger equation

$$[T(R) + w_e(R) - w_e(R_0) + V(R)]\psi_{en}(R) = w_{en}\psi_{en}(R), \qquad (2.8.21)$$

which describes the motion of the nuclei constrained by an effective potential energy arising from the nuclear–nuclear interactions $V(R)$ together with the difference between the electronic energy $w_e(R)$ at some general nuclear configuration and the electronic energy $w_e(R_0)$ at the equilibrium nuclear configuration, the molecule being in some adiabatic electronic eigenstate $\psi_e(r, R)$.

If the variation of $\psi_e(r, R)$ with R is sufficiently small that $T(R)\psi_e(r, R)$ can be neglected, we can use (2.8.20) and (2.8.21) to write the complete Schrödinger equation (2.8.18) as

$$[T(r) + T(R) + V(r, R) + V(R)]\psi_e(r, R)\psi_{en}(R)$$
$$= [T(R) + w_e(R) + V(R)]\psi_e(r, R)\psi_{en}(R)$$
$$= [w_e(R_0) + w_{en}]\psi_e(r, R)\psi_{en}(R). \qquad (2.8.22)$$

Thus the energy eigenvalue for the state represented by the adiabatic eigenfunction (2.8.19) is

$$W_{en} = w_e(R_0) + w_{en}, \qquad (2.8.23)$$

which is the electronic energy at the equilibrium nuclear configuration plus the energy due to the nuclear motion. The justification for the adiabatic approximation lies in the slow nuclear motion compared with the electronic motion resulting from the large disparity between the nuclear and the electronic masses, so that the nuclear motion constitutes an adiabatic perturbation of the electronic quantum state.

In general, the nuclear motion has vibrational, rotational and translational contributions which can be separated to a good approximation. Translational motion is eliminated by working in a molecule-fixed set of axes. The adiabatic eigenfunction and energy eigenvalue now become

$$\Psi_{evr} = \psi_e(r, Q)\psi_{ev}(Q)\psi_{evr}(\theta, \phi, \chi), \qquad (2.8.24a)$$

$$W_{evr} = w_e(Q_0) + w_{ev} + w_{evr}, \qquad (2.8.25b)$$

where subscripts v and r denote vibrational and rotational quantum states, Q denotes the particular set of internal nuclear coordinates known as normal vibrational coordinates, and θ, ϕ, χ are the Euler angles that specify the orientation of the molecule-fixed axes relative to space-fixed axes.

In *ket* notation, the jth electronic–nuclear state, prior to invoking the Born–Oppenheimer approximation, we write as $|j\rangle = |e_j v_j r_j\rangle$; after invoking the approximation this can be written as $|e_j\rangle|v_j\rangle|r_j\rangle$, provided that the electronic part is not orbitally degenerate. The complete specification of the vibrational part is rather messy because it is necessary to specify the number of vibrational quanta in each normal mode. Thus the vibrational part of the jth state is written

$$|v_j\rangle \equiv |n_{1_j}, n_{2_j}, \cdots n_{p_j}, \cdots n_{(3N-6)_j}\rangle,$$

where n_{p_j} is the number of vibrational quanta in the normal mode associated with the normal coordinate Q_{p_j}, there being $3N-6$ normal modes in all in a nonlinear molecule. We shall often use simplified notations that are clear from the particular context. For example, $|1_j\rangle$ is used to denote a vibrational state associated with an electronic state $|e_j\rangle$ in which one of the normal modes contains one quantum and all the others no quanta. We do not bother to specify which normal mode is excited because this is usually clear, as in the vibrational transition moment $\langle 1_j|Q_p|0\rangle$ for example. Elsewhere, $|1_p\rangle$ is used to denote a singly-excited vibrational state corresponding to the normal coordinate Q_p and associated with the ground electronic state.

The fundamental approximation (2.8.19) is only valid when the electronic function $\psi_e(r, R)$ is orbitally nondegenerate at all points in the relevant R space. The extension to orbitally-degenerate states is outlined in Section 2.8.4 in the simplified context of the 'crude' adiabatic approximation.

2.8.3 The vibrational Raman transition tensors in Placzek's approximation

The adiabatic approximation can be used to simplify the Raman transition tensors derived in Section 2.8.1. Since we are concerned only with vibrational Raman scattering, the adiabatic wavefunction and energy (2.8.24) are employed first to isolate the vibrational parts of the general Raman transition polarizabilities in (2.8.6). Up until now we have used n, j and m to denote initial, intermediate and final quantum states: we now append these as subscripts to e, v and r to specify the corresponding electronic, vibrational and rotational parts. Thus in the adiabatic approximation, the general molecular eigenstate is written as a product of separate electronic, vibrational and rotational parts:

$$|j\rangle = |e_j v_j r_j\rangle = |e_j\rangle|v_j\rangle|r_j\rangle, \tag{2.8.25a}$$

the second equality holding only if $|e_j\rangle$ is not orbitally degenerate, with energy

$$W_{e_j v_j r_j} = w_{e_j} + w_{v_j} + w_{r_j}. \tag{2.8.25b}$$

This means that the frequency separation of two general molecular eigenstates ψ_j and ψ_n can be written as the sum of the frequency separations of the electronic, vibrational and rotational parts:

$$\omega_{e_j v_j r_j e_n v_n r_n} = \omega_{e_j e_n} + \omega_{v_j v_n} + \omega_{r_j r_n}. \tag{2.8.26}$$

The real transition polarizability (2.8.6*b*), for example, now becomes

$$(\alpha_{\alpha\beta})_{e_m v_m r_m e_n v_n r_n} = \frac{1}{2\hbar} \sum_{\substack{e_j v_j r_j \neq \\ e_n v_n r_n, \\ e_m v_m r_m}} \frac{1}{(\omega_{e_j v_j r_j e_n v_n r_n} - \omega)(\omega_{e_j v_j r_j e_m v_m r_m} + \omega)}$$

$$\times [(\omega_{e_j v_j r_j e_n v_n r_n} + \omega_{e_j v_j r_j e_m v_m r_m}) \text{Re} (\langle e_m v_m r_m | \mu_\alpha | e_j v_j r_j \rangle \langle e_j v_j r_j | \mu_\beta | e_n v_n r_n \rangle$$

$$+ \langle e_m v_m r_m | \mu_\beta | e_j v_j r_j \rangle \langle e_j v_j r_j | \mu_\alpha | e_n v_n r_n \rangle)$$

$$+ (2\omega + \omega_{e_n v_n r_n e_m v_m r_m}) \text{Re} (\langle e_m v_m r_m | \mu_\alpha | e_j v_j r_j \rangle \langle e_j v_j r_j | \mu_\beta | e_n v_n r_n \rangle$$

$$- \langle e_m v_m r_m | \mu_\beta | e_j v_j r_j \rangle \langle e_j v_j r_j | \mu_\alpha | e_n v_n r_n \rangle)].$$

$$(2.8.27)$$

For incident radiation at transparent frequencies, it is a good approximation to neglect the rotational contributions to the transition frequencies except for the terms with $e_j v_j = e_n v_n = 00$. These terms, which involve pure rotational virtual excited states, will only be significant for incident radiation at microwave frequencies. In the remaining terms, we can therefore invoke the closure theorem with respect to the complete set of rotational states associated with every electronic–vibrational level:

$$\sum_{r_j} |e_j v_j r_j\rangle\langle e_j v_j r_j| = |e_j v_j\rangle\langle e_j v_j|. \qquad (2.8.28)$$

Neglecting the microwave term, the transition polarizability (2.8.27) can now be written

$$(\alpha_{\alpha\beta})_{e_m v_m r_m e_n v_n r_n} = \langle r_m | (\alpha_{\alpha\beta})_{e_m v_m e_n v_n} | r_n \rangle, \qquad (2.8.29)$$

where $(\alpha_{\alpha\beta})_{e_m v_m e_n v_n}$ is simply (2.8.27) with all rotational states and energies removed. The same approximation should also be good at absorbing frequencies if the lifetimes of the excited states are taken into account. If we were interested in rotational Raman scattering, we would relate the space-fixed axes $\alpha, \beta, \cdots$ to molecule-fixed axes $\alpha', \beta', \cdots$ using direction cosines such as $l_{\alpha\alpha'}$ between the α and the α' axis and write

$$(\alpha_{\alpha\beta})_{e_m v_m r_m e_n v_n r_n} = (\alpha_{\alpha'\beta'})_{e_m v_m e_n v_n} \langle r_m | l_{\alpha\alpha'} l_{\beta\beta'} | r_n \rangle \qquad (2.8.30)$$

since only direction cosine operators can effect pure rotational transitions. However, since we are concerned only with vibrational Raman scattering from fluids and solids, the rotational states are dropped henceforth. In the case of fluids, isotropic averages of intensity expressions are ultimately taken: this gives results identical with those that would be obtained by retaining the complete transition polarizability

(2.8.30) and ultimately summing the intensity expressions over the complete set of initial and final rotational states (Van Vleck, 1932; Bridge and Buckingham, 1966).

At ordinary temperatures a molecule is usually in a quantum state belonging to the lowest electronic level, taken here to be e_n, so for vibrational Raman scattering we need only consider the vibrational transition polarizability $(\alpha_{\alpha\beta})_{e_n v_m e_n v_n}$ given by

$$
(\alpha_{\alpha\beta})_{e_n v_m e_n v_n} = \frac{1}{2\hbar} \sum_{\substack{e_j v_j \neq \\ e_n v_n, \\ e_n v_m}} \frac{1}{(\omega_{e_j v_j e_n v_n} - \omega)(\omega_{e_j v_j e_n v_m} + \omega)}
$$

$$
\times [(\omega_{e_j v_j e_n v_n} + \omega_{e_j v_j e_n v_m}) \mathrm{Re}(\langle e_n v_m | \mu_\alpha | e_j v_j \rangle \langle e_j v_j | \mu_\beta | e_n v_n \rangle
$$

$$
+ \langle e_n v_m | \mu_\beta | e_j v_j \rangle \langle e_j v_j | \mu_\alpha | e_n v_n \rangle)
$$

$$
+ (2\omega + \omega_{e_n v_n e_n v_m}) \mathrm{Re}(\langle e_n v_m | \mu_\alpha | e_j v_j \rangle \langle e_j v_j | \mu_\beta | e_n v_n \rangle
$$

$$
- \langle e_n v_m | \mu_\beta | e_j v_j \rangle \langle e_j v_j | \mu_\alpha | e_n v_n \rangle)]. \tag{2.8.31}
$$

We now split the summation over e_j into two parts corresponding to the two cases $e_j = e_n$ and $e_j \neq e_n$. For the latter part at transparent frequencies it is a good approximation to neglect vibrational contributions $\omega_{v_j v_n}$ and $\omega_{v_j v_m}$ to the virtual transition frequencies $\omega_{e_j v_j e_n v_n}$ and $\omega_{e_j v_j e_n v_m}$, in which case (2.8.31) becomes

$$
(\alpha_{\alpha\beta})_{e_n v_m e_n v_n} = \frac{1}{2\hbar} \sum_{\substack{v_j \neq \\ v_n, v_m}} \frac{1}{(\omega_{v_j v_n} - \omega)(\omega_{v_j v_m} + \omega)}
$$

$$
\times [(\omega_{v_j v_n} + \omega_{v_j v_m}) \mathrm{Re}(\langle e_n v_m | \mu_\alpha | e_n v_j \rangle \langle e_n v_j | \mu_\beta | e_n v_n \rangle
$$

$$
+ \langle e_n v_m | \mu_\beta | e_n v_j \rangle \langle e_n v_j | \mu_\alpha | e_n v_n \rangle)
$$

$$
+ (2\omega + \omega_{v_n v_m}) \mathrm{Re}(\langle e_n v_m | \mu_\alpha | e_n v_j \rangle \langle e_n v_j | \mu_\beta | e_n v_n \rangle
$$

$$
- \langle e_n v_m | \mu_\beta | e_n v_j \rangle \langle e_n v_j | \mu_\alpha | e_n v_n \rangle)]
$$

$$
+ \frac{1}{2\hbar} \sum_{e_j \neq e_n} \frac{1}{(\omega_{e_j e_n}^2 - \omega^2)}
$$

$$
\times \left[2\omega_{e_j e_n} \sum_{v_j} \mathrm{Re}(\langle e_n v_m | \mu_\alpha | e_j v_j \rangle \langle e_j v_j | \mu_\beta | e_n v_n \rangle \right.
$$

$$
+ \langle e_n v_m | \mu_\beta | e_j v_j \rangle \langle e_j v_j | \mu_\alpha | e_n v_n \rangle)
$$

$$
+ 2\omega \sum_{v_j} \mathrm{Re}(\langle e_n v_m | \mu_\alpha | e_j v_j \rangle \langle e_j v_j | \mu_\beta | e_n v_n \rangle
$$

$$
\left. - \langle e_n v_m | \mu_\beta | e_j v_j \rangle \langle e_j v_j | \mu_\alpha | e_n v_n \rangle) \right], \tag{2.8.32}
$$

where $\omega_{v_j v_n}$, $\omega_{v_j v_m}$ and $\omega_{v_n v_m}$ in the first part denote transition frequencies between vibrational states belonging to the lowest electronic level. In the antisymmetric term of the second part, we have neglected the vibrational Raman transition frequency $\omega_{v_m v_n}$ relative to ω.

In these expressions, the electric dipole moment operator μ_α is a function of both electronic and nuclear coordinates; and the matrix elements are to be formed with complete adiabatic wavefunctions of the form (2.8.25a). To simplify these expressions, we now introduce the adiabatic permanent electric dipole moment and adiabatic dynamic polarizability of the molecule in the lowest electronic level, the nuclei being held fixed in a configuration Q so that only the electrons are free to move. Both quantities are evidently functions of Q and will be denoted by $\mu_\alpha(Q)$ and $\alpha_{\alpha\beta}(Q)$ respectively. Thus

$$\mu_\alpha(Q) = \langle \psi_0(r, Q)|\mu_\alpha|\psi_0(r, Q)\rangle, \tag{2.8.33}$$

and from (2.6.27a)

$$\alpha_{\alpha\beta}(Q) = \frac{2}{\hbar} \sum_{e_j \neq e_n} \frac{\omega_{e_j e_n}}{\omega_{e_j e_n}^2 - \omega^2} \text{Re}(\langle \psi_0(r, Q)|\mu_\alpha|\psi_{e_j}(r, Q)\rangle$$
$$\times \langle \psi_{e_j}(r, Q)|\mu_\beta|\psi_0(r, Q)\rangle), \tag{2.8.34}$$

which is only valid at transparent frequencies.

Using (2.8.33), the first part of the vibrational transition polarizability (2.8.32) becomes

$$(\alpha_{\alpha\beta})_{v_m v_n}^{(\text{ionic})} = \frac{1}{2\hbar} \sum_{v_j \neq v_n} \frac{1}{(\omega_{v_j v_n} - \omega)(\omega_{v_j v_m} + \omega)}$$
$$\times [(\omega_{v_j v_n} + \omega_{v_j v_m})\text{Re}(\langle v_m|\mu_\alpha(Q)|v_j\rangle\langle v_j|\mu_\beta(Q)|v_n\rangle$$
$$+ \langle v_m|\mu_\beta(Q)|v_j\rangle\langle v_j|\mu_\alpha(Q)|v_n\rangle)$$
$$+ (2\omega + \omega_{v_n v_m})\text{Re}(\langle v_m|\mu_\alpha(Q)|v_j\rangle\langle v_j|\mu_\beta(Q)|v_n\rangle$$
$$- \langle v_m|\mu_\beta(Q)|v_j\rangle\langle v_j|\mu_\alpha(Q)|v_n\rangle)]. \tag{2.8.35}$$

This is usually known as the ionic part of the vibrational transition polarizability and describes Raman scattering through virtual excited vibrational states alone, the molecule remaining in the ground electronic state. Except when the frequency of the exciting light is in the infrared region or below, this term can be ignored.

In the second part of the vibrational transition polarizability (2.8.32) the closure theorem with respect to the complete set of vibrational states can be invoked, leaving

$$(\alpha_{\alpha\beta})_{v_m v_n}^{(\text{electronic})} = \frac{2}{\hbar} \sum_{e_j \neq e_n} \frac{\omega_{e_j e_n}}{\omega_{e_j e_n}^2 - \omega^2} \langle v_m|\text{Re}(\langle e_n|\mu_\alpha|e_j\rangle\langle e_j|\mu_\beta|e_n\rangle)|v_n\rangle$$
$$= \langle v_m|\alpha_{\alpha\beta}(Q)|v_n\rangle, \tag{2.8.36}$$

where we have assumed that ψ_{v_m} and ψ_{v_n} are real and have introduced the adiabatic dynamic polarizability (2.8.34). Notice that in this approximation the antisymmetric part vanishes (because the real part of a quantity minus its complex conjugate, which is pure imaginary, is specified). The vibrational part $(\alpha'_{\alpha\beta})_{v_m v_n}$ of the imaginary transition polarizability may be developed in a similar fashion except that, since $\alpha'_{\alpha\beta}$ is time odd, it is necessary to consider its dependence on the conjugate momentum $\dot{Q}$ of the normal coordinate rather than on Q itself (see Section 8.2). Now the symmetric part vanishes (because the imaginary part of a quantity plus its complex conjugate, which is pure real, is specified).

The discussion in this section has followed that given by Born and Huang (1954) and Placzek (1934), and is usually known as Placzek's approximation. Thus within this approximation, when visible or ultraviolet exciting light far from any electronic absorption frequency of the molecule is used, the real and imaginary vibrational parts of the complex transition polarizability (2.8.7*b*) are written

$$(\alpha_{\alpha\beta})_{v_m v_n} = \langle v_m|\alpha_{\alpha\beta}(Q)|v_n\rangle = (\alpha_{\beta\alpha})_{v_m v_n}, \tag{2.8.37a}$$

$$(\alpha'_{\alpha\beta})_{v_m v_n} = \langle v_m|\alpha'_{\alpha\beta}(\dot{Q})|v_n\rangle = -(\alpha'_{\beta\alpha})_{v_m v_n}, \tag{2.8.37b}$$

and are pure symmetric and pure antisymmetric, respectively.

Similar developments are possible for the vibrational transition optical activity tensors leading to, for example,

$$(G'_{\alpha\beta})_{v_m v_n} = \langle v_m|G'_{\alpha\beta}(Q)|v_n\rangle, \tag{2.8.38a}$$

$$(A_{\alpha,\beta\gamma})_{v_m v_n} = \langle v_m|A_{\alpha,\beta\gamma}(Q)|v_n\rangle = (A_{\alpha,\gamma\beta})_{v_m v_n}. \tag{2.8.38b}$$

As in the case of Rayleigh scattering, there is no meaningful separation of the vibrational transition optical activity tensors into symmetric and antisymmetric parts at transparent frequencies in Placzek's approximation.

2.8.4 Vibronic interactions: the Herzberg–Teller approximation

The adiabatic wavefunction and energy (2.8.24) were arrived at through the neglect of coupling between electronic and nuclear motions. This coupling can have important consequences: of relevance here is vibrational–electronic ('vibronic') coupling, which is responsible for certain vibrational Raman transitions, and also gives rise to some of the vibrational structure of electronic absorption and circular dichroism bands. Vibronic coupling can be taken into account at various levels of sophistication, as discussed in reviews such as Longuet–Higgins (1961), Englman (1972), Özkan and Goodman (1979) and Ballhausen (1979).

Here we shall be content with the grossest approximation, due to Herzberg and Teller (1933), which starts by invoking the *crude adiabatic approximation* in which the complete molecular wavefunction is written

$$\Psi_{ev}(r, Q) = \psi_e(r, Q_0)\psi_{ev}^{(CA)}(Q). \tag{2.8.39}$$

Apart from our neglect of the rotational part and the particularization to normal vibrational coordinates, this differs from (2.8.19) in that the electronic factors $\psi_e(r, Q_0)$ now apply to the equilibrium nuclear configuration Q_0. The Q-dependence is now contained solely in the vibrational functions $\psi_{ev}^{(CA)}(Q)$, which are not quite the same as the $\psi_{ev}(Q)$, being obtained as solutions of

$$[T(Q) + \langle\psi_e(r, Q_0)|H_e(r, Q)|\psi_e(r, Q_0)\rangle - w_e(Q_0) + V(Q)]\psi_{ev}^{(CA)}(Q)$$
$$= w_{ev}^{(CA)}\psi_{ev}^{(CA)}(Q), \tag{2.8.40a}$$

rather than (2.8.21), where

$$H_e(r, Q) = T(r) + V(r, Q) \tag{2.8.41}$$

is the electronic Hamiltonian whose functional dependence on Q may be represented by an expansion in the nuclear displacements around Q_0:

$$H_e(Q) = (H_e)_0 + \sum_p \left(\frac{\partial H_e}{\partial Q_p}\right)_0 Q_p + \tfrac{1}{2}\sum_{p,q}\left(\frac{\partial^2 H_e}{\partial Q_p \partial Q_q}\right)_0 Q_p Q_q + \cdots. \tag{2.8.42}$$

Notice that, since the total vibronic energy is $W_{ev} = w_e(Q_0) + w_{ev}$, we can write (2.8.40a) more simply as

$$[T(Q) + \langle\psi_e(r, Q_0)|H_e(r, Q)|\psi_e(r, Q_0)\rangle + V(Q)]\psi_{ev}^{(CA)}(Q)$$
$$= W_{ev}^{(CA)}\psi_{ev}^{(CA)}(Q). \tag{2.8.40b}$$

Although the numerical aspects of the crude adiabatic approximation are not even qualitatively correct (Özkan and Goodman, 1979), we shall persist with the Herzberg–Teller method because it provides a simple framework on which to hang symmetry arguments, which is our main concern.

The perturbation that mixes the electronic states is taken to be the second and higher terms in the expansion (2.8.42) of the electronic Hamiltonian in the normal coordinates. Taking as our electronic basis the set of electronic functions $\psi_e(r, Q_0)$ at the equilibrium nuclear configuration, the Q-dependence of the electronic functions is considered to arise from the vibrational perturbation mixing the $\psi_e(r, Q_0)$; that is

$$\psi_{e_j}(Q) = \psi_{e_j}'(Q_0)$$
$$= \psi_{e_j}(Q_0) + \sum_{e_k \neq e_j} \frac{\langle\psi_{e_k}(Q_0)|\sum_p (\partial H_e/\partial Q_p)_0 Q_p|\psi_{e_j}(Q_0)\rangle}{w_{e_j} - w_{e_k}} \psi_{e_k}(Q_0) + \cdots. \tag{2.8.43}$$

In the case that the nuclear motion mixes orbitally-degenerate electronic states (Jahn–Teller effect) or near-degenerate states (pseudo Jahn–Teller effect) the electronic and nuclear motions are closely intertwined and the Herzberg–Teller approach must be reformulated. A general vibronic wavefunction for a doubly-degenerate electronic state, for example, is written

$$\Psi(r, Q) = \psi_{e_1}(r, Q_0)\psi_{e_1v}^{(CA)}(Q) + \psi_{e_2}(r, Q_0)\psi_{e_2v}^{(CA)}(Q), \qquad (2.8.44)$$

where $\psi_{e_1}(r, Q_0)$ and $\psi_{e_2}(r, Q_0)$ are the two electronic wavefunctions which are degenerate at some preselected symmetrical configuration Q_0, and $\psi_{e_1v}^{(CA)}(Q)$ and $\psi_{e_2v}^{(CA)}(Q)$ are vibrational wavefunctions, or component vibrational amplitudes, associated with the two electronic states. In fact, Q_0 is not necessarily an equilibrium configuration, but must have sufficient symmetry for the degeneracy to be non-accidental. Thus (2.8.44) allows heavy mixing of $\psi_{e_1}(r, Q_0)$ and $\psi_{e_2}(r, Q_0)$, but ignores mixing with all other electronic states.

The pair of degenerate electronic wavefunctions are solutions of the Schröodinger equation

$$H_e(r, Q_0)\psi_e(r, Q_0) = w_e(Q_0)\psi_e(r, Q_0), \qquad (2.8.45)$$

where $H_e(r, Q_0)$ is the electronic Hamiltonian at the symmetric configuration Q_0, being the first term in the expansion (2.8.42). Taking the second and higher terms in (2.8.42) to be the perturbation that mixes the degenerate wavefunctions, degenerate perturbation theory yields the following secular determinant:

$$\begin{vmatrix} H_{11} - w_e(Q) & H_{12} \\ H_{21} & H_{22} - w_e(Q) \end{vmatrix} = 0, \qquad (2.8.46)$$

where $H_{ij} = \langle \psi_{e_i}(r, Q_0)|H_e(r, Q)|\psi_{e_j}(r, Q)\rangle$. If there are nonzero off-diagonal matrix elements linear in Q_p, that is arising from the operator $\Sigma_p(\partial H_e/\partial Q_p)_0 Q_p$ in the expansion of $H_e(r, Q)$, the electronic degeneracy will be lifted so that Q_0 is not an equilibrium configuration. The solution of (2.8.46) to determine the electronic potential energy surfaces constitutes the *static* aspect of the Jahn–Teller effect. In the *dynamic* aspect the vibrational wavefunctions, here regarded as amplitudes of the degenerate electronic wavefunctions in (2.8.44), are determined by the coupled equations

$$\begin{pmatrix} T(Q)+H_{11}-W_{ev}^{(CA)}+V(Q) & H_{12} \\ H_{21} & T(Q)+H_{22}-W_{ev}^{(CA)}+V(Q) \end{pmatrix} \begin{pmatrix} \psi_{e_1v}^{(CA)}(Q) \\ \psi_{e_2v}^{(CA)}(Q) \end{pmatrix} = 0,$$

$$(2.8.47)$$

which can be viewed as a generalization of (2.8.40b).

3

Molecular scattering of polarized light

Edward VII to Lord Rayleigh and Augustine Birrell at a palace party:
'Well, Lord Rayleigh, discovering something I suppose? You know, he's
always at it.'

(Diana Cooper in her autobiography, reporting conversations with Augustine Birrell)

3.1 Introduction

This chapter constitutes the heart of the book. In it, the theoretical material developed in Chapter 2 is used to calculate explicit expressions, in terms of molecular property tensors, for the polarization and intensity of light scattered into any direction from a collection of molecules. These expressions, which are applied in detail in subsequent chapters, therefore contain the basic equations for all of the optical activity phenomena under discussion.

Polarization phenomena have always been an important part of light scattering studies. For example, Tyndall's early investigations with aerosols (1869) showed that linear polarization was an important feature of light scattered at right angles, and he pointed out that (quoted by Kerker, 1969) 'The blue colour of the sky, and the polarization of skylight . . . constitute, in the opinion of our most eminent authorities, the two great standing enigmas of meteorology.' This enigma was resolved by Lord Rayleigh (1871) who showed that the intensity of light scattered by a uniform sphere much smaller than the wavelength is proportional to $1/\lambda^4$, the component scattered at right angles being completely linearly polarized perpendicular to the scattering plane, indicating that the colour and polarization of skylight originates in the scattering of sunlight by air molecules. In fact imperfections *are* observed in the polarization of skylight scattered at right angles, and these were ascribed at first to factors such as dust and multiple scattering, but since imperfections are also observed in dust-free molecular gases, it was realized that departure from spherical symmetry in the optical properties of the molecules is also an important factor.

3.2 Molecular scattering of light

When an electromagnetic wave encounters an obstacle, bound charges are set into oscillation and secondary waves are scattered in all directions. Within a medium, the 'obstacles' responsible for light scattering can be gross inclusions of foreign matter such as impurities in crystals, droplets of water or dust particles in the atmosphere, and colloidal matter suspended in liquids. But light scattering also occurs in transparent materials completely free of contaminants on account of in-homogeneities at the molecular level. As indicated previously, molecular scattering in which the frequency is essentially unchanged is known as *Rayleigh scattering*, whereas molecular scattering with well defined frequency shifts is known as *Raman scattering*. These frequency shifts in scattered light were first observed by Raman and Krishnan in 1928, and shortly afterwards, quite independently, by Landsberg and Mandelstam (1928). In the Russian literature, Raman scattering is often referred to as *combination scattering*.

It should be realized that a perfectly transparent and homogeneous medium does not scatter light. Consider a plane wave propagating in a medium in which identical numbers of molecules of one type are found in equivalent volume elements at any instant. If the dimension of each volume element is small compared with the wavelength of the incident light, the waves scattered by different parts of any one volume element have the same phase. The part of the wave scattered from a particular volume element V at an angle ξ to the direction of propagation of the incident beam can be assigned a particular amplitude and phase. Since the medium is completely uniform, a second volume V' at a distance l from V along the incident plane wavefront can always be found which radiates a wave in the same direction, with the same amplitude, but with opposite phase, as illustrated in Fig. 3.1. The condition for phase opposition is that the two scattered waves have a path length difference of $\lambda/2$, so that

$$l = \frac{\lambda}{2\sin\xi}. \tag{3.2.1}$$

Consequently, for any ξ except $\xi = 0$, one can find within the plane wavefront two volumes radiating waves which destructively interfere, so there is no light scattered away from the forward direction in a perfectly homogeneous medium. Only forward scattering survives, and gives rise to refraction through interference with the unscattered component of the incident wave.

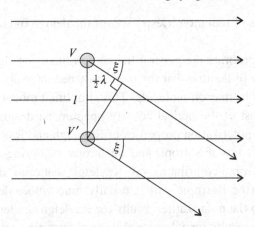

Fig. 3.1 Destructive interference of waves scattered from volume elements V and V'.

Rayleigh scattering in pure transparent samples arises because no medium can be perfectly homogeneous. For example, in the limiting case of a rarified gas, the molecules execute disordered thermal motions with a mean free path length much greater than the wavelength of the light. The phase difference between the waves scattered by any one pair of molecules is as often positive as negative, so on the average destructive interference occurs half the time and constructive interference the rest of the time, and the total scattered intensity is the sum of the individual scattered intensities. This conclusion is due to Lord Rayleigh. A more general theory, applicable to dense media, was developed by Smoluchowski and Einstein, and explains the origin of light scattering in terms of optical inhomogeneities arising from local density fluctuations: the number of scatterers in the volume element V is constant only on the average, so the destructive interference with waves in phase opposition from a second volume V' is not complete at any instant. Einstein's equation for the scattered intensity reduces to that of Rayleigh for an ideal gas. The total scattering power per molecule decreases with increasing density, and in a liquid it can be an order of magnitude smaller than in a gas. The fluctuation theory was extended by Cabannes to include light scattering by anisotropy fluctuations produced by irregularities in the orientations of the molecules. For theories of light scattering in dense media, we refer to works such as Bhagavantam (1942), Landau and Lifshitz (1960) and Fabelinskii (1968).

On the other hand, vibrational Raman scattering is completely incoherent (except in the case of stimulated Raman scattering in intense laser beams), and the total vibrational Raman scattered intensity from N molecules is simply N times that from a single molecule at all sample densities. This is because the phase of a Raman scattered wave depends on the phase of the molecular vibration, which, to a good approximation, varies arbitrarily from molecule to molecule, so the molecules act as independent sources of radiation irrespective of the degree of correlation between their positions.

We shall not incorporate the general fluctuation theory of light scattering into our treatment, but will assume that the total scattered intensity is the sum of the scattered intensities from each molecule. This simplified model actually provides valid results for most of the optical activity phenomena discussed in this book. Expressions for the polarization properties of Rayleigh and Raman scattered light involve quotients of sums of isotropic and anisotropic scattering contributions. The isotropic and anisotropic contributions to Rayleigh scattering depend differently on sample density (the isotropic part is usually much more dependent than the anisotropic part), so the polarization results for Rayleigh scattering apply only to ideal gases. But the results for vibrational Raman scattering are true for all samples since both isotropic and anisotropic vibrational Raman scattering are always incoherent.

That part of the scattered light in the forward direction is fully coherent at all sample densities, and there is complete constructive interference from all of the scatterers. Consequently, the polarization results for birefringence effects such as optical rotation, derived from a model involving interference between the forward-scattered and unscattered components of the incident light wave, are basically correct at all sample densities. We need only correct for the modification of the optical fields by the internal fields of the sample using a relation such as

$$\mathbf{E}' = \tfrac{1}{3}(n^2 + 2)\mathbf{E}, \qquad (3.2.2)$$

where $\tfrac{1}{3}(n^2 + 2)$ is the *Lorentz factor*.

3.3 Radiation by induced oscillating molecular multipole moments

We consider the origin of scattered light to be the characteristic radiation fields generated by the oscillating electric and magnetic multipole moments induced in a molecule by the electromagnetic fields of the incident light wave. Equation (2.4.43a) gives the electric field radiated by time-dependent multipole moments at distances large compared with the wavelength. The scattered electric field detected in the wave zone at a point d at a distance R from the molecular origin is therefore the real part of

$$\tilde{E}_\alpha^{\mathrm{d}} = \frac{\omega^2 \mu_0}{4\pi R} \mathrm{e}^{\mathrm{i}\omega(R/c-t)} \Big[\big(\tilde{\mu}_\alpha^{(0)} - n_\alpha^{\mathrm{d}} n_\beta^{\mathrm{d}} \tilde{\mu}_\beta^{(0)}\big)$$
$$- \frac{1}{c} \varepsilon_{\alpha\beta\gamma} n_\beta^{\mathrm{d}} \tilde{m}_\gamma^{(0)} - \frac{\mathrm{i}\omega}{3c} \big(n_\beta^{\mathrm{d}} \tilde{\Theta}_{\alpha\beta}^{(0)} - n_\alpha^{\mathrm{d}} n_\beta^{\mathrm{d}} n_\gamma^{\mathrm{d}} \tilde{\Theta}_{\beta\gamma}^{(0)}\big) + \cdots \Big], \qquad (3.3.1)$$

where $\mathbf{n}^d$ is the propagation vector in the direction of the detected wave. The terms in $n_\alpha^d n_\beta^d \tilde{\mu}_\beta^{(0)}$ and $n_\alpha^d n_\beta^d n_\gamma^d \tilde{\Theta}_{\beta\gamma}^{(0)}$ ensure that the wave is transverse. Since this wave is travelling in the free space between the molecules, μ and ϵ are taken as unity and $\mathbf{n}^d$ is now a unit propagation vector. The complex induced moments are given in terms of the dynamic molecular property tensors by (2.6.43); these moments are best written in terms of the electric vector of the incident plane wave light beam, so the required amplitudes are

$$\tilde{\mu}_\alpha^{(0)} = \left(\tilde{\alpha}_{\alpha\beta} + \frac{i\omega}{3c} n_\gamma^i \tilde{A}_{\alpha,\gamma\beta} + \frac{1}{c} \varepsilon_{\gamma\delta\beta} n_\delta^i \tilde{G}_{\alpha\gamma} + \cdots \right) \tilde{E}_\beta^{(0)}, \qquad (3.3.2a)$$

$$\tilde{\Theta}_{\alpha\beta}^{(0)} = (\tilde{\mathscr{A}}_{\gamma,\alpha\beta} + \cdots) \tilde{E}_\gamma^{(0)}, \qquad (3.3.2b)$$

$$\tilde{m}_\alpha^{(0)} = (\tilde{\mathscr{G}}_{\alpha\beta} + \cdots) \tilde{E}_\beta^{(0)}, \qquad (3.3.2c)$$

where $\mathbf{n}^i$ is the propagation vector of the incident wave. Equation (3.3.1) can now be written

$$\tilde{E}_\alpha^d = \frac{\omega^2 \mu_0}{4\pi R} e^{i\omega(R/c-t)} \tilde{a}_{\alpha\beta} \tilde{E}_\beta^{(0)}, \qquad (3.3.3)$$

where $\tilde{a}_{\alpha\beta}$ is a scattering tensor for particular incident and scattered directions given by the unit vectors $\mathbf{n}^i$ and $\mathbf{n}^d$:

$$\tilde{a}_{\alpha\beta} = \tilde{\alpha}_{\alpha\beta} + \frac{i\omega}{3c} \left(n_\gamma^i \tilde{A}_{\alpha,\gamma\beta} - n_\gamma^d \tilde{\mathscr{A}}_{\beta,\gamma\alpha} \right)$$

$$+ \frac{1}{c} \left(\varepsilon_{\gamma\delta\beta} n_\delta^i \tilde{G}_{\alpha\gamma} + \varepsilon_{\gamma\delta\alpha} n_\delta^d \tilde{\mathscr{G}}_{\gamma\beta} \right) - n_\alpha^d n_\gamma^d \tilde{\alpha}_{\gamma\beta} - \frac{i\omega}{3c} n_\alpha^d n_\gamma^d \left(n_\delta^i \tilde{A}_{\gamma,\delta\beta} - n_\delta^d \tilde{\mathscr{A}}_{\beta,\delta\gamma} \right)$$

$$- \frac{1}{c} n_\alpha^d n_\gamma^d \varepsilon_{\epsilon\delta\beta} n_\delta^i \tilde{\mathscr{G}}_{\gamma\epsilon} + \cdots \qquad (3.3.4)$$

The last three terms are required for calculations of electric field gradient-induced birefringence (Section 3.4.5) and related phenomena, and also for the dependence of Rayleigh and Raman scattering phenomena on the finite cone of collection.

A penetrating discussion of the realm of validity of this 'local multipole' approximation to spatial dispersion in molecular light scattering has been given by Baranova and Zel'dovich (1979b).

3.4 Polarization phenomena in transmitted light

3.4.1 Refraction as a consequence of light scattering

The polarization changes in a light beam passing through a transparent medium are usually accounted for in terms of circular and linear birefringence which refer, respectively, to different refractive indices for right- and left-circularly polarized light and light linearly polarized in two perpendicular directions. Additional polarization changes can occur in an attenuating medium, and are usually described in terms of circular and linear dichroism, which refer to different absorption coefficients for the corresponding polarized light.

Lord Rayleigh pointed out that the refraction of light is a consequence of light scattering. Modern treatments are given in the books by van de Hulst (1957), Newton (1966) and Jenkins and White (1976). The individual molecules scatter a small part of the incident light, and the forward parts of the resulting spherical waves combine and interfere with the primary wave, resulting in a phase change which is equivalent to an alteration of the wave velocity. We call this process *refringent scattering*. Very little of the nonforward scattered light is actually lost from the transmitted wave if the medium is optically homogeneous on account of destructive interference: in contrast, waves scattered into the forward direction from any point in the medium interfere constructively. It is therefore natural to formulate a molecular theory of 'refringent polarization effects' directly from Lord Rayleigh's scattering model, without introducing an index of refraction. Kauzmann (1957) was the first to present such a scattering theory of optical rotation, but this was restricted to small angles of rotation at transparent wavelengths. We consider a light beam of arbitrary azimuth, ellipticity and degree of polarization incident on an infinitesimal lamina of a dilute molecular medium which may be oriented and absorbing. Expressions in terms of components of dynamic molecular property tensors are derived for the infinitesimal changes in azimuth, ellipticity, degree of polarization and intensity of the emergent light beam. Integration of these infinitesimal changes over a finite optical path provides the standard equations for the finite polarization and intensity changes in well known phenomena such as natural and magnetic optical rotation and the Kerr and Cotton-Mouton effects, together with some newer effects such as magnetochiral birefringence and dichroism

The conventional theories of refringent polarization effects start from the circular and linear birefringence and dichroism description. The transition to a molecular theory is made by relating the refractive index to the bulk electric polarization and magnetization of the medium, which are related in turn to an appropriate sum of the electric and magnetic multipole moments induced in individual molecules by the light wave. Although such use of an index of refraction has proved invaluable for deriving expressions for refringent polarization effects, it can obscure some of the fundamental processes responsible. The infinitesimal scattering theory automatically includes the general case of circular and linear birefringence and dichroism existing simultaneously, together with changes in the degree of polarization, all of which can be interdependent. The refractive index theories can accommodate this general stituation within the Mueller or Jones matrix techniques. The Mueller calculus (Mueller, 1948) describes the effects of particular optical elements on a polarized light beam characterized by the four Stokes parameters: the properties of an optical element are represented by a real four-by-four matrix, the elements of which are functions of refractive index components, which multiplies the input real Stokes four-vector; and by applying successively matrices corresponding to infinitesimal optical elements, the effect of a medium showing simultaneous circular and linear birefringence and dichroism can be calculated. The Jones calculus (Jones, 1948) is similar, but involves complex two-by-two matrices operating on the complex Jones two-vector. Since the Jones vector can only describe a pure-polarized beam, whereas the Stokes vector can accomodate partial polarization, only the Mueller calculus can incorporate changes in the degree of polarization. For further discussion of the Mueller and Jones methods, we refer to Ramachandran and Ramaseshan (1961).

It should be mentioned that there is a procedure intermediate between the basic scattering theory used in this book and the refractive index theory outlined above. Instead of calculating the refractive indices through the bulk electric polarization and magnetization, the refractive indices for linearly and circularly polarized light can be calculated using Lord Rayleigh's scattering model, and the results used in the Mueller or Jones matrices.

3.4.2 Refringent scattering of polarized light

Consider a quasi-monochromatic light wave propagating along z and incident on an infinitely wide lamina in the xy plane in a dilute molecular medium, as shown in Fig. 3.2. The thickness of the lamina is infinitesimal relative to the wavelength of the light. If only a small fraction of the wave is scattered, the disturbance reaching a point f at R_0 a large distance from the lamina in the forward direction is essentially

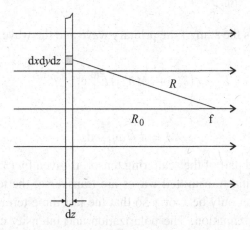

Fig. 3.2 Geometry for forward scattering by a thin lamina.

the original wave plus a small contribution due to scattering by the molecules in the lamina. From (3.3.3) the electric field of the scattered wave at f from a volume element $dxdydz$ at $(x, y, 0)$ in the lamina is

$$\tilde{E}_\alpha^f = \frac{N\omega^2\mu_0\,dxdydz}{4\pi R}e^{i\omega(R/c-t)}\tilde{a}_{\alpha\beta}\tilde{E}_\beta^{(0)},\tag{3.4.1}$$

where N is the number density of molecules. Only molecules within the base of a narrow cone with apex at f will contribute effectively to forward scattering, since waves scattered from molecules outside this area tend to interfere destructively at f. This means that we may calculate the total scattered electric vector at f by integrating (3.4.1) over the infinite surface of the lamina since only those molecules close to the axis of the cone will contribute coherently. The propagation vector in the direction of the detected wave may be written

$$\mathbf{n}^d = (\mathbf{n}^d \cdot \mathbf{i})\mathbf{i} + (\mathbf{n}^d \cdot \mathbf{j})\mathbf{j} + (\mathbf{n}^d \cdot \mathbf{k})\mathbf{k} = -\frac{x}{R}\mathbf{i} - \frac{y}{R}\mathbf{j} + \frac{R_0}{R}\mathbf{k}\tag{3.4.2a}$$

which, for $R_0 \gg x$ or y, may be approximated by

$$\mathbf{n}^d \approx -\frac{x}{R_0}\mathbf{i} - \frac{y}{R_0}\mathbf{j} + \mathbf{k}.\tag{3.4.2b}$$

For simplicity we shall consider explicitly only the contributions to $\mathbf{n}^d$ depending on the third term, $\mathbf{k}$, in (3.4.2b). Writing

$$R = \left[R_0^2 + (x^2 + y^2)\right]^{1/2} \approx R_0 + \frac{1}{2R_0}(x^2 + y^2),$$

the required integral is

$$\frac{1}{R_0}\int_{-\infty}^{\infty}\int_{-\infty}^{\infty} dxdy\,e^{i\omega(x^2+y^2)/2R_0c} = \frac{i2\pi c}{\omega}.\tag{3.4.3}$$

The total wave at f is the sum of the primary wave and the wave scattered from the lamina:

$$\tilde{E}_\alpha^f = \left(\delta_{\alpha\beta} + iM\tilde{a}_{\alpha\beta}^f\right)\tilde{E}_\beta^{(0)}e^{i\omega(R_0/c-t)},\tag{3.4.4a}$$

where

$$M = \tfrac{1}{2}N\omega\mu_0cdz\tag{3.4.4b}$$

and $\tilde{a}_{\alpha\beta}^f$, the forward part of the scattering tensor, is given by (3.3.4) with $\mathbf{n}^d = \mathbf{n}^i$. Since the incident and transmitted waves are transverse, the tensor subscripts α and β in (3.4.4a) can only be x or y so that the last three terms of (3.3.4) vanish in the present approximation. The polarization and intensity changes associated with refringent scattering arise from cross products of the first and second terms in

(3.4.4*a*). Since i in the second term may be replaced by $\exp(i\pi/2)$, (3.4.4*a*) reveals that the net plane wave front in the forward direction obtained by summing the scattered wavelets from molecules in the lamina is shifted in phase by $\pi/2$ relative to the transmitted wave. This phase shift is crucial in generating the polarization changes characteristic of linear birefringence, optical rotation, etc. developed below.

First, however, it is useful to derive an expression for the complex refractive index $\tilde{n} = n + in'$ of the dilute molecular medium from this refringent scattering formalism. Taking $z = 0$ at the front face of the thin lamina, we can take account of the retardation resulting from propagation of the light wave through the lamina by writing the electric field at f, for linearly polarized light (along x, say), in the form

$$\tilde{E}_x^f = \tilde{E}_x^{(0)} e^{i\omega\{[\tilde{n}dz + (R_0 - dz)]/c - t\}}$$
$$= \tilde{E}_x^{(0)} e^{i\omega(\tilde{n}-1)dz/c} e^{i\omega(R_0/c - t)}. \tag{3.4.5}$$

Since $\exp(x) = 1 + x + \cdots$, (3.4.4*a*) can be rewritten as

$$\tilde{E}_x^f = e^{iM\tilde{a}_{xx}^f} \tilde{E}_x^{(0)} e^{i\omega(R_0/c - t)}. \tag{3.4.6}$$

Comparing this with (3.4.5), the complex refractive index is found to be

$$\tilde{n} \approx 1 + \tfrac{1}{2} N\mu_0 c^2 \tilde{a}_{xx}^f. \tag{3.4.7}$$

The refractive index n and absorption index n' are then given by the dispersive and absorptive parts, respectively, of the property tensors within $\tilde{a}_{xx}^f$:

$$n \approx 1 + \tfrac{1}{2} N\mu_0 c^2 \tilde{a}_{xx}^f(f), \tag{3.4.8a}$$
$$n' \approx \tfrac{1}{2} N\mu_0 c^2 \tilde{a}_{xx}^f(g). \tag{3.4.8b}$$

The Stokes parameters of the transmitted wave can be found in terms of the scattering tensor components and the Stokes parameters of the incident wave by substituting (3.4.4) into (2.3.6). Since very little scattering occurs, $Ma \ll 1$, and we can neglect terms in $M^2 a^2$. For example, the first Stokes parameter is

$$\begin{aligned}
S_0^f &= \tilde{E}_x^f \tilde{E}_x^{f*} + \tilde{E}_y^f \tilde{E}_y^{f*} \\
&= \big[(\delta_{x\beta} + iM\tilde{a}_{x\beta}^f)(\delta_{x\gamma} - iM\tilde{a}_{x\gamma}^{f*}) \\
&\quad + (\delta_{y\beta} + iM\tilde{a}_{y\beta}^f)(\delta_{y\gamma} - iM\tilde{a}_{yy}^{f*})\big] \tilde{E}_\beta \tilde{E}_\gamma \\
&\approx \tilde{E}_x \tilde{E}_x^* + \tilde{E}_y \tilde{E}_y^* \\
&\quad - 2M\,\mathrm{Im}\big(\tilde{a}_{xx}^f \tilde{E}_x \tilde{E}_x^* + \tilde{a}_{yy}^f \tilde{E}_y \tilde{E}_y^* + \tilde{a}_{xy}^f \tilde{E}_y \tilde{E}_x^* + \tilde{a}_{yx}^f \tilde{E}_x \tilde{E}_y^*\big) \\
&\approx S_0 - M\,\mathrm{Im}\big[(\tilde{a}_{xx}^f + \tilde{a}_{yy}^f)S_0 + (\tilde{a}_{xx}^f - \tilde{a}_{yy}^f)S_1 \\
&\quad - (\tilde{a}_{xy}^f + \tilde{a}_{yx}^f)S_2 - i(\tilde{a}_{xy}^f - \tilde{a}_{yx}^f)S_3\big]. \tag{3.4.9a}
\end{aligned}$$

The others are found to be

$$S_1^f = \tilde{E}_x^f \tilde{E}_x^{f*} - \tilde{E}_y^f \tilde{E}_y^{f*}$$
$$\approx S_1 - M \operatorname{Im}\big[(\tilde{a}_{xx}^f - \tilde{a}_{yy}^f)S_0 + (\tilde{a}_{xx}^f + \tilde{a}_{yy}^f)S_1$$
$$- (\tilde{a}_{xy}^f - \tilde{a}_{yx}^f)S_2 - \mathrm{i}(\tilde{a}_{xy}^f + \tilde{a}_{yx}^f)S_3\big], \qquad (3.4.9b)$$

$$S_2^f = -\big(\tilde{E}_x^f \tilde{E}_y^{f*} + \tilde{E}_y^f \tilde{E}_x^{f*}\big)$$
$$\approx S_2 + M \operatorname{Im}\big[(\tilde{a}_{xy}^f + \tilde{a}_{yx}^f)S_0 - (\tilde{a}_{xy}^f - \tilde{a}_{yx}^f)S_1$$
$$- (\tilde{a}_{xx}^f + \tilde{a}_{yy}^f)S_2 + \mathrm{i}(\tilde{a}_{xx}^f - \tilde{a}_{yy}^f)S_3\big], \qquad (3.4.9c)$$

$$S_3^f = -\mathrm{i}\big(\tilde{E}_x^f \tilde{E}_y^{f*} - \tilde{E}_y^f \tilde{E}_x^{f*}\big)$$
$$\approx S_3 + M \operatorname{Re}\big[(\tilde{a}_{xy}^f - \tilde{a}_{yx}^f)S_0 - (\tilde{a}_{xy}^f + \tilde{a}_{yx}^f)S_1$$
$$- (\tilde{a}_{xx}^f - \tilde{a}_{yy}^f)S_2 + \mathrm{i}(\tilde{a}_{xx}^f + \tilde{a}_{yy}^f)S_3\big]. \qquad (3.4.9d)$$

The intensity, azimuth, ellipticity and degree of polarization of the transmitted wave are now found by using (3.4.9) in (2.3.9) and (2.3.18e). The corresponding changes are effectively infinitesimal so we can write $I^f - I = \mathrm{d}I$, etc. The changes as a function of the azimuth, ellipticity and degree of polarization of the incident wave are found to be

$$\mathrm{d}I \approx I M\big[\operatorname{Im}(\tilde{a}_{xx}^f + \tilde{a}_{yy}^f) + \operatorname{Im}(\tilde{a}_{xx}^f - \tilde{a}_{yy}^f)P\cos 2\eta \cos 2\theta$$
$$- \operatorname{Im}(\tilde{a}_{xy}^f + \tilde{a}_{yx}^f)P\cos 2\eta \sin 2\theta - \operatorname{Re}(\tilde{a}_{xy}^f - \tilde{a}_{yx}^f)P\sin 2\eta\big], \quad (3.4.10a)$$

$$\mathrm{d}\theta \approx \tfrac{1}{2}M\big\{\big[\operatorname{Re}(\tilde{a}_{xx}^f - \tilde{a}_{yy}^f)\cos 2\theta - \operatorname{Re}(\tilde{a}_{xy}^f + \tilde{a}_{yx}^f)\sin 2\theta\big]\tan 2\eta$$
$$+ \big[\operatorname{Im}(\tilde{a}_{xx}^f - \tilde{a}_{yy}^f)\sin 2\theta + \operatorname{Im}(\tilde{a}_{xy}^f + \tilde{a}_{yx}^f)\cos 2\theta\big]/(P\cos 2\eta)$$
$$- \operatorname{Im}(\tilde{a}_{xy}^f - \tilde{a}_{yx}^f)\big\}, \qquad (3.4.10b)$$

$$\mathrm{d}\eta \approx \tfrac{1}{2}M\big\{-\operatorname{Re}(\tilde{a}_{xx}^f - \tilde{a}_{yy}^f)\sin 2\theta - \operatorname{Re}(\tilde{a}_{xy}^f + \tilde{a}_{yx}^f)\cos 2\theta$$
$$+ \big[\operatorname{Im}(\tilde{a}_{xx}^f - \tilde{a}_{yy}^f)\cos 2\theta - \operatorname{Im}(\tilde{a}_{xy}^f + \tilde{a}_{yx}^f)\sin 2\theta\big]\sin 2\eta/P$$
$$+ \operatorname{Re}(\tilde{a}_{xy}^f - \tilde{a}_{yx}^f)\cos 2\eta/P\big\}, \qquad (3.4.10c)$$

$$\mathrm{d}P \approx M(P^2 - 1)\big\{\big[\operatorname{Im}(\tilde{a}_{xx}^f - \tilde{a}_{yy}^f)\cos 2\theta - \operatorname{Im}(\tilde{a}_{xy}^f + \tilde{a}_{yx}^f)\sin 2\theta\big]\cos 2\eta$$
$$- \operatorname{Re}(\tilde{a}_{xy}^f - \tilde{a}_{yx}^f)\sin 2\eta\big\}. \qquad (3.4.10d)$$

In obtaining the azimuth and ellipticity changes we used the relations

$$\tan 2\theta^f - \tan 2\theta \approx 2\mathrm{d}\theta/\cos^2 2\theta,$$
$$\tan 2\eta^f - \tan 2\eta \approx 2\mathrm{d}\eta/\cos^2 2\eta.$$

In developing these equations for the refringent intensity and polarization changes in terms of explicit dynamic molecular property tensors, it is convenient to group together appropriate components of the optical activity tensors $\tilde{G}_{\alpha\beta}$ and

$\tilde{A}_{\alpha,\beta\gamma}$ into a single third-rank tensor defined by

$$\tilde{\xi}_{\alpha\beta\gamma} = \frac{1}{c}\left[\tfrac{1}{3}i\omega(\tilde{A}_{\alpha,\beta\gamma} - \tilde{\mathscr{A}}_{\beta,\alpha\gamma}) + \varepsilon_{\delta\gamma\beta}\tilde{G}_{\alpha\delta} + \varepsilon_{\delta\gamma\alpha}\tilde{\mathscr{G}}_{\delta\beta}\right]. \qquad (3.4.11)$$

Like $\tilde{\alpha}_{\alpha\beta}$, $\tilde{\xi}_{\alpha\beta\gamma}$ can be decomposed into symmetric and antisymmetric parts with respect to the first two suffixes:

$$\tilde{\xi}_{\alpha\beta\gamma} = \xi_{\alpha\beta\gamma} - i\xi'_{\alpha\beta\gamma}, \qquad (3.4.12)$$

where

$$\xi_{\alpha\beta\gamma} = \frac{1}{c}\left[\tfrac{1}{3}\omega(A'_{\alpha,\beta\gamma} + A'_{\beta,\alpha\gamma}) + \varepsilon_{\delta\gamma\alpha}G_{\beta\delta} + \varepsilon_{\delta\gamma\beta}G_{\alpha\delta}\right], \qquad (3.4.13a)$$

$$\xi'_{\alpha\beta\gamma} = -\frac{1}{c}\left[\tfrac{1}{3}\omega(A_{\alpha,\beta\gamma} - A_{\beta,\alpha\gamma}) + \varepsilon_{\delta\gamma\alpha}G'_{\beta\delta} - \varepsilon_{\delta\gamma\beta}G'_{\alpha\delta}\right]. \qquad (3.4.13b)$$

The forward part of the scattering tensor (3.3.4) in the present approximation now simplifies to

$$\tilde{a}^{\mathrm{f}}_{\alpha\beta} = \tilde{\alpha}_{\alpha\beta} + \tilde{\xi}_{\alpha\beta\gamma}n_{\gamma} + \cdots, \qquad (3.4.14)$$

where **n** is the unit vector in the propagation direction of the incident light beam.

Using (2.6.42) to write these property tensors in terms of dispersive and absorptive parts, we have

$$\mathrm{Re}\,\tilde{a}^{\mathrm{f}}_{\alpha\beta} = \alpha_{\alpha\beta}(f) + \xi_{\alpha\beta\gamma}(f)n_{\gamma} + \alpha'_{\alpha\beta}(g) + \xi'_{\alpha\beta\gamma}(g)n_{\gamma} + \cdots, \qquad (3.4.15a)$$

$$\mathrm{Im}\,\tilde{a}^{\mathrm{f}}_{\alpha\beta} = -\alpha'_{\alpha\beta}(f) - \xi'_{\alpha\beta\gamma}(f)n_{\gamma} + \alpha_{\alpha\beta}(g) + \xi_{\alpha\beta\gamma}(g)n_{\gamma} + \cdots. \qquad (3.4.15b)$$

Using these results in (3.4.10), we obtain finally the following expressions for the rate of change of intensity, azimuth, ellipticity and degree of polarization of a quasi-monochromatic light wave on traversing a dilute optically active birefringent absorbing medium:

$$\begin{aligned}
\frac{dI}{dz} &\approx -\tfrac{1}{2}IN\omega\mu_0\, c\{\alpha_{xx}(g) + \alpha_{yy}(g) + \xi_{xxz}(g) + \xi_{yyz}(g) \\
&\quad + [(\alpha_{xx}(g) - \alpha_{yy}(g) + \xi_{xxz}(g) - \xi_{yyz}(g))\cos 2\theta \\
&\quad - 2(\alpha_{xy}(g) + \xi_{xyz}(g))\sin 2\theta]P\cos 2\eta \\
&\quad - 2(\alpha'_{xy}(g) + \xi'_{xyz}(g))P\sin 2\eta\}, \qquad (3.4.16a)
\end{aligned}$$

$$\begin{aligned}
\frac{d\theta}{dz} &\approx \tfrac{1}{4}N\omega\mu_0\, c\{2(\alpha'_{xy}(f) + \xi'_{xyz}(f)) \\
&\quad + [(\alpha_{xx}(f) - \alpha_{yy}(f) + \xi_{xxz}(f) - \xi_{yyz}(f))\cos 2\theta \\
&\quad - 2(\alpha_{xy}(f) + \xi_{xyz}(f))\sin 2\theta]\tan 2\eta \\
&\quad + [(\alpha_{xx}(g) - \alpha_{yy}(g) + \xi_{xxz}(g) - \xi_{yyz}(g))\sin 2\theta \\
&\quad + 2(\alpha_{xy}(g) + \xi_{xyz}(g))\cos 2\theta]/(P\cos 2\eta)\}, \qquad (3.4.16b)
\end{aligned}$$

$$\frac{\mathrm{d}\eta}{\mathrm{d}z} \approx \tfrac{1}{4} N\omega\mu_0\, c\{-(\alpha_{xx}(f) - \alpha_{yy}(f) + \zeta_{xxz}(f) - \zeta_{yyz}(f))\sin 2\theta$$

$$- 2(\alpha_{xy}(f) + \zeta_{xyz}(f))\cos 2\theta + 2(\alpha'_{xy}(g) + \zeta'_{xyz}(g))\cos 2\eta/P$$

$$+ [(\alpha_{xx}(g) - \alpha_{yy}(g) + \zeta_{xxz}(g) - \zeta_{yyz}(g))\cos 2\theta$$

$$- 2(\alpha_{xy}(g) + \zeta_{xyz}(g))\sin 2\theta]\sin 2\eta/P\}, \qquad (3.4.16c)$$

$$\frac{\mathrm{d}P}{\mathrm{d}z} \approx \tfrac{1}{2} N\omega\mu_0\, c(P^2 - 1)\{[(\alpha_{xx}(g) - \alpha_{yy}(g) + \zeta_{xxz}(g) - \zeta_{yyz}(g))\cos 2\theta$$

$$- 2(\alpha_{xy}(g) + \zeta_{xyz}(g))\sin 2\theta]\cos 2\eta$$

$$- 2(\alpha'_{xy}(g) + \zeta'_{xyz}(g))\sin 2\eta\}. \qquad (3.4.16d)$$

Notice that equation (3.4.16a) for the differential change in intensity contains only the absorptive (or antiHermitian) parts of the dynamic molecular property tensors, as required. Also equation (3.4.16d) for the differential change in the degree of polarization shows that if the incident beam is completely polarized, the transmitted beam is also completely polarized under all the circumstances relevant to this model: any change can only occur in the direction of an increase in the degree of polarization, and then only at absorbing frequencies.

We shall not apply these equations in detail to every one of the large number of phenomena which they embrace, but will use them to obtain the macroscopic polarization and intensity changes for the basic refringent optical activity phenomena, together with a few other related effects. The criteria for deciding whether or not a particular component of a particular property tensor can contribute to a certain polarization or intensity change are elaborated in detail in subsequent chapters, particularly Chapter 4 in which symmetry classifications are developed.

3.4.3 Simple absorption

The simplest application of these results is to an unpolarized light beam (or linearly polarized taking, for convenience, the azimuth $\theta = 0$) traversing a system of randomly oriented molecules that can support only components of the real polarizability $\alpha_{\alpha\beta}$. This would obtain in the case of a fluid composed of achiral molecules in the absence of applied magnetic fields. Using the unit vector average (4.2.48), we obtain the isotropic averages

$$\langle\alpha_{xx}\rangle = \alpha_{\alpha\beta}\langle i_\alpha i_\beta\rangle = \tfrac{1}{3}\alpha_{\alpha\alpha} = \langle\alpha_{yy}\rangle,$$

$$\langle\alpha_{xy}\rangle = \alpha_{\alpha\beta}\langle i_\alpha i_\beta\rangle = 0,$$

so that (3.4.16) reduce to

$$\frac{\mathrm{d}\theta}{\mathrm{d}z} = \frac{\mathrm{d}\eta}{\mathrm{d}z} = \frac{\mathrm{d}P}{\mathrm{d}z} \approx 0, \qquad (3.4.17a)$$

$$\frac{\mathrm{d}I}{\mathrm{d}z} \approx -\tfrac{1}{3}IN\omega\mu_0 c\alpha_{\alpha\alpha}(g). \tag{3.4.17b}$$

The only change is therefore a reduction in intensity due to absorption, being a function of that part of the polarizability tensor involving the absorption lineshape function g, in agreement with the conclusions at the end of Section 2.6.3. Integration over a finite path length l provides the following expression for the final attenuated intensity:

$$I_l \approx I_0 \, \mathrm{e}^{-\frac{1}{3}N\omega\mu_0 cl\alpha_{\alpha\alpha}(g)}, \tag{3.4.18}$$

where I_0 is the initial intensity. Comparing this result with (1.2.12), the absorption index is found to be

$$n' \approx \tfrac{1}{6}N\mu_0 c^2\alpha_{\alpha\alpha}(g). \tag{3.4.19}$$

3.4.4 Linear dichroism and birefringence (the Kerr effect)

If the molecules, while still supporting only components of the real polarizability $\alpha_{\alpha\beta}$, are now completely oriented, as in a crystal, or partially oriented, as in a fluid in a static external field, there is the possibility of polarization changes through linear dichroism and birefringence. It is convenient to take the incident light beam to be completely linearly polarized with an azimuth $\theta = \pi/4$, in which case (3.4.16) reduce to

$$\frac{\mathrm{d}I}{\mathrm{d}z} \approx -\tfrac{1}{2}IN\omega\mu_0 c(\alpha_{xx}(g) + \alpha_{yy}(g) - 2\alpha_{xy}(g)), \tag{3.4.20a}$$

$$\frac{\mathrm{d}\theta}{\mathrm{d}z} \approx \tfrac{1}{4}N\omega\mu_0 c(\alpha_{xx}(g) - \alpha_{yy}(g)), \tag{3.4.20b}$$

$$\frac{\mathrm{d}\eta}{\mathrm{d}z} \approx -\tfrac{1}{4}N\omega\mu_0 c(\alpha_{xx}(f) - \alpha_{yy}(f)), \tag{3.4.20c}$$

$$\frac{\mathrm{d}P}{\mathrm{d}z} \approx 0. \tag{3.4.20d}$$

The first equation describes absorption via the absorptive parts of the appropriate dynamic polarizability tensor components; the second describes an azimuth change due to linear dichroism brought about through a differential absorption of the two linearly polarized components of the incident light beam resolved along the x and y directions; the third describes the corresponding ellipticity change due to linear birefringence; and the fourth shows that the beam suffers no depolarization.

We now develop (3.4.20c) to obtain an expression for the ellipticity change in the Kerr effect in which a static uniform electric field is applied to a fluid perpendicular to the propagation direction, and at 45° to the azimuth, of an incident linearly polarized light beam (so here the electric field is applied along the x direction). But first we note that, since the ellipticity change at transparent frequencies does not depend on the initial ellipticity, the macroscopic ellipticity (in radians) developed along a finite path length l (in metres) is simply

$$\eta \approx -\tfrac{1}{4}N\omega\mu_0 cl(\alpha_{xx}(f) - \alpha_{yy}(f)). \tag{3.4.21}$$

The electric field generates anisotropy in the fluid on account of a partial orientation of the molecular electric dipole moments, both permanent and induced. According to (2.6.4a), the electric dipole moment in the presence of a static uniform electric field is

$$\mu_\alpha = \mu_{0\alpha} + \alpha_{\alpha\beta}E_\beta + \cdots,$$

where $\alpha_{\alpha\beta}$ is the *static* polarizability. There is no need to specify $(E_\beta)_0$, the field at the molecular origin, because the field here is uniform. A further contribution to the Kerr effect originates in the perturbation of the *dynamic* molecular polarizability by the electric field, in accordance with (2.7.1):

$$\alpha_{\alpha\beta}(\mathbf{E}) = \alpha_{\alpha\beta} + \alpha_{\alpha\beta,\gamma}^{(\mu)}E_\gamma + \tfrac{1}{2}\alpha_{\alpha\beta,\gamma\delta}^{(\mu\mu)}E_\gamma E_\delta + \cdots.$$

Thus in (3.4.20c), a weighted average must be taken of the polarizability tensor components perturbed by the static electric field.

For our purposes, the classical Boltzmann average for a system in thermodynamic equilibrium at the temperature T is adequate:

$$\overline{X(\Omega)} = \int d\Omega X(\Omega)e^{-V(\Omega)/kT} \bigg/ \int d\Omega e^{-V(\Omega)/kT}, \tag{3.4.22}$$

where $X(\Omega)$ is the value of a particular component, in space-fixed axes, of a molecular property tensor when the molecule is at some orientation Ω to the field, and $V(\Omega)$ is the corresponding potential energy of the molecule in the field. If $V(\Omega)$ is much smaller than kT, we can use the expansion

$$\overline{X(\Omega)} = \langle X(\Omega)\rangle - \frac{1}{kT}[\langle X(\Omega)V(\Omega)\rangle - \langle X(\Omega)\rangle\langle V(\Omega)\rangle]$$

$$+ \frac{1}{k^2T^2}\big[\tfrac{1}{2}\langle X(\Omega)V(\Omega)^2\rangle - \tfrac{1}{2}\langle X(\Omega)\rangle\langle V(\Omega)^2\rangle$$

$$- \langle X(\Omega)V(\Omega)\rangle\langle V(\Omega)\rangle\big] + \cdots. \tag{3.4.23}$$

The potential energy here is the interaction between the static field and the permanent and induced molecular electric dipole moments, so from (2.6.1) and (2.6.4),

$$V(\Omega) = -\mu_{0x}E_x - \tfrac{1}{2}\alpha_{xx}E_x^2 + \cdots. \tag{3.4.24}$$

Using the unit vector averages (4.2.53), we obtain terms such as

$$\langle(\alpha_{xx}(f) - \alpha_{yy}(f))\alpha_{xx}\rangle = \alpha_{\alpha\beta}(f)\alpha_{\gamma\delta}\langle i_\alpha i_\beta i_\gamma i_\delta - j_\alpha j_\beta i_\gamma i_\delta\rangle$$
$$= \tfrac{1}{15}(3\alpha_{\alpha\beta}(f)\alpha_{\alpha\beta} - \alpha_{\alpha\alpha}(f)\alpha_{\beta\beta}),$$

and the complete expression for the ellipticity is found to be (Buckingham and Pople, 1955)

$$\eta \approx -\tfrac{1}{120}\omega\mu_0 clNE_x^2\Big[3\alpha_{\alpha\beta,\alpha\beta}^{(\mu\mu)}(f) - \alpha_{\alpha\alpha,\beta\beta}^{(\mu\mu)}(f)$$
$$+ \frac{2}{kT}(3\alpha_{\alpha\beta,\alpha}^{(\mu)}(f)\mu_{0\beta} - \alpha_{\alpha\alpha,\beta}^{(\mu)}(f)\mu_{0\beta}) + \frac{1}{kT}(3\alpha_{\alpha\beta}(f)\alpha_{\alpha\beta} - \alpha_{\alpha\alpha}(f)\alpha_{\beta\beta})$$
$$+ \frac{1}{k^2T^2}(3\alpha_{\alpha\beta}(f)\mu_{0\alpha}\mu_{0\beta} - \alpha_{\alpha\alpha}(f)\mu_{0\beta}\mu_{0\beta})\Big]. \tag{3.4.25}$$

It is stressed that this result for the macroscopic ellipticity is strictly valid only at transparent frequencies. To facilitate comparison with standard molecular expressions for the Kerr birefringence (Buckingham and Pople, 1955; Buckingham, 1962), note that the phase difference between transmitted light waves linearly polarized along x and y is

$$\delta = \frac{2\pi l}{\lambda}(n^x - n^y) \tag{3.4.26}$$

and that the ellipticity (using the present sign convention) is equal to $-\tan(\delta/2)$ (Fredericq and Houssier, 1973) so that, for small ellipticities,

$$\eta \approx -\frac{\pi l}{\lambda}(n^x - n^y). \tag{3.4.27}$$

Buckingham (1962) has discussed the detailed application of this equation at absorbing frequencies. However, such discussions of the frequency dependence of the Kerr effect only apply to ellipticity changes that are effectively infinitesimal for, once an ellipticity develops, (3.4.16c) shows that additional changes can be generated through terms in $\alpha_{\alpha\beta}(g)$ since these depend on $\sin 2\eta$. Furthermore, as outlined below, the simultaneous presence of linear dichroism can lead to additional complexity.

The development of an expression for the azimuth change at absorbing frequencies due to Kerr linear dichroism proceeds in an analogous fashion. However, integration over a finite path length to derive an expression for a macroscopic azimuth change is no longer trivial because, according to (3.4.16b), the differential azimuth change depends on both the ellipticity and azimuth of the light beam incident on the lamina. We refer to Kuball and Singer (1969) for further discussion of this complicated situation.

Similar expressions can be developed for the Cotton–Mouton effect, with a static uniform magnetic field replacing the electric field.

3.4.5 Electric field gradient-induced birefringence: measurement of molecular electric quadrupole moments and the problem of origin invariance

It is of considerable interest to extend the development of linear birefringence in the previous section to allow for a static electric field gradient. This provides the theoretical background for the experimental determination of molecular quadrupole moments in fluids. Although taking us a little outside the realm of optical activity phenomena, this example reveals the power and generality of the refringent scattering formalism and provides a glimpse of one of the great achievements of molecular optics. A similar treatment has been given independently by Raab and de Lange (2003).

The static electric field is now taken to be inhomogeneous with gradient $E_{xx} = -E_{yy}$. By allowing for the perturbation of the dynamic molecular polarizability by this static electric field gradient,

$$\alpha_{\alpha\beta}(\nabla \mathbf{E}) = \tfrac{1}{3}\alpha^{(\Theta)}_{\alpha\beta,\gamma\delta} E_{\gamma\delta} + \cdots,$$

and adding the interaction between the static field gradient and the permanent electric quadrupole moment, namely $-\tfrac{1}{3}(\Theta_{0_{xx}} - \Theta_{0_{yy}})E_{xx}$, to the orientation-dependent potential energy (3.4.24), the following additional contribution to the Kerr ellipticity (3.4.25) is found (Buckingham, 1958):

$$\mu \approx -\tfrac{1}{30}\omega\mu_0 c l N E_{xx}\big(\alpha^{(\Theta)}_{\alpha\beta,\,\alpha\beta} + \alpha_{\alpha\beta}\Theta_{0_{\alpha\beta}}/kT\big). \qquad (3.4.28)$$

The perceptive reader will notice a problem with this result: if the quadrupolar molecule also possesses a permanent electric dipole moment, then according to (2.4.9) the electric quadrupole moment will be origin dependent. This situation is unsatisfactory, for it requires a bulk observable, the electric field gradient-induced birefringence, to depend on an arbitrary molecular origin. The problem was resolved by Buckingham and Longuet-Higgins (1968) who realized that, in addition to the partial alignment of the quadrupolar molecules by the electric field gradient, there will be a nonuniform distribution of dipolar molecules as a result of the interaction of their permanent electric dipole moments with a position-dependent electric field that is proportional to the displacement of the molecule along x or y from the z axis where the field is zero. The associated temperature-dependent birefringence then arises from a combination of electric dipole scattering by the aligned quadrupolar molecules with magnetic dipole plus electric quadrupole scattering from molecules with locally oriented electric dipoles displaced slightly from the z axis. There is also a temperature-independent contribution from the electric dipole–magnetic dipole and electric dipole–electric quadrupole tensors $G'_{\alpha\beta}$ and $A_{\alpha,\beta\gamma}$ perturbed by the position-dependent electric field, again from molecules displaced slightly from the axis.

To accommodate these features, the refringent scattering formalism of Section 3.4.2 must be extended: specifically, the last three terms of the scattering tensor (3.3.4) must be retained along with the small components along x and y in the propagation vector (3.4.2b) for the waves scattered from the molecules within the thin lamina. The present treatment is equivalent to the original treatment of Buckingham and Longuet-Higgins (1968) who also employed a molecular scattering approach. Because of the additional complexity we shall not calculate the ellipticity directly via the Stokes parameters, but instead will calculate the refractive index difference for light linearly polarized along the x and y directions. We therefore require the following components of the scattering tensor:

$$\tilde{a}_{xx} = \tilde{\alpha}_{xx} + \frac{i\omega}{3c}(\tilde{A}_{x,zx} - \mathscr{A}_{x,zx}) + \frac{1}{c}(\tilde{G}_{xy} + \mathscr{G}_{yx})$$

$$+ \frac{x}{R_0}\tilde{\alpha}_{zx} + \frac{i\omega}{3cR_0}(x\tilde{A}_{z,zx} - x\mathscr{A}_{x,zz} + x\mathscr{A}_{x,xx} + y\mathscr{A}_{x,yx})$$

$$+ \frac{1}{cR_0}(x\tilde{G}_{zy} + y\mathscr{G}_{zx}) + \cdots, \qquad (3.4.29a)$$

$$\tilde{a}_{yy} = \tilde{\alpha}_{yy} + \frac{i\omega}{3c}(\tilde{A}_{y,zy} - \mathscr{A}_{y,zy}) - \frac{1}{c}(\tilde{G}_{yx} + \mathscr{G}_{xy})$$

$$+ \frac{y}{R_0}\tilde{\alpha}_{zy} + \frac{i\omega}{3cR_0}(y\tilde{A}_{z,zy} - y\mathscr{A}_{y,zz} + y\mathscr{A}_{y,yy} + x\mathscr{A}_{y,xy})$$

$$- \frac{1}{cR_0}(y\tilde{G}_{zx} + x\mathscr{G}_{zy}) + \cdots. \qquad (3.4.29b)$$

It is now necessary to consider the arrangement by which the electric field gradient is generated in the experiment. Typically, the sample is contained in a long tube within which are two fine wires running parallel to the axis of the tube. When a potential difference is set up between the walls of the tube and the wires (which are at the same potential), an inhomogeneous electric field is established between the wires. The probe light beam is directed along the tube between the two wires. If the z axis is taken to be the axis of the tube and the wires lie along the lines $(x = a, y = 0)$ and $(x = -a, y = 0)$, the nonzero electric field components near the z axis are $E_x = qx$ and $E_y = -qy$, and the nonzero electric field gradient components are $E_{xx} = q$ and $E_{yy} = -q$ (Buckingham and Longuet-Higgins, 1968). In these expressions q is proportional to the associated line charge (charge per unit length). The potential energy of a molecule at $(x, y, 0)$ is then

$$V(x, y, 0) = -\mu_{0\alpha}E_\alpha - \tfrac{1}{3}\Theta_{0\alpha\beta}E_{\alpha\beta} + \cdots$$

$$= -q\left(\mu_{0x}x - \mu_{0y}y + \tfrac{1}{3}\Theta_{0xx} - \tfrac{1}{3}\Theta_{0yy} + \cdots\right). \qquad (3.4.30)$$

The integration over the surface of the lamina of the scattered waves, detected at f, from molecules within the lamina must now take account of the probability distribution of molecules in the xy plane. We make the artificial assumption that the molecules remain in a fixed orientation: the rotational averaging will be performed at the end. The probability that there is a molecule in the volume element $dxdydz$ at equilibrium in small fields is then

$$P(x, y, z)\,dxdydz = Ne^{-V(x,y,z)/kT}dxdydz$$

$$= N\left[1 + \frac{q}{kT}\left(\mu_{0x}x - \mu_{0y}y + \tfrac{1}{3}\Theta_{0xx} - \tfrac{1}{3}\Theta_{0yy} + \cdots\right)\right]dxdydz, \quad (3.4.31)$$

where N is the number density of molecules in the absence of the field. This expression replaces $N\,dxdydz$ in (3.4.1). The required electric fields of the scattered waves at f are now

$$\tilde{E}_x^f = \frac{N\omega^2\mu_0\,dxdydz}{4\pi R_0}\left[1 + \frac{q}{kT}\left(\mu_{0x}x - \mu_{0y}y + \tfrac{1}{3}\Theta_{0xx} - \tfrac{1}{3}\Theta_{0yy} + \cdots\right)\right]$$

$$\times \left[1 + \tilde{\alpha}_{xx} + \frac{i\omega}{3c}(\tilde{A}_{x,zx} - \tilde{\mathscr{A}}_{x,zx}) + \frac{1}{c}(\tilde{G}_{xy} + \tilde{\mathscr{G}}_{yx})\right.$$

$$+ \frac{x}{R_0}\tilde{\alpha}_{zx} + \frac{i\omega}{3cR_0}(x\tilde{A}_{z,zx} - x\tilde{\mathscr{A}}_{x,zz} + x\tilde{\mathscr{A}}_{x,xx} + y\tilde{\mathscr{A}}_{x,yx})$$

$$\left. + \frac{1}{cR_0}(x\tilde{G}_{zy} + y\,\tilde{\mathscr{G}}_{zx}) + \cdots\right]\tilde{E}_x^{(0)}e^{i\omega(x^2+y^2)/2R_0c}e^{i\omega(R_0/c-t)}, \quad (3.4.32a)$$

$$\tilde{E}_y^f = \frac{N\omega^2\mu_0 dxdydz}{4\pi R_0}\left[1 + \frac{q}{kT}\left(\mu_{0x}x - \mu_{0y}y + \tfrac{1}{3}\Theta_{0xx} - \tfrac{1}{3}\Theta_{0yy} + \cdots\right)\right]$$

$$\times \left[1 + \tilde{\alpha}_{yy} + \frac{i\omega}{3c}(\tilde{A}_{y,zy} - \tilde{\mathscr{A}}_{y,zy}) - \frac{1}{c}(\tilde{G}_{yx} + \tilde{\mathscr{G}}_{xy})\right.$$

$$+ \frac{y}{R_0}\tilde{\alpha}_{zy} + \frac{i\omega}{3cR_0}(y\tilde{A}_{z,zy} - y\tilde{\mathscr{A}}_{y,zz} + y\tilde{\mathscr{A}}_{y,yy} + x\tilde{\mathscr{A}}_{y,xy})$$

$$\left. - \frac{1}{cR_0}(y\tilde{G}_{zx} + x\,\tilde{\mathscr{G}}_{zy}) + \cdots\right]\tilde{E}_y^{(0)}e^{i\omega(x^2+y^2)/2R_0c}e^{i\omega(R_0/c-t)}. \quad (3.4.32b)$$

Using the integral (3.4.3) together with

$$\frac{1}{R_0^2}\int_{-\infty}^{\infty}\int_{-\infty}^{\infty}dxdy\,x^2e^{i\omega(x^2+y^2)/2R_0c} = -\frac{2\pi c^2}{\omega^2} \quad (3.4.33)$$

and $\exp(x) = 1 + x + \cdots$, and comparing with (3.4.5), the temperature-dependent birefringence is found to be

$$n^x - n^y \approx \tfrac{1}{2}N\mu_0c^2\left\{\frac{q}{3kT}(\Theta_{0xx} - \Theta_{0yy})(\tilde{\alpha}_{xx} - \tilde{\alpha}_{yy})\right.$$

$$- \frac{q}{kT}\left[\tfrac{1}{3}(\mu_{0x}\tilde{A}_{z,zx} - \mu_{0x}\tilde{\mathscr{A}}_{x,zz} + \mu_{0x}\tilde{\mathscr{A}}_{x,xx} - \mu_{0y}\tilde{\mathscr{A}}_{x,yx}\right.$$

$$+ \mu_{0y}\tilde{A}_{z,zy} - \mu_{0y}\tilde{\mathscr{A}}_{y,zz} + \mu_{0y}\tilde{\mathscr{A}}_{y,yy} - \mu_{0x}\tilde{\mathscr{A}}_{y,xy})$$

$$\left.\left. - \frac{i}{\omega}(\mu_{0x}\tilde{G}_{zy} - \mu_{0y}\tilde{\mathscr{G}}_{zx} - \mu_{0y}\tilde{G}_{zx} + \mu_{0x}\tilde{\mathscr{G}}_{zy})\right] + \cdots\right\}, \quad (3.4.34)$$

where we have retained only those terms which provide nonzero averages over all molecular orientations. As shown in Chapter 4, in the absence of a static magnetic field, $\tilde{A}_{\alpha,\beta\gamma} = A_{\alpha,\beta\gamma}$, $\tilde{\mathscr{A}}_{\alpha,\beta\gamma} = A_{\alpha,\beta\gamma}$, $i\tilde{G}_{\alpha\beta} = G'_{\alpha\beta}$, $i\mathscr{G}_{\alpha\beta} = -G'_{\beta\alpha}$. Performing the orientational averaging using the unit vector averages (4.2.49) and (4.2.53), the following expression for the temperature-dependent contribution to the birefringence is obtained:

$$n^x - n^y \approx \frac{N\mu_0 c^2 q}{15kT}\left[\Theta_{0\alpha\beta}\alpha_{\alpha\beta} - \mu_{0\alpha}\left(A_{\beta,\alpha\beta} + \frac{5}{\omega}\varepsilon_{\alpha\beta\gamma}G'_{\beta\gamma}\right)\right]. \quad (3.4.35)$$

By allowing for the perturbation of the dynamic molecular property tensors in the scattered electric fields (3.4.32) by the electric field and field gradient,

$$\tilde{\alpha}_{\alpha\beta}(\mathbf{E}, \nabla\mathbf{E}) = \tilde{\alpha}_{\alpha\beta} + \tilde{\alpha}^{(\mu)}_{\alpha\beta,\gamma}\tilde{E}_\gamma + \tfrac{1}{3}\tilde{\alpha}^{(\Theta)}_{\alpha\beta,\gamma\delta}\tilde{E}_{\gamma\delta} + \cdots, \quad (3.4.36a)$$

$$\tilde{A}_{\alpha,\beta\gamma}(\mathbf{E}) = \tilde{A}_{\alpha,\beta\gamma} + \tilde{A}^{(\mu)}_{\alpha,\beta\gamma,\delta}\tilde{E}_\delta + \cdots, \quad (3.4.36b)$$

$$\tilde{G}_{\alpha\beta}(\mathbf{E}) = \tilde{G}_{\alpha\beta} + \tilde{G}^{(\mu)}_{\alpha\beta,\gamma}\tilde{E}_\delta + \cdots, \quad (3.4.36c)$$

and similarly for $\tilde{\mathscr{A}}_{\alpha,\beta\gamma}(\mathbf{E})$ and $\mathscr{G}_{\alpha\beta}(\mathbf{E})$, the temperature-independent contribution is produced. The final complete result for the birefringence due to the electric field gradient is

$$n^x - n^y \approx \frac{N\mu_0 c^2 q}{15}\left\{\alpha^{(\Theta)}_{\alpha\beta,\alpha\beta} - A^{(\mu)}_{\alpha,\beta\alpha,\beta} - \frac{5}{\omega}\varepsilon_{\alpha\beta\gamma}G'^{(\mu)}_{\alpha\beta,\gamma}\right.$$
$$\left. + \frac{1}{kT}\left[\Theta_{0\alpha\beta}\alpha_{\alpha\beta} - \mu_{0\alpha}\left(A_{\beta,\alpha\beta} + \frac{5}{\omega}\varepsilon_{\alpha\beta\gamma}G'_{\beta\gamma}\right)\right]\right\}, \quad (3.4.37)$$

which is equivalent to the result of Buckingham and Longuet-Higgins (1968). de Lange and Raab (2004) have recently shown how a very different theory based on the solution of a wave equation derived from Maxwell's macroscopic equations may be refined to give the same result, thereby resolving a long-standing puzzle.

Using (2.6.35) for the origin dependencies of $\alpha_{\alpha\beta}$, $A_{\alpha,\beta\gamma}$ and $G'_{\alpha\beta}$, it is readily verified that this expression is independent of the choice of molecular origin, as required. The point at which the origin-dependent vector

$$A_{\beta,\alpha\beta} + \frac{5}{\omega}\varepsilon_{\alpha\beta\gamma}G'_{\beta\gamma} = 0 \quad (3.4.38)$$

is called the *effective quadrupole centre*. Hence the apparent electric quadrupole moment given by (3.4.28) has its origin at the point which satisfies (3.4.38).

3.4.6 Natural optical rotation and circular dichroism

To determine the natural optical activity contributions to the refringent intensity and polarization changes, we retain only terms in $G'_{\alpha\beta}$ and $A_{\alpha,\beta\gamma}$ since in Chapter 4

(Section 4.4.4) it is shown that only chiral molecules can support the appropriate components in most situations. These tensors always contribute to refringent scattering in the antisymmetric combination (3.4.13*b*). The required component for light propagating along z is

$$\zeta'_{xyz} = -\frac{1}{c}\left[\tfrac{1}{3}\omega(A_{x,yz} - A_{y,xz}) + G'_{xx} + G'_{yy}\right]. \qquad (3.4.39)$$

According to (2.6.35), general components of $A_{\alpha,\beta\gamma}$ and $G'_{\alpha\beta}$ are origin dependent. However, it is easily verified that the combination of components in (3.4.39) is such that ζ'_{xyz} is independent of the choice of origin, as required for a term contributing to observables such as optical rotation and circular dichroism.

In isotropic samples such as fluids in the absence of static fields, the unweighted average of ζ'_{xyz} over all molecular orientations must be taken. Using the unit vector averages (4.2.48) and (4.2.49), we find

$$\langle\zeta'_{xyz}\rangle = -\frac{1}{c}\left[\tfrac{1}{3}\omega A_{\alpha,\beta\gamma}\langle i_\alpha k_\beta j_\gamma - j_\alpha k_\beta i_\gamma\rangle + G'_{\alpha\beta}\langle i_\alpha i_\beta + j_\alpha j_\beta\rangle\right]$$

$$= -\frac{2}{3c}G'_{\alpha\alpha} \qquad (3.4.40)$$

since $A_{\alpha,\beta\gamma} = A_{\alpha,\gamma\beta}$. Thus only electric dipole–magnetic dipole scattering contributes to the natural optical rotation and circular dichroism of isotropic samples, the electric dipole–electric quadrupole contribution averaging to zero. Although, according to (2.6.35), a general component of $G'_{\alpha\beta}$ is origin dependent, the trace is independent of origin and so can contribute by itself to optical rotation in an isotropic sample.

It should be mentioned that the results of this section give the complete polarization changes only for nonmagnetic samples which are isotropic in the plane perpendicular to the direction of propagation. Thus they are valid for light propagating along the optic axis of uniaxial crystals and, after averaging, to fluids. For other propagation directions in anisotropic media additional terms can contribute.

Thus (3.4.16*b*) indicates that a chiral medium generates an azimuth change which depends on the dispersion lineshape function f:

$$\frac{d\theta}{dz} \approx \tfrac{1}{2}\omega\mu_0 cN\zeta'_{xyz}(f). \qquad (3.4.41)$$

Since this is independent of the polarization of the light beam incident on the lamina dz, the macroscopic natural optical rotation (in radians) for a finite path length l (in metres) along the z direction in an oriented medium can be written immediately as (Buckingham and Dunn, 1971)

$$\Delta\theta \approx -\tfrac{1}{2}\omega\mu_0 lN\left[\tfrac{1}{3}\omega(A_{x,yz}(f) - A_{y,xz}(f)) + G'_{xx}(f) + G'_{yy}(f)\right]. \qquad (3.4.42)$$

In an isotropic sample, we use the average (3.4.40) and so recover the celebrated *Rosenfeld equation* (Rosenfeld, 1928) for natural optical rotation:

$$\Delta\theta \approx -\tfrac{1}{3}\omega\mu_0 l N G'_{\alpha\alpha}(f).\tag{3.4.43}$$

From (3.4.16c) we see that a chiral medium generates an ellipticity change which depends on the absorption lineshape function g and on the ellipticity and degree of polarization of the light beam incident on the lamina:

$$\frac{d\eta}{dz} \approx \tfrac{1}{2}\omega\mu_0 c N \zeta'_{xyz}(g)\frac{1}{P}\cos 2\eta.\tag{3.4.44}$$

Assuming that the degree of polarization remains unity, the macroscopic ellipticity change is obtained from an integral of the form

$$\int_{\eta_0}^{\eta_l} \sec 2\eta \, d\eta = C \int_0^l dz,$$

where $C = \tfrac{1}{2}\omega\mu_0 c N \zeta'_{xyz}(g)$ and η_0 and η_1 are the initial and final ellipticities. If the incident light is linearly polarized, $\eta_0 = 0$ and

$$\eta_l = \tan^{-1} e^{2Cl} - \pi/4 = \tan^{-1}\tanh Cl.$$

The macroscopic ellipticity developed over the path length l is thus

$$\eta \approx \tan^{-1}\tanh\left(\tfrac{1}{2}\omega\mu_0 c l N \zeta'_{xyz}(g)\right).\tag{3.4.45}$$

For very small ellipticities, this reduces to

$$\eta = \tfrac{1}{2}\omega\mu_0 c l N \zeta'_{xyz}(g).\tag{3.4.46}$$

Equation (3.4.16a) shows that, in addition to the usual absorption due to $\alpha_{\alpha\beta}(g)$, a chiral medium can generate a loss of intensity which depends on the absorption lineshape function and on the ellipticity and degree of polarization of the incident light:

$$\frac{dI}{dz} \approx -\tfrac{1}{2}I\omega\mu_0 c N(\alpha_{xx}(g) + \alpha_{yy}(g) - 2\zeta'_{xyz}(g)P\sin 2\eta).\tag{3.4.47}$$

If the degree of polarization remains unity, the macroscopic loss of intensity is obtained from an integral of the form

$$\int_{I_0}^{I_l} \frac{dI}{I} = \int_0^l (C' + 2C\sin 2\eta)\, dz$$

$$= \int_0^l [C' + 2C\sin 2(\tan^{-1}\tanh Cz)]\, dz,$$

where $C' = -\frac{1}{2}\omega\mu_0 cN(\alpha_{xx}(g) + \alpha_{yy}(g))$. We have assumed that the incident light is linearly polarized and used (3.4.45). Therefore

$$I_l = I_0 e^{C'l} \cosh 2Cl,$$

and the final attenuated intensity is given by

$$I_l \approx I_0 \, e^{-\frac{1}{2}\omega\mu_0 clN(\alpha_{xx}(g)+\alpha_{yy}(g))} \cosh\left(\omega\mu_0 clN\zeta'_{xyz}(g)\right) \qquad (3.4.48)$$

which is a generalization to oriented samples of the modified Beer–Lambert law for the passage of an initially linearly polarized light beam through an absorbing chiral medium (Velluz, Legrand and Grosjean, 1965). If the incident light beam is right- or left-circularly polarized, (3.4.44) shows that no further change in ellipticity can occur, and the final attenuated intensity is found from (3.4.47) to be

$$I_{L_l}^R \approx I_{L_0}^R e^{-\frac{1}{2}\omega\mu_0 clN(\alpha_{xx}(g)+\alpha_{yy}(g)\mp 2\zeta'_{xyz}(g))}. \qquad (3.4.49)$$

Using (1.2.11), this last result immediately provides an expression, in terms of the absorptive parts of the dynamic molecular property tensors, for Kuhn's dissymmetry factor (1.2.15):

$$g = \frac{\epsilon^L - \epsilon^R}{\frac{1}{2}(\epsilon^L + \epsilon^R)} = \frac{4\zeta'_{xyz}(g)}{\alpha_{xx}(g) + \alpha_{yy}(g)}. \qquad (3.4.50)$$

From (3.4.16d) we see that the degree of polarization increases in an absorbing chiral medium:

$$\frac{dP}{dz} \approx -\omega\mu_0 \, cN\zeta'_{xyz}(g)(P^2 - 1)\sin 2\eta. \qquad (3.4.51)$$

The macroscopic change in the degree of polarization is obtained from an integral of the form

$$\int_{P_0}^{P_l} \frac{dP}{P^2 - 1} = -2C \int_0^l \sin 2\eta \, dz.$$

If the incident light is unpolarized, we can take $\sin 2\eta = \pm 1$ (the sign being given by the sign of C) since (3.4.44) shows that the polarized component that is acquired is circular. The final degree of polarization is therefore

$$P_l = \left|\tanh\left(\omega\mu_0 clN\zeta'_{xyz}(g)\right)\right|. \qquad (3.4.52)$$

Notice that an equivalent result is obtained by calculating directly the degree of circularity of the transmitted light using (3.4.49):

$$\frac{S_3}{S_0} = \frac{I_{R_l} - I_{L_l}}{I_{R_l} + I_{L_l}} = \tanh\left(\omega\mu_0 clN\zeta'_{xyz}(g)\right). \qquad (3.4.53)$$

In his early experiments on circular dichroism, Cotton actually found that unpolarized light becomes partially circularly polarized in an absorbing chiral medium (Lowry, 1935). Measurement of the degree of circular polarization of transmitted light could be useful in situations where it is not possible to prepare the polarization state of the incident light; for example, in the search for resolved chiral molecules in interstellar gas clouds by looking for circular polarization, at characteristic absorption frequencies of the particular molecules, in light transmitted from a star behind the gas cloud. Of course other possible sources of circular polarization, such as magnetic fields and light scattering by dust particles (Whittet, 1992), would have to be investigated carefully.

3.4.7 Magnetic optical rotation and circular dichroism

In the general expressions (3.4.16) for the refringent intensity and polarization changes, it is seen that the imaginary dynamic polarizability tensor component α'_{xy} contributes in just the same way as the natural optical activity tensor component ζ'_{xyz}. However, as discussed in Chapter 4, $\alpha'_{\alpha\beta}$ is time odd and therefore requires the presence of some other time-odd influence such as a static magnetic field in order to contribute to refringent scattering; although it can contribute to incoherent phenomena such as the nonrefringent antisymmetric scattering discussed in Chapter 8. In the Faraday effect, parity arguments (Section 1.9.3) require the magnetic field to be applied along the direction of propagation of the light beam. A fluid, for example, then becomes effectively a uniaxial medium.

Thus all the basic results of the previous section apply if ζ'_{xyz} is replaced by α'_{xy}, so magnetic optical rotation is given by

$$\Delta\theta \approx \tfrac{1}{2}\omega\mu_0 c l N\alpha'_{xy}(f), \qquad (3.4.54)$$

and the ellipticity associated with circular dichroism by

$$\eta = \tan^{-1}\tanh\left(\tfrac{1}{2}\omega\mu_0 c l N\alpha'_{xy}(g)\right). \qquad (3.4.55)$$

We must now bring the magnetic field into these expressions. Clearly we seek a linear dependence on B_z. This could come about through a partial orientation of any permanent molecular magnetic moments (but, unlike the Kerr effect, not of magnetic moments induced by the field since such contributions would be quadratic in B_z), and also through a linear perturbation of α'_{xy}:

$$\alpha'_{xy}(\mathbf{B}) = \alpha'_{xy} + \alpha'^{(m)}_{xy,z}B_z + \cdots. \qquad (3.4.56)$$

We consider first the Faraday effect in a fluid. Applying the classical Boltzmann average (3.4.23) with a potential energy

$$V(\Omega) = -m_{0_z} B_z + \cdots,$$

we find, using the unit vector average (4.2.49),

$$\begin{aligned}
\overline{\alpha'_{xy}} &= \left(\alpha''^{(m)}_{\alpha\beta,\gamma} + \alpha'_{\alpha\beta}m_{0\gamma}/kT\right)\langle i_\alpha j_\beta k_\gamma\rangle B_z \\
&= \tfrac{1}{6} B_z \varepsilon_{\alpha\beta\gamma}\left(\alpha''^{(m)}_{\alpha\beta,\gamma} + m_{0\alpha}\alpha'_{\beta\gamma}/kT\right) + \cdots.
\end{aligned} \tag{3.4.57}$$

This shows that only a field along the propagation direction can generate nonzero contributions after spatial averaging, which is consistent with the parity arguments. This result is now used in (3.4.54) taking the dynamic molecular property tensors as functions of the dispersion lineshape f, and in (3.4.55) taking the tensors as functions of the absorption lineshape g. Thus the Faraday optical rotation, for example, becomes

$$\Delta\theta \approx \tfrac{1}{12}\omega\mu_0 c l N B_z \varepsilon_{\alpha\beta\gamma}\left(\alpha''^{(m)}_{\alpha\beta,\gamma}(f) + m_{0\alpha}\alpha'_{\beta\gamma}(f)/kT\right). \tag{3.4.58}$$

If the quantum-mechanical expressions for $\alpha''^{(m)}_{\alpha\beta}$ (the magnetic analogue of (2.7.8)) and $\alpha'_{\alpha\beta}$ are introduced, the standard expressions for Faraday optical rotation and circular dichroism in fluids are recovered (Buckingham and Stephens, 1966). But we shall not write them out explicitly until Chapter 6.

Unlike permanent electric dipole moments, permanent magnetic dipole moments are not necessarily tied to a molecule's frame and can exist, for example, in free atoms, and in atomic ions in molecular complexes (the first excited state of the hydrogen atom is the only atomic system showing a permanent electric dipole moment on account of the accidental near degeneracy of electronic states of opposite parity). Consequently, a uniform static magnetic field can induce anisotropy in a collection of ionic or molecular magnetic moments in a crystal. It is now necessary to use a quantum-statistical average in place of the classical Boltzmann average (3.4.22) since it is the relative populations of quantum states with nonzero spin or orbital angular momentum projections onto the magnetic field direction that determines the induced magnetic anisotropy. Consider a molecule in a quantum state ψ_n, where n specifies a complete set of quantum numbers including the magnetic quantum number defining the projection of any nonzero angular momentum vector (so that ψ_n could be one component of a degenerate set). If the system is perturbed, the number of molecules per unit volume in the perturbed state ψ'_n is related to the number in the unperturbed state ψ_n by

$$N'_n = N_n e^{-(W_{n'}-W_n)/kT}.$$

In the case of a weak magnetic field and 'high' temperatures,

$$W'_n - W_n = -m_{n_z} B_z \ll kT$$

so that

$$N'_n = N_n(1 + m_{n_z} B_z/kT + \cdots). \tag{3.4.59}$$

In equations (3.4.54) and (3.4.55) for Faraday optical rotation and circular dichroism, we replace N by (3.4.59), and for α'_{xy}, which pertains to a molecule in the quantum state ψ_n, we use the expansion (3.4.56) in the magnetic field. If ψ_n is a component eigenstate of a degenerate set we must sum the contributions from all such components. The Faraday optical rotation, for example, then becomes

$$\Delta\theta \approx \tfrac{1}{2}\omega\mu_0 cl \left(\frac{N}{d_n}\right) B_z \sum_n (\alpha''^{(m)}_{xy,z}(f) + m_{n_z}\alpha'_{xy}(f)/kT), \tag{3.4.60}$$

where d_n is the degeneracy and $N = N_n d_n$ is the total number of molecules per unit volume in the degenerate set. The molecules themselves may be completely oriented as in a crystal; if in a fluid, an average over all orientations produces an expression equivalent to (3.4.58) derived from a consideration of a collection of classical magnetic moments in a fluid, namely

$$\Delta\theta \approx \tfrac{1}{12}\omega\mu_0 cl \left(\frac{N}{d_n}\right) B_z \varepsilon_{\alpha\beta\gamma} \sum_n (\alpha''^{(m)}_{\alpha\beta,\gamma}(f) + m_{n\alpha}\alpha'_{\beta\gamma}(f)/kT). \tag{3.4.61}$$

3.4.8 Magnetochiral birefringence and dichroism

Equations (3.4.16) for the general refringent intensity and polarization changes contain contributions from components of the symmetric tensor $\zeta_{\alpha\beta\gamma}$ defined in (3.4.13a). This contains components of $A'_{\alpha,\beta\gamma}$, the imaginary part of the electric dipole–electric quadrupole dynamic property tensor, together with $G_{\alpha\beta}$, the real part of the electric dipole–magnetic dipole dynamic property tensor. As shown in Chapter 4, both of these tensors are time odd and so $\zeta_{\alpha\beta\gamma}$ can only contribute in the presence of a time-odd influence such as a magnetic field. As elaborated in this section, $\zeta_{\alpha\beta\gamma}$ is responsible for magnetochiral phenomena.

In the expression (3.4.16a) for the rate of change of intensity of a light beam traversing an absorbing dilute molecular medium along z, terms in $\zeta_{xxz}(g) + \zeta_{yyz}(g)$ are specified which are completely independent of the polarization state of the incident light beam. These generate magnetochiral dichroism. If the incident light beam is unpolarized, and we assume it remains so over the sample path length, only the conventional absorption terms and the magnetochiral terms survive. Integration over a finite path length l then provides the following expression for the final attenuated intensity:

$$I_l \approx I_0 e^{-\frac{1}{2}\omega\mu_0 c l N[\alpha_{xx}(g) + \alpha_{yy}(g) + \zeta_{xxz}(g) + \zeta_{yyz}(g)]}. \qquad (3.4.62)$$

Comparing this result with (1.2.12), the associated absorption index is found to be

$$n' \approx \tfrac{1}{4}\mu_0 c^2 N[\alpha_{xx}(g) + \alpha_{yy}(g) + \zeta_{xxz}(g) + \zeta_{yyz}(g)]. \qquad (3.4.63)$$

The same result may be deduced directly from the expression (3.4.8b) for the absorption index in linearly polarized light by taking unpolarized light to be an incoherent superposition of light beams linearly polarized along x and y. The expression (3.4.8a) for the refractive index may similarly be used to deduce the following result for the associated refractive index in unpolarized light:

$$n \approx 1 + \tfrac{1}{4}\mu_0 c^2 N[\alpha_{xx}(f) + \alpha_{yy}(f) + \zeta_{xxz}(f) + \zeta_{yyz}(f)]. \qquad (3.4.64)$$

In isotropic samples such as fluids in the absence of static fields, the magnetochiral terms ζ_{xxz} and ζ_{yyz} give zero when averaged over all orientations. They do, however, give nonzero averages in the presence of a static magnetic field along z. This may be seen by using (3.4.13a) to write them out in terms of components of $G_{\alpha\beta}$ and $A_{\alpha,\beta\gamma}$,

$$\zeta_{xxz} = \frac{2}{c}\left(\tfrac{1}{3}\omega A'_{x,xz} + G_{xy}\right), \qquad (3.4.65a)$$

$$\zeta_{yyz} = \frac{2}{c}\left(\tfrac{1}{3}\omega A'_{y,yz} - G_{yx}\right), \qquad (3.4.65b)$$

and considering a perturbation linear in **B**:

$$A'_{x,xz}(\mathbf{B}) = A'_{x,xz} + A'^{(m)}_{x,xz,z}B_z + \cdots, \qquad (3.4.66a)$$

$$G_{xy}(\mathbf{B}) = G_{xy} + G^{(m)}_{xy,z}B_z + \cdots, \qquad (3.4.66b)$$

and similarly for $A'_{y,yz}(\mathbf{B})$ and $G_{yx}(\mathbf{B})$. The classical Boltzmann average (3.4.22) with potential energy

$$V(\Omega) = -m_{0_z}B_z + \cdots$$

is then applied, and using the unit vector averages (4.2.49) and (4.2.53) we find that the magnetochiral terms give the following average:

$$\begin{aligned}
\bar{\zeta}_{xxz} + \bar{\zeta}_{yyz} = \frac{2}{c}B_z \Big\{ &\tfrac{1}{45}\omega\Big[3A'^{(m)}_{\alpha,\alpha\beta,\beta} - A'^{(m)}_{\alpha,\beta\beta,\alpha} \\
&+ (3A'_{\alpha,\alpha\beta}m_{0\beta} - A'_{\alpha,\beta\beta}m_{0\alpha})/kT\Big] \\
&+ \tfrac{1}{3}\varepsilon_{\alpha\beta\gamma}\big(G^{(m)}_{\alpha\beta,\gamma} + G_{\alpha\beta}m_{0\gamma}/kT\big) + \cdots \Big\}. \qquad (3.4.67)
\end{aligned}$$

This expression reverses sign if the direction of **B** is reversed relative to the propagation direction of the light beam. Hence the required magnetochiral birefringence is found to be (Barron and Vrbancich, 1984)

$$n^{\uparrow\uparrow} - n^{\uparrow\downarrow} \approx \mu_0 cN B_z \left\{ \tfrac{1}{45}\omega\left[3A_{\alpha,\alpha\beta,\beta}^{\prime(m)}(f) - A_{\alpha,\beta\beta,\alpha}^{\prime(m)}(f)\right.\right.$$
$$+ (3A_{\alpha,\alpha\beta}'(f)m_{0\beta} - A_{\alpha,\beta\beta}'(f)m_{0\alpha})/kT\Big]$$
$$+ \tfrac{1}{3}\varepsilon_{\alpha\beta\gamma}\left(G_{\alpha\beta,\gamma}^{(m)}(f) + G_{\alpha\beta}(f)m_{0\gamma}/kT\right)\Bigg\}, \qquad (3.4.68)$$

where, as defined in Section 1.7, $n^{\uparrow\uparrow}$ and $n^{\uparrow\downarrow}$ are the refractive indices for an unpolarized light beam (or a beam of arbitrary polarization) propagating parallel and antiparallel to the static magnetic field. A similar expression obtains for the magnetochiral dichroism $n'^{\uparrow\uparrow} - n'^{\uparrow\downarrow}$ in which the dispersion lineshape function f is replaced by the absorption lineshape function g.

It was indicated in Chapter 1 that magnetochiral birefringence and dichroism require chiral samples. This is discussed in more detail in Chapter 6, where it is shown that the components of $A_{\alpha,\beta\gamma}'$ and $G_{\alpha\beta}$ specified in (3.4.68) are supported only by chiral molecules.

Using (2.6.35) for the origin dependencies of $A_{\alpha,\beta\gamma}'$ and $G_{\alpha\beta}$, it may be verified that (3.4.68) is independent of the choice of molecular origin (Coriani *et al.*, 2002).

3.4.9 Nonreciprocal (gyrotropic) birefringence

Equations (3.4.16) for the general refringent intensity and polarization changes contain contributions from components of the symmetric tensor $\zeta_{\alpha\beta\gamma}$ defined in (3.4.13*a*). This contains components of $A_{\alpha,\beta\gamma}'$, the imaginary part of the electric dipole–electric quadrupole dynamic property tensor, together with $G_{\alpha\beta}$, the real part of the electric dipole–magnetic dipole dynamic property tensor. As shown in Chapter 4, both these tensors are time odd and so $\zeta_{\alpha\beta\gamma}$ can only contribute in the presence of a time-odd influence such as a magnetic field.

Brown, Shtrikman and Treves (1963), and Birss and Shrubsall (1967), suggested that certain magnetic crystals could show an effect called *non-reciprocal* or *gyrotropic birefringence*, the origin of which Hornreich and Shtrikman (1968) ascribed to property tensors equivalent to our $G_{\alpha\beta}$ and $A_{\alpha,\beta\gamma}'$. Thus $\zeta_{\alpha\beta\gamma}$ generates gyrotropic birefringence, and it is seen from (3.4.16) that $\zeta_{\alpha\beta\gamma}$ contributes to polarization and intensity changes in just the same way as the real symmetric dynamic polarizability $\alpha_{\alpha\beta}$ which is responsible for conventional linear birefringence. So, like linear birefringence, gyrotropic birefringence can only exist in oriented media, and the associated polarization changes are subject to all the complications indicated in Section 3.4.4. But in addition, since there must be a static magnetic field, or bulk magnetization in the case of a magnetic crystal, parallel to the light beam, any polarization changes associated with gyrotropic birefringence add on reflecting the light beam back through the sample: this contrasts with polarization effects associated with linear birefringence, which cancel.

3.4.10 The Jones birefringence

In the development of his optical calculus, mentioned in Section 3.4.1, Jones (1948) predicted the existence of a new kind of linear birefringence together with its corresponding dichroism. These two new properties arose from the two-by-two matrix that Jones derived for determining the effect of a nondepolarizing medium on a polarized monochromatic light beam incident on it in certain directions. Having four complex elements, the Jones matrix represents in general eight distinct optical effects, namely refraction, absorption, circular birefingence and circular dichroism, linear birefringence and linear dichroism with respect to a pair of orthogonal axes, and linear birefringence and linear dichroism with respect to a second pair of orthogonal axes that bisect the first.

The last two properties were the new ones. They have since been predicted to occur naturally in certain magnetic and nonmagnetic crystals, and in fluids by the simultaneous application of uniform static electric and magnetic fields parallel to each other and transverse to the light beam (Graham and Raab, 1983; Ross, Sherbourne and Stedman, 1989). Observation of the Jones birefringence in crystals is hampered by the presence of conventional birefringence, but is favourable in fluids due to its dependence on EB whereas the conventional birefringence depends on E^2 for the Kerr effect and B^2 for the Cotton–Mouton effect. This magnetoelectric Jones birefringence has been observed by Roth and Rikken (2000) in paramagnetic molecules such as the organometallic complex methylcyclopentadienyl-manganese-tricarbonyl, $C_9H_7MnO_3$, in the neat liquid state.

A molecular theory of the magnetoelectric Jones birefringence in fluids has been given by Graham and Raab (1983), who showed that it depends on $G_{\alpha\beta}$, the real part of the electric dipole–magnetic dipole dynamic property tensor and $A'_{\alpha,\beta\gamma}$, the imaginary part of the electric dipole–electric quadrupole dynamic property tensor, perturbed by the static electric and magnetic fields simultaneously. The magnetoelectric Jones birefringence therefore shares a kinship with magnetochiral birefringence and nonreciprocal birefringence since in all three cases an essential element is the activation of the same time-odd property tensors by a time-odd influence, a static magnetic field.

There is another distinct magnetoelectric birefringence, this time induced by perpendicular static electric and magnetic fields transverse to the light beam. It has been observed by Roth and Rikken (2002) in fluids and compared with the Jones magnetoelectric birefringence where it was found to have the same magnitude, as predicted (Graham and Raab, 1984; Ross, Sherbourne and Stedman, 1989).

A third distinct magnetoelectric optical phenomenon, an anisotropy in the refractive index for an unpolarized light beam, propagating parallel and antiparallel to $\mathbf{E} \times \mathbf{B}$ using perpendicular static electric and magnetic fields transverse to the light beam, has also been observed (Rikken, Strohm and Wyder, 2002). This effect is related to the Cotton–Mouton effect through special relativity.

It has been shown that origin invariance of the expressions describing Jones birefringence requires the diamagnetic contribution to the magnetic dipole moment interaction with the static magnetic field in (2.5.1) to be retained in developing $G_{\alpha\beta}$ (Rizzo and Coriani, 2003). This may also be necessary for other phenomena that depend on $G_{\alpha\beta}$, although it is not required in the particular case of magnetochiral birefringence of fluids, described by (3.4.68), since the extra terms vanish on averaging over all orientations (Rizzo and Coriani, 2003).

3.4.11 Electric optical rotation (electrogyration) and circular dichroism

The simple pictorial symmetry arguments in Section 1.7.3 demonstrate that no *direct* electric analogue of the Faraday effect exists in fluids, even of chiral molecules. An electric analogue of the Faraday effect can exist in certain crystals, however, and we refer to Buckingham, Graham and Raab (1971), Gunning and Raab (1997) and Kaminsky (2000) for further details.

It is easy to understand one particular source of linear electric optical rotation. In Section 3.4.7, the Faraday effect was formulated in terms of a linear perturbation of the imaginary dynamic polarizability component α'_{xy} by a magnetic field along the z direction. So one source of an electric analogue is the activation of the same tensor component by an electric field along z in crystals which exhibit the magneto-electric effect, which is the generation of a small magnetization in the direction of an applied electric field. The electrically-induced magnetization may be regarded as arising from an imbalance in the fluctuations associated with the two equal and opposite spin lattices in antiferromagnetic crystals (Hornreich and Shtrikman, 1967).

Returning briefly to fluids, it is easy to show that an additional optical rotation and circular dichroism can exist in an isotropic collection of *chiral* molecules in perpendicular electric and magnetic fields, at right angles to the direction of propagation, that varies linearly with the strength of each field (Baranova, Bogdanov and Zel'dovich, 1977; Buckingham and Shatwell, 1978). This effect originates in the simultaneous electric and magnetic field perturbation of $\alpha'_{\alpha\beta}$.

3.5 Polarization phenomena in Rayleigh and Raman scattered light

3.5.1 Nonrefringent scattering of polarized light

We now consider polarization effects in light scattering processes other than those involving interference between the forward-scattered and the unscattered

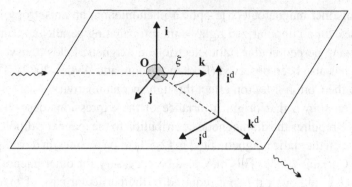

Fig. 3.3 The system of unit vectors used to describe the incident $(\mathbf{i}, \mathbf{j}, \mathbf{k})$ and scattered $(\mathbf{i}^{d}, \mathbf{j}^{d}, \mathbf{k}^{d})$ waves. ξ is the scattering angle.

components. These include Rayleigh and Raman scattering in any nonforward direction, and also Raman scattering in the forward direction since interference with the unscattered wave does not occur on account of the different frequencies.

Figure 3.3 shows a molecule at the origin $\mathbf{O}$ of a right-handed coordinate system x, y, z associated with unit vectors $\mathbf{i}, \mathbf{j}, \mathbf{k}$ in an incident quasi-monochromatic light wave propagating along $\mathbf{n}^{i} = \mathbf{k}$. We require the polarization and intensity in the wave zone of light scattered at an arbitrary angle ξ away from the forward direction. The unit vectors $\mathbf{i}, \mathbf{j}, \mathbf{k}$ are chosen so that the scattered direction is always in the $\mathbf{jk}$ plane, called the scattering plane. If a unit vector $\mathbf{k}^{d}$ is assigned to the direction of the propagation vector $\mathbf{n}^{d}$ of the detected wave, the characteristics of the detected plane wave in the wave zone can be specified in terms of a coordinate system x^{d}, y^{d}, z^{d} associated with unit vectors $\mathbf{i}^{d}, \mathbf{j}^{d}, \mathbf{k}^{d}$. From Fig. 3.3, the two sets of unit vectors are related by

$$\mathbf{i}^{d} = \mathbf{i}, \tag{3.5.1a}$$

$$\mathbf{j}^{d} = \mathbf{j} \cos \xi - \mathbf{k} \sin \xi, \tag{3.5.1b}$$

$$\mathbf{k}^{d} = \mathbf{k} \cos \xi + \mathbf{j} \sin \xi. \tag{3.5.1c}$$

The Stokes parameters of the scattered electric vector $\tilde{\mathbf{E}}^{d}$ in the x^{d}, y^{d}, z^{d} system are

$$S_{0}^{d} = \tilde{E}_{x^{d}}^{d} \tilde{E}_{x^{d}}^{d*} + \tilde{E}_{y^{d}}^{d} \tilde{E}_{y^{d}}^{d*},$$

$$S_{1}^{d} = \tilde{E}_{x^{d}}^{d} \tilde{E}_{x^{d}}^{d*} - \tilde{E}_{y^{d}}^{d} \tilde{E}_{y^{d}}^{d*},$$

$$S_{2}^{d} = -\left(\tilde{E}_{x^{d}}^{d} \tilde{E}_{y^{d}}^{d*} + \tilde{E}_{y^{d}}^{d} \tilde{E}_{x^{d}}^{d*}\right),$$

$$S_{3}^{d} = -\mathrm{i}\left(\tilde{E}_{x^{d}}^{d} \tilde{E}_{y^{d}}^{d*} - \tilde{E}_{y^{d}}^{d} \tilde{E}_{x^{d}}^{d*}\right).$$

We require these parameters in the system x, y, z used to describe the incident wave; from (3.5.1),

$$S_0^d = \tilde{E}_x^d \tilde{E}_x^{d*} + \tilde{E}_y^d \tilde{E}_y^{d*} \cos^2 \xi + \tilde{E}_z^d \tilde{E}_z^{d*} \sin^2 \xi$$
$$- \left(\tilde{E}_y^d \tilde{E}_z^{d*} + \tilde{E}_z^d \tilde{E}_y^{d*} \right) \cos \xi \sin \xi, \tag{3.5.2a}$$

$$S_1^d = \tilde{E}_x^d \tilde{E}_x^{d*} - \tilde{E}_y^d \tilde{E}_y^{d*} \cos^2 \xi - \tilde{E}_z^d \tilde{E}_z^{d*} \sin^2 \xi$$
$$+ \left(\tilde{E}_y^d \tilde{E}_z^{d*} + \tilde{E}_z^d \tilde{E}_y^{d*} \right) \cos \xi \sin \xi, \tag{3.5.2b}$$

$$S_2^d = -\left(\tilde{E}_x^d \tilde{E}_y^{d*} \cos \xi - \tilde{E}_x^d \tilde{E}_z^{d*} \sin \xi \right.$$
$$\left. + \tilde{E}_y^d \tilde{E}_x^{d*} \cos \xi - \tilde{E}_z^d \tilde{E}_x^{d*} \sin \xi \right), \tag{3.5.2c}$$

$$S_3^d = -i\left(\tilde{E}_x^d \tilde{E}_y^{d*} \cos \xi - \tilde{E}_x^d \tilde{E}_z^{d*} \sin \xi \right.$$
$$\left. - \tilde{E}_y^d \tilde{E}_x^{d*} \cos \xi + \tilde{E}_z^d \tilde{E}_x^{d*} \sin \xi \right). \tag{3.5.2d}$$

The electric vector of the scattered wave is given by (3.3.3) in terms of the scattering tensor and the electric vector of the incident wave, so using

$$\tilde{E}_\alpha^d \tilde{E}_\beta^{d*} = \left(\frac{\omega^2 \mu_0}{4\pi R} \right)^2 \tilde{a}_{\alpha\gamma} \tilde{a}_{\beta\delta}^* \tilde{E}_\gamma^{(0)} \tilde{E}_\delta^{(0)*},$$

the Stokes parameters (3.5.2) of the scattered wave can be written in terms of the scattering tensor and the Stokes parameters of the incident wave:

$$S_0^d = \frac{1}{2} \left(\frac{\omega^2 \mu_0}{4\pi R} \right)^2 \{ (|\tilde{a}_{xx}|^2 + |\tilde{a}_{xy}|^2) S_0 + (|\tilde{a}_{xx}|^2 - |\tilde{a}_{xy}|^2) S_1$$
$$- 2\mathrm{Re}(\tilde{a}_{xx}\tilde{a}_{xy}^*) S_2 - 2\mathrm{Im}(\tilde{a}_{xx}\tilde{a}_{xy}^*) S_3$$
$$+ [(|\tilde{a}_{yx}|^2 + |\tilde{a}_{yy}|^2) S_0 + (|\tilde{a}_{yx}|^2 - |\tilde{a}_{yy}|^2) S_1$$
$$- 2\mathrm{Re}(\tilde{a}_{yx}\tilde{a}_{yy}^*) S_2 - 2\mathrm{Im}(\tilde{a}_{yx}\tilde{a}_{yy}^*) S_3] \cos^2 \xi$$
$$+ [(|\tilde{a}_{zx}|^2 + |\tilde{a}_{zy}|^2) S_0 + (|\tilde{a}_{zx}|^2 - |\tilde{a}_{zy}|^2) S_1$$
$$- 2\mathrm{Re}(\tilde{a}_{zx}\tilde{a}_{zy}^*) S_2 - 2\mathrm{Im}(\tilde{a}_{zx}\tilde{a}_{zy}^*) S_3] \sin^2 \xi$$
$$- 2[\mathrm{Re}(\tilde{a}_{yx}\tilde{a}_{zx}^* + \tilde{a}_{yy}\tilde{a}_{zy}^*) S_0 + \mathrm{Re}(\tilde{a}_{yx}\tilde{a}_{zx}^* - \tilde{a}_{yy}\tilde{a}_{zy}^*) S_1$$
$$- \mathrm{Re}(\tilde{a}_{yx}\tilde{a}_{zy}^* + \tilde{a}_{zx}\tilde{a}_{yy}^*) S_2 - \mathrm{Im}(\tilde{a}_{yx}\tilde{a}_{zy}^* + \tilde{a}_{zx}\tilde{a}_{yy}^*) S_3] \cos \xi \sin \xi \},$$
$$\tag{3.5.3a}$$

$$S_1^d = \frac{1}{2} \left(\frac{\omega^2 \mu_0}{4\pi R} \right)^2 \{ (|\tilde{a}_{xx}|^2 + |\tilde{a}_{xy}|^2) S_0 + (|\tilde{a}_{xx}|^2 - |\tilde{a}_{xy}|^2) S_1$$
$$- 2\mathrm{Re}(\tilde{a}_{xx}\tilde{a}_{xy}^*) S_2 - 2\mathrm{Im}(\tilde{a}_{xx}\tilde{a}_{xy}^*) S_3$$
$$- [(|\tilde{a}_{yx}|^2 + |\tilde{a}_{yy}|^2) S_0 + (|\tilde{a}_{yx}|^2 - |\tilde{a}_{yy}|^2) S_1$$
$$- 2\mathrm{Re}(\tilde{a}_{yx}\tilde{a}_{yy}^*) S_2 - 2\mathrm{Im}(\tilde{a}_{yx}\tilde{a}_{yy}^*) S_3] \cos^2 \xi$$
$$- [(|\tilde{a}_{zx}|^2 + |\tilde{a}_{zy}|^2) S_0 + (|\tilde{a}_{zx}|^2 - |\tilde{a}_{zy}|^2) S_1$$

$$- 2\mathrm{Re}(\tilde{a}_{zx}\tilde{a}_{zy}^*)S_2 - 2\mathrm{Im}(\tilde{a}_{zx}\tilde{a}_{zy}^*)S_3]\sin^2\xi$$

$$+ 2[\mathrm{Re}(\tilde{a}_{yx}\tilde{a}_{zx}^* + \tilde{a}_{yy}\tilde{a}_{zy}^*)S_0 + \mathrm{Re}(\tilde{a}_{yx}\tilde{a}_{zx}^* - \tilde{a}_{yy}\tilde{a}_{zy}^*)S_1$$

$$- \mathrm{Re}(\tilde{a}_{yx}\tilde{a}_{zy}^* + \tilde{a}_{zx}\tilde{a}_{yy}^*)S_2 - \mathrm{Im}(\tilde{a}_{yx}\tilde{a}_{zy}^* + \tilde{a}_{zx}\tilde{a}_{yy}^*)S_3]\cos\xi\sin\xi\},$$

$$\hspace{10cm}(3.5.3b)$$

$$S_2^{\mathrm{d}} = -\left(\frac{\omega^2\mu_0}{4\pi R}\right)^2 \{[\mathrm{Re}(\tilde{a}_{xx}\tilde{a}_{yx}^* + \tilde{a}_{xy}\tilde{a}_{yy}^*)S_0 + \mathrm{Re}(\tilde{a}_{xx}\tilde{a}_{yx}^* - \tilde{a}_{xy}\tilde{a}_{yy}^*)S_1$$

$$- \mathrm{Re}(\tilde{a}_{xx}\tilde{a}_{yy}^* + \tilde{a}_{yx}\tilde{a}_{xy}^*)S_2 - \mathrm{Im}(\tilde{a}_{xx}\tilde{a}_{yy}^* + \tilde{a}_{yx}\tilde{a}_{xy}^*)S_3]\cos\xi$$

$$- [\mathrm{Re}(\tilde{a}_{xx}\tilde{a}_{zx}^* + \tilde{a}_{xy}\tilde{a}_{zy}^*)S_0 + \mathrm{Re}(\tilde{a}_{xx}\tilde{a}_{zx}^* - \tilde{a}_{xy}\tilde{a}_{zy}^*)S_1$$

$$- \mathrm{Re}(\tilde{a}_{xx}\tilde{a}_{zy}^* + \tilde{a}_{zx}\tilde{a}_{xy}^*)S_2 - \mathrm{Im}(\tilde{a}_{xx}\tilde{a}_{zy}^* + \tilde{a}_{zx}\tilde{a}_{xy}^*)S_3]\sin\xi\}, \hspace{1cm}(3.5.3c)$$

$$S_3^{\mathrm{d}} = \left(\frac{\omega^2\mu_0}{4\pi R}\right)^2 \{[\mathrm{Im}(\tilde{a}_{xx}\tilde{a}_{yx}^* + \tilde{a}_{xy}\tilde{a}_{yy}^*)S_0 + \mathrm{Im}(\tilde{a}_{xx}\tilde{a}_{yx}^* - \tilde{a}_{xy}\tilde{a}_{yy}^*)S_1$$

$$- \mathrm{Im}(\tilde{a}_{xx}\tilde{a}_{yy}^* - \tilde{a}_{yx}\tilde{a}_{xy}^*)S_2 + \mathrm{Re}(\tilde{a}_{xx}\tilde{a}_{yy}^* - \tilde{a}_{yx}\tilde{a}_{xy}^*)S_3]\cos\xi$$

$$- [\mathrm{Im}(\tilde{a}_{xx}\tilde{a}_{zx}^* + \tilde{a}_{xy}\tilde{a}_{zy}^*)S_0 + \mathrm{Im}(\tilde{a}_{xx}\tilde{a}_{zx}^* - \tilde{a}_{xy}\tilde{a}_{zy}^*)S_1$$

$$- \mathrm{Im}(\tilde{a}_{xx}\tilde{a}_{zy}^* - \tilde{a}_{zx}\tilde{a}_{xy}^*)S_2 + \mathrm{Re}(\tilde{a}_{xx}\tilde{a}_{zy}^* - \tilde{a}_{zx}\tilde{a}_{xy}^*)S_3]\sin\xi\}. \hspace{1cm}(3.5.3d)$$

By considering the various contributions to the complex scattering tensor $\tilde{a}_{\alpha\beta}$ specified by (3.3.4), these equations can be used to derive explicit expressions, in terms of dynamic molecular property tensors, for the intensity and polarization of light scattered into any direction from an incident beam of arbitrary polarization by a gaseous, liquid or solid medium which can be transparent or absorbing, oriented or isotropic, and also optically active. However, such general expressions are of overwhelming complexity, so we shall extract explicit expressions for particular situations as required. Notice that the last three terms in (3.3.4) do not contribute here since the scattered waves are purely transverse, and we are not considering a finite cone of collection.

Most Rayleigh scattered intensity originates in $\alpha_{\alpha\beta}$, the real symmetric dynamic polarizability, so the dominant polarization effects which it generates are discussed first. Polarization effects arising from other tensors are then discussed in turn, and expressions presented which must be added to those in $\alpha_{\alpha\beta}$ since the additional effects usually have to be measured in the presence of the dominant contributions from $\alpha_{\alpha\beta}$. We consider explicitly only fluid samples that are isotropic in the absence of external fields.

The same expressions apply to Raman scattering if the dynamic molecular property tensors are replaced by the corresponding transition tensors, so terms in the real symmetric and imaginary antisymmetric property tensors $\alpha_{\alpha\beta}$ and $\alpha'_{\alpha\beta}$ apply equally well to scattering through the real symmetric and imaginary antisymmetric transition tensors $(\alpha_{\alpha\beta})^{\mathrm{s}}_{mn}$ and $(\alpha'_{\alpha\beta})^{\mathrm{a}}_{mn}$ defined in (2.8.8b) and (2.8.8e). But there is also the possibility of scattering through the real antisymmetric and imaginary symmetric transition tensors $(\alpha_{\alpha\beta})^a_{mn}$ and $(\alpha'_{\alpha\beta})^s_{mn}$, defined in (2.8.8c) and (2.8.8d): as shown in Chapter 8, these can be important in resonance Raman scattering.

Furthermore, the same expressions apply to scattering at both transparent and absorbing frequencies. In the former case the dynamic molecular property tensors or transition tensors are written as functions of just the dispersion lineshape f, whereas in the latter case the complete complex lineshape $f + ig$ must be used (for an isolated transition we can then write, for example, $\alpha_{\alpha\beta} G'^*_{\alpha\beta} = \alpha^*_{\alpha\beta} G'_{\alpha\beta}$). Discussion of the variation of the scattered intensity with the frequency of the incident light in the region of an electronic absorption frequency, known as an *excitation profile* in the case of resonance Raman scattering, is postponed until Chapter 8 since the internal molecular mechanism generating the scattering tensor exerts a considerable influence.

3.5.2 Symmetric scattering

The situation most commonly encountered is Rayleigh or Raman scattering at transparent frequencies from randomly oriented achiral molecules in the absence of external static fields. The scattering is then usually dominated by the real dynamic polarizability $\alpha_{\alpha\beta}$, which is always symmetric, or by $(\alpha_{\alpha\beta})^s_{mn}$, the symmetric part of the real transition polarizability. In equations (3.5.3) for the Stokes parameters of the scattered wave, the specified products of tensor components must be averaged over all orientations of the molecule. Using the unit vector averages (4.2.53), together with $\alpha_{\alpha\beta} = \alpha_{\beta\alpha}$, we find the following types of nonzero average:

$$\langle \alpha_{xx}\alpha^*_{xx}\rangle = \alpha_{\alpha\beta}\alpha^*_{\gamma\delta}\langle i_\alpha i_\beta i_\gamma i_\delta\rangle$$
$$= \tfrac{1}{15}(\alpha_{\alpha\alpha}\alpha^*_{\beta\beta} + 2\alpha_{\alpha\beta}\alpha^*_{\alpha\beta}), \tag{3.5.4a}$$
$$\langle \alpha_{xx}\alpha^*_{yy}\rangle = \alpha_{\alpha\beta}\alpha^*_{\gamma\delta}\langle i_\alpha i_\beta j_\gamma j_\delta\rangle$$
$$= \tfrac{1}{15}(2\alpha_{\alpha\alpha}\alpha^*_{\beta\beta} - \alpha_{\alpha\beta}\alpha^*_{\alpha\beta}), \tag{3.5.4b}$$
$$\langle \alpha_{xy}\alpha^*_{xy}\rangle = \alpha_{\alpha\gamma}\alpha^*_{\beta\delta}\langle i_\alpha i_\beta j_\gamma j_\delta\rangle$$
$$= \tfrac{1}{30}(3\alpha_{\alpha\beta}\alpha'_{\alpha\beta} - \alpha_{\alpha\alpha}\alpha^*_{\beta\beta}). \tag{3.5.4c}$$

The Stokes parameters for Rayleigh light scattered into the forward direction ($\xi = 0°$) are then

$$S^d_0(0°) = K(7\alpha_{\alpha\beta}\alpha^*_{\alpha\beta} + \alpha_{\alpha\alpha}\alpha^*_{\beta\beta}), \tag{3.5.5a}$$
$$S^d_1(0°) = K(3\alpha_{\alpha\alpha}\alpha^*_{\beta\beta} + \alpha_{\alpha\beta}\alpha^*_{\alpha\beta})P\cos 2\eta\cos 2\theta, \tag{3.5.5b}$$
$$S^d_2(0°) = K(3\alpha_{\alpha\alpha}\alpha^*_{\beta\beta} + \alpha_{\alpha\beta}\alpha^*_{\alpha\beta})P\cos 2\eta\sin 2\theta, \tag{3.5.5c}$$
$$S^d_3(0°) = 5K(\alpha_{\alpha\alpha}\alpha^*_{\beta\beta} - \alpha_{\alpha\beta}\alpha^*_{\alpha\beta})P\sin 2\eta; \tag{3.5.5d}$$

for light scattered at right angles ($\xi = 90°$),

$$S_0^d(90°) = \tfrac{1}{2} K[(13\alpha_{\alpha\beta}\alpha_{\alpha\beta}^* - \alpha_{\alpha\alpha}\alpha_{\beta\beta}^*)$$
$$+ (\alpha_{\alpha\beta}\alpha_{\alpha\beta}^* + 3\alpha_{\alpha\alpha}\alpha_{\beta\beta}^*)P\cos 2\eta\cos 2\theta], \tag{3.5.6a}$$

$$S_1^d(90°) = \tfrac{1}{2} K[(\alpha_{\alpha\beta}\alpha_{\alpha\beta}^* + 3\alpha_{\alpha\alpha}\alpha_{\beta\beta}^*)(1 + P\cos 2\eta\cos 2\theta), \tag{3.5.6b}$$

$$S_2^d(90°) = 0, \tag{3.5.6c}$$

$$S_3^d(90°) = 0; \tag{3.5.6d}$$

and for backward-scattered light ($\xi = 180°$),

$$S_0^d(180°) = S_0^d(0°), \tag{3.5.7a}$$

$$S_1^d(180°) = S_1^d(0°), \tag{3.5.7b}$$

$$S_2^d(180°) = -S_2^d(0°), \tag{3.5.7c}$$

$$S_3^d(180°) = -S_3^d(0°), \tag{3.5.7d}$$

where

$$K = \frac{1}{30}\left(\frac{\omega^2\mu_0 E^{(0)}}{4\pi R}\right)^2 \tag{3.5.8}$$

and P, θ and η specify the polarization of the incident beam.

The same Stokes parameters apply to Raman scattering if $\alpha_{\alpha\beta}$ is replaced by $(\alpha_{\alpha\beta})_{mn}^s$. In fact the Stokes parameters (3.5.5) apply only to the Raman case; *forward* Rayleigh scattering is not meaningful since forward-scattered waves with the same frequency as the incident wave interfere with the transmitted wave and generate refraction and birefringence phenomena. However, we can talk about *near-forward* Rayleigh scattering.

A significant quantity in measurements on scattered light is the *depolarization ratio*, defined as the ratio of intensities linearly polarized parallel and perpendicular to the scattering plane. For 90° scattering,

$$\rho = \frac{I_z}{I_x} = \frac{I_{y^d}}{I_{x^d}} = \frac{S_0^d(90°) - S_1^d(90°)}{S_0^d(90°) + S_1^d(90°)}$$

$$= \frac{6\beta(\alpha)^2}{45\alpha^2 + 7\beta(\alpha)^2 + [45\alpha^2 + \beta(\alpha)^2]P\cos 2\eta\cos 2\theta}, \tag{3.5.9}$$

where the isotropic and anisotropic invariants

$$\alpha^2 = \tfrac{1}{9}\alpha_{\alpha\alpha}\alpha_{\beta\beta}^*, \tag{3.5.10a}$$

$$\beta(\alpha)^2 = \tfrac{1}{2}(3\alpha_{\alpha\beta}\alpha_{\alpha\beta}^* - \alpha_{\alpha\alpha}\alpha_{\beta\beta}^*), \tag{3.5.10b}$$

which are discussed in Section 4.2.6, are the only combinations of components of $\alpha_{\alpha\beta}\alpha_{\gamma\delta}^*$ that can contribute to light scattering in an isotropic sample. Equation

(3.5.9) generates the standard expressions for the depolarization ratio in incident light of particular polarizations (Placzek, 1934). Thus for unpolarized incident light ($P = 0$), and right-or left-circularly polarized incident light ($P = 1, \eta = \pm\pi/4$),

$$\rho(n) = \frac{6\beta(\alpha)^2}{45\alpha^2 + 7\beta(\alpha)^2};$$ (3.5.11)

for incident light linearly polarized perpendicular to the scattering plane ($P = 1$, $\eta = 0, \theta = 0$),

$$\rho(x) = \frac{3\beta(\alpha)^2}{45\alpha^2 + 4\beta(\alpha)^2};$$ (3.5.12)

and for incident light linearly polarized parallel to the scattering plane ($P = 1, \eta = 0, \theta = \pi/2$),

$$\rho(y) = 1.$$ (3.5.13)

In the Raman case ρ depends on the effective symmetry of the molecule and the symmetry species of the molecular vibration. Thus $\rho(x)$ can vary between 0 for a totally symmetric vibration spanned only by the isotropic polarizability α (as in the cubic point groups, for example), and $\frac{3}{4}$ for nontotally symmetric vibrations spanned only by the anisotropic polarizability $\beta(\alpha)$.

Another quantity of interest is the circularly polarized component of the scattered light. This is given by S_3^d, and from (3.5.3d) it can be seen that, for randomly oriented archiral molecules, a circularly polarized component only exists in the scattered light if the incident light has a circularly polarized component and the scattering angle is other than 90°. In the forward direction (or near-forward for Rayleigh scattering) the fraction of the scattered light that is circularly polarized (the degree of circularity) is found from (3.5.5) to be

$$\frac{S_3^d(0°)}{S_0^d(0°)} = \frac{5[9\alpha^2 - \beta(\alpha)^2]}{45\alpha^2 + 7\beta(\alpha)^2} P \sin 2\eta.$$ (3.5.14)

Thus if the incident beam is completely circularly polarized, the near-forward Rayleigh component is also completely circularly polarized in the same sense if the molecule is isotropically polarizable; polarizability anisotropy reduces the circularly polarized component. The Raman light scattered into the forward direction from circularly polarized incident light is completely circularly polarized in the same sense if the vibration is spanned only by α, and is partially circularly polarized in the opposite sense (with a degree of circularity $\frac{5}{7}$) if the vibration is spanned only by $\beta(\alpha)$. Equations (3.5.7) show that in the backward direction the degree of circularity of the scattered light is the same as (3.5.14) for the forward direction, but with opposite sign.

The use of circularly polarized light in conventional Rayleigh and Raman scattering was discussed by Placzek (1934), who defined a *reversal coefficient R* as the ratio of the intensity of the component circularly polarized in the same sense as the incident beam to that polarized in the reverse sense. Thus for backward scattering in right-circularly polarized incident light, for example,

$$R(180°) = \frac{I_R}{I_L} = \frac{S_0^d(180°) + S_3^d(180°)}{S_0^d(180°) - S_3^d(180°)}$$

$$= \frac{6\beta(\alpha)^2}{45\alpha^2 + \beta(\alpha)^2} = \frac{2\rho(x)}{1 - \rho(x)}, \qquad (3.5.15)$$

where I_R and I_L are the scattered intensities with right- and left-circular polarization, and for forward scattering

$$R(0°) = \frac{S_0^d(0°) + S_3^d(0°)}{S_0^d(0°) - S_3^d(0°)} = \frac{1}{R(180°)}. \qquad (3.5.16)$$

This technique enables totally symmetric and nontotally symmetric Raman bands to be distinguished in striking fashion since they have opposite signs in an $I_R - I_L$ spectrum (Clark *et al.*, 1974). But, as discussed in the next section, in the absence of antisymmetric scattering it gives no more information than the depolarization ratio.

3.5.3 Antisymmetric scattering

Rayleigh and Raman scattering can occur also through the imaginary dynamic polarizability $\alpha'_{\alpha\beta}$, which is always antisymmetric, through $(\alpha'_{\alpha\beta})^a_{mn}$, the antisymmetric part of the imaginary transition polarizability, and through $(\alpha_{\alpha\beta})^a_{mn}$, the antisymmetric part of the real transition polarizability. Since $\alpha'_{\alpha\beta} = -\alpha'_{\beta\alpha}$, the only type of nonzero average is now

$$\langle \alpha'_{xy}\alpha'^*_{xy} \rangle = \tfrac{1}{6}\alpha'_{\alpha\beta}\alpha'^*_{\alpha\beta}. \qquad (3.5.17)$$

The corresponding contributions to be added to the Rayleigh Stokes parameters (3.5.5) to (3.5.7) are

$$S_0^d(0°) = 5K\alpha'_{\alpha\beta}\alpha'^*_{\alpha\beta}, \qquad (3.5.18a)$$

$$S_1^d(0°) = -5K\alpha'_{\alpha\beta}\alpha'^*_{\alpha\beta}P\cos 2\eta \cos 2\theta, \qquad (3.5.18b)$$

$$S_2^d(0°) = -5K\alpha'_{\alpha\beta}\alpha'^*_{\alpha\beta}P\cos 2\eta \sin 2\theta, \qquad (3.5.18c)$$

$$S_3^d(0°) = 5K\alpha'_{\alpha\beta}\alpha'^*_{\alpha\beta}P\sin 2\eta; \qquad (3.5.18d)$$

$$S_0^d(90°) = \tfrac{5}{2}K\alpha'_{\alpha\beta}\alpha'^*_{\alpha\beta}(3 - P\cos 2\eta \cos 2\theta), \qquad (3.5.19a)$$

$$S_1^d(90°) = -\tfrac{5}{2}K\alpha'_{\alpha\beta}\alpha'^*_{\alpha\beta}(1 + P\cos 2\eta \cos 2\theta), \qquad (3.5.19b)$$

$$S_2^d(90°) = 0, \qquad (3.5.19c)$$

$$S_3^d(90°) = 0; \qquad (3.5.19d)$$

$$S_0^d(180°) = S_0^d(0°), \qquad (3.5.20a)$$

$$S_1^d(180°) = S_1^d(0°), \qquad (3.5.20b)$$

$$S_2^d(180°) = -S_2^d(0°), \qquad (3.5.20c)$$

$$S_3^d(180°) = -S_3^d(0°). \qquad (3.5.20d)$$

The general depolarization ratio for pure antisymmetric scattering at 90° is therefore

$$\rho = \frac{2}{1 - P\cos 2\eta \cos 2\theta}. \qquad (3.5.21)$$

Thus if the incident light is unpolarized or circularly polarized,

$$\rho(n) = 2; \qquad (3.5.22)$$

if linearly polarized perpendicular to the scattering plane,

$$\rho(x) = \infty; \qquad (3.5.23)$$

and if linearly polarized parallel to the scattering plane,

$$\rho(y) = 1. \qquad (3.5.24)$$

The phenomenon described by (3.5.23) is called *inverse polarization* and was first predicted by Placzek (1934).

The degree of circularity for pure antisymmetric scattering in the forward direction is

$$\frac{S_3^d(0°)}{S_0^d(0°)} = P\sin 2\eta. \qquad (3.5.25)$$

In the backward direction, the degree of circularity is the same but with opposite sign. Thus, as for pure isotropic scattering, if the incident beam is completely circularly polarized, the near-forward Rayleigh and the forward Raman components arising from pure antisymmetric scattering are also completely circularly polarized in the same sense. The corresponding reversal coefficient is

$$R(0°) = \frac{1}{R(180°)} = \infty. \qquad (3.5.26)$$

In fact antisymmetric scattering is usually encountered in the form of *anomalous* polarization ($\infty > \rho(x) > \tfrac{3}{4}$), rather than pure inverse polarization ($\rho(x) = \infty$).

This arises because symmetric and antisymmetric scattering contribute to the same band.

Antisymmetric Rayleigh scattering can produce large 'anomalies' in the depolarization ratio of light scattered from atoms (such as sodium) in spin-degenerate ground states when the incident frequency is in the vicinity of an electronic absorption frequency. Antisymmetric resonance Rayleigh and Raman scattering is also possible from molecules in degenerate states; but it can also arise without degeneracy in resonance Raman scattering associated with modes of vibration that transform as components of axial vectors. These questions are discussed in detail in Chapters 4 and 8.

If isotropic, anisotropic and antisymmetric scattering contribute simultaneously to the same Raman band, it is necessary to measure both the depolarization ratio in 90° scattering and the degree of circularity or reversal coefficient in 0° or 180° scattering in order to separate them (Placzek, 1934; McClain, 1971; Hamaguchi, 1985). General expressions for the depolarization ratio (incident light linearly polarized perpendicular to the scattering plane) and the reversal coefficient (backward scattering) are

$$\rho(x) = \frac{3\beta(\alpha)^2 + 5\beta(\alpha')^2}{45\alpha^2 + 4\beta(\alpha)^2}, \tag{3.5.27}$$

$$R(180°) = \frac{6\beta(\alpha)^2}{45\alpha^2 + \beta(\alpha)^2 + 5\beta(\alpha')^2}, \tag{3.5.28}$$

where α^2 and $\beta(\alpha)^2$ are the isotropic and anisotropic invariants (3.5.10), and

$$\beta(\alpha')^2 = \tfrac{3}{2}\alpha'_{\alpha\beta}\alpha'^*_{\alpha\beta} \tag{3.5.29}$$

is the corresponding antisymmetric invariant.

Thus the relative magnitudes of α^2, $\beta(\alpha)^2$ and $\beta(\alpha')^2$ can be determined from the three independent expressions given by the following three intensity measurements:

1. Intensity of light scattered at 90° and linearly polarized parallel to the scattering plane, in incident light linearly polarized perpendicular to the scattering plane: $[3\beta(\alpha)^2 + 5\beta(\alpha')^2]$.
2. Intensity of light scattered at 90° and linearly polarized perpendicular to the scattering plane, in incident light linearly polarized perpendicular to the scattering plane: $[45\alpha^2 + 4\beta(\alpha)^2]$.
3. Intensity of the component of light scattered at 180° with the same sense of circular polarization as the incident light: $6\beta(\alpha)^2$.

Complete polarization measurements such as these have been reported for resonance Raman scattering from ferrocytochrome c (Pézolet, Nafie and Peticolas,

1973; Nestor and Spiro, 1973), and provide information about the effective symmetry of the haem group.

Since the shapes of the Raman bands generated by isotropic, anisotropic and antisymmetric scattering are different, it is worth noting that the relative contributions to a particular Raman band could be determined just from 90° scattering by decomposing the lineshape into the three characteristic parts.

3.5.4 Natural Rayleigh and Raman optical activity

Rayleigh and Raman scattering from chiral samples can show additional polarization effects that originate in the slight difference in response to right- and left-circularly polarized light. The main contribution to 'optically active' Rayleigh scattering arises from interference between waves generated by $\alpha_{\alpha\beta}$ and waves generated by $G'_{\alpha\beta}$ plus $A_{\alpha,\beta\gamma}$. Similarly, the main contribution to optically active Raman scattering arises from interference between waves generated by $(\alpha_{\alpha\beta})^s_{mn}$ and waves generated by $(G'_{\alpha\beta})_{mn}$ plus $(A_{\alpha,\beta\gamma})_{mn}$. The averages over all molecular orientations of products of components of $\alpha_{\alpha\beta}$ with components of $G'_{\alpha\beta}$ are similar to (3.5.4), but in addition we must use the unit vector average (4.2.54) to obtain the following type of nonzero average:

$$\langle \alpha_{zx} A^*_{z,zy} \rangle = \alpha_{\gamma\alpha} A^*_{\delta,\epsilon\beta} \langle i_\alpha j_\beta k_\gamma k_\delta k_\epsilon \rangle$$
$$= \tfrac{1}{30}(\varepsilon_{\alpha\beta\gamma}\alpha_{\gamma\alpha}A^*_{\delta,\delta\beta} + \varepsilon_{\alpha\beta\delta}\alpha_{\gamma\alpha}A^*_{\delta,\gamma\beta} + \varepsilon_{\alpha\beta\epsilon}\alpha_{\gamma\alpha}A^*_{\gamma,\epsilon\beta}). \quad (3.5.30)$$

The first and third terms of this expression are in fact zero because $\alpha_{\alpha\beta} = \alpha_{\beta\alpha}$ and $A_{\alpha,\beta\gamma} = A_{\alpha,\gamma\beta}$.

The corresponding contributions to be added to the Rayleigh Stokes parameters (3.5.5) to (3.5.7) are

$$S^d_0(0°) = \frac{4K}{c}\left(3\alpha_{\alpha\alpha}G'^*_{\beta\beta} + \alpha_{\alpha\beta}G'^*_{\alpha\beta} - \tfrac{1}{3}\omega\alpha_{\alpha\beta}\varepsilon_{\alpha\gamma\delta}A^*_{\gamma,\delta\beta}\right)P\sin 2\eta, \quad (3.5.31a)$$

$$S^d_1(0°) = 0, \quad (3.5.31b)$$

$$S^d_2(0°) = 0, \quad (3.5.31c)$$

$$S^d_3(0°) = \frac{4K}{c}\left(3\alpha_{\alpha\alpha}G'^*_{\beta\beta} + \alpha_{\alpha\beta}G'^*_{\alpha\beta} - \tfrac{1}{3}\omega\alpha_{\alpha\beta}\varepsilon_{\alpha\gamma\delta}A^*_{\gamma,\delta\beta}\right); \quad (3.5.31d)$$

$$S^d_0(90°) = \frac{K}{c}\left(13\alpha_{\alpha\beta}G'^*_{\alpha\beta} - \alpha_{\alpha\alpha}G'^*_{\beta\beta} - \tfrac{1}{3}\omega\alpha_{\alpha\beta}\varepsilon_{\alpha\gamma\delta}A^*_{\gamma,\delta\beta}\right)P\sin 2\eta, \quad (3.5.32a)$$

$$S^d_1(90°) = \frac{K}{c}\left(3\alpha_{\alpha\alpha}G'^*_{\beta\beta} + \alpha_{\alpha\beta}G'^*_{\alpha\beta} + \omega\alpha_{\alpha\beta}\varepsilon_{\alpha\gamma\delta}A^*_{\gamma,\delta\beta}\right)P\sin 2\eta, \quad (3.5.32b)$$

$$S^d_2(90°) = 0, \quad (3.5.32c)$$

$$S_3^d(90°) = \frac{K}{c}\Big[13\alpha_{\alpha\beta}G'^*_{\alpha\beta} - \alpha_{\alpha\alpha}G'^*_{\beta\beta} - \tfrac{1}{3}\omega\alpha_{\alpha\beta}\varepsilon_{\alpha\gamma\delta}A^*_{\gamma,\delta\beta}$$
$$+ (3\alpha_{\alpha\alpha}G'^*_{\beta\beta} + \alpha_{\alpha\beta}G'^*_{\alpha\beta} + \omega\alpha_{\alpha\beta}\varepsilon_{\alpha\gamma\delta}A^*_{\gamma,\delta\beta})P\cos 2\eta\cos 2\theta\Big];$$

$$(3.5.32d)$$

$$S_0^d(180°) = \frac{8K}{c}\big(3\alpha_{\alpha\beta}G'^*_{\alpha\beta} - \alpha_{\alpha\alpha}G'^*_{\beta\beta} + \tfrac{1}{3}\omega\alpha_{\alpha\beta}\varepsilon_{\alpha\gamma\delta}A^*_{\gamma,\delta\beta}\big)P\sin 2\eta,$$

$$(3.5.33a)$$

$$S_1^d(180°) = 0, \tag{3.5.33b}$$

$$S_2^d(180°) = 0, \tag{3.5.33c}$$

$$S_3^d(180°) = \frac{8K}{c}\big(3\alpha_{\alpha\beta}G'^*_{\alpha\beta} - \alpha_{\alpha\alpha}G'^*_{\beta\beta} + \tfrac{1}{3}\omega\alpha_{\alpha\beta}\varepsilon_{\alpha\gamma\delta}A^*_{\gamma,\delta\beta}\big). \tag{3.5.33d}$$

These equations show that the optically active contribution to the scattered intensity depends on $P\sin 2\eta$ (Atkins and Barron, 1969), and is therefore zero if the incident light is unpolarized or linearly polarized; they also show that optical activity gives rise to a circularly polarized component in the scattered light. Notice that in the forward and backward directions there is no change in azimuth, and the azimuth of the light scattered at 90° is always perpendicular to the scattering plane (although optical rotation of the scattered light leaving an optically active sample can occur subsequently).

An appropriate experimental quantity in Rayleigh and Raman optical activity is a dimensionless circular intensity difference

$$\Delta = \frac{I^R - I^L}{I^R + I^L}, \tag{1.4.1}$$

where I^R and I^L are the scattered intensities in right- and left-circularly polarized incident light. From (3.5.31) to (3.5.33) and (3.5.5) to (3.5.7) we find the following Δs for scattering at 0°, 180°, and 90° (Barron and Buckingham, 1971):

$$\Delta(0°) = \frac{4\big(3\alpha_{\alpha\alpha}G'^*_{\beta\beta} + \alpha_{\alpha\beta}G'^*_{\alpha\beta} - \tfrac{1}{3}\omega\alpha_{\alpha\beta}\varepsilon_{\alpha\gamma\delta}A^*_{\gamma,\delta\beta}\big)}{c(7\alpha_{\lambda\mu}\alpha^*_{\lambda\mu} + \alpha_{\lambda\lambda}\alpha^*_{\mu\mu})}; \tag{3.5.34}$$

$$\Delta(180°) = \frac{8\big(3\alpha_{\alpha\beta}G'^*_{\alpha\beta} - \alpha_{\alpha\alpha}G'^*_{\beta\beta} + \tfrac{1}{3}\omega\alpha_{\alpha\beta}\varepsilon_{\alpha\gamma\delta}A^*_{\gamma,\delta\beta}\big)}{c(7\alpha_{\lambda\mu}\alpha^*_{\lambda\mu} + \alpha_{\lambda\lambda}\alpha^*_{\mu\mu})}; \tag{3.5.35}$$

$$\Delta_x(90°) = \frac{2\big(7\alpha_{\alpha\beta}G'^*_{\alpha\beta} + \alpha_{\alpha\alpha}\,G'^*_{\beta\beta} + \tfrac{1}{3}\omega\alpha_{\alpha\beta}\varepsilon_{\alpha\gamma\delta}A^*_{\gamma,\delta\beta}\big)}{c(7\alpha_{\lambda\mu}\alpha^*_{\lambda\mu} + \alpha_{\lambda\lambda}\alpha^*_{\mu\mu})}, \tag{3.5.36a}$$

$$\Delta_z(90°) = \frac{4\big(3\alpha_{\alpha\beta}G'^*_{\alpha\beta} - \alpha_{\alpha\alpha}\,G'^*_{\beta\beta} - \tfrac{1}{3}\omega\alpha_{\alpha\beta}\varepsilon_{\alpha\gamma\delta}A^*_{\gamma,\delta\beta}\big)}{2c(3\alpha_{\lambda\mu}\alpha^*_{\lambda\mu} - \alpha_{\lambda\lambda}\alpha^*_{\mu\mu})}. \tag{3.5.36b}$$

Only in scattering at 90° is it meaningful to define components polarized perpendicular and parallel to the scattering plane and we refer to $\Delta_x(90°)$ and $\Delta_z(90°)$ as the *polarized* and *depolarized* circular intensity differences; the circular intensity difference with no analyzer in the scattered beam is obtained by adding the numerators and denominators in (3.5.36a) and (3.5.36b). Notice that the degree of circularity of the scattered light wave gives information equivalent to that from the circular intensity difference. For example, $S_3^d(90°)/S_0^d(90°)$ equals (3.5.36a) if the incident light is linearly polarized perpendicular to the scattering plane, and equals (3.5.36b) if linearly polarized parallel to the scattering plane.

The symmetry requirements for optically active Rayleigh and Raman scattering are discussed in detail in Chapter 7. For the moment, we note that only chiral molecules can support such scattering. This is because the same components of $\alpha_{\alpha\beta}$, a second-rank polar tensor, and $G'_{\alpha\beta}$, a second-rank axial tensor, are specified in each cross term, and polar and axial tensors of the same rank only have the same transformation properties in the chiral point groups. Furthermore, although $A_{\alpha,\beta\gamma}$ does not transform the same as $G'_{\alpha\beta}$, it always occurs in the cross terms with $\alpha_{\alpha\beta}$ in the form $\varepsilon_{\alpha\gamma\delta}A_{\gamma,\delta\beta}$, which has transformation properties identical with $G'_{\alpha\beta}$. Notice that although $A_{\alpha,\beta\gamma}$ only contributes to birefringent optical activity phenomena such as optical rotation and circular dichroism in oriented media, it contributes to natural Rayleigh and Raman optical activity even in isotropic media where its contributions are of the same order of magnitude as those from $G'_{\alpha\beta}$.

Contributions to the Stokes parameters of the Rayleigh and Raman scattered light from terms in G'^2 and A^2 can be calculated from the general equations (3.5.3), but are not written down explicitly here since they are expected to be about 10^{-6} times terms in α^2 and 10^{-3} times terms in $\alpha G'$ and αA, which is probably too small to be detected at present. Furthermore, they do not describe optically active scattering since they do not have the circular polarization dependence of the $\alpha G'$ and αA terms; this also makes them more difficult to isolate from the dominant α^2 terms. In addition, the molecules do not necessarily need to be chiral to support such scattering; but if the molecules do happen to be chiral, racemic collections would show the same G'^2 and A^2 scattering as resolved collections since it is independent of the sign of the optical activity tensor (Pomeau, 1973). This topic has been revisited in the context of fluctuations in achiral, rather than racemic, systems which generate fleeting chiral configurations (Harris, 2001).

Although the results for optically active scattering presented above apply to most Raman scattering situations, we have not included contributions from cross terms between the real antisymmetric transition polarizability $(\alpha_{\alpha\beta})_{mn}^a$ and the transition optical activity tensors $(G'_{\alpha\beta})_{mn}$ and $(A_{\alpha,\beta\gamma})_{mn}$. Such cross terms could be important in certain resonance Raman scattering situations. There is also the possibility of optically active scattering involving cross terms between $\alpha'_{\alpha\beta}$ and $G_{\alpha\beta}$ plus $A'_{\alpha,\beta\gamma}$, which could be important in resonance Rayleigh and Raman scattering from odd-electron chiral molecules.

3.5.5 Magnetic Rayleigh and Raman optical activity

Just as all samples in a static magnetic field parallel to the incident light beam can show Faraday optical rotation and circular dichroism, so all samples in a static magnetic field can show Rayleigh and Raman optical activity. The main contribution to optically active magnetic Rayleigh scattering arises from interference between waves generated by $\alpha_{\alpha\beta}$ unperturbed by the external magnetic field and waves generated by $\alpha'_{\alpha\beta}$ perturbed to first order in the external magnetic field; and *vice versa*. Similarly, the analogous contribution to optically active magnetic Raman scattering arises from interference between waves generated by $(\alpha_{\alpha\beta})^s_{mn}$ unperturbed by the external magnetic field and waves generated by $(\alpha'_{\alpha\beta})^a_{mn}$ perturbed to first order in the external magnetic field; and *vice versa*. It is emphasized that the perturbation must arise from an external magnetic field: although a magnetic perturbation arising within a molecule can generate nonzero components of $\alpha'_{\alpha\beta}$, this does not give rise to optically active scattering in isotropic samples.

The complex dynamic polarizability is written as a power series in the external magnetic field:

$$\tilde{\alpha}_{\alpha\beta}(\mathbf{B}) = \alpha_{\alpha\beta} - i\alpha'_{\alpha\beta} + \alpha^{(m)}_{\alpha\beta,\gamma}B_\gamma - i\alpha'^{(m)}_{\alpha\beta,\gamma}B_\gamma + \cdots. \qquad (3.5.37)$$

The quantum mechanical expressions for $\alpha^{(m)}_{\alpha\beta,\gamma}$ and $\alpha'^{(m)}_{\alpha\beta,\gamma}$ are the magnetic analogues of (2.7.8). In view of the almost overwhelming complexity of some of the results of this section, we shall omit the superscripts (m) and the commas separating tensor subscripts. We now apply a weighted Boltzmann average in the form (3.4.23), with $V(\Omega) = -m_{n\gamma}B_\gamma$, and obtain expressions such as

$$\overline{\tilde{\alpha}_{xx}\tilde{\alpha}^*_{xy}} = iB_\gamma\Big(\alpha_{xx}\alpha'^*_{xyy} - \alpha'_{xx}\alpha^*_{xyy} + \alpha_{xx\gamma}\alpha'^*_{xy}$$

$$- \alpha'_{xx\gamma}\alpha^*_{xy} + \frac{1}{kT}(\alpha_{xx}\alpha'^*_{xy}m_{n\gamma} - \alpha'_{xx}\alpha^*_{xy}m_{n\gamma}) + \cdots\Big). \qquad (3.5.38)$$

Nonzero terms occur here only for $B_\gamma = B_z$; for example,

$$\langle\alpha_{xx}\alpha'^*_{xyz}\rangle = \alpha_{\gamma\delta}\alpha'^*_{\epsilon\alpha\beta}\langle j_\alpha k_\beta i_\gamma i_\delta i_\epsilon\rangle$$

$$= \tfrac{1}{30}(2\alpha_{\alpha\beta}\varepsilon_{\alpha\gamma\delta}\alpha'^*_{\beta\gamma\delta} + \alpha_{\alpha\alpha}\varepsilon_{\beta\gamma\delta}\alpha'^*_{\gamma\delta\beta}). \qquad (3.5.39)$$

The corresponding contributions to be added to the Rayleigh Stokes parameters (3.5.5) to (3.5.7) are

$$S_0^d(0°) = -2KB_z\Big[2\alpha_{\alpha\beta}\varepsilon_{\alpha\gamma\delta}\alpha'^*_{\beta\gamma\delta} + \alpha_{\alpha\alpha}\varepsilon_{\beta\gamma\delta}\alpha'^*_{\gamma\delta\beta} + 2\alpha'_{\alpha\beta}\varepsilon_{\alpha\gamma\delta}\alpha^*_{\gamma\beta\delta}$$
$$+ \alpha'_{\alpha\beta}\varepsilon_{\alpha\beta\gamma}\alpha^*_{\delta\delta\gamma} + \frac{1}{kT}(2\alpha_{\alpha\beta}\varepsilon_{\alpha\gamma\delta}\alpha'^*_{\beta\gamma}m_{n\delta}$$
$$+ \alpha_{\alpha\alpha}\varepsilon_{\beta\gamma\delta}\alpha'^*_{\gamma\delta}m_{n\beta}\Big]P\sin 2\eta, \tag{3.5.40a}$$

$$S_1^d(0°) = 0, \tag{3.5.40b}$$

$$S_2^d(0°) = 0, \tag{3.5.40c}$$

$$S_3^d(0°) = -2KB_z\Big[2\alpha_{\alpha\beta}\varepsilon_{\alpha\gamma\delta}\alpha'^*_{\beta\gamma\delta} + \alpha_{\alpha\alpha}\varepsilon_{\beta\gamma\delta}\alpha'^*_{\gamma\delta\beta} + 2\alpha'_{\alpha\beta}\varepsilon_{\alpha\gamma\delta}\alpha^*_{\gamma\beta\delta}$$
$$+ \alpha'_{\alpha\beta}\varepsilon_{\alpha\beta\gamma}\alpha^*_{\delta\delta\gamma} + \frac{1}{kT}(2\alpha_{\alpha\beta}\varepsilon_{\alpha\gamma\delta}\alpha'^*_{\beta\gamma}m_{n\delta} + \alpha_{\alpha\alpha}\varepsilon_{\beta\gamma\delta}\alpha'^*_{\gamma\delta}m_{n\beta}\Big];$$
$$\tag{3.5.40d}$$

$$S_0^d(90°) = -KB_z\Big[4\alpha_{\alpha\beta}\varepsilon_{\alpha\gamma\delta}\alpha'^*_{\beta\gamma\delta} + \alpha_{\alpha\alpha}\varepsilon_{\beta\gamma\delta}\alpha'^*_{\gamma\delta\beta} - 2\alpha_{\alpha\beta}\varepsilon_{\alpha\gamma\delta}\alpha'^*_{\gamma\delta\beta}$$
$$+ 4\alpha'_{\alpha\beta}\varepsilon_{\alpha\gamma\delta}\alpha^*_{\beta\gamma\delta} - \alpha'_{\alpha\beta}\varepsilon_{\alpha\gamma\beta}\alpha^*_{\delta\delta\gamma} - 2\alpha'_{\alpha\beta}\varepsilon_{\alpha\beta\gamma}\alpha^*_{\gamma\delta\delta}$$
$$+ \frac{1}{kT}(4\alpha_{\alpha\beta}\varepsilon_{\alpha\gamma\delta}\alpha'^*_{\beta\gamma}m_{n\delta} + \alpha_{\alpha\alpha}\varepsilon_{\beta\gamma\delta}\alpha'^*_{\gamma\delta}m_{n\beta}$$
$$- 2\alpha_{\alpha\beta}\varepsilon_{\alpha\gamma\delta}\alpha'^*_{\gamma\delta}m_{n\beta})\Big]P\sin 2\eta, \tag{3.5.41a}$$

$$S_1^d(90°) = -KB_z\Big[\alpha_{\alpha\alpha}\varepsilon_{\beta\gamma\delta}\alpha'^*_{\gamma\delta\beta} + 2\alpha_{\alpha\beta}\varepsilon_{\alpha\gamma\delta}\alpha'^*_{\gamma\delta\beta}$$
$$- \alpha'_{\alpha\beta}\varepsilon_{\alpha\gamma\beta}\alpha^*_{\delta\delta\gamma} + 2\alpha'_{\alpha\beta}\varepsilon_{\alpha\beta\gamma}\alpha^*_{\gamma\delta\delta}$$
$$+ \frac{1}{kT}(\alpha_{\alpha\alpha}\varepsilon_{\beta\gamma\delta}\alpha'^*_{\gamma\delta}m_{n\beta} + 2\alpha_{\alpha\beta}\varepsilon_{\alpha\gamma\delta}\alpha'^*_{\gamma\delta}m_{n\beta})\Big]P\sin 2\eta, \tag{3.5.41b}$$

$$S_2^d(90°) = 0, \tag{3.5.41c}$$

$$S_3^d(90°) = 0; \tag{3.5.41d}$$

$$S_0^d(180°) = S_0^d(0°), \tag{3.5.42a}$$

$$S_1^d(180°) = 0, \tag{3.5.42b}$$

$$S_2^d(180°) = 0, \tag{3.5.42c}$$

$$S_3^d(180°) = -S_3^d(0°). \tag{3.5.42d}$$

Notice that magnetic optical activity does not lead to a circularly polarized component in the light scattered at 90° if the magnetic field is parallel to the incident beam; it follows from (3.5.3*d*) that such a component is only generated by a magnetic field parallel to the scattered beam. On the other hand, the intensity of the scattered light is only dependent on the degree of circularity of the incident light (which leads to a circular intensity difference) if the magnetic field is parallel to the incident beam. This contrasts with natural optical activity in which the light scattered at any angle shows a circularly polarized component and a circular intensity difference simultaneously.

From these equations, together with (3.5.5) to (3.5.7) and (3.5.18) to (3.5.20), we find the following magnetic circular intensity differences for scattering at 0°, 180° and 90° (Barron and Buckingham, 1972):

$$\Delta(0°) = -2B_z\Bigg[2\alpha_{\alpha\beta}\varepsilon_{\alpha\gamma\delta}\alpha'^*_{\beta\gamma\delta} + \alpha_{\alpha\alpha}\varepsilon_{\beta\gamma\delta}\alpha'^*_{\gamma\delta\beta}$$

$$+ 2\alpha'_{\alpha\beta}\varepsilon_{\alpha\gamma\delta}\alpha^*_{\gamma\beta\delta} + \alpha'_{\alpha\beta}\varepsilon_{\alpha\beta\gamma}\alpha^*_{\delta\delta\gamma}$$

$$+ \frac{1}{kT}(2\alpha_{\alpha\beta}\varepsilon_{\alpha\gamma\delta}\alpha'^*_{\beta\gamma}m_{n\delta} + \alpha_{\alpha\alpha}\varepsilon_{\beta\gamma\delta}\alpha'^*_{\gamma\delta}m_{n\beta})\Bigg]\Bigg/$$

$$(7\alpha_{\lambda\mu}\alpha^*_{\lambda\mu} + \alpha_{\lambda\lambda}\alpha^*_{\mu\mu} + 5\alpha'_{\lambda\mu}\alpha'^*_{\lambda\mu}); \tag{3.5.43}$$

$$\Delta(180°) = \Delta(0°); \tag{3.5.44}$$

$$\Delta_x(90°) = -2B_z\Bigg[2\alpha_{\alpha\beta}\varepsilon_{\alpha\gamma\delta}\alpha'^*_{\beta\gamma\delta} + \alpha_{\alpha\alpha}\varepsilon_{\beta\gamma\delta}\alpha'^*_{\gamma\delta\beta}$$

$$+ 2\alpha'_{\alpha\beta}\varepsilon_{\alpha\gamma\delta}\alpha^*_{\beta\gamma\delta} + \alpha'_{\alpha\beta}\varepsilon_{\alpha\beta\gamma}\alpha^*_{\delta\delta\gamma}$$

$$+ \frac{1}{kT}(2\alpha_{\alpha\beta}\varepsilon_{\alpha\gamma\delta}\alpha'^*_{\beta\gamma}m_{n\delta} + \alpha_{\alpha\alpha}\varepsilon_{\beta\gamma\delta}\alpha'^*_{\gamma\delta}m_{n\beta})\Bigg]\Bigg/$$

$$(7\alpha_{\lambda\mu}\alpha^*_{\lambda\mu} + \alpha_{\lambda\lambda}\alpha^*_{\mu\mu} + 5\alpha'_{\lambda\mu}\alpha'^*_{\lambda\mu}), \tag{3.5.45a}$$

$$\Delta_z(90°) = -2B_z\Bigg[\alpha_{\alpha\beta}\varepsilon_{\alpha\gamma\delta}\alpha'^*_{\beta\gamma\delta} - \alpha_{\alpha\beta}\varepsilon_{\alpha\gamma\delta}\alpha'^*_{\gamma\delta\beta}$$

$$+ \alpha'_{\alpha\beta}\varepsilon_{\alpha\gamma\delta}\alpha^*_{\beta\gamma\delta} - \alpha'_{\alpha\beta}\varepsilon_{\alpha\beta\gamma}\alpha^*_{\gamma\delta\delta}$$

$$+ \frac{1}{kT}(\alpha_{\alpha\beta}\varepsilon_{\alpha\gamma\delta}\alpha'^*_{\beta\gamma}m_{n\delta} - \alpha_{\alpha\beta}\varepsilon_{\alpha\gamma\delta}\alpha'^*_{\gamma\delta}m_{n\beta})\Bigg]\Bigg/$$

$$(3\alpha_{\lambda\mu}\alpha^*_{\lambda\mu} - \alpha_{\lambda\lambda}\alpha^*_{\mu\mu} + 5\alpha'_{\lambda\mu}\alpha'^*_{\lambda\mu}). \tag{3.5.45b}$$

The symmetry aspects of magnetic Rayleigh and Raman optical activity are discussed in detail in Chapter 8. For the moment, we note that all molecules can support such scattering, because the components of the unperturbed and perturbed dynamic polarizabilities specified in each temperature-independent term always have the same transformation properties. For example, in $\alpha_{\alpha\beta}\varepsilon_{\alpha\gamma\delta}\alpha'^*_{\gamma\delta,\beta}$ both $\alpha_{\alpha\beta}$ and $\varepsilon_{\alpha\gamma\delta}\alpha'^*_{\gamma\delta,\beta}$ are symmetric second-rank polar tensors; and in $\alpha'_{\alpha\beta}\varepsilon_{\alpha\gamma\beta}\alpha^*_{\delta\delta,\gamma}$ both $\alpha'_{\alpha\beta}$ and $\varepsilon_{\alpha\gamma\beta}\alpha^*_{\delta\delta,\gamma}$ are antisymmetric second-rank polar tensors.

So far, we have not included cross terms containing the real antisymmetric transition polarizability $(\alpha_{\alpha\beta})^a_{mn}$, and cross terms containing the imaginary symmetric transition polarizability $(\alpha'_{\alpha\beta})^s_{mn}$. Such terms are important in certain resonance Raman scattering situations. The Stokes parameter contributions (3.5.40) to (3.5.42) are therefore generalized to

$$S_0^d(0°) = -2KB_z\left[4\alpha_{\alpha\beta}^s\varepsilon_{\alpha\gamma\delta}\alpha_{\beta\gamma\delta}^{\prime s*} + 2\alpha_{\alpha\beta}^s\varepsilon_{\alpha\gamma\delta}\alpha_{\beta\gamma\delta}^{\prime a*} + \alpha_{\alpha\alpha}^s\varepsilon_{\beta\gamma\delta}\alpha_{\gamma\delta\beta}^{\prime a*}\right.$$

$$- 2\alpha_{\alpha\beta}^a\varepsilon_{\alpha\gamma\delta}\alpha_{\gamma\beta\delta}^{\prime s*} - \alpha_{\alpha\beta}^a\varepsilon_{\alpha\beta\gamma}\alpha_{\delta\delta\gamma}^{\prime s*} - 4\alpha_{\alpha\beta}^{\prime s}\varepsilon_{\alpha\gamma\delta}\alpha_{\beta\gamma\delta}^{s*}$$

$$- 2\alpha_{\alpha\beta}^{\prime s}\varepsilon_{\alpha\gamma\delta}\alpha_{\beta\gamma\delta}^{a*} - \alpha_{\alpha\alpha}^{\prime s}\varepsilon_{\beta\gamma\delta}\alpha_{\gamma\delta\beta}^{a*} + 2\alpha_{\alpha\beta}^{\prime a}\varepsilon_{\alpha\gamma\delta}\alpha_{\gamma\beta\delta}^{s*}$$

$$+ \alpha_{\alpha\beta}^{\prime a}\varepsilon_{\alpha\beta\gamma}\alpha_{\delta\delta\gamma}^{s*} + \frac{1}{kT}\left(4\alpha_{\alpha\beta}^s\varepsilon_{\alpha\gamma\delta}\alpha_{\beta\gamma}^{\prime s*}m_{n\delta} + 2\alpha_{\alpha\beta}^s\varepsilon_{\alpha\gamma\delta}\alpha_{\beta\gamma}^{\prime a*}m_{n\delta}\right.$$

$$\left.\left. + \alpha_{\alpha\alpha}^s\varepsilon_{\beta\gamma\delta}\alpha_{\gamma\delta}^{\prime a*}m_{n\beta} - 2\alpha_{\alpha\beta}^{\prime s}\varepsilon_{\alpha\gamma\delta}\alpha_{\beta\gamma}^{a*}m_{n\delta} - \alpha_{\alpha\alpha}^{\prime s}\varepsilon_{\beta\gamma\delta}\alpha_{\gamma\delta}^{a*}m_{n\beta}\right)\right]P\sin 2\eta,$$

$$(3.5.46a)$$

$$S_1^d(0°) = 0, \tag{3.5.46b}$$

$$S_2^d(0°) = 0, \tag{3.5.46c}$$

$$S_3^d(0°) = -2KB_z\left[-4\alpha_{\alpha\beta}^s\varepsilon_{\alpha\gamma\delta}\alpha_{\beta\gamma\delta}^{\prime s*} + 2\alpha_{\alpha\beta}^s\varepsilon_{\alpha\gamma\delta}\alpha_{\beta\gamma\delta}^{\prime a*} + \alpha_{\alpha\alpha}^s\varepsilon_{\beta\gamma\delta}\alpha_{\gamma\delta\beta}^{\prime a*}\right.$$

$$- 2\alpha_{\alpha\beta}^a\varepsilon_{\alpha\gamma\delta}\alpha_{\gamma\beta\delta}^{\prime s*} - \alpha_{\alpha\beta}^a\varepsilon_{\alpha\beta\gamma}\alpha_{\delta\delta\gamma}^{\prime s*} + 4\alpha_{\alpha\beta}^{\prime s}\varepsilon_{\alpha\gamma\delta}\alpha_{\beta\gamma\delta}^{s*}$$

$$- 2\alpha_{\alpha\beta}^{\prime s}\varepsilon_{\alpha\gamma\delta}\alpha_{\beta\gamma\delta}^{a*} - \alpha_{\alpha\alpha}^{\prime s}\varepsilon_{\beta\gamma\delta}\alpha_{\gamma\delta\beta}^{a*} + 2\alpha_{\alpha\beta}^{\prime a}\varepsilon_{\alpha\gamma\delta}\alpha_{\gamma\beta\delta}^{s*}$$

$$+ \alpha_{\alpha\beta}^{\prime a}\varepsilon_{\alpha\beta\gamma}\alpha_{\delta\delta\gamma}^{s*} + \frac{1}{kT}\left(-4\alpha_{\alpha\beta}^s\varepsilon_{\alpha\gamma\delta}\alpha_{\beta\gamma}^{\prime s*}m_{n\delta} + 2\alpha_{\alpha\beta}^s\varepsilon_{\alpha\gamma\delta}\alpha_{\beta\gamma}^{\prime a*}m_{n\delta}\right.$$

$$\left.\left. + \alpha_{\alpha\alpha}^s\varepsilon_{\beta\gamma\delta}\alpha_{\gamma\delta}^{\prime a*}m_{n\beta} - 2\alpha_{\alpha\beta}^{\prime s}\varepsilon_{\alpha\gamma\delta}\alpha_{\beta\gamma}^{a*}m_{n\delta} - \alpha_{\alpha\alpha}^{\prime s}\varepsilon_{\beta\gamma\delta}\alpha_{\gamma\delta}^{a*}m_{n\beta}\right)\right];$$

$$(3.5.46d)$$

$$S_0^d(90°) = -KB_z\left[6\alpha_{\alpha\beta}^s\varepsilon_{\alpha\gamma\delta}\alpha_{\beta\gamma\delta}^{\prime s*} + 4\alpha_{\alpha\beta}^s\varepsilon_{\alpha\gamma\delta}\alpha_{\beta\gamma\delta}^{\prime a*} + \alpha_{\alpha\alpha}^s\varepsilon_{\beta\gamma\delta}\alpha_{\gamma\delta\beta}^{\prime a*}\right.$$

$$- 4\alpha_{\alpha\beta}^a\varepsilon_{\alpha\gamma\delta}\alpha_{\gamma\beta\delta}^{\prime s*} + \alpha_{\alpha\beta}^a\varepsilon_{\alpha\gamma\beta}\alpha_{\delta\delta\gamma}^{\prime s*} + 2\alpha_{\alpha\beta}^a\varepsilon_{\alpha\beta\gamma}\alpha_{\delta\gamma\delta}^{\prime s*}$$

$$+ 2\alpha_{\alpha\beta}^a\varepsilon_{\alpha\beta\gamma}\alpha_{\delta\gamma\delta}^{\prime a*} + 2\alpha_{\alpha\beta}^s\varepsilon_{\alpha\gamma\delta}\alpha_{\delta\gamma\beta}^{\prime a*} - 2\alpha_{\alpha\beta}^a\varepsilon_{\alpha\gamma\delta}\alpha_{\delta\gamma\beta}^{\prime a*}$$

$$- 2\alpha_{\alpha\beta}^a\varepsilon_{\alpha\gamma\delta}\alpha_{\beta\gamma\delta}^{\prime a*} - 6\alpha_{\alpha\beta}^{\prime s}\varepsilon_{\alpha\gamma\delta}\alpha_{\beta\gamma\delta}^{s*} - 4\alpha_{\alpha\beta}^{\prime s}\varepsilon_{\alpha\gamma\delta}\alpha_{\beta\gamma\delta}^{a*}$$

$$- \alpha_{\alpha\alpha}^{\prime s}\varepsilon_{\beta\gamma\delta}\alpha_{\gamma\delta\beta}^{a*} + 4\alpha_{\alpha\beta}^{\prime a}\varepsilon_{\alpha\gamma\delta}\alpha_{\gamma\beta\delta}^{s*} - \alpha_{\alpha\beta}^{\prime a}\varepsilon_{\alpha\gamma\beta}\alpha_{\delta\delta\gamma}^{s*}$$

$$- 2\alpha_{\alpha\beta}^{\prime a}\varepsilon_{\alpha\beta\gamma}\alpha_{\delta\gamma\delta}^{s*} - 2\alpha_{\alpha\beta}^{\prime a}\varepsilon_{\alpha\beta\gamma}\alpha_{\delta\gamma\delta}^{a*} - 2\alpha_{\alpha\beta}^{\prime s}\varepsilon_{\alpha\gamma\delta}\alpha_{\delta\gamma\beta}^{a*}$$

$$+ 2\alpha_{\alpha\beta}^{\prime a}\varepsilon_{\alpha\gamma\delta}\alpha_{\delta\gamma\beta}^{a*} + 2\alpha_{\alpha\beta}^{\prime a}\varepsilon_{\alpha\gamma\delta}\alpha_{\beta\gamma\delta}^{a*} + \frac{1}{kT}\left(6\alpha_{\alpha\beta}^s\varepsilon_{\alpha\gamma\delta}\alpha_{\beta\gamma}^{\prime s*}m_{n\delta}\right.$$

$$+ 4\alpha_{\alpha\beta}^s\varepsilon_{\alpha\gamma\delta}\alpha_{\beta\gamma}^{\prime a*}m_{n\delta} + \alpha_{\alpha\alpha}^s\varepsilon_{\beta\gamma\delta}\alpha_{\gamma\delta}^{\prime a*}m_{n\beta} - 4\alpha_{\alpha\beta}^{\prime s}\varepsilon_{\alpha\gamma\delta}\alpha_{\beta\gamma}^{a*}m_{n\delta}$$

$$- \alpha_{\alpha\alpha}^{\prime s}\varepsilon_{\beta\gamma\delta}\alpha_{\gamma\delta}^{a*}m_{n\beta} + 2\alpha_{\alpha\beta}^a\varepsilon_{\alpha\beta\gamma}\alpha_{\delta\gamma}^{\prime s*}m_{n\delta} + 2\alpha_{\alpha\beta}^a\varepsilon_{\alpha\beta\gamma}\alpha_{\delta\gamma}^{\prime a*}m_{n\delta}$$

$$\left.\left. + 2\alpha_{\alpha\beta}^s\varepsilon_{\alpha\gamma\delta}\alpha_{\delta\gamma}^{\prime a*}m_{n\beta} - 2\alpha_{\alpha\beta}^a\varepsilon_{\alpha\gamma\delta}\alpha_{\delta\gamma}^{\prime a*}m_{n\beta} - 2\alpha_{\alpha\beta}^a\varepsilon_{\alpha\gamma\delta}\alpha_{\beta\gamma}^{\prime a*}m_{n\delta}\right)\right]P\sin 2\eta,$$

$$(3.5.47a)$$

$$S_1^d(90°) = -KB_z\left[2\alpha_{\alpha\beta}^s\varepsilon_{\alpha\gamma\delta}\alpha_{\beta\gamma\delta}^{\prime s*} + \alpha_{\alpha\alpha}^s\varepsilon_{\beta\gamma\delta}\alpha_{\gamma\delta\beta}^{\prime a*} + \alpha_{\alpha\beta}^a\varepsilon_{\alpha\gamma\beta}\alpha_{\delta\delta\gamma}^{\prime s*}\right.$$

$$- 2\alpha_{\alpha\beta}^a\varepsilon_{\alpha\beta\gamma}\alpha_{\delta\gamma\delta}^{\prime s*} - 2\alpha_{\alpha\beta}^a\varepsilon_{\alpha\beta\gamma}\alpha_{\delta\gamma\delta}^{\prime a*} - 2\alpha_{\alpha\beta}^s\varepsilon_{\alpha\gamma\delta}\alpha_{\delta\gamma\beta}^{\prime a*}$$

$$+ 2\alpha_{\alpha\beta}^a\varepsilon_{\alpha\gamma\delta}\alpha_{\delta\gamma\beta}^{\prime a*} + 2\alpha_{\alpha\beta}^a\varepsilon_{\alpha\gamma\delta}\alpha_{\beta\gamma\delta}^{\prime a*} - 2\alpha_{\alpha\beta}^{\prime s}\varepsilon_{\alpha\gamma\delta}\alpha_{\beta\gamma\delta}^{s*}$$

$$-\alpha'^{s}_{\alpha\alpha}\varepsilon_{\beta\gamma\delta}\alpha'^{a*}_{\gamma\delta\beta} - \alpha^{a}_{\alpha\beta}\varepsilon_{\alpha\gamma\beta}\alpha^{s*}_{\delta\delta\gamma} + 2\alpha'^{a}_{\alpha\beta}\varepsilon_{\alpha\beta\gamma}\alpha^{s*}_{\delta\gamma\delta}$$

$$+2\alpha'^{a}_{\alpha\beta}\varepsilon_{\alpha\beta\gamma}\alpha^{a*}_{\delta\gamma\delta} + 2\alpha'^{s}_{\alpha\beta}\varepsilon_{\alpha\gamma\delta}\alpha^{a*}_{\delta\gamma\beta} - 2\alpha'^{a}_{\alpha\beta}\varepsilon_{\alpha\gamma\delta}\alpha^{a*}_{\delta\gamma\beta}$$

$$-2\alpha'^{a}_{\alpha\beta}\varepsilon_{\alpha\gamma\delta}\alpha^{a*}_{\beta\gamma\delta} + \frac{1}{kT}\big(2\alpha^{s}_{\alpha\beta}\varepsilon_{\alpha\gamma\delta}\alpha'^{s*}_{\beta\gamma}m_{n\delta} + \alpha^{s}_{\alpha\alpha}\varepsilon_{\beta\gamma\delta}\alpha'^{a*}_{\gamma\delta}m_{n\beta}$$

$$-\alpha'^{s}_{\alpha\alpha}\varepsilon_{\beta\gamma\delta}\alpha^{a*}_{\gamma\delta}m_{n\beta} - 2\alpha^{a}_{\alpha\beta}\varepsilon_{\alpha\beta\gamma}\alpha'^{s*}_{\delta\gamma}m_{n\delta} - 2\alpha^{a}_{\alpha\beta}\varepsilon_{\alpha\beta\gamma}\alpha'^{a*}_{\delta\gamma}m_{n\delta}$$

$$-2\alpha^{s}_{\alpha\beta}\varepsilon_{\alpha\gamma\delta}\alpha'^{a*}_{\delta\gamma}m_{n\beta} + 2\alpha^{a}_{\alpha\beta}\varepsilon_{\alpha\gamma\delta}\alpha'^{a*}_{\delta\gamma}m_{n\beta} + 2\alpha^{a}_{\alpha\beta}\varepsilon_{\alpha\gamma\delta}\alpha'^{a*}_{\beta\gamma}m_{n\delta}\big)\Big]P\sin 2\eta,$$

$$(3.5.47b)$$

$$S_{2}^{d}(90°) = 0, \qquad\qquad (3.5.47c)$$

$$S_{3}^{d}(90°) = 0; \qquad\qquad (3.5.47d)$$

$$S_{0}^{d}(180°) = S_{0}^{d}(0°), \qquad\qquad (3.5.48a)$$

$$S_{1}^{d}(180°) = 0, \qquad\qquad (3.5.48b)$$

$$S_{2}^{d}(180°) = 0, \qquad\qquad (3.5.48c)$$

$$S_{3}^{d}(180°) = -S_{3}^{d}(0°), \qquad\qquad (3.5.48d)$$

where again, for simplicity, we have written $(\alpha_{\alpha\beta})^{s}_{mn}$ etc. as $\alpha^{s}_{\alpha\beta}$ etc. The superscripts s and a in $\alpha_{\alpha\beta\gamma}$ and $\alpha'_{\alpha\beta\gamma}$ refer to the symmetry with respect to interchange of the first two tensor subscripts.

The generalized magnetic circular intensity differences follow immediately, but we shall not write them down explicitly because of their complexity. However, it is shown in Chapter 8 that they simplify considerably when applied to a specific situation.

3.5.6 Electric Rayleigh and Raman optical activity

It was shown in Section 1.9.3 that, except in certain magnetic crystals, there is no *simple* electric analogue of the Faraday effect (optical rotation and circular dichroism induced by, and proportional to, a static electric field parallel to the light beam) because such an effect would violate parity and reversality. However, as demonstrated in Section 1.9.4, Rayleigh and Raman optical activity is allowed for light scattered at 90° from all molecules in a static electric field perpendicular to both the incident and scattered directions. The circular intensity difference changes sign if one of the following is reversed: the electric field direction, the incident beam direction or the direction of observation. No electric Rayleigh or Raman optical activity exists for light scattered in the forward or backward direction.

Electric Rayleigh optical activity depends on cross terms between the unperturbed $\alpha_{\alpha\beta}$ and $G'_{\alpha\beta}$ plus $A_{\alpha,\beta\gamma}$ perturbed to first order in the electric field, together with cross terms between $\alpha_{\alpha\beta}$ perturbed to first order in the electric field with the unperturbed $G'_{\alpha\beta}$ plus $A_{\alpha,\beta\gamma}$, the calculation proceeding in an analogous fashion to the magnetic case discussed in the previous section. Similarly for electric Raman optical activity with the dynamic molecular property tensors replaced by corresponding transition tensors. The resulting contributions to the Stokes parameters are complicated and are not given here. We refer to Buckingham and Raab (1975) for explicit expressions for the electric Rayleigh circular intensity difference in a number of molecular symmetries; although it is worth quoting here the expressions for the temperature-dependent contribution in highly polar molecules in isotropic media since this is likely to be the most important case:

$$\Delta_x(90°) = 2(E_x\mu_\alpha/kT)\big[\varepsilon_{\alpha\beta\gamma}\alpha_{\beta\delta}G'^*_{\delta\gamma} - 3\varepsilon_{\beta\gamma\delta}\alpha_{\alpha\beta}G'^*_{\gamma\delta} - \varepsilon_{\alpha\beta\gamma}\alpha_{\delta\delta}G'^*_{\beta\gamma}$$
$$+ \tfrac{1}{3}\omega(\alpha_{\beta\gamma}A^*_{\alpha,\beta\gamma} - \alpha_{\alpha\beta}A^*_{\gamma,\beta\gamma} + \alpha_{\beta\gamma}A^*_{\beta,\gamma\alpha} - \alpha_{\beta\beta}A^*_{\gamma,\gamma\alpha})\big]/$$
$$c(7\alpha_{\lambda\mu}\alpha^*_{\lambda\mu} + \alpha_{\lambda\lambda}\alpha^*_{\mu\mu}), \tag{3.5.49a}$$
$$\Delta_z(90°) = (E_x\mu_\alpha/kT)\big[\varepsilon_{\alpha\beta\gamma}\alpha_{\gamma\delta}G'^*_{\beta\delta} - \varepsilon_{\beta\gamma\delta}\alpha_{\alpha\beta}G'^*_{\gamma\delta}$$
$$+ \tfrac{1}{3}\omega(2\alpha_{\beta\gamma}A^*_{\alpha,\beta\gamma} - 2\alpha_{\alpha\beta}A^*_{\gamma,\beta\gamma} - \alpha_{\beta\gamma}A^*_{\beta,\gamma\alpha} + \alpha_{\beta\beta}A^*_{\gamma,\gamma\alpha})\big]/$$
$$c(3\alpha_{\lambda\mu}\alpha^*_{\lambda\mu} - \alpha_{\lambda\lambda}\alpha^*_{\mu\mu}). \tag{3.5.49b}$$

Although the natural optical activity tensors $G'_{\alpha\beta}$ and $A_{\alpha,\beta\gamma}$ are involved, it is emphasized that the molecules do not need to be chiral to show electric Rayleigh and Raman optical activity.

De Figueiredo and Raab (1981) have given a molecular theory of a number of other differential light scattering effects that are of the same order as electric Rayleigh optical activity.

4

Symmetry and optical activity

Ubi materia, ibi geometria. Johannes Kepler

4.1 Introduction

This chapter is a rambling affair. It collects together a number of disparate topics, all of which have some bearing on the application of symmetry arguments to molecular properties in general and optical activity in particular.

Optical activity is a splendid subject for the application of symmetry principles. As well as conventional point group symmetry, the fundamental symmetries of space inversion, time reversal and even charge conjugation have something to say about optical activity at all levels: the experiments that show up optical activity observables, the objects generating these observables and the nature of the quantum states that these objects must be able to support. There are also technical matters such as the simplification and evaluation of matrix elements using irreducible tensor methods, a topic of great importance in magnetic optical activity. One topic set apart from the others is the application of permutation symmetry to ligand sites on molecular skeletons: this generates an imposing algebra based on 'chirality functions' which gives mathematical insight into the phenomenon of molecular chirality.

4.2 Cartesian tensors

In this book, considerable use is made of a cartesian tensor notation, and the symmetry aspects of various phenomena discussed in terms of the transformation properties of the corresponding molecular property tensors. A review of the relevant parts of the theory of cartesian tensors is therefore appropriate. More complete accounts can be found in works such as Jeffreys (1931), Milne (1948), Temple (1960) and Bourne and Kendall (1977). A knowledge of elementary vector algebra is assumed.

170

4.2.1 Scalars, vectors and tensors

A *scalar* physical quantity, such as density or temperature, is not associated in any way with a direction and is specified by a single number.

A *vector* physical quantity, such as velocity or electric field strength, is associated with a single direction and is specified by a scalar magnitude and the direction. A vector is specified analytically by resolving the components along three mutually perpendicular directions defined by unit vectors. Thus if $\mathbf{i}, \mathbf{j}, \mathbf{k}$ are unit vectors along the axes x, y, z and V_x, V_y, V_z are the corresponding components of a vector $\mathbf{V}$, we write

$$\mathbf{V} = V_x\mathbf{i} + V_y\mathbf{j} + V_z\mathbf{k}. \tag{4.2.1}$$

Alternative representations include the triad

$$\mathbf{V} = (V_x, V_y, V_z), \tag{4.2.2}$$

which is not meant to be a row matrix, and the column matrix

$$\mathbf{V} = \begin{pmatrix} V_x \\ V_y \\ V_z \end{pmatrix}, \tag{4.2.3a}$$

with the row matrix

$$\mathbf{V}^{\mathrm{T}} = (V_x\, V_y\, V_z) \tag{4.2.3b}$$

as its transpose. Thus in matrix notation, the *scalar product* of two vectors $\mathbf{V}$ and $\mathbf{W}$ is

$$\mathbf{V} \cdot \mathbf{W} = \mathbf{V}^{\mathrm{T}}\mathbf{W} = V_x W_x + V_y W_y + V_z W_z. \tag{4.2.4}$$

The magnitude of a vector is defined as

$$|\mathbf{V}| = V = \left(V_x^2 + V_y^2 + V_z^2\right)^{\frac{1}{2}}. \tag{4.2.5}$$

A physical quantity associated with two or more directions is called a *tensor.* Thus the electric polarizability α of a molecule is a tensor since it relates the induced electric dipole moment vector to the applied electric field vector through

$$\mu = \alpha \cdot \mathbf{E}. \tag{4.2.6}$$

The directions of the influence $\mathbf{E}$ and the response μ are not necessarily the same on account of anisotropy in the electrical properties of the molecule. If μ and $\mathbf{E}$ are written in the form (4.2.1), then α must be written as the *dyad*

$$\alpha = \alpha_{xx}\mathbf{ii} + \alpha_{xy}\mathbf{ij} + \alpha_{xz}\mathbf{ik} + \alpha_{yx}\mathbf{ji} + \alpha_{yy}\mathbf{jj}$$
$$+ \alpha_{yz}\mathbf{jk} + \alpha_{zx}\mathbf{ki} + \alpha_{zy}\mathbf{kj} + \alpha_{zz}\mathbf{kk}. \tag{4.2.7}$$

If the vectors $\boldsymbol{\mu}$ and $\mathbf{E}$ are written in the column matrix form (4.2.3a), then $\boldsymbol{\alpha}$ must be written as the square matrix

$$\boldsymbol{\alpha} = \begin{pmatrix} \alpha_{xx} & \alpha_{xy} & \alpha_{xz} \\ \alpha_{yx} & \alpha_{yy} & \alpha_{yz} \\ \alpha_{zx} & \alpha_{zy} & \alpha_{zz} \end{pmatrix}. \tag{4.2.8}$$

Whatever representation is used, if the components of (4.2.6) are written out explicitly, the same result obtains:

$$\mu_x = \alpha_{xx} E_x + \alpha_{xy} E_y + \alpha_{xz} E_z,$$
$$\mu_y = \alpha_{yx} E_x + \alpha_{yy} E_y + \alpha_{yz} E_z,$$
$$\mu_z = \alpha_{zx} E_x + \alpha_{zy} E_y + \alpha_{zz} E_z. \tag{4.2.9}$$

Tensor manipulations are simplified considerably by the use of the following notation. The set of equations (4.2.9) can be written

$$\mu_\alpha = \sum_{\beta = x,y,z} \alpha_{\alpha\beta} E_\beta, \quad \alpha = x, y, z. \tag{4.2.10}$$

The summation sign is now omitted, and the *Einstein summation convention* introduced: when a Greek suffix occurs twice in the same term, summation with respect to that suffix is understood. Thus (4.2.10) is now written

$$\mu_\alpha = \alpha_{\alpha\beta} E_\beta. \tag{4.2.11}$$

In these equations, α is called a free suffix and β a dummy suffix. μ_α denotes the array of three numbers that specifies the vector $\boldsymbol{\mu}$, and $\alpha_{\alpha\beta}$ denotes the array of nine numbers that specifies the tensor $\boldsymbol{\alpha}$. In this book, Greek letters are used for free or dummy suffixes, whereas Roman letters or numerals are used for suffixes which denote specific tensor components.

Although the word tensor is often reserved for physical quantities associated with two or more directions, we shall see that it is more systematic to generalize the definition of a tensor so as to include scalars and vectors. Thus a scalar is a tensor of rank zero, being specified by a number unrelated to any axis. A vector is a tensor of the first rank, being specified by three numbers, each of which is associated with one coordinate axis. A tensor of the second rank is specified by nine numbers, each of which is associated with two coordinate axes. Tensors of higher rank may be introduced as natural extensions: thus a third-rank tensor is specified by 27 numbers which form, not a square array as in (4.2.8), but a cubic array. Notice that, except in the case of a tensor of zero rank, the actual values of the numbers in the array specifying a tensor will change as the coordinate axes are rotated because they are associated with both the axes and with the tensor quantity itself, which is a physical entity that retains its identity however the axes are changed. We shall see that a study of the relationships between the components of a tensor in one coordinate system and those in another will provide an indication of the essential character of a particular tensor.

The operation, implied in the Einstein summation convention, of putting two suffixes equal in a tensor and summing, is known as *contraction* and gives a new tensor whose rank is less by two than that of the original tensor. Thus contraction is the tensor equivalent of the scalar product in vector analysis. So we can write the scalar product of two vectors **V** and **W** as

$$\mathbf{V} \cdot \mathbf{W} = V_\alpha W_\alpha = V_x W_x + V_y W_y + V_z W_z. \tag{4.2.12}$$

Hence contraction of the second-rank tensor **VW** (a dyadic product) has given a tensor of rank zero (a scalar).

A tensor with components that satisfy

$$T_{\alpha\beta} = T_{\beta\alpha} \tag{4.2.13}$$

for all α and β is said to be *symmetric*. On the other hand, if

$$T_{\alpha\beta} = -T_{\beta\alpha} \tag{4.2.14}$$

for all α and β, the tensor is said to be *antisymmetric*. Clearly the diagonal elements of a second-rank antisymmetric tensor are zero. This definition may be extended to tensors of higher rank, the symmetry or antisymmetry being defined with respect to a particular pair of suffixes. Notice that any second-rank tensor can be represented as a sum of a symmetric tensor $T_{\alpha\beta}^{s}$ and an antisymmetric tensor $T_{\alpha\beta}^{a}$:

$$T_{\alpha\beta} = T_{\alpha\beta}^{s} + T_{\alpha\beta}^{a}, \tag{4.2.15a}$$

$$T_{\alpha\beta}^{s} = \tfrac{1}{2}(T_{\alpha\beta} + T_{\beta\alpha}), \tag{4.2.15b}$$

$$T_{\alpha\beta}^{a} = \tfrac{1}{2}(T_{\alpha\beta} - T_{\beta\alpha}). \tag{4.2.15c}$$

This decomposition is a step towards the construction of irreducible tensorial sets, to be encountered later.

4.2.2 Rotation of axes

Consider two sets of cartesian axes x, y, z and x', y', z' with a common origin **O**. The relative orientation of the two sets may be specified by a set of nine *direction cosines* $l_{\lambda'\alpha}$ where, for example, $\cos^{-1} l_{x'y}$ is the angle between the x' and the y axes. (Although summation would not be implied in a direction cosine such as

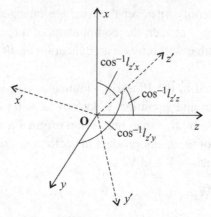

Fig. 4.1 The direction cosines specifying the orientation of a rotated axis system (primed) relative to the original axis system.

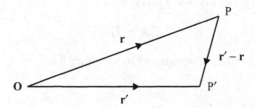

Fig. 4.2 The relative positions of two points P and P'.

$l_{\alpha'\alpha}$ because α' and α are components of different axis systems, in order to avoid possible confusion we shall use suffixes $\lambda', \mu', \nu' \dots$ for the primed axis system, and suffixes $\alpha, \beta, \gamma \dots$ for the unprimed system.) Thus the direction cosines of a particular axis z' with respect to x, y, z are $l_{z'x}, l_{z'y}, l_{z'z}$ (Fig. 4.1); and the direction cosines of z with respect to x', y', z' are $l_{x'z}, l_{y'z}, l_{z'z}$.

Important relations exist between direction cosines. Consider a point P defined by a position vector $\mathbf{r}$ from the origin. Denoting the direction cosines between $\mathbf{r}$ and the x, y, z axes by l, m, n, we can write

$$l = x/r, m = y/r, n = z/r. \qquad (4.2.16)$$

Since $r^2 = x^2 + y^2 + z^2$, we have

$$l^2 + m^2 + n^2 = 1, \qquad (4.2.17)$$

which can be regarded as a normalization relation. Now consider a second point P' defined by a position vector $\mathbf{r}'$ with orientation specified by the direction cosines l', m', n' and making an angle θ with $\mathbf{r}$. Applying the cosine rule with reference to Fig. 4.2, taking $r = r' = 1$ and using (4.2.16) and (4.2.17), we find

$$
\begin{aligned}
\cos\theta &= (r^2 + r'^2 - |\mathbf{r}' - \mathbf{r}|^2)/2rr' = 1 - \tfrac{1}{2}|\mathbf{r}' - \mathbf{r}|^2 \\
&= 1 - \tfrac{1}{2}[(x' - x)^2 + (y' - y)^2 + (z' - z)^2] \\
&= 1 - \tfrac{1}{2}[(l' - l)^2 + (m' - m)^2 + (n' - n)^2] \\
&= 1 - \tfrac{1}{2}[(l'^2 + m'^2 + n'^2) + (l^2 + m^2 + n^2) - 2(ll' + mm' + nn')] \\
&= ll' + mm' + nn'.
\end{aligned}
\tag{4.2.18}
$$

Thus if $\mathbf{r}$ and $\mathbf{r}'$ are perpendicular to each other, the following orthogonality relation obtains:

$$
ll' + mm' + nn' = 0.
\tag{4.2.19}
$$

We now apply these relations to the direction cosines that specify the relative orientation of the two sets of axes x, y, z and x', y', z'. From (4.2.17), the sum of the squares of the direction cosines relating a particular axis in one coordinate system to the three axes in the other system are unity. Thus, concentrating on each of the axes x', y', z' we obtain

$$
\begin{aligned}
l_{x'x}^2 + l_{x'y}^2 + l_{x'z}^2 &= 1, \\
l_{y'x}^2 + l_{y'y}^2 + l_{y'z}^2 &= 1, \\
l_{z'x}^2 + l_{z'y}^2 + l_{z'z}^2 &= 1.
\end{aligned}
\tag{4.2.20}
$$

Also, since x', y', z' are mutually perpendicular, we can use (4.2.19) to write

$$
\begin{aligned}
l_{x'x}l_{y'x} + l_{x'y}l_{y'y} + l_{x'z}l_{y'z} &= 0, \\
l_{y'x}l_{z'x} + l_{y'y}l_{z'y} + l_{y'z}l_{z'z} &= 0, \\
l_{z'x}l_{x'x} + l_{z'y}l_{x'y} + l_{z'z}l_{x'z} &= 0.
\end{aligned}
\tag{4.2.21}
$$

The six equations (4.2.20) and (4.2.21) are called the *orthonormality relations*. Equivalent orthonormality relations can be obtained by concentrating on the other set of axes x, y, z:

$$
\begin{aligned}
l_{x'x}^2 + l_{y'x}^2 + l_{z'x}^2 &= 1, \\
l_{x'y}^2 + l_{y'y}^2 + l_{z'y}^2 &= 1, \\
l_{x'z}^2 + l_{y'z}^2 + l_{z'z}^2 &= 1;
\end{aligned}
\tag{4.2.22}
$$

$$
\begin{aligned}
l_{x'x}l_{x'y} + l_{y'x}l_{y'y} + l_{z'x}l_{z'y} &= 0, \\
l_{x'y}l_{x'z} + l_{y'y}l_{y'z} + l_{z'y}l_{z'z} &= 0, \\
l_{x'z}l_{x'x} + l_{y'z}l_{y'x} + l_{z'z}l_{z'x} &= 0.
\end{aligned}
\tag{4.2.23}
$$

Introducing the *Kronecker delta* defined by

$$\delta_{\alpha\beta} = \begin{cases} 0 & \text{when } \alpha \neq \beta, \\ 1 & \text{when } \alpha = \beta, \end{cases} \tag{4.2.24}$$

together with the summation convention, the orthonormality relations can be embodied in the single equations

$$l_{\lambda'\alpha}l_{\mu'\alpha} = \delta_{\lambda'\mu'}, \tag{4.2.25a}$$

$$l_{\lambda'\alpha}l_{\lambda'\beta} = \delta_{\alpha\beta}. \tag{4.2.25b}$$

For example, taking $\lambda' = \mu' = x'$, (4.2.25a) becomes

$$l_{x'x}^2 + l_{x'y}^2 + l_{x'z}^2 = 1$$

as in (4.2.20); and taking $\lambda' = x'$, $\mu' = y'$, (4.2.25a) becomes

$$l_{x'x}l_{y'x} + l_{x'y}l_{y'y} + l_{x'z}l_{y'z} = 0$$

as in (4.2.21).

The direction cosines enable the components $(V_{x'}, V_{y'}, V_{z'})$ of a vector **V** expressed in a new coordinate system x', y', z' to be written immediately in terms of its components (V_x, V_y, V_z) expressed in an original axis system x, y, z. Thus resolving each of V_x, V_y, V_z along each of x', y', z' in turn, we obtain

$$V_{x'} = l_{x'x}V_x + l_{x'y}V_y + l_{x'z}V_z,$$
$$V_{y'} = l_{y'x}V_x + l_{y'y}V_y + l_{y'z}V_z, \tag{4.2.26}$$
$$V_{z'} = l_{z'x}V_x + l_{z'y}V_y + l_{z'z}V_z.$$

Using the summation convention, these equations can be written

$$V_{\lambda'} = l_{\lambda'\alpha}V_\alpha. \tag{4.2.27a}$$

The corresponding inverse transformation is

$$V_\alpha = l_{\lambda'\alpha}V_{\lambda'}. \tag{4.2.27b}$$

These equations show how the components of a vector transform under a rotation of the axes.

We can now go on and write the components of a second-rank tensor expressed in the new axis system x', y', z' in terms of the components in the original system x, y, z. For example, the defining equation (4.2.11) for the polarizability tensor can be written in the x', y', z' system as

$$\mu_{\lambda'} = \alpha_{\lambda'\mu'}E_{\mu'}. \tag{4.2.28}$$

Successive application of (4.2.27*a*) and (4.2.27*b*) yields

$$\mu_{\lambda'} = l_{\lambda'\alpha}\mu_\alpha = l_{\lambda'\alpha}\alpha_{\alpha\beta}E_\beta = l_{\lambda'\alpha}\alpha_{\alpha\beta}l_{\mu'\beta}E_{\mu'}, \tag{4.2.29}$$

and comparing (4.2.29) with (4.2.28) gives

$$\alpha_{\lambda'\mu'} = l_{\lambda'\alpha}l_{\mu'\beta}\alpha_{\alpha\beta}. \tag{4.2.30}$$

This result illustrates the economy of the dummy suffix notation since the single equation (4.2.30) represents nine equations, each with nine terms on the right hand side.

Direction cosines should not be confused with second-rank tensors. Although $l_{\lambda'\alpha}$ and $\alpha_{\alpha\beta}$ are both arrays of nine numbers, they are very different quantities. The $l_{\lambda'\alpha}$ relate two sets of axes, whereas the $\alpha_{\alpha\beta}$ represent a physical quantity referred to one particular set of axes. It would be meaningless to speak of transforming the $l_{\lambda'\alpha}$ to another set of axes.

The existence of the transformation law (4.2.27*a*) for the components of a vector and (4.2.30) for the components of a second-rank tensor, together with the fact that a scalar is invariant under a rotation of the axes, suggests that a tensor be defined as a quantity which transforms according to

$$T_{\lambda'\mu'\nu'}\ldots = l_{\lambda'\alpha}l_{\mu'\beta}l_{\nu'\gamma}\ldots T_{\alpha\beta\gamma}\ldots. \tag{4.2.31}$$

The number of suffixes attached to $T_{\alpha\beta\gamma\ldots}$ determines the rank of the tensor. This is the reason behind our earlier statement that scalars and vectors are to be regarded as tensors of rank zero and rank one, respectively. It is emphasized that although, according to this definition, we cannot describe a tensor without reference to some coordinate system, the tensor itself is to be distinguished from any one of its descriptions. No meaning attaches to asking whether a particular set of numbers constitutes a tensor or not. It is only when we are given a rule for obtaining the corresponding set of numbers in any other coordinate system that we can compare the rule with (4.2.31) and so answer the question.

4.2.3 Polar and axial tensors

In order to generalize further the transformation law (4.2.31), it is necessary to distinguish between polar and axial tensors. We saw in Section 1.9.2 that a polar vector such as a position vector changes sign under space inversion, whereas an axial vector such as angular momentum does not. If, instead of actually inverting the vectors, we invert the coordinate axes, then the components of a polar vector will change sign and the components of an axial vector will not. An inversion of the axes, or reflecting them in a plane, is equivalent to changing the hand of the axes, as illustrated in Fig. 4.3. Consequently, the generalization of the tensor transformation

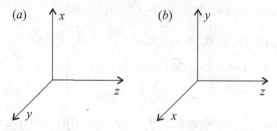

Fig. 4.3 (*a*) a right-handed and (*b*) a left-handed axis system.

law (4.2.31) to a *polar* tensor of any rank is

$$P_{\lambda'\mu'\nu'} \ldots = l_{\lambda'\alpha} l_{\mu'\beta} l_{\nu'\gamma} \ldots P_{\alpha\beta\gamma} \ldots, \qquad (4.2.32a)$$

and to an *axial* tensor of any rank is

$$A_{\lambda'\mu'\nu'} \ldots = (\pm) l_{\lambda'\alpha} l_{\mu'\beta} l_{\nu'\gamma} \ldots A_{\alpha\beta\gamma} \ldots . \qquad (4.2.32b)$$

The negative sign in (4.2.32b) is taken for transformations such as reflections and inversions which change the sign of the axes (*improper rotations*), and the positive sign for transformations which do not change the sign of the axes (*proper rotations*).

As an illustration, we use these transformation laws to determine the effect of an inversion of the coordinate axes on a polar and on an axial vector. The direction cosine corresponding to an inversion is $l_{\lambda'\alpha} = \cos \pi = -1$, since the angle between the new axis λ' and the old axis α is 180°. Thus, applying (4.2.32a), the new components $P_{\lambda'}$ of a polar vector after inversion are related to the original components P_α by

$$P_{\lambda'} = l_{\lambda'\alpha} P_\alpha = -P_\alpha;$$

applying (4.2.32b), the new components of an axial vector are

$$A_{\lambda'} = -l_{\lambda'\alpha} A_\alpha = A_\alpha,$$

as required.

The scalar product $P_\alpha P'_\alpha$ of two polar vectors, or $A_\alpha A'_\alpha$ of two axial vectors, is clearly a number that does not change sign under inversion: a *scalar*. But the scalar product $P_\alpha A_\alpha$ of a polar and an axial vector is a number that does change sign under inversion, and this is called a *pseudoscalar.*

Vector analysis defines the vector product $\mathbf{P} \times \mathbf{P}'$ of two polar vectors $\mathbf{P}$ and $\mathbf{P}'$ as a vector with magnitude equal to the area of the parallelogram defined by the two vectors and with direction $\mathbf{n}$ perpendicular to the parallelogram in the direction which makes $\mathbf{P}, \mathbf{P}', \mathbf{n}$ a right-handed set. Thus if $\mathbf{i}, \mathbf{j}, \mathbf{k}$ are unit vectors associated with a right-handed axis system, the vector product is written analytically as

$$\mathbf{P} \times \mathbf{P}' = (P_y P_z' - P_z P_y')\mathbf{i} + (P_z P_x' - P_x P_z')\mathbf{j}$$
$$+ (P_x P_y' - P_y P_x')\mathbf{k} \equiv \mathbf{A}. \tag{4.2.33}$$

The components of this axial vector $\mathbf{A}$ are equivalent to the components of the second-rank antisymmetric polar tensor

$$P_{\alpha\beta} = P_\alpha P_\beta' - P_\beta P_\alpha' = -P_{\beta\alpha}. \tag{4.2.34}$$

The explicit components are

$$\begin{pmatrix} P_{xx} & P_{xy} & P_{xz} \\ P_{yx} & P_{yy} & P_{yz} \\ P_{zx} & P_{zy} & P_{zz} \end{pmatrix} =$$

$$\begin{pmatrix} 0 & (P_x P_y' - P_y P_x') & (P_x P_z' - P_z P_x') \\ -(P_x P_y' - P_y P_x') & 0 & (P_y P_z' - P_z P_y') \\ -(P_x P_z' - P_z P_x') & -(P_y P_z' - P_z P_y') & 0 \end{pmatrix}$$

$$= \begin{pmatrix} 0 & A_z & -A_y \\ -A_z & 0 & A_x \\ A_y & -A_x & 0 \end{pmatrix}. \tag{4.2.35}$$

In general, an axial tensor can be represented by an antisymmetric polar tensor of higher rank, which usually provides a more fundamental description of the corresponding physical entity.

Vector products are formulated in tensor notation by means of the *alternating* or *Levi-Città tensor* $\varepsilon_{\alpha\beta\gamma}$ which is a completely antisymmetric unit axial tensor of the third rank. The only nonvanishing components of $\varepsilon_{\alpha\beta\gamma}$ are those with three different suffixes. We set $\varepsilon_{xyz} = 1$ and the other nonvanishing components are either $+1$ or -1 depending on whether the sequence $\alpha\beta\gamma$ can be brought to the order xyz by a cyclic or a noncyclic permutation. Thus we define

$$\varepsilon_{xyz} = \varepsilon_{zxy} = \varepsilon_{yzx} = 1,$$
$$\varepsilon_{xzy} = \varepsilon_{yxz} = \varepsilon_{zyx} = -1, \tag{4.2.36}$$
$$\text{all other components zero.}$$

This definition applies whether x, y, z are axes of a right- or left-handed coordinate system, because components of axial vectors of odd rank do not change sign under an inversion of the coordinate axes. The tensor formulation of the vector product (4.2.33) is therefore

$$A_\alpha = \varepsilon_{\alpha\beta\gamma} P_\beta P_\gamma'. \tag{4.2.37}$$

For example, taking $\alpha = x$, summing over pairs of repeated suffixes, and remembering the definition (4.2.36), we find

$$
\begin{aligned}
A_x &= \varepsilon_{xxx} P_x P'_x + \varepsilon_{xxy} P_x P'_y + \varepsilon_{xxz} P_x P'_z \\
&\quad + \varepsilon_{xyx} P_y P'_x + \varepsilon_{xyy} P_y P'_y + \varepsilon_{xyz} P_y P'_z \\
&\quad + \varepsilon_{xzx} P_z P'_x + \varepsilon_{xzy} P_z P'_y + \varepsilon_{xzz} P_z P'_z \\
&= P_y P'_z - P_z P'_y.
\end{aligned}
$$

An important term in the theory of optical activity is $\varepsilon_{\alpha\gamma\delta} A_{\gamma,\delta\beta}$, where $A_{\gamma,\delta\beta}$ is the electric dipole–electric quadrupole tensor (2.6.27c) (it is unfortunate that we use the same symbol for a general axial tensor). Since $\varepsilon_{\alpha\gamma\delta}$ is a third-rank axial tensor and $A_{\gamma,\delta\beta}$ is a third-rank polar tensor, and contraction with respect to two pairs of suffixes is specified, the complete term transforms as a second-rank axial tensor just like the electric dipole–magnetic dipole tensor $G'_{\alpha\beta}$ (2.6.27f).

4.2.4 Some algebra of unit tensors

The Kronecker delta $\delta_{\alpha\beta}$ defined in (4.2.24) is a symmetric unit polar tensor of the second rank. The alternating tensor $\varepsilon_{\alpha\beta\gamma}$ defined in (4.2.36) is a completely antisymmetric unit axial tensor of the third rank. We now collect together a few useful relationships involving $\delta_{\alpha\beta}$ and $\varepsilon_{\alpha\beta\gamma}$.

Consider first contraction with respect to the two suffixes in the delta tensor:

$$
\delta_{\alpha\alpha} = \delta_{xx} + \delta_{yy} + \delta_{zz} = 3. \tag{4.2.38}
$$

This is actually equivalent to the product

$$
\begin{aligned}
\delta_{\alpha\beta}\delta_{\alpha\beta} &= \delta_{xx}\delta_{xx} + \delta_{xy}\delta_{xy} + \delta_{xz}\delta_{xz} \\
&\quad + \delta_{yx}\delta_{yx} + \delta_{yy}\delta_{yy} + \delta_{yz}\delta_{yz} \\
&\quad + \delta_{zx}\delta_{zx} + \delta_{zy}\delta_{zy} + \delta_{zz}\delta_{zz} \\
&= 3. \tag{4.2.39}
\end{aligned}
$$

In view of (4.2.38),

$$
\delta_{\alpha\alpha}\delta_{\beta\beta} = 9. \tag{4.2.40}
$$

Since a component of the alternating tensor having any two subscripts the same is zero, we find

$$
\delta_{\alpha\beta}\varepsilon_{\alpha\beta\gamma} = \varepsilon_{\alpha\alpha\gamma} = 0. \tag{4.2.41}
$$

A most useful relation between the alternating and delta tensors is

$$
\varepsilon_{\alpha\beta\gamma}\varepsilon_{\delta\lambda\gamma} = \delta_{\alpha\delta}\delta_{\beta\lambda} - \delta_{\alpha\lambda}\delta_{\beta\delta}. \tag{4.2.42}
$$

This may be established as follows. If $\alpha = \beta$ or $\delta = \lambda$ both sides of (4.2.42) vanish independently. Without loss of generality we may now choose $\alpha = x$ and $\beta = y$. The left-hand side of (4.2.42) then becomes

$$\varepsilon_{xyx}\varepsilon_{\delta\lambda x} + \varepsilon_{xyy}\varepsilon_{\delta\lambda y} + \varepsilon_{xyz}\varepsilon_{\delta\lambda z} = \varepsilon_{\delta\lambda z}.$$

The right-hand side becomes

$$\delta_{x\delta}\delta_{y\lambda} - \delta_{x\lambda}\delta_{y\delta} = \Delta, \text{ say.}$$

As $\delta \neq \lambda$ there are just the following possibilities: $\delta = z$, in which case $\Delta = 0$ for all λ; $\lambda = z$, in which case $\Delta = 0$ for all δ; $\delta = x$, $\lambda = y$, giving $\Delta = 1$; $\delta = y$, $\lambda = x$, giving $\Delta = -1$. Hence $\Delta = \varepsilon_{\delta\lambda z}$, and the identity (4.2.42) is proved. Notice that this identity is the tensor equivalent of the vector identity

$$(\mathbf{T} \times \mathbf{U}) \cdot (\mathbf{V} \times \mathbf{W}) = (\mathbf{T} \cdot \mathbf{V})(\mathbf{U} \cdot \mathbf{W}) - (\mathbf{T} \cdot \mathbf{W})(\mathbf{U} \cdot \mathbf{V}).$$

By contraction of (4.2.42) we have

$$\varepsilon_{\alpha\beta\gamma}\varepsilon_{\delta\beta\gamma} = \delta_{\alpha\delta}\delta_{\beta\beta} - \delta_{\alpha\beta}\delta_{\beta\delta}$$
$$= 3\delta_{\alpha\delta} - \delta_{\alpha\delta} = 2\delta_{\alpha\delta}. \tag{4.2.43}$$

Further contraction yields

$$\varepsilon_{\alpha\beta\gamma}\varepsilon_{\alpha\beta\gamma} = \delta_{\alpha\alpha}\delta_{\beta\beta} - \delta_{\alpha\beta}\delta_{\alpha\beta} = 9 - 3 = 6. \tag{4.2.44}$$

Notice that the components of the unit tensors $\delta_{\alpha\beta}$ and $\varepsilon_{\alpha\beta\gamma}$ are identical in all coordinate systems. Such tensors are called *isotropic tensors*, or *tensor invariants*, and play a fundamental role in the study of isotropic materials such as fluids. This is because in a collection of freely rotating molecules, all proper transformations between molecule-fixed and space-fixed axes are possible so that, on the average, only the tensor invariants survive. General higher-rank tensor invariants are written in terms of $\delta_{\alpha\beta}$ and $\varepsilon_{\alpha\beta\gamma}$: thus fourth-rank and fifth-rank tensor invariants are linear combinations of products such as $\delta_{\alpha\beta}\delta_{\gamma\delta}$ and $\varepsilon_{\alpha\beta\gamma}\delta_{\delta\varepsilon}$, respectively. We shall see in the next section that the isotropic averages of tensor components are always expressed in terms of isotropic tensors.

4.2.5 *Isotropic averages of tensor components*

A problem encountered frequently in the theory of light scattering from isotropic collections of molecules such as fluids is the evaluation of isotropic averages of tensor components. This problem reduces to the evaluation of products of direction cosines, between particular pairs of axes in a molecule-fixed and a space-fixed coordinate system, averaged over all possible relative orientations of the two coordinate systems. Thus an expression for an observable, such as a polarization change, is first written in terms of molecular property tensor components specified in space-fixed axes: since we want to relate the observable to intrinsic molecular properties, we must transform to a set of axes fixed to the molecule's frame. Then if the molecule is tumbling freely, the expressions must be averaged over all orientations.

If the primed suffixes refer to space-fixed and unprimed to molecule-fixed axes we have, from the polar tensor transformation law (4.2.32*a*), the following general expression for the isotropic average of a general tensor component:

$$\langle P_{\lambda'\mu'\nu'} \ldots \rangle = \langle l_{\lambda'\alpha} l_{\mu'\beta} l_{\nu'\gamma} \ldots \rangle P_{\alpha\beta\gamma} \ldots \qquad (4.2.45)$$

Notice that we do not need to invoke the axial tensor transformation law (4.2.32*b*), even when averaging axial tensor components, because no improper rotations are involved. The first few averages can be obtained from a simple trigonometric analysis. It is now necessary to consider explicit tensor components, and in order to produce results in a notation conforming with that used in the rest of the book, we shall take x, y, z and X, Y, Z as the space-fixed and molecule-fixed coordinate systems, respectively (while still using general Greek subscripts $\lambda', \mu', \nu' \ldots$ and $\alpha, \beta, \gamma \ldots$ for the former and latter). It is also convenient to use the replacements $i_\alpha = l_{x\alpha}, j_\alpha = l_{y\alpha}, k_\alpha = l_{z\alpha}$ for direction cosines between the space fixed axes x, y, z and a molecule fixed axis α, where, as usual, $\mathbf{i}, \mathbf{j}, \mathbf{k}$ are unit vectors along x, y, z. Thus the isotropic average of a tensor component such as P_{xyzy}, for example, would be written

$$\langle P_{xyzy} \rangle = \langle i_\alpha j_\beta k_\gamma j_\delta \rangle P_{\alpha\beta\gamma\delta},$$

for which the problem reduces to evaluating $\langle i_\alpha j_\beta k_\gamma j_\delta \rangle$.

It is first necessary to note the form of the average of certain trigonometric functions over a sphere. If we denote the angle between a space-fixed and a molecule-fixed axis by θ, and identify θ with the polar angle in spherical coordinates, then isotropic averages of products of the same direction cosine take the form

$$\langle \cos^n \theta \rangle = \frac{\int_0^{2\pi} d\phi \int_0^{\pi} \cos^n \theta \sin \theta \, d\theta}{\int_0^{2\pi} d\phi \int_0^{\pi} \sin \theta \, d\theta}$$

since the volume element in spherical coordinates is $\sin\theta \, d\theta \, d\phi$, where ϕ is the azimuthal angle. On integration, the following general result obtains:

$$\langle \cos^n \theta \rangle = \begin{cases} \dfrac{1}{2k+1} & \text{for } n = 2k; \\ 0 & \text{for } n = 2k+1; \text{ with } k = 0, 1, 2, 3 \ldots. \end{cases} \qquad (4.2.46)$$

Consider first the isotropic average of a single direction cosine, say i_X. Taking the angle between the x axis and the X axis to be θ, we have

$$\langle i_X \rangle = \langle \cos \theta \rangle = 0$$

since, according to (4.2.46), the average of $\cos \theta$ over a sphere is zero. The same result obtains for any single direction cosine, so we may write

$$\langle i_\alpha \rangle = \langle j_\alpha \rangle = \langle k_\alpha \rangle = 0. \tag{4.2.47}$$

Consider next the isotropic average of a product of two direction cosines. If the two are the same, say i_X, we have from (4.2.46)

$$\langle i_X^2 \rangle = \langle \cos^2 \theta \rangle = \tfrac{1}{3}.$$

The same result obtains for any pair of identical direction cosines. Notice that the same result can be deduced by writing out a scalar product in the X, Y, Z coordinate system of a unit vector in the x, y, z coordinate system with itself and averaging both sides: for example, from $i_\alpha i_\alpha = 1$ we can write

$$\langle i_X^2 \rangle + \langle i_Y^2 \rangle + \langle i_Z^2 \rangle = 1,$$

and since the three averages are all equal, each has the value $\tfrac{1}{3}$. The isotropic average of any pair of different direction cosines is zero. For example, from $i_\alpha j_\alpha = 0$ we can write

$$\langle i_X j_X \rangle + \langle i_Y j_Y \rangle + \langle i_Z j_Z \rangle = 0,$$

and since the three averages are all equal, they must separately be zero. This analysis can be summarized neatly in terms of the *second-rank tensor invariant* $\delta_{\alpha\beta}$:

$$\langle i_\alpha i_\beta \rangle = \langle j_\alpha j_\beta \rangle = \langle k_\alpha k_\beta \rangle = \tfrac{1}{3}\delta_{\alpha\beta}, \tag{4.2.48}$$

with all other types of average equal to zero.

Consider now the isotropic average of a product of three direction cosines. This can be deduced by considering expressions such as $(\mathbf{i} \times \mathbf{j}) \cdot \mathbf{k} = 1$. Writing this out in terms of components in the X, Y, Z coordinate system, we have

$$(i_Y j_Z - i_Z j_Y)k_X + (i_Z j_X - i_X j_Z)k_Y + (i_X j_Y - i_Y j_X)k_Z = 1.$$

Averaging both sides, and recognizing that the averages of the three terms are all equal, yields the result

$$\langle i_Y j_Z k_X \rangle = -\langle i_Z j_Y k_X \rangle = \langle i_Z j_X k_Y \rangle = -\langle i_X j_Z k_Y \rangle = \langle i_X j_Y k_Z \rangle = -\langle i_Y j_X k_Z \rangle = \tfrac{1}{6}.$$

By considering expressions such as $(\mathbf{i} \times \mathbf{j}) \cdot \mathbf{j} = 0$ it can be shown that all other types of isotropic average are zero. These results can be summarized in terms of the *third-rank tensor invariant* $\varepsilon_{\alpha\beta\gamma}$;

$$\langle i_\alpha j_\beta k_\gamma \rangle = \tfrac{1}{6}\varepsilon_{\alpha\beta\gamma}, \tag{4.2.49}$$

with all other types of average equal to zero.

We turn now to the isotropic average of a product of four direction cosines. If the four are the same, say i_X, we have from (4.2.46)

$$\langle i_X^4 \rangle = \langle \cos^4 \theta \rangle = \tfrac{1}{5}. \tag{4.2.50}$$

Similarly for the product of any other four identical direction cosines. We can obtain the isotropic averages of products of pairs of identical direction cosines from the orthonormality relations (4.2.25). For example, taking a particular normalization relation such as

$$i_X^2 + i_Y^2 + i_Z^2 = 1,$$

squaring both sides and averaging, gives

$$3\langle i_X^4 \rangle + 6\langle i_X^2 i_Y^2 \rangle = 1$$

since $\langle i_X^4 \rangle = \langle i_Y^4 \rangle = \langle i_Z^4 \rangle$ and $\langle i_X^2 i_Y^2 \rangle = \langle i_X^2 i_Z^2 \rangle = \langle i_Y^2 i_Z^2 \rangle$. Using (4.2.50), we then obtain averages such as

$$\langle i_X^2 i_Y^2 \rangle = \tfrac{1}{6}\left(1 - \tfrac{3}{5}\right) = \tfrac{1}{15}. \tag{4.2.51}$$

Similarly, starting with

$$i_X^2 + j_X^2 + k_X^2 = 1,$$

we obtain averages such as

$$\langle i_X^2 j_X^2 \rangle = \tfrac{1}{15}. \tag{4.2.52}$$

Taking a product such as

$$(i_X^2 + j_X^2 + k_X^2)(i_Y^2 + j_Y^2 + k_Y^2) = 1,$$

we can write

$$3\langle i_X^2 i_Y^2 \rangle + 6\langle i_X^2 j_Y^2 \rangle = 1,$$

and using (4.2.51), we obtain averages such as

$$\langle i_X^2 j_Y^2 \rangle = \tfrac{1}{6}\left(1 - \tfrac{3}{15}\right) = \tfrac{2}{15}.$$

Finally, taking a particular orthogonality relation such as

$$i_X j_X + i_Y j_Y + i_Z j_Z = 0$$

and squaring both sides enables us to write

$$3\langle i_X^2 j_X^2 \rangle + 6\langle i_X j_X i_Y j_Y \rangle = 0$$

from which, using (4.2.52), we obtain averages such as

$$\langle i_X j_X i_Y j_Y \rangle = \tfrac{1}{6}\left(-\tfrac{3}{15}\right) = -\tfrac{1}{30}.$$

All other types of isotropic average are zero. These results can be summarized in terms of the *fourth-rank tensor invariant* $\delta_{\alpha\beta}\delta_{\gamma\delta}$ (Buckingham and Pople, 1955; Kielich, 1961):

$$\langle i_\alpha i_\beta i_\gamma i_\delta \rangle = \langle j_\alpha j_\beta j_\gamma j_\delta \rangle = \langle k_\alpha k_\beta k_\gamma k_\delta \rangle$$

$$= \tfrac{1}{15}(\delta_{\alpha\beta}\delta_{\gamma\delta} + \delta_{\alpha\gamma}\delta_{\beta\delta} + \delta_{\alpha\delta}\delta_{\beta\gamma}), \qquad (4.2.53a)$$

$$\langle i_\alpha i_\beta j_\gamma j_\delta \rangle = \langle j_\alpha j_\beta k_\gamma k_\delta \rangle = \langle i_\alpha i_\beta k_\gamma k_\delta \rangle$$

$$= \tfrac{1}{30}(4\delta_{\alpha\beta}\delta_{\gamma\delta} - \delta_{\alpha\gamma}\delta_{\beta\delta} - \delta_{\alpha\delta}\delta_{\beta\gamma}), \qquad (4.2.53b)$$

with all other types of average zero.

The last isotropic average required in this book is of a product of five direction cosines. This is obtained by considering expressions such as $(\mathbf{i} \times \mathbf{j}) \cdot \mathbf{k}(\mathbf{i} \cdot \mathbf{i}) = 1$. However, the trigonometrical analysis now becomes very complicated, and we simply quote the general result in terms of the *fifth-rank tensor invariant* $\varepsilon_{\alpha\beta\gamma}\delta_{\delta\epsilon}$ (Kielich, 1968/69):

$$\langle i_\alpha j_\beta k_\gamma k_\delta k_\varepsilon \rangle = \langle k_\alpha i_\beta j_\gamma j_\delta j_\varepsilon \rangle = \langle j_\alpha k_\beta i_\gamma i_\delta i_\varepsilon \rangle$$

$$= \tfrac{1}{30}(\varepsilon_{\alpha\beta\gamma}\delta_{\delta\epsilon} + \varepsilon_{\alpha\beta\delta}\delta_{\gamma\epsilon} + \varepsilon_{\alpha\beta\epsilon}\delta_{\gamma\delta}), \qquad (4.2.54)$$

with all other types of average equal to zero. Boyle and Matthews (1971) have provided a general discussion of fifth-rank tensor invariants and isotropic averages.

4.2.6 *Principal axes*

It was shown in Section 3.5.2 that isotropic averages such as $\langle \alpha_{xx}\alpha_{xx}^* \rangle$, $\langle \alpha_{xx}\alpha_{yy}^* \rangle$ and $\langle \alpha_{xy}\alpha_{xy}^* \rangle$ contribute to conventional light scattering in fluids, and equations (3.5.4) gave these averages in terms of $\alpha_{\alpha\alpha}\alpha_{\beta\beta}$ and $\alpha_{\alpha\beta}\alpha_{\alpha\beta}$. All resulting expressions for observables such as the depolarization ratio can be written in terms of α^2 and $\beta(\alpha)^2$, where

$$\alpha^2 = \tfrac{1}{9}\alpha_{\alpha\alpha}\alpha_{\beta\beta} = \tfrac{1}{9}(\alpha_{XX} + \alpha_{YY} + \alpha_{ZZ})^2, \qquad (4.2.55a)$$

$$\beta(\alpha)^2 = \tfrac{1}{2}(3\alpha_{\alpha\beta}\alpha_{\alpha\beta} - \alpha_{\alpha\alpha}\alpha_{\beta\beta})$$

$$= \tfrac{1}{2}\big[(\alpha_{XX} - \alpha_{YY})^2 + (\alpha_{XX} - \alpha_{ZZ})^2 + (\alpha_{YY} - \alpha_{ZZ})^2 \qquad (4.2.55b)$$

$$+ 6\big(\alpha_{XY}^2 + \alpha_{XZ}^2 + \alpha_{YZ}^2\big)\big].$$

These are effectively the invariants of the fourth-rank tensor $\alpha_{\alpha\beta}\alpha_{\gamma\delta}$, being the only combinations of components that contribute in isotropic media. The mean polarizability itself, $\alpha = \tfrac{1}{3}\alpha_{\alpha\alpha}$, is the invariant of the second-rank tensor $\alpha_{\alpha\beta}$. Although X, Y, Z refer to a particular set of axes attached to the molecule's frame, the values of α^2 and $\beta(\alpha)^2$ are invariant to a rotation of these axes.

A famous theorem, too long for proof here, is that for any second-rank symmetric cartesian tensor, it is always possible to choose a set of axes, called *principal* axes, such that only the diagonal components are nonzero (Nye, 1985). The anisotropy invariant then takes the simple form

$$\beta(\alpha)^2 = \tfrac{1}{2}[(\alpha_{XX} - \alpha_{YY})^2 + (\alpha_{XX} - \alpha_{ZZ})^2 + (\alpha_{YY} - \alpha_{ZZ})^2]. \qquad (4.2.56)$$

The principal axes are associated with any symmetry elements present in a molecule. Thus a proper rotation axis is always a principal axis and a reflection plane always contains two of the principal axes and is perpendicular to the third.

Consider, for example, an axially-symmetric molecule. This has a threefold or higher proper rotation axis (which we take to be the Z axis), and its physical properties are isotropic with respect to rotations about this axis. This isotropy in the plane perpendicular to the principal rotation axis is obvious if the molecule is linear ($C_{\infty v}$ or $D_{\infty h}$), but is not immediately apparent in the case of a symmetric top molecule such as NH_3 having C_{3v} symmetry. A simple argument for this case runs as follows: if it is accepted that a reflection plane always contains two principal axes and is perpendicular to the third, the fact that there are three vertical reflection planes at $120°$ to each other is only consistent with the presence of two principal axes (which we take to be X and Y) at $90°$ to each other if X and Y can have any orientation in the plane perpendicular to Z.

The polarizability tensor of an axially-symmetric molecule can be written as follows in terms of components referred to principal axes:

$$\alpha_{\alpha\beta} = \alpha_\perp \delta_{\alpha\beta} + (\alpha_\| - \alpha_\perp)K_\alpha K_\beta, \qquad (4.2.57)$$

where $\alpha_\perp = \alpha_{XX} = \alpha_{YY}$ and $\alpha_\| - \alpha_{ZZ}$ denote polarizability components perpendicular and parallel to the threefold or higher rotation axis Z, and $\mathbf{K}$ is the unit vector along Z. It is convenient to write (4.2.57) in the form

$$\alpha_{\alpha\beta} = \alpha(1 - \kappa)\delta_{\alpha\beta} + 3\alpha\kappa K_{\alpha}K_{\beta}, \tag{4.2.58a}$$

where

$$\alpha = \tfrac{1}{3}(2\alpha_{\perp} + \alpha_{\parallel}) \tag{4.2.58b}$$

is the usual mean polarizability, and

$$\kappa = \frac{\alpha_{\parallel} - \alpha_{\perp}}{3\alpha} \tag{4.2.58c}$$

is a *dimensionless polarizability anisotropy*. It will be seen in subsequent chapters that the polarizability written in the form (4.2.58) facilitates the development of theories of optical activity and Rayleigh and Raman scattering in molecules composed of axially symmetric bonds or groups.

We can also write useful expressions for $G'_{\alpha\beta}$ and $A_{\alpha,\beta\gamma}$ in certain cases of axial symmetry. For example, for the point groups C_{4v} and C_{6v}, Tables 4.2, developed later in this chapter, tell us that the only nonzero components are $G'_{XY} = -G'_{YX}$, $A_{Z,ZZ}$ and $A_{Z,XX} = A_{Z,YY}$. Writing $A_{\parallel} = A_{Z,ZZ}$ and $A_{\perp} = A_{Z,XX}$, we have

$$G'_{\alpha\beta} = G'_{XY}\varepsilon_{\alpha\beta\gamma}K_{\gamma}, \tag{4.2.59}$$

$$A_{\alpha,\beta\gamma} = \left(\tfrac{3}{2}A_{\parallel} - 2A_{\perp}\right)K_{\alpha}K_{\beta}K_{\gamma}$$
$$+ A_{\perp}(K_{\beta}\delta_{\alpha\gamma} + K_{\gamma}\delta_{\alpha\beta}) - \tfrac{1}{2}A_{\parallel}K_{\alpha}\delta_{\beta\gamma}. \tag{4.2.60}$$

These results also apply to linear dipolar molecules ($C_{\infty v}$), and were first derived for this case by Buckingham and Longuet-Higgins (1968). Furthermore, (4.2.59) also applies to C_{3v}, but (4.2.60) does not because additional components of $A_{\alpha,\beta\gamma}$ can be nonzero.

4.3 Inversion symmetry in quantum mechanics

The classification of quantum states and operators with respect to space inversion and time reversal is a cornerstone of atomic and molecular physics. Here we review some aspects that are relevant to optical activity and light scattering.

4.3.1 Space inversion

We introduce a *parity operator* P that changes the sign of the space coordinates in the wavefunction:

$$P\psi(\mathbf{r}) = \psi(-\mathbf{r}). \tag{4.3.1a}$$

P is a linear unitary operator with eigenvalues p determined by

$$P\psi(\mathbf{r}) = p\psi(\mathbf{r}). \qquad (4.3.2)$$

The eigenvalues are found by noticing that a double application amounts to the identity:

$$P^2\psi(\mathbf{r}) = p^2\psi(\mathbf{r}) = \psi(\mathbf{r})$$

so that

$$p^2 = 1, \quad p = \pm 1. \qquad (4.3.3)$$

The wavefunction (and the corresponding state) is said to have even or odd parity depending on whether $p = +1$ or -1. Thus for even and odd wavefunctions $\psi(+)$ and $\psi(-)$ we have

$$P\psi(+) = \psi(+), \quad P(\psi)(-) = -\psi(-). \qquad (4.3.4)$$

It is emphasized that P is an inversion with respect to space-fixed axes and can be applied to all systems. It should not be confused with the inversion operation with respect to molecule-fixed axes in systems with a centre of inversion which leads to the 'g' or 'u' classification of quantum states.

The development so far refers to *orbital* parity which describes the symmetry properties of motions of particles. But in order to understand the processes in which elementary particles are created and destroyed, it has been found necessary to introduce the notion of the *intrinsic* or *internal* parity of a particle (see, for example, Gibson and Pollard, 1976; Berestetskii, Lifshitz and Pitaevskii, 1982). This is incorporated by generalizing the transformation law (4.3.1a) to

$$P\psi(\mathbf{r}) = \eta\psi(-\mathbf{r}), \qquad (4.3.1b)$$

where η is the intrinsic parity of the particle described by the wavefunction $\psi(\mathbf{r})$. Since two inversions restore the original coordinate system, and

$$P^2\psi(\mathbf{r}) = \eta P\psi(-\mathbf{r}) = \eta^2\psi(\mathbf{r}),$$

η^2 may at the most be a phase factor of unit magnitude, and it can be shown that $\eta^2 = +1$ or ± 1 depending on whether the spin of the particle is integral or half odd-integral. Thus for particles with integral spin, $\eta = \pm 1$; and for particles with half odd-integral spin, $\eta = \pm 1$ or $\pm i$. Particles for which $\eta = +1$ and -1 are said to be scalar and pseudoscalar, respectively. The intrinsic parity of the photon is defined absolutely from theoretical considerations to be negative; whereas the intrinsic parities of electrons and nucleons are relative and are taken by convention to be positive (with negative parities for the corresponding antiparticles). Notice that intrinsic parities can be assigned to particle wavefunctions that are not eigenfunctions of the parity operator.

If two eigenfunctions $\psi(+)$ and $\psi(-)$ of opposite orbital parity have energy eigenvalues that are degenerate, or nearly so, the system can exist in states of *mixed parity* with wavefunctions

$$\psi_1 = \frac{1}{\sqrt{2}}[\psi(+) + \psi(-)], \qquad (4.3.5a)$$

$$\psi_2 = P\psi_1 = \frac{1}{\sqrt{2}}[\psi(+) - \psi(-)]. \qquad (4.3.5b)$$

Since the conventional Hamiltonian operator is unaffected by inversion of the coordinates we can write

$$H = PHP^{-1} \text{ or } PH - HP = 0. \qquad (4.3.6)$$

It then follows from a consideration of the time derivative of an operator that the expectation value of P is constant in time (Landau and Lifshitz, 1977). Equation (4.3.6) expresses the *law of conservation of parity*: if the state of a closed system has definite parity, that parity is conserved. It follows that definite parity states $\psi(\pm)$ are stationary states with constant energy $W(\pm)$.

All observables can be classified as even or odd depending on whether their operators do not or do change sign under inversion of the coordinates. Even and odd operators $A(+)$ and $A(-)$ are thus defined by

$$PA(+)P^{-1} = A(+), \quad PA(-)P^{-1} = -A(-). \qquad (4.3.7)$$

Since integrals taken over all space are only nonzero for totally symmetric integrands, the expectation values of these operators in a state such as (4.3.5a) are

$$\langle\psi_1|A(+)|\psi_1\rangle = \tfrac{1}{2}[\langle\psi(+)|A(+)|\psi(+)\rangle$$
$$+ \langle\psi(-)|A(+)|\psi(-)\rangle], \qquad (4.3.8a)$$

$$\langle\psi_1|A(-)|\psi_1\rangle = \tfrac{1}{2}[\langle\psi(+)|A(-)|\psi(-)\rangle$$
$$+ \langle\psi(-)|A(-)|\psi(+)\rangle], \qquad (4.3.8b)$$

from which we deduce immediately that the expectation value of any odd observable vanishes in any state of definite parity, that is, a state for which either $\psi(+)$ or $\psi(-)$ is zero. It also follows that the expectation values of an odd parity operator have equal magnitude but opposite sign for the pair of mixed parity states ψ_1 and $\psi_2 = P\psi_1$. Consequently, a system in a state of *definite* parity can possess only observables with *even* parity, examples being electric charge, magnetic dipole moment, electric quadrupole moment, etc.; whereas a system in a state of *mixed* parity can, in addition, possess observables with *odd* parity, examples being linear momentum, electric dipole moment, etc. (Kaempffer, 1965). A well known example of a system with states of mixed parity is the hydrogen atom. Here the special dynamical symmetry leads to degenerate eigenstates of opposite parity: for example, the states with $n = 2$, $l = 0$ and $n = 2$, $l = 1$ are degenerate, and since the parity of the spherical harmonic function Y_{lm} is $(-1)^l$, the first excited state of the hydrogen atom has mixed parity and therefore supports a permanent electric dipole moment, evidenced by a first order Stark effect (Buckingham, 1972). In fact these states are not exactly degenerate because of a small relativistic splitting, and in very weak electric fields only a second-order Stark effect is observed (Woolley, 1975*b*).

We saw in Section 1.9.2 that the natural optical rotatory parameter is a pseudoscalar and so has odd parity. It was shown, furthermore, that the optical rotation experiment conserves parity, because if one inverts the entire experiment (light beam plus active medium) the resulting experiment is also realized in nature. Consequently, resolved chiral molecules exist in quantum states of mixed parity.

The origin of the mixed parity states of a chiral molecule can be appreciated by considering the vibrational wavefunctions (associated with the inversion coordinate) of a molecule such as NH_3 which is said to invert between two equivalent configurations, as shown in Fig. 4.4., although this motion does not in fact correspond to inversion through the centre of mass (Townes and Schawlow, 1955). If the planar configuration were the most stable, the adiabatic potential energy function would have the parabolic form shown on the left with simple harmonic vibrational energy levels equally spaced. If a potential hill is raised gradually in the middle, the two pyramidal configurations become the most stable and the energy levels approach each other in pairs. For an infinitely high potential hill, the pairs of energy levels are exactly degenerate, as shown on the right. The rise of the central potential hill modifies the wavefunctions as shown, but does not destroy their parity. The even and odd parity wavefunctions $\psi^{(0)}(+)$ and $\psi^{(0)}(-)$ describe stationary states in all circumstances. On the other hand, the wavefunctions ψ_L and ψ_R, corresponding to the system in its lowest state of oscillation and localized completely in the left and right wells, respectively, are not true stationary states. They are obtained from the following combinations of the even and odd parity

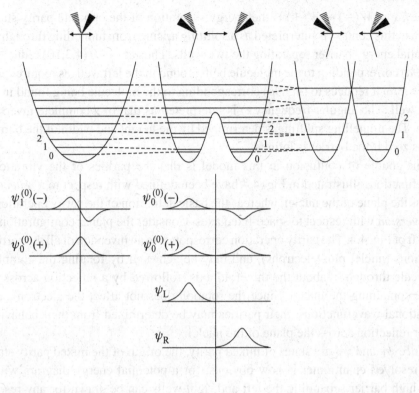

Fig. 4.4 The vibrational states of a molecule that can invert between two equivalent configurations. $\psi^{(0)}(+)$ and $\psi^{(0)}(-)$ are two definite parity states for which there is complete uncertainty, and ψ_L and ψ_R are two mixed parity states for which there is complete certainty, about whether the molecule is in the left or right well.

wavefunctions:

$$\psi_L = \frac{1}{\sqrt{2}}[\psi^{(0)}(+) + \psi^{(0)}(-)], \qquad (4.3.9a)$$

$$\psi_R = \frac{1}{\sqrt{2}}[\psi^{(0)}(+) - \psi^{(0)}(-)], \qquad (4.3.9b)$$

which provide explicit physical realizations of the mixed parity wavefunctions (4.3.5). The wavefunctions (4.3.9) are in fact specializations of the general time-dependent wavefunction of a degenerate two-state system (see Section 4.3.4 below). To be precise, we assume that the system is in the left well at $t = 0$. Then at a later time t we have

$$\psi(t) = \frac{1}{\sqrt{2}}[\psi^{(0)}(+)e^{-iW(+)t/\hbar} + \psi^{(0)}(-)e^{-iW(-)t/\hbar}]$$

$$= \frac{1}{\sqrt{2}}[\psi^{(0)}(+) + \psi^{(0)}(-)e^{-i\omega t}]e^{-iW(+)t/\hbar}, \qquad (4.3.10)$$

where $\hbar\omega = W(-) - W(+)$ is the energy separation of the opposite parity states, which in this context is interpreted as a splitting arising from tunnelling through the potential energy barrier separating the two wells. Thus at $t = 0$ (4.3.10) reduces to (4.3.9a) corresponding to the molecule being found in the left well, as required; and at $t = \pi/\omega$ it reduces to (4.3.9b) corresponding to the molecule being found in the right well. The angular frequency ω is interpreted as that of a complete inversion cycle. The tunnelling splitting is determined by the height and width of the barrier, and is zero if the barrier is infinite.

One source of confusion in this model is that the parities of the vibrational wavefunctions illustrated in Fig. 4.4 have been defined with respect to a *reflection* across the plane of the nuclei, whereas the basic definition of the parity operation is an *inversion* with respect to space-fixed axes. Consider the planar configuration on the left of Fig. 4.4. The parity operation corresponds to an inversion of all the particle positions (nuclei plus electrons), and may be achieved by rotating the complete molecule through π about the threefold axis, followed by a reflection across the plane containing the nuclei. Since the rotation does not affect the electronic and vibrational wavefunctions, their parities may be determined from their behaviour under reflection across the plane of the nuclei.

Since ψ_L and ψ_R are states of mixed parity, the origin of the mixed parity states of a resolved enantiomer is now obvious, for a potential energy diagram with a very high barrier separating the left and right wells can be drawn for any resolvable chiral molecule: the horizontal axis might represent the position of an atom above a plane containing three different atoms, or the torsion coordinate in a chiral biphenyl, or some more complicated collective coordinate of the molecule. If such a state is prepared, but the tunnelling splitting is finite, its energy will be indefinite because it is a superposition of two opposite parity states of different energy. From the discussion above (or, more generally, using $\Delta W = \hbar/t$, where t is the average lifetime and ΔW is the width of the level corresponding to a quasi-stationary state), the splitting of the two definite parity states is seen to be inversely proportional to the left–right conversion time. A crucial point is therefore the relation between the time scale of the optical activity measurement and the lifetime of the resolved enantiomer. A manifestation of the uncertainty principle appears to arise here which has been stated loosely as follows (Barron, 1979a): 'If, for the duration of the measurement, there is complete certainty about the enantiomer, there is complete uncertainty about the parity of its quantum state; whereas if there is complete uncertainty about the enantiomer, there is complete certainty about the parity of its quantum state.' Thus experimental resolution of the definite parity states of an enantiomer of tartaric acid, say, which has a lifetime probably greater than the age of the universe, is impossible unless the duration of the experiment is virtually infinite; whereas for a nonresolvable chiral molecule such as H_2O_2, spectroscopic transitions between states of definite parity are observed routinely.

4.3.2 Time reversal

Although it is possible to classify time-even and time-odd Hermitian operators and the corresponding time-even and time-odd observables, depending on whether they do not or do change sign under time reversal, the division of quantum states into even and odd reversality, by analogy with the division of states into even and odd parity, is obscure for several reasons.

Since the Hamiltonian is time even, the simple time reversal operation of replacing t by $-t$ everywhere changes the time-dependent Schrodinger equation

$$H\psi(\mathbf{r}, t) = i\hbar\frac{\partial}{\partial t}\psi(\mathbf{r}, t) \tag{4.3.11a}$$

into

$$H\psi(\mathbf{r}, -t) = -i\hbar\frac{\partial}{\partial t}\psi(\mathbf{r}, -t), \tag{4.3.11b}$$

which is not satisfactory because $\psi(\mathbf{r}, t)$ and $\psi(\mathbf{r}, -t)$ do not obey the same equation. However, by taking the complex conjugate of both sides as well, for real H and ignoring any spin variables, we obtain

$$H\psi^*(\mathbf{r}, -t) = i\hbar\frac{\partial\psi^*}{\partial t}(\mathbf{r}, -t). \tag{4.3.12}$$

This shows that, if $\psi(\mathbf{r}, t)$ is a solution of the Schrödinger equation, then so is $\psi^*(\mathbf{r}, -t)$. For example, the stationary state eigenfunction

$$\psi_n(\mathbf{r}, t) = \psi_n^{(0)}e^{-iW_n t/\hbar}$$

gives $H\psi_n^{(0)} = W_n\psi_n^{(0)}$ from (4.3.11a); and

$$\psi_n^*(\mathbf{r}, -t) = \psi_n^{(0)*}e^{-iW_n t/\hbar}$$

gives $H\psi_n^{(0)*} = W_n\psi_n^{(0)*}$ from (4.3.12). Hence $\psi_n^{(0)}$ and $\psi_n^{(0)*}$ belong to the same energy level with energy eigenvalue W_n.

Considerations such as these lead to the following definition of the time reversal operator in quantum mechanics:

$$\Theta = UK, \tag{4.3.13}$$

where U is a unitary operator and K is the operator of complex conjugation (Wigner, 1959; Abragam and Bleaney, 1970; Sachs, 1987). The unitarity condition on U follows from the requirement that the probability of finding a particle must be conserved under time reversal, that is $\langle \psi | \psi \rangle = \langle \Theta\psi | \Theta\psi \rangle$. This is only true if $U^\dagger U = 1$:

$$\langle \Theta\psi | \Theta\psi \rangle = \langle UK\psi | UK\psi \rangle = \langle K\psi | U^\dagger U | K\psi \rangle$$
$$= \langle \psi^* | \psi^* \rangle = \langle \psi | \psi \rangle. \tag{4.3.14}$$

For the case of a spinless particle, U is the unit operator so that Θ is simply the operation of complex conjugation, as may be verified by applying K to both sides of the Schrödinger equation (4.3.11a):

$$K[H\psi(\mathbf{r}, t)] = K\left[i\hbar \frac{\partial}{\partial t} \psi(\mathbf{r}, t) \right],$$

$$HK\psi(\mathbf{r}, t) = -i\hbar \frac{\partial}{\partial t} K\psi(\mathbf{r}, t).$$

Replacing t by $-t$, this may be rewritten as

$$HK\psi(\mathbf{r}, -t) = -i\hbar \frac{\partial}{\partial t} K\psi(\mathbf{r}, -t), \tag{4.3.15}$$

which is the same as (4.3.12). For the more complicated case of a particle with spin, $U = i\sigma_y$ where

$$\sigma_y = \begin{pmatrix} 0 & -i \\ i & 0 \end{pmatrix}$$

is one of the Pauli spin matrices (Wigner, 1959; Abragam and Bleaney, 1970; Sachs, 1987). This is consistent with the result (4.3.22) below if a spinor representation in the form of a column matrix is used for the spin states α and β. Since K is an *antilinear* operator and U is a *unitary* operator, Θ is called an *antiunitary* operator.

We recall here for convenience a few definitions. Two functions are linearly dependent if they are proportional to each other: a set of functions is therefore linearly independent if the identity

$$c_1\psi_1 + c_2\psi_2 + \cdots c_n\psi_n = 0$$

can only be satisfied by taking $c_1 = c_2 = \cdots = c_n = 0$. A unitary operator A is one whose inverse A^{-1} equals its Hermitian conjugate $A^\dagger$, the latter being the complex conjugate of the transpose. Linear and antilinear operators A and B have the properties

$$A(a\psi_1 + b\psi_2) = aA\psi_1 + bA\psi_2, \tag{4.3.16a}$$
$$B(a\psi_1 + b\psi_2) = a^*B\psi_1 + b^*B\psi_2. \tag{4.3.16b}$$

Thus, whereas a linear unitary operator satisfies

$$\langle A\psi | A\phi \rangle = \langle \psi | A^\dagger A | \phi \rangle = \langle \psi | \phi \rangle, \tag{4.3.17a}$$

an antiunitary operator satisfies

$$\langle B\psi | B\phi \rangle = \langle \psi | \phi \rangle^* = \langle \phi | \psi \rangle. \tag{4.3.17b}$$

We shall see below that, unlike the parity operator, it is not possible to characterize a quantum state by an eigenvalue of the operator Θ. We can, however, characterize a quantum state by the eigenvalue of the linear unitary operator Θ^2. This follows from the physical requirement that the operation Θ applied twice should result in the same state (within a phase factor):

$$\Theta^2 \psi = \epsilon \psi; \quad |\epsilon| = 1. \tag{4.3.18}$$

Since $K^2 = 1$, we may use (4.3.18) to write

$$\Theta^2 = UKUK = UU^* K^2 = UU^* = \epsilon. \tag{4.3.19}$$

Furthermore, since U is unitary, we have

$$U^{-1} = U^{T*}$$

where the superscript T denotes the transpose, so that the last equality in (4.3.19) may be written

$$U^* = \epsilon U^{T*}$$

which, transposing again, becomes

$$U^{T*} = \epsilon U^* = \epsilon^2 U^{T*}.$$

This can only be true if

$$\epsilon^2 = 1, \quad \epsilon = \pm 1. \tag{4.3.20}$$

The possible eigenvalues of Θ^2 are therefore $+1$ and -1.

It can be shown quite generally (see, for example. Wigner, 1959; Heine, 1960; Kaempffer, 1965; or Abragam and Bleaney, 1970) that, for a system containing an even number of electrons (or a system with an integral total angular momentum quantum number J), the quantum state belongs to the eigenvalue $\epsilon = +1$ of Θ^2, and for an odd number of electrons (or half odd-integral J) the quantum state belongs to the eigenvalue $\epsilon = -1$:

$$\Theta^2 \psi = \psi \ \text{(integral } J), \tag{4.3.21a}$$

$$\Theta^2 \psi = -\psi \ \text{(half odd-integral } J). \tag{4.3.21b}$$

Equation (4.3.21b) leads to an important theorem. Consider a system with an odd number of electrons, and assume that the Hamiltonian commutes with Θ (as in the absence of a magnetic field, for the kinetic and potential energy, as well as spin–spin and spin–orbit interactions, are invariant under time reversal). Then if ψ is an eigenstate with energy W, the function $\phi = \Theta\psi$ is also an eigenstate with the same energy. For this to lead to a degeneracy, we have to show that ψ and ϕ are linearly independent. Suppose that

$$\Theta\psi = \lambda\psi,$$

where λ is some constant. Then

$$\Theta^2\psi = \Theta\lambda\psi = \lambda^*\Theta\psi = \lambda^*\lambda\psi.$$

For a system with an odd number of electrons, this contradicts (4.3.21b) since $\lambda^*\lambda$ must be positive. Thus $\Theta\psi \neq \lambda\psi$ and so ψ and $\Theta\psi$ are linearly independent. Since $\Theta^2\psi = -\psi$, the degeneracy of every energy level is even. Hence *Kramers' theorem*: in the presence of any electric potential but in the absence of an external magnetic field, every energy level of a system with an odd number of electrons is m-fold degenerate, where m is an even number (not necessarily the same for each level). $\Theta\psi$ is called the *Kramers conjugate* of ψ.

Notice that, if $\alpha = |\frac{1}{2} - \frac{1}{2}\rangle$ and $\beta = |\frac{1}{2} - \frac{1}{2}\rangle$ are the two orthogonal spin states $|sm_s\rangle$ for a single electron, the following statements, for a particular choice of phase, are consistent with the foregoing:

$$\Theta\alpha = \beta, \quad \Theta\beta = -\alpha. \tag{4.3.22}$$

We now develop the eigenstates of Θ^2 a little more. Consider first eigenfunctions $\psi(+)$ with eigenvalue $\epsilon = +1$,

$$\Theta^2\psi(+) = \psi(+). \tag{4.3.23}$$

First notice that $\Theta\psi(+)$ is also an eigenfunction of Θ^2 with eigenvalue $+1$; and so is

$$\psi_{\text{even}} = c[\psi(+) + \Theta\psi(+)]. \tag{4.3.24}$$

If c is real, and $\Theta\psi(+) \neq -\psi(+)$, we have

$$\Theta\psi_{\text{even}} = \psi_{\text{even}}. \tag{4.3.25}$$

If it should happen that, for a particular state, $\Theta\psi'(+) = -\psi'(+)$, we can choose

$$\psi_{\text{even}} = i\psi'(+) \tag{4.3.26}$$

and (4.3.25) is again valid. Similarly, we can construct

$$\psi_{\text{odd}} = c[\psi(+) - \Theta\psi(+)] \tag{4.3.27}$$

where, if c is real and $\Theta\psi(+) \neq \psi(+)$,

$$\Theta\psi_{odd} = -\psi_{odd}; \tag{4.3.28}$$

and if it happens that $\Theta\psi'(+) = \psi'(+)$, we can choose

$$\psi_{odd} = i\psi'(+). \tag{4.3.29}$$

This possibility of finding even and odd states under time reversal is not equivalent to labelling the states by a physically meaningful quantum number, characteristic of Θ, such as parity in the case of space inversion P, because an even state can be transformed into an odd state, and *vice versa*, simply by multiplication with a physically unobservable phase factor i. The quantum number ϵ, characteristic of Θ^2, however, is not affected by such a phase change.

Now consider eigenfunctions $\psi(-)$ with eigenvalue $\epsilon = -1$:

$$\Theta^2\psi(-) = -\psi(-). \tag{4.3.30}$$

It follows that $\Theta\psi(-)$ is also an eigenfunction of Θ^2 with eigenvalue $\epsilon = -1$. However, one feature that did not arise in the $\epsilon = +1$ case is that $\Theta\psi(-)$ is always orthogonal to $\psi(-)$ (like α and β in (4.3.22)). This follows from

$$\langle\psi(-)|\Theta\psi(-)\rangle = \langle\Theta^2\psi(-)|\Theta\psi(-)\rangle$$
$$= -\langle\psi(-)|\Theta\psi(-)\rangle = 0, \tag{4.3.31}$$

where (4.3.17b) provides the first equality and (4.3.30) the second. This is equivalent to the earlier demonstration that, for odd electron systems, a state and its Kramers conjugate are linearly independent, leading to an even-fold degeneracy. Unlike the $\epsilon = +1$ case, it does not appear to be possible to construct even and odd states with respect to Θ. However, it is shown below that states can be constructed for which the expectation values with time-even and time-odd operators, respectively, vanish.

Operators can be classified as time even or time odd according to the following criteria (Abragam and Bleaney, 1970):

$$\Theta A(+)\Theta^{-1} = A(+)^\dagger, \quad \Theta A(-)\Theta^{-1} = -A(-)^\dagger. \tag{4.3.32}$$

This follows from a development similar to (4.3.34) below. Some important statements about matrix elements can now be made.

(*a*) If $\epsilon = -1$, a time-even operator has no matrix elements between Kramers conjugate states:

$$\langle\psi|A(+)|\Theta\psi\rangle = \langle\psi|A(+)\Theta\psi\rangle = \langle\Theta A(+)\Theta\psi|\Theta\psi\rangle$$
$$= \langle\Theta A(+)\Theta^{-1}\Theta^2\psi|\Theta\psi\rangle = -\langle A(+)^\dagger\psi|\Theta\psi\rangle$$
$$= -\langle\psi|A(+)|\Theta\psi\rangle = 0. \tag{4.3.33}$$

If $\epsilon = +1$, a time-odd operator has no matrix elements between Kramers conjugate states, the proof being similar to (4.3.33).

(b) For both $\epsilon = +1$ and -1, a time-even operator has the same expectation value in two Kramers conjugate states:

$$\langle\psi|A(+)|\psi\rangle = \langle\psi|A(+)\psi\rangle = \langle\Theta A(+)\psi|\Theta\psi\rangle$$
$$= \langle\Theta A(+)\Theta^{-1}\Theta\psi|\Theta\psi\rangle = \langle A(+)^{\dagger}\Theta\psi|\Theta\psi\rangle$$
$$= \langle\Theta\psi|A(+)|\Theta\psi\rangle. \tag{4.3.34}$$

(c) For both $\epsilon = +1$ and -1, a time-odd operator has opposite expectation values in Kramers conjugate states. The proof is similar to (4.3.34).

We can now see that for the invariant states ψ_{even} (4.3.24) and ψ_{odd} (4.3.27), which can always be constructed when $\epsilon = +1$, the expectation value of a time-odd operator vanishes. This is true for any general invariant state for which $\Theta\psi = e^{i\alpha}\psi$ with arbitrary α. Invariant states cannot be constructed for the $\epsilon = -1$ case, but states such as

$$\psi' = c[\psi(-) + i\Theta\psi(-)], \tag{4.3.35a}$$
$$\psi'' = c[\psi(-) - i\Theta\psi(-)], \tag{4.3.35b}$$

can be constructed for which the expectation values of *Hermitian* time-odd and time-even operators, respectively, vanish (see Kaempffer, 1965, p. 110 for the proof).

Although the well known selection rules for matrix elements between component states of *different* levels are unchanged whatever the behaviour under time reversal of the operators and eigenfunctions, the selection rules must be modified for matrix elements between component states of the *same* degenerate level (Griffith, 1961; Landau and Lifshitz, 1977; Abragam and Bleaney, 1970; Stedman and Butler, 1980). We follow Abragam and Bleaney (1970), who consider matrix elements of the form $\langle\Theta\psi_j|V|\psi_k\rangle$, where ψ_j and ψ_k are component eigenfunctions of the same basis set spanning an irreducible representation Γ. According to these authors, since the $\Theta\psi_j$ form a set of orthonormal functions spanning the same manifold as the ψ_k, the vanishing of all of the $\langle\Theta\psi_j|V|\psi_k\rangle$ implies that of the $\langle\psi_j|V|\psi_k\rangle$, and *vice versa*. We now transform $\langle\Theta\psi_j|V|\psi_k\rangle$ as follows:

$$\langle\Theta\psi_j|V|\psi_k\rangle = \langle\Theta\psi_j|V\psi_k\rangle = \langle\Theta V\psi_k|\Theta^2\psi_j\rangle$$
$$= \langle\Theta V\Theta^{-1}\Theta\psi_k|\Theta^2\psi_j\rangle = \epsilon\lambda\langle V^{\dagger}\Theta\psi_k|\psi_j\rangle$$
$$= \epsilon\lambda\langle\Theta\psi_k|V|\psi_j\rangle, \tag{4.3.36}$$

where λ equals $+1$ or -1 depending on whether the operator V is time even or time odd, and ϵ, being the eigenvalue of Θ^2, equals $+1$ or -1 depending on whether there are an even or odd number of electrons. We can now write

$$\langle\Theta\psi_j|V|\psi_k\rangle = \tfrac{1}{2}(\langle\Theta\psi_j|V|\psi_k\rangle + \epsilon\lambda\langle\Theta\psi_k|V|\psi_j\rangle). \qquad (4.3.37)$$

Thus depending on whether $\epsilon\lambda$ is positive or negative, $\langle\psi_j|V|\psi_k\rangle$ belongs to the representation $[\Gamma^2] \times \Gamma_V$ or $\{\Gamma^2\} \times \Gamma_V$, the square and curly brackets denoting the symmetric and antisymmetric parts of the direct product. It is stressed that, in the odd electron case, the representations refer to the appropriate *double* group. The result (4.3.37) applies whether V is Hermitian, antiHermitian or nonHermitian.

A simple but important example of the application of the generalized selection rule (4.3.37) is to the existence of permanent electric and magnetic dipole moments in atoms and molecules (Landau and Lifshitz, 1977). Since the electric dipole moment $\boldsymbol{\mu}$ is a time-even polar vector, it follows that a permanent electric dipole moment can exist in a system with an even number of electrons (or integral J) in a state ψ_j if $[\Gamma_j^2] \times \Gamma_P$ contains the totally symmetric irreducible representation, and in a system with an odd number of electrons (or half odd-integral J) if $\{\Gamma_j^2\} \times \Gamma_P$ contains the totally symmetric irreducible representation, where Γ_P is a representation spanned by a polar vector component. Similarly, since the magnetic dipole moment $\mathbf{m}$ is a time-odd axial vector, a permanent magnetic dipole moment can exist in a system with an even number of electrons (or integral J) if $\{\Gamma_j^2\} \times \Gamma_A$ contains the totally symmetric irreducible representation, and in a system with an odd number of electrons (or half odd-integral J) if $[\Gamma_j^2] \times \Gamma_A$ contains the totally symmetric irreducible representation, where Γ_A is a representation spanned by an axial vector component.

Consider an octahedral molecule. Its electronic states can be classified with respect to the single point group O if it has an even number of electrons, and the double point group O^* if it has an odd number of electrons. Looking first at the even electron case, since both (P_x, P_y, P_z) and (A_x, A_y, A_z) span T_1, and $[E^2] = A_1 + E$, $\{E^2\} = A_2$, $[T_1^2] = A_1 + E + T_2$, $\{T_1^2\} = T_1$, $[T_2^2] = A_1 + E + T_2$ and $\{T_2^2\} = T_1$, we conclude that a permanent electric dipole moment is not supported by any of the electronic states, but a permanent magnetic dipole moment is supported by states belonging to the T_1 and T_2 sets. Turning to the odd electron case, since $[E'^2] = T_1$, $\{E'^2\} = A_1$, $[E''^2] = T_1$, $\{E''^2\} = A_1$, $[U'^2] = A_2 + 2T_1 + T_2$ and $\{U'^2\} = A_1 + E + T_2$, we conclude that a permanent electric dipole moment is not supported by any of the electronic states, but a permanent magnetic dipole moment is supported by states belonging to the E', E'' and U' sets.

It will prove useful in later applications to have an expression for the effect of the time reversal operator on a general atomic state of the form $|JM\rangle$. Our derivation is based on one given by Judd and Runciman (1976), and we adhere to their choice of phase factors. The effect of time reversal on pure spin and orbital states is straightforward. For a spin eigenstate $|s \, m_s\rangle$ we have, using the phase choice of (4.3.22),

$$\Theta|s \, m_s\rangle = (-1)^{s-m_s}|s -m_s\rangle. \qquad (4.3.38)$$

The part $\pm m_s$ in the exponent of the phase factor is crucial; whereas the part s is included merely to avoid complex phase factors. (It is worth noting that Heine (1960) writes $\Theta\alpha = -\beta$ and $\Theta\beta = \alpha$, which implies a phase factor of $(-1)^{s+m_s}$, but this is unconventional). For an orbital eigenstate $|l m_l\rangle$, the phase factor chosen in

$$\Theta|l m_l\rangle = (-1)^{l-m_l}|l -m_l\rangle \qquad (4.3.39)$$

is consistent with that in (4.3.38). We now investigate how the coupled states $|jm\rangle$ behave under time reversal by performing an uncoupling and using the following property of the real vector coupling coefficients (Edmonds, 1960):

$$\langle j_1 m_1 j_2 m_2|j_3 m_3\rangle = (-1)^{j_1+j_2-j_3}\langle j_1 -m_1 j_2 -m_2|j_3 -m_3\rangle. \qquad (4.3.40)$$

Thus

$$\Theta|jm\rangle = \Theta \sum_{m_s,m_l} \langle s m_s l m_l|jm\rangle|s m_s\rangle|l m_l\rangle$$

$$= \sum_{m_s,m_l} \langle s m_s l m_l|jm\rangle(-1)^{s+l-m}|s -m_s\rangle|l -m_l\rangle$$

$$= \sum_{m_s,m_l} \langle s -m_s \, l -m_l|j -m\rangle(-1)^{j-m}|s -m_s\rangle|l -m_l\rangle$$

$$= (-1)^{j-m}|j -m\rangle. \qquad (4.3.41)$$

In the second line we have used $m = m_s + m_l$, and in the third line we have used the fact that, since the phase factor is real, it satisfies $(-1)^{s+l-m} = (-1)^{-(s+l-m)}$. By considering a sequence of couplings we obtain the following result for a many electron system whose various spin and orbital angular momenta are coupled to a resultant J:

$$\Theta|JM\rangle = (-1)^{J-M}|J -M\rangle. \qquad (4.3.42)$$

This result presupposes that the orbital functions behave as in (4.3.39). But in fact the usual form used for orbital functions is that of the spherical harmonics Y_{lm} with the following phase convention of Condon and Shortley (1935):

$$Y_{l-m} = (-1)^m Y_{lm}^*. \qquad (4.3.43)$$

Since the action of Θ on a spherical harmonic is equivalent to complex conjugation, we have

$$\Theta Y_{lm} = (-1)^{-m} Y_{l-m}. \tag{4.3.44}$$

Compared with (4.3.39), which was used to derive (4.3.42), there is a missing part $(-1)^l$ in the phase; so if the orbital part of our general atomic state $|JM\rangle$ is based on spherical harmonics, (4.3.42) must be changed to

$$\Theta|JM\rangle = (-1)^{J-M+p}|J-M\rangle, \tag{4.3.45}$$

where p is the sum of the individual orbital quantum numbers l of all the electrons in the atom.

4.3.3 The parity and reversality classification of optical activity observables

By considering the helical pattern of the electric field vectors of a linearly polarized light beam established in an optical rotatory medium, it was deduced in Section 1.9.2 that the optical rotation observable is a time-even pseudoscalar. This classification seems reasonable for natural optical rotation in an isotropic collection of chiral molecules because the direction of propagation of the light beam is immaterial. But the classification becomes slippery when we apply it to magnetic optical rotation because the direction of the light beam relative to the magnetic field is crucial.

In order to properly classify the natural and magnetic optical rotation observables we must get away from the approach used in Section 1.9.3 in which the *complete* experiment was subjected to space inversion and time reversal (this was to demonstrate that the laws of conservation of parity and reversality are obeyed by the natural and magnetic optical rotation experiments). Now we leave the observer and his linearly polarized probe light beam alone and apply space inversion and time reversal to just the sample and any applied fields.

Under space inversion, an isotropic collection of chiral molecules is replaced by a collection of the enantiomeric molecules, and the observer will measure an equal and opposite optical rotation. This indicates that the observable has odd parity, and it is easy to deduce that it is a pseudoscalar (rather than, say, a polar vector) because it is invariant with respect to any proper rotation in space of the complete sample. Under time reversal, an isotropic collection of chiral molecules is unchanged (even if paramagnetic), and so the optical rotation is unchanged. Thus the *natural* optical rotation observable in an isotropic sample is a *time-even pseudoscalar.*

Consider now a collection of achiral molecules in a static uniform magnetic field. Under space inversion, the molecules and the magnetic field direction are unchanged, so the same magnetic optical rotation will be observed. This indicates that the observable has even parity, and we can further deduce that it is an axial vector (rather than a scalar) by noticing that a proper rotation of the complete sample, including the magnetic field, through π about any axis perpendicular to the field reverses the relative directions of the magnetic field and the probe beam and so changes the sign of the observable. Under time reversal, the collection of molecules can be regarded as unchanged provided it is isotropic in the absence of the field (even though individual paramagnetic molecules will change to their Kramers conjugates, there will be the same number of Kramers conjugate pairs before and after), but again the relative directions of the magnetic field and the probe beam are reversed and so the optical rotation changes sign. Thus the *magnetic* optical rotation observable is a *time-odd axial vector.*

These conclusions accord with the explicit expressions for the optical rotation angle obtained in Chapter 3:

$$\Delta\theta \approx -\tfrac{1}{3}\omega\mu_0 l N G'_{\alpha\alpha}(f), \tag{3.4.43}$$

$$\Delta\theta_z \approx \tfrac{1}{2}\omega\mu_0 c l N \alpha'_{xy}(f). \tag{3.4.54}$$

Thus (3.4.43) shows that the natural optical rotation in an isotropic sample is proportional to $G'_{\alpha\alpha}$, which transforms as a time-even pseudoscalar; and (3.4.54) shows that the magnetic optical rotation for light propagating in the α direction is proportional to $\varepsilon_{\alpha\beta\gamma}\alpha'_{\beta\gamma}$, which transforms as a time-odd axial vector (this classification of molecular tensors is discussed in Section 4.4.1).

Similar arguments may be used to demonstrate that the *magnetochiral birefringence* observable transforms as a *time-odd polar vector* (Barron and Vrbancich, 1984).

In order to apply quantum mechanical arguments to the symmetry classification of natural and magnetic optical rotation observables, it is necessary to specify corresponding operators with well defined behaviour under space inversion using (4.3.7), and time reversal using (4.3.32). A good start has already been made with the introduction of effective polarizability and optical activity operators in (2.8.14). Consider first the product of two noncommuting Hermitian operators A and B:

$$AB = \tfrac{1}{2}(AB + BA) + \tfrac{1}{2}(AB - BA) = p + q. \tag{4.3.46}$$

Recalling that a Hermitian operator satisfies $A = A^\dagger$; an antiHermitian operator satisfies $A = -A^\dagger$; and that $(AB)^\dagger = B^\dagger A^\dagger$; it is clear that $p = \tfrac{1}{2}(AB + BA)$ is Hermitian and $q = \tfrac{1}{2}(AB - BA)$ is antiHermitian. Extending this to the product of three Hermitian operators, it is found that

$$(AB)C \pm C(BA) = pC + qC \pm Cp \mp Cq, \tag{4.3.47}$$

where $pC + Cp$ and $qC - Cq$ are Hermitian, and $pC - Cp$ and $qC + Cq$ are antiHermitian. Since μ_α and μ_β are Hermitian and have odd parity, and O^s and O^a are Hermitian and have even parity, it follows that both $\hat{\alpha}^s_{\alpha\beta}$ and $\hat{\alpha}^a_{\alpha\beta}$ have even parity, but the first is Hermitian and the second is antiHermitian. To determine the behaviour under time reversal, it is necessary to appreciate that the product of two noncommuting Hermitian operators A and B of well defined reversality does not itself have well defined reversality but is the sum of a time-even and a time-odd operator (Abragam and Bleaney, 1970). This can be seen by developing p and q in (4.3.46) as follows:

$$\Theta p \Theta^{-1} = \tfrac{1}{2}(\Theta A \Theta^{-1} \Theta B \Theta^{-1} + \Theta B \Theta^{-1} \Theta A \Theta^{-1})$$

$$= \tfrac{1}{2}(A^\dagger B^\dagger + B^\dagger A^\dagger) = \tfrac{1}{2}(AB + BA)^\dagger = p^\dagger, \qquad (4.3.48a)$$

$$\Theta q \Theta^{-1} = \tfrac{1}{2}(\Theta A \Theta^{-1} \Theta B \Theta^{-1} - \Theta B \Theta^{-1} \Theta A \Theta^{-1})$$

$$= \tfrac{1}{2}(A^\dagger B^\dagger - B^\dagger A^\dagger) = -\tfrac{1}{2}(AB - BA)^\dagger = -q^\dagger. \qquad (4.3.48b)$$

Thus p is time even but q is time odd! By extending these considerations to (4.3.47) and using the fact that μ_α, μ_β, O^s and O^a are time even, we deduce that $\hat{\alpha}^s_{\alpha\beta}$ is time even and $\hat{\alpha}^a_{\alpha\beta}$ is time odd.

Consider next the effective optical activity operators $\hat{G}_{\alpha\beta}$ and $\hat{A}_{\alpha,\beta\gamma}$, also given in (2.8.14). Repeating the same procedure as for $\hat{\alpha}_{\alpha\beta}$ but now using the fact that m_β is Hermitian, has even parity and is time odd, we deduce that $\hat{G}^s_{\alpha\beta}$ is Hermitian, has odd parity and is time odd; and that $\hat{G}^a_{\alpha\beta}$ is antiHermitian, has odd parity and is time even. Similarly, since $\Theta_{\beta\gamma}$ is Hermitian, has even parity and is time even, we deduce that $\hat{A}^s_{\alpha,\beta\gamma}$ is Hermitian, has odd parity and is time even, and that $\hat{A}^a_{\alpha,\beta\gamma}$ is antiHermitian, has odd parity and is time odd.

Finally, we generate the required natural and magnetic optical activity tensors by taking diagonal matrix elements of the appropriate operators:

$$G'_{\alpha\beta} = i\langle n|\hat{G}^a_{\alpha\beta}|n\rangle, \qquad (4.3.49a)$$

$$\alpha'_{\alpha\beta} = i\langle n|\hat{\alpha}^a_{\alpha\beta}|n\rangle. \qquad (4.3.49b)$$

Since the operators are antiHermitian, the expectation values are pure imaginary (Bohm, 1951). These results are consistent with the symmetry classifications introduced earlier: thus natural optical rotation, being a time-even pseudoscalar observable, is generated by a time-even odd-parity operator; and magnetic optical rotation, being a time-odd axial vector, is generated by a time-odd even-parity operator. The magnetic result is also consistent with the statement given in the previous section that, for both even and odd electron systems, a time-odd operator has opposite expectation values in Kramers conjugate states: hence the result that Kramers conjugate states generate equal and opposite magnetic optical rotation.

It was shown in Section 4.3.1 that, since the natural optical rotation observable has odd parity, a resolved chiral molecule must exist in a state of mixed parity. Now that it has been shown that the magnetic optical rotation observable has even parity, we can understand why an atomic state such as $|JM\rangle$, which has definite parity, can also show optical rotation, provided the degeneracy with its Kramers conjugate state $\Theta|JM\rangle$ is lifted by a magnetic field (or a pure $|JM\rangle$ state is prepared in, say, a molecular beam). But notice that a state such as $|JM\rangle$ does not have definite reversality because $\Theta|JM\rangle$ is a new state orthogonal to $|JM\rangle$. For an even-electron system, it is always possible using (4.3.24) and (4.3.27) to write $|JM\rangle$ as a combination of states that have definite reversality; but this is not possible for odd electron systems since invariant states cannot be constructed for them. Thus natural optical rotation is supported only by systems in states with indefinite *space* parity, and magnetic optical rotation is supported only by systems in states with indefinite *time* parity.

It was shown in Section 4.3.1 that both natural optical rotation and a permanent space fixed electric dipole moment are odd-parity observables and so require mixed parity quantum states. We are now in a position to appreciate that time reversal invariance provides a fundamental quantum mechanical distinction between these two different odd-parity observables. It is well known in elementary particle and atomic physics that both parity conservation and time reversal invariance lead independently to the vanishing of a permanent electric dipole moment in a stationary state (see, for example, Sandars, 1968, 2001; or Gibson and Pollard, 1976). Taking an atom as a simple example, this means that observation of a permanent electric dipole moment in, say, a pure $|JM\rangle$ state would violate both P and T.

Since $|JM\rangle$ is a state of definite parity $(-1)^p$, where p is the sum of the individual orbital quantum numbers l of all the electrons in the atom (and using the standard convention that the intrinsic parity of an electron spin state is $+1$ (Heine, 1960)), the vanishing of the electric dipole moment through P invariance follows from the discussion in Section 4.3.1 which shows that the expectation value of any odd-parity observable vanishes in any state of definite parity. In the present context we can use

$$P|JM\rangle = (-1)^p|JM\rangle$$

and

$$P\mu_\alpha P^{-1} = -\mu_\alpha,$$

together with $P^{-1}P = 1$ and $P^{\dagger}P = 1$ (since P is unitary) to write

$$\mu_\alpha = \langle JM|\mu_\alpha|JM\rangle$$
$$= \langle JM|P^{\dagger}(P\mu_\alpha P^{-1})P|JM\rangle$$
$$= -\langle JM|\mu_\alpha|JM\rangle = 0. \tag{4.3.50}$$

The argument showing that T invariance also requires the electric dipole moment to vanish is less straightforward. Since the electric dipole moment operator is time even,

$$\Theta\mu_\alpha\Theta^{-1} = \mu_\alpha^{\dagger} = \mu_\alpha.$$

Therefore, using the methods of the previous section, we can write the expectation value of the electric dipole moment operator with respect to a time reversed state $\Theta|JM\rangle$ as

$$\langle\Theta JM|\mu_\alpha|\,\Theta JM\rangle = \langle\Theta JM|\mu_\alpha\Theta JM\rangle$$
$$= \langle\Theta\mu_\alpha\Theta JM|\Theta^2 JM\rangle = \langle\Theta\mu_\alpha\Theta^{-1}\Theta^2 JM|\Theta^2 JM\rangle$$
$$= \langle\mu_\alpha^{\dagger} JM|JM\rangle = \langle JM|\mu_\alpha|JM\rangle. \tag{4.3.51}$$

We now invoke a unitary operator R which rotates the axes through π about an axis perpendicular to the quantization axis z, say the y axis. This operation therefore retains the handedness of the axis system with $x \to -x, y \to y, z \to -z$, and it can be shown, using Wigner rotation matrices (Silver, 1976), that it has the same effect on $|JM\rangle$ as time reversal, given in (4.3.45), at least within an inessential phase factor. Thus we can write

$$\langle\Theta JM|\mu_z|\Theta JM\rangle = \langle RJM|\mu_z|RJM\rangle$$
$$= \langle JM|R^{-1}\mu_z R|JM\rangle = -\langle JM|\mu_z|JM\rangle \tag{4.3.52}$$

because a rotation through π about y changes the sign of the polar vector operator component μ_z. Since we can choose z arbitrarily in the absence of external fields, (4.3.52) and (4.3.51) are only compatible if $\mu_z = 0$.

The effective optical activity operator $\hat{G}_{\alpha\alpha}^{a}$ responsible for natural optical rotation in isotropic collections of chiral molecules is, like μ_α, time even, and so a development of its expectation value for an atomic state can be written analogous to (4.3.51). But since $\hat{G}_{\alpha\alpha}^{a}$ is a pseudoscalar operator, rather than a polar vector operator, it is invariant to the rotation by π about y and so

$$\langle\Theta JM|\hat{G}_{\alpha\alpha}^{a}|\Theta JM\rangle = \langle JM|R^{-1}\hat{G}_{\alpha\alpha}^{a}R|JM\rangle$$
$$= \langle JM|\hat{G}_{\alpha\alpha}^{a}|JM\rangle. \tag{4.3.53}$$

Thus T invariance does not prohibit natural optical rotation in atoms! A somewhat different proof has been given by Bouchiat and Bouchiat (1974). What does prevent natural optical rotation, of course, is P invariance. This confirms that the tiny optical rotations observed in free atoms, described in Section 1.9.6, are manifestations of P but not T violation.

So far, the discussion of parity and reversality requirements has been confined to atoms. In order to discuss the effect of parity and reversality on rotating molecules, we must ascertain the behaviour of the molecule-fixed axes when the space-fixed axes are transformed. Consider the simple case of a linear dipolar molecule (with zero angular momentum about its symmetry axis) and use the polar tensor transformation law (4.2.32a) to write the molecule-fixed electric dipole moment in terms of space-fixed axes:

$$\mu_{\lambda'} = l_{\lambda'\alpha}\mu_\alpha,$$

where primed and unprimed components now refer to space-fixed and molecule-fixed axes. Then in place of (4.3.50) we have

$$\mu_{\lambda'} = \langle J M ev|l_{\lambda'\alpha}\mu_\alpha|J M ev\rangle$$
$$= \langle J M|l_{\lambda'\alpha}|J M\rangle(\mu_\alpha)_{ev},$$

where $(\mu_\alpha)_{ev} = \langle ev|\mu_\alpha|ev\rangle$ is the molecule-fixed electric dipole moment for a given internal vibrational–electronic state, and so only the direction cosine part of the electric dipole moment operator affects the rotational states. We take the internal axes to be X, Y, Z, with Z parallel to the symmetry axis and pointing in the direction of the electric dipole moment. As a result of the inversion operation, the directions of the space-fixed axes are reversed, thereby changing the handedness. The system X, Y, Z must also change its handedness, but since the Z axis is rigidly connected to the nuclei, it retains its former direction. Hence the direction of either one of the axes X, Y must be reversed. Thus the operation of inversion of the space fixed axes must be accompanied in the molecule-fixed (rotating) axes by a reflection in a plane passing through the symmetry axis of the molecule. This important point is elaborated in Landau and Lifshitz (1977) p. 307, and in Judd (1975) p. 134. Having seen that all the λ' change sign under inversion whereas Z does not, we can now write

$$Pl_{\lambda'Z}P^{-1} = l_{-\lambda'Z} = -l_{\lambda'Z},$$

and so

$$\mu_{\lambda'} = \langle J M|l_{\lambda'Z}|J M\rangle(\mu_Z)_{ev}$$
$$= \langle J M|P^\dagger(Pl_{\lambda'Z}P^{-1})P|J M\rangle(\mu_Z)_{ev}$$
$$= \langle J M|l_{\lambda'Z}|J M\rangle(\mu_Z)_{ev} = 0.$$

Thus parity prevents a dipolar molecule in a rotational quantum state $|J M\rangle$ from showing a space-fixed electric dipole moment. The arguments embodied in (4.3.51, 2) can be extended in a similar fashion to show that this is also prohibited by reversality. On the other hand, it is easy to see that a molecule in a rotational quantum state $|J M\rangle$ is allowed by reversality, and by parity if it is chiral, to show natural optical rotation.

These arguments cover linear tops in nondegenerate electronic states, and asymmetric tops, because their rotational quantum states depend on only the two quantum numbers J and M. Symmetric tops, on the other hand, have an additional quantum number K and a degeneracy of states with angular momentum $\pm K\hbar$ about the molecular symmetry axis. Since the symmetric top wave function has the form (see, for example, Eyring, Watter and Kimball, 1944)

$$\Psi_{JKM} = \Theta_{JKM}(\theta)e^{iM\phi}e^{iK\chi}, \tag{4.3.54}$$

where θ, ϕ, χ are the Euler angles and Θ_{JKM} is a complicated function of θ (and is not to be confused with the time reversal operator), it follows that the time reversal operator transforms a state $|JKM\rangle$ into a different state $|J-K-M\rangle$ (times an inessential phase factor). Since this state cannot be generated by any spatial symmetry operation that changes the sign of μ_α (for example, inversion followed by a rotation through π about the y axis transforms $|JKM\rangle$ into $|J-K-M\rangle$, but does not change the sign of μ_α), arguments of the sort embodied in (4.3.51, 52) cannot be applied, and so reversality does not prohibit a space-fixed electric dipole moment in a symmetric top (unless $K = 0$). The parity operator transforms $|JKM\rangle$ into $|J-KM\rangle$, which therefore has mixed parity, and so parity does not prohibit a space-fixed electric dipole moment either. Thus only dipolar symmetric tops with $K \neq 0$ can show first-order Stark effects: despite the fact that many asymmetric tops and certain linear tops have molecule-fixed permanent electric dipole moments, they usually do not show first-order Stark effects. But it should not be thought that $|JKM\rangle$ and $|J-KM\rangle$ are enantiomeric states, for neither can show natural optical rotation: this follows because time reversal generates the same state from $|JKM\rangle$ as inversion followed by a rotation through π about y, namely $|J-K-M\rangle$, yet time reversal does not change the sign of $\hat{G}^a_{\alpha\alpha}$ while the second combined operation does. States $|JKM\rangle$ and $|J-KM\rangle$ will, however, generate the same magnetic optical rotation, which is equal and opposite to that generated by $|JK-M\rangle$ and $|J-K-M\rangle$.

4.3.4 Optical enantiomers, two-state systems and parity violation

We saw in Section 4.3.1 how the mixed parity states of a resolved chiral molecule can be pictured in terms of a double well potential. This aspect is now developed further by considering the quantum mechanics of a degenerate two-state system in order to gain an insight into the apparent paradox of the stability of optical enantiomers which was recognized at the beginning of the quantum era since the existence of optical enantiomers was difficult to reconcile with basic quantum mechanics. In the words of Hund (1927):

If a molecule admits two different nuclear configurations being the mirror images of each other, then the stationary states do not correspond to a motion around one of these two equilibrium configurations. Rather, each stationary state is composed of left-handed and right-handed configurations in equal shares ... The fact that the right-handed or left-handed configuration of a molecule is not a quantum state (eigenstate of the Hamiltonian) might appear to contradict the existence of optical isomers.

Similarly Rosenfeld (1928):

A system (state) with sharp energy is optically inactive.

And Born and Jordan (1930):

Since each molecule consists of point charges interacting via Coulomb's law, the energy function (Hamiltonian) is always invariant with respect to space inversion. Consequently there could not exist any optically active molecules, which contradicts experience.

These translated quotations are taken from a critical review by Pfeifer (1980).

Hund's resolution of the 'paradox' involves arguments of the type given in Section 4.3.1, namely that typical chiral molecules have such high barriers to inversion that the lifetime of a prepared enantiomer is virtually infinite. In this section Hund's approach is brought up to date by injecting a small parity-violating term into the Hamiltonian, which results in the two enantiomeric states becoming the true stationary states (Harris and Stodolsky, 1978).

We start by reviewing the perturbation treatment for two interacting degenerate states ψ_1 and ψ_2, following a treatment given by Bohm (1951) which is particularly appropriate. The usual result for the perturbed energy is

$$W_\pm = W \pm |V_{12}|, \tag{4.3.55}$$

where W is the unperturbed energy shared by ψ_1 and ψ_2 and V is the perturbation Hamiltonian (for simplicity we have assumed that $V_{11} = V_{22} = 0$). The amplitudes of the corresponding perturbed wavefunctions can be written

$$\psi_\pm^{(0)} = \frac{1}{\sqrt{2}}(\psi_1 \pm e^{i\alpha}\psi_2), \tag{4.3.56}$$

with $V_{12} = |V_{12}|e^{-i\alpha}$ so that, if V_{12} is real and positive, $\alpha = 0$, and, if V_{12} is real and negative, $\alpha = \pi$. These approximate wavefunctions have two important properties: they are orthogonal, and the matrix elements of V between $\psi_+^{(0)}$ and $\psi_-^{(0)}$ vanish. (Do not confuse the subscripts $\pm$, which denote higher and lower energy solutions, with the notation ($\pm$) used in Section 4.3.1 to denote even- and odd-parity wavefunctions.)

We now consider how the wavefunction changes with time. Take V_{12} to be real and negative so that $\alpha = \pi$. The amplitudes of the two perturbed wavefunctions are then

$$\psi_\pm^{(0)} = \frac{1}{\sqrt{2}}(\psi_1 \mp \psi_2). \tag{4.3.57}$$

The $\psi_\pm^{(0)}$ are the amplitudes of stationary states with time-dependent wavefunctions

$$\psi_\pm(t) = \psi_\pm^{(0)}e^{-i(W\pm|V_{12}|)t/\hbar}. \tag{4.3.58}$$

The general time-dependent wavefunction for the two-state system is now given by the sum of the two stationary state wavefunctions:

$$\psi(t) = \frac{1}{\sqrt{2}}\left(\psi_+^{(0)}e^{-i|V_{12}|t/\hbar} + \psi_-^{(0)}e^{i|V_{12}|t/\hbar}\right)e^{-iWt/\hbar}. \tag{4.3.59}$$

This can be rewritten in terms of ψ_1 and ψ_2:

$$\psi(t) = [\psi_1 \cos(|V_{12}|t/\hbar) + i\psi_2 \sin(|V_{12}|t/\hbar)]e^{-iWt/\hbar}. \tag{4.3.60}$$

Thus at $t = 0$ the system is entirely in the state ψ_1 and at $t = \pi\hbar/2|V_{12}|$ it is entirely in the state ψ_2 which is seen to have a phase $e^{-i\pi/2}$ relative to ψ_1. This oscillation of amplitude between the two states ψ_1 and ψ_2 is formally similar to that between two resonant classical harmonic oscillators, such as pendulums, that are weakly coupled. If just one of the pendulums is made to swing, the energy is transferred back and forth between the two pendulums at a rate proportional to the strength of the coupling force. But if the two pendulums are made to swing simultaneously with identical energies, two possible states of stationary oscillation are possible (stationary in the sense that each pendulum retains constant energy) corresponding to the in-phase and out-of-phase local oscillations. The transformation from a description in terms of local pendulum coordinates to the stationary combinations of the local coordinates is simply a transformation to the normal coordinates of the vibrating system: the local coordinates are not 'diagonal' in the sense that they couple with each other; whereas there is no coupling between the normal coordinates so they oscillate independently of each other. Likewise the set of quantum states (ψ_1, ψ_2) couple with each other whereas the set ($\psi_+^{(0)}$, $\psi_-^{(0)}$) do not and are true stationary states.

Thus if no external perturbation is applied to a two-state system, any 'perturbation' which couples ψ_1 and ψ_2 is internal and is simply an 'artifact' of the chosen representation: the Hamiltonian is the same for (ψ_1, ψ_2) and ($\psi_+^{(0)}$, $\psi_-^{(0)}$). It might be appropriate in some situations to set up the problem in terms of perturbation theory, as above, if the chosen representation is 'almost diagonal' in the sense that the coupling is weak, or indeed if an external perturbation is present. But for a general two-state system (not necessarily degenerate) the exact energy eigenvalues and eigenfunctions are, in place of (4.3.55) and (4.3.56),

$$W_\pm = \tfrac{1}{2}(H_{11} + H_{22}) \pm \tfrac{1}{2}[(H_{11} - H_{22})^2 + 4|H_{12}|^2]^{\frac{1}{2}}, \qquad (4.3.61)$$

$$\psi_+^{(0)} = \cos\phi\,\psi_1^{(0)} + \sin\phi\,\psi_2^{(0)}, \qquad (4.3.62a)$$

$$\psi_-^{(0)} = -\sin\phi\,\psi_1^{(0)} + \cos\phi\,\psi_2^{(0)}, \qquad (4.3.62b)$$

with

$$\tan 2\phi = 2|H_{12}|/(H_{11} - H_{22}). \qquad (4.3.62c)$$

If ψ_1 and ψ_2 happen to be degenerate and are interconverted by a particular symmetry operation of the Hamiltonian, $\psi_+^{(0)}$ and $\psi_-^{(0)}$ transform according to one or other of the irreducible representations of the group comprising the identity and the operation in question. Thus if a two-state system is prepared in a nonstationary state, it might appear (falsely) to be influenced by a time-dependent perturbation lacking some fundamental symmetry of the internal Hamiltonian of the system.

We now identify the two enantiomeric states ψ_L and ψ_R of a chiral molecule with ψ_1 and ψ_2. Since these states are interconverted by a fundamental symmetry operation of the Hamiltonian, the inversion, they couple with each other; whereas the stationary states $\psi_+^{(0)}$ and $\psi_-^{(0)}$ transform according to one or other of the irreducible representations of the inversion group, $\psi_+^{(0)} \equiv \psi^{(0)}(-)$ having odd parity and energy $W_+ \equiv W(-)$, and $\psi_-^{(0)} \equiv \psi^{(0)}(+)$ having even parity and energy $W_- \equiv W(+)$. This identification enables (4.3.10) to be recovered from (4.3.59). The Born–Oppenheimer approximation is invoked in order to envisage this coupling in terms of an overlap of ψ_L and ψ_R due to tunnelling through the barrier in the double well potential (Fig. 4.4), but it is emphasized that this is a convenience: the coupling is independent of any model of molecular structure. It happens that, because we are able to distinguish the left- and right-handed forms of a chiral object, we can prepare a chiral molecule in a state ψ_L or ψ_R, but these are not the stationary states (neglecting for the moment a small parity-violating term in the Hamiltonian): having prepared ψ_L or ψ_R, if the molecule is isolated from all external influences, it will oscillate forever between ψ_L and ψ_R in accordance with (4.3.60).

The natural optical activity observables shown by this oscillating system are time dependent and are given by the expectation values of the effective optical activity operators $\hat{G}_{\alpha\beta}$ and $\hat{A}_{\alpha,\beta\gamma}$, defined in (2.8.14), for the general time-dependent wavefunction (4.3.60). Isotropic optical rotation, for example, is proportional to the imaginary part of

$$\langle\psi|\hat{G}^a_{\alpha\alpha}|\psi\rangle = \langle\psi_L|\hat{G}^a_{\alpha\alpha}|\psi_L\rangle\cos^2(\delta t/\hbar)$$
$$+\langle\psi_R|\hat{G}^a_{\alpha\alpha}|\psi_R\rangle\sin^2(\delta t/\hbar)$$
$$+i\left[\langle\psi_L|\hat{G}^a_{\alpha\alpha}|\psi_R\rangle - \langle\psi_R|\hat{G}^a_{\alpha\alpha}|\psi_L\rangle\right]\cos(\delta t/\hbar)\sin(\delta t/\hbar), \quad (4.3.63)$$

where $\delta = |\langle\psi_L|H|\psi_R\rangle|$. Using the fact that $\hat{G}^a_{\alpha\beta}$ has odd parity, together with $P^{-1}P = 1$ and $P^\dagger P = 1$, we find

$$\langle\psi_L|\hat{G}^a_{\alpha\alpha}|\psi_L\rangle = \langle\psi_L|P^\dagger(P\hat{G}^a_{\alpha\alpha}P^{-1})P|\psi_L\rangle$$
$$= \langle P\psi_L|P\hat{G}^a_{\alpha\alpha}P^{-1}|P\psi_L\rangle = -\langle\psi_R|\tilde{G}^a_{\alpha\alpha}|\psi_R\rangle. \quad (4.3.64a)$$

Similarly,

$$\langle\psi_L|\hat{G}^a_{\alpha\alpha}|\psi_R\rangle = -\langle\psi_R|\hat{G}^a_{\alpha\alpha}|\psi_L\rangle. \quad (4.3.64b)$$

But since $\hat{G}^a_{\alpha\beta}$ is antiHermitian, we also have

$$\langle\psi_L|\hat{G}^a_{\alpha\alpha}|\psi_R\rangle = -\langle\psi_R|\hat{G}^a_{\alpha\alpha}|\psi_L\rangle^*. \quad (4.3.64c)$$

The last two results show that $\langle\psi_L|\hat{G}^a_{\alpha\alpha}|\psi_R\rangle$ is real; whereas time reversal arguments in Section 4.4.3 below show that it is imaginary (at least for even electron systems). We therefore conclude that both real and imaginary parts of $\langle\psi_L|\hat{G}^a_{\alpha\alpha}|\psi_R\rangle$ are zero. Thus (4.3.63) becomes

$$\langle\psi|\hat{G}^a_{\alpha\alpha}|\psi\rangle = \langle\psi_L|\hat{G}^a_{\alpha\alpha}|\psi_L\rangle\cos(2\delta t/\hbar) \quad (4.3.65)$$

and so the time-averaged natural optical rotation angle is zero.

We now introduce a small parity-violating term into the Hamiltonian of the chiral molecule that lifts the exact degeneracy of the mirror image enantiomers, as described in Section 1.9.6. The weak neutral current interaction generates parity violating interactions between electrons, and between electrons and nucleons. The latter leads to the following electron–nucleus contact interaction (in atomic units where $\hbar = e = m_e = 1$ in atoms and molecules (Bouchiat and Bouchiat, 1974; Hegstrom, Rein and Sandars, 1980):

$$V^{PV}_{eN} = \frac{G\alpha}{4\sqrt{2}}Q_W\{\sigma_e.\mathbf{p}_e, \rho_N(\mathbf{r}_e)\}, \quad (4.3.66)$$

where { } denotes an anticommutator, G is the Fermi weak coupling constant, α is the fine structure constant, σ_e and $\mathbf{p}_e$ are the Pauli spin operator and linear momentum operator of the electron, $\rho_N(\mathbf{r}_e)$ is a normalized nuclear density function and

$$Q_W = Z(1 - 4\sin^2\theta_W) - N$$

is an effective weak charge which depends on the proton and neutron numbers Z and N together with the Weinberg electroweak mixing angle θ_W which relates the weak and electromagnetic unit charges g and e through $g \sin \theta_W = e$. The much smaller electron–electron interaction is usually neglected. Since σ_e and $\mathbf{p}_e$ are time-odd axial and polar vectors, respectively, and all the other factors are time-even scalars, V_{eN}^{PV} transforms as a time-even pseudoscalar, as required, and so can mix even- and odd-parity electronic states at the nucleus. Hence

$$P V_{eN}^{PV} P^{-1} = -V_{eN}^{PV}, \qquad (4.3.67)$$

so parity violation shifts the energies of the enantiomeric states in opposite directions:

$$\langle \psi_L | V_{eN}^{PV} | \psi_L \rangle = -\langle \psi_R | V_{eN}^{PV} | \psi_R \rangle = \epsilon. \qquad (4.3.68)$$

Attempts to calculate ϵ are faced with the following difficulty. The electronic coordinate part of V_{eN}^{PV} in (4.3.66) is linear in $\mathbf{p}_e$ and is therefore purely imaginary. Since, in the absence of external magnetic fields, the molecular wavefunction may always be chosen to be real, V_{eN}^{PV} has zero expectation values. Also the presence of σ_e means that only matrix elements between different spin states survive. Consequently, it is necessary to invoke a magnetic perturbation of the wavefunction that involves spin, such as spin–orbit coupling. This leads to a tractable method for detailed quantum chemical calculations of the tiny parity-violating energy differences between enantiomers. Results at the time of writing are summarized by Quack (2002) and Wesendrup *et al.* (2003).

Since, on account of parity violation, the two enantiomeric states of the chiral molecule are no longer degenerate, the energies and wavefunctions of the two stationary states $\psi_+^{(0)}$ and $\psi_-^{(0)}$ are given by the general two-state results (4.3.61) and (4.3.62) with H now containing V_{eN}^{PV} from which it follows that (Harris and Stodolsky, 1978; Harris, 1980)

$$W_+ - W_- = 2(\epsilon^2 + \delta^2)^{\frac{1}{2}}, \qquad (4.3.69a)$$

$$\tan 2\phi = \delta/\epsilon. \qquad (4.3.69b)$$

When $\epsilon = 0$, $W_+ - W_- = 2\delta$ and is interpreted as the tunnelling splitting $W(-) - W(+)$ between the definite parity states $\psi^{(0)}(-)$ and $\psi^{(0)}(+)$, as discussed in Section 4.3.1. When $\epsilon \neq 0$, the Hamiltonian lacks inversion symmetry so the stationary states $\psi_+^{(0)}$ and $\psi_-^{(0)}$ may no longer be identified with the definite parity states $\psi^{(0)}(-)$ and $\psi^{(0)}(+)$, respectively. Thus $\psi_+^{(0)}$ and $\psi_-^{(0)}$ are no longer equal combinations of ψ_L and ψ_R. If the system is prepared in ψ_L, say, it will never become completely ψ_R: the optical activity oscillates asymmetrically. This can be shown explicitly by inverting (4.3.62a and b) (and multiplying each stationary state amplitude by its exponential time factor),

$$\psi_L = \cos\phi\ \psi_+^{(0)} e^{-iW_+ t/\hbar} - \sin\phi\ \psi_-^{(0)} e^{-iW_- t/\hbar}, \qquad (4.3.70a)$$

$$\psi_R = \cos\phi\ \psi_-^{(0)} e^{-iW_- t/\hbar} + \sin\phi\ \psi_+^{(0)} e^{-iW_+ t/\hbar}, \qquad (4.3.70b)$$

and working out the appropriate expectation value. Thus, for a system prepared in ψ_L, the time dependence of the isotropic optical rotation is proportional to the imaginary part of

$$\langle\psi_L|\hat{G}_{\alpha\alpha}^a|\psi_L\rangle = \langle\psi_L^{(0)}|\hat{G}_{\alpha\alpha}^a|\psi_L^{(0)}\rangle \left\{ \frac{\epsilon^2 + \delta^2 \cos[2(\delta^2 + \epsilon^2)^{\frac{1}{2}} t/\hbar]}{(\delta^2 + \epsilon^2)} \right\}. \quad (4.3.71)$$

As discussed just after (4.3.64), terms in $\langle\psi_L^{(0)}|\hat{G}_{\alpha\alpha}^a|\psi_R^{(0)}\rangle$ are zero, at least for even electron systems. Taking the time average, we can write

$$\frac{\overline{\Delta\theta}}{\Delta\theta_{max}} = \frac{\epsilon^2}{(\delta^2 + \epsilon^2)}. \qquad (4.3.72)$$

Thus parity violation causes a shift away from zero of the time-averaged optical rotation angle $\overline{\Delta\theta}$.

It follows from (4.3.61) and (4.3.62) that, as $\delta/\epsilon \to 0$, ψ_L and ψ_R become the true stationary states. In fact for typical chiral molecules, δ corresponds to tunnelling times of the order of millions of years: Harris and Stodolsky (1978) have estimated ϵ to correspond to times of the order of seconds to days, so at low temperature (to prevent thermal 'hopping' over the barrier) and in a vacuum (to minimize interaction with the environment) a prepared enantiomer will retain its handedness essentially for ever. These considerations therefore suggest that the ultimate answer to the 'paradox' of the stability of optical enantiomers lies in the weak interactions. However, the situation is more complicated because the influence of the environment must also be considered (Harris and Stodolsky, 1981).

Because any observable quantities are expected to be so very small, the detection of manifestations of parity violation in chiral molecules and the measurement of the parity-violating energy differences between enantiomers remains a major challenge for molecular physics. There has been much discussion of possible experimental strategies that exploit different aspects of the quantum mechanics of the two-state system perturbed by parity violation (see, for example, Quack, 2002 and Harris, 2002).

4.3.5 Symmetry breaking and symmetry violation

The appearance of parity-violating phenomena is interpreted in quantum mechanics by saying that, contrary to what had been previously supposed, the Hamiltonian lacks inversion symmetry due to the presence of pseudoscalar terms such as the weak neutral current interaction (4.3.66). This means that P and H no longer commute, so the associated law of conservation of parity no longer holds. Such symmetry *violation* must be distinguished from symmetry *breaking*: current usage in the physics literature applies the latter term to the situation which arises when a system displays a lower symmetry than that of its Hamiltonian (Anderson, 1972, 1983; Michel, 1980; Blaizot and Ripka, 1986). More specifically, a state has broken symmetry if it cannot be classified according to an irreducible representation of the symmetry group of the Hamiltonian or, equivalently, if it does not carry the quantum numbers of the eigenstates of the Hamiltonian, such as parity, angular momentum, etc. Natural optical activity is therefore a phenomenon arising from parity breaking since, as we have seen, a resolved chiral molecule displays a lower symmetry than that of its associated Hamiltonian. If the small parity-violating term in the Hamiltonian is neglected, the symmetry operation that the Hamiltonian possesses but the chiral molecule lacks is parity, and it is the parity operation that interconverts the two enantiomeric parity-broken states. In the context of nuclear physics, broken symmetry states are often called deformed states (Blaizot and Ripka, 1986).

A symmetry violation may often be conceptualized as a symmetry breaking with respect to some new and previously unsuspected deeper symmetry operation of the Hamiltonian. For example, parity violation was found to imply a violation of charge conjugation symmetry, with the combined CP symmetry being conserved overall (Section 1.9.6). Hence the P *violation* that lifts the degeneracy of the P-enantiomers of a chiral molecule is associated with a symmetry *breaking* with respect to CP, since CP generates a distinguishable system (the mirror-image molecule composed of antiparticles) with identical energy to the original. Likewise, assuming CPT is conserved, CP violation is associated with symmetry breaking with respect to CPT, although now the physical interpretation is more subtle. For example, a process which violates CP, such as the decay of the neutral K-meson where CP violation is manifest as an asymmetry in the decay rates to the two sets of CP-enantiomeric states (Section 1.9.6), will be invariant under CPT. This means that the rate from the initial state to the final state will be identical to the rate for the reverse process from the final state to the initial state but now with all the particles replaced by their CP-enantiomers.

The conventional view, formulated in terms of the double well model in Section 4.3.1, is that parity violation plays no part in the stabilization of chiral molecules. The natural optical activity remains observable only so long as the observation time is short compared with the interconversion time between enantiomers, which is proportional to the inverse of the tunnelling splitting. Such parity-breaking optical activity therefore averages to zero over a sufficiently long observation time. These considerations lead us to an important criterion for distinguishing between natural optical activity generated through parity breaking from that generated through parity violation. The former is time dependent and averages to zero, at least in isolated chiral molecules; whereas the latter is constant in time (recall from the previous section that the handed states become the stationary states when $\delta/\epsilon \to 0$). Since it is due entirely to parity violation, the tiny natural optical rotation shown by a free atomic vapour is constant in time.

There is considerable interest in the development of quantitative measures of the *degree of chirality* of individual chiral molecules (Mislow, 1999). While such measures are of mathematical interest in the context of static geometry and topology and may have practical applications in chemistry, it should be clear from the discussion above that the degree of chirality of individual molecular structures in the form of some fundamental time-even pseudoscalar quantity analogous to, say energy (a time-even scalar) is a Will o' the wisp (Barron, 1996). This is because the degree of chirality evaporates under close quantum mechanical scrutiny: neglecting parity violation, chiral molecules are not in stationary states of the Hamiltonian so any pseudoscalar quantity will average to zero on an appropriate timescale.

In condensed matter physics symmetry breaking is associated with phase transitions in which large numbers of particles cooperate to produce sudden transitions between symmetric and asymmetric states of the complete macroscopic sample, as in ferromagnetism. The Hamiltonian of an iron crystal is invariant under spatial rotations. However the ground state of a magnetized sample, in which all the microscopic magnetic dipole moments are aligned in the same direction, is not invariant: it distinguishes a specific direction in space, the direction of magnetization. This nonzero magnetization in zero applied field also breaks time reversal symmetry. When the temperature is raised above the Curie point, the magnetization disappears and the rotational and time reversal symmetries become manifest. A vestige of the rotational symmetry still survives in the ferromagnetic phase in that the sense of magnetization with respect to space-fixed axes is arbitrary. Temperature is a central feature here, because behaviour reflecting the full symmetry of the Hamiltonian can be recovered at sufficiently high temperature. Molecules behave rather differently from macroscopic systems in that they do not support sharp phase transitions between symmetric and asymmetric states (Anderson, 1972, 1983). There has been much discussion on the relationship between the microscopic and macroscopic aspects of the broken-parity states of chiral systems (see, for example, Woolley, 1975b, 1982; Quack, 1989; Vager, 1997).

The expression 'spontaneous symmetry breaking' is usually employed in macroscopic systems (ideally in the limit of an infinite number of particles) to describe phase transitions to less symmetric states (Binney *et al.*, 1992). This expression is derived from 'spontaneous magnetization' in the case of ferromagnetism. An analogous type of spontaneous symmetry breaking occurs in gauge theories of elementary particles (Gottfried and Weisskopf, 1984; Weinberg, 1996). The broken symmetry phase is described by an *order parameter*, indicating that this phase possesses the lower symmetry and hence greater order. The order parameter in the case of ferromagnetism is the magnetization, which transforms as a time-odd axial vector. A phase transition from an achiral (racemic) state to a chiral state of a macroscopic system would be characterized by an order parameter transforming as a time-even pseudoscalar.

4.3.6 CP violation and molecular physics

Heisenberg (1966) once made remarks to the effect that elementary particles are much more akin to molecules than to atoms. This insight gains force from a consideration of the curious behaviour of the neutral K-meson (Gibson and Pollard, 1976; Gottfried and Weisskopf, 1984; Sachs, 1987). The neutral K-meson displays four distinct states: particle and antiparticle states $|K^0\rangle$ and $|K^{0*}\rangle$ which are interconverted by the operation CP, and two mixed states $|K_1\rangle = (|K^0\rangle + |K^{0*}\rangle)/\sqrt{2}$ and $|K_2\rangle = (|K^0\rangle - |K^{0*}\rangle)/\sqrt{2}$ which have different energies because of coupling between $|K^0\rangle$ and $|K^{0*}\rangle$ via the weak force. This means that $|K_1\rangle$ and $|K_2\rangle$ are even and odd eigenstates with respect to CP, and that $|K^0\rangle$ and $|K^{0*}\rangle$ are mixed (symmetry broken) with respect to CP. Wigner (1965) has therefore likened these four distinct states of the neutral K-meson to the four possible states of a chiral molecule in the real world, namely the even- and odd-parity states $\psi(+)$ and $\psi(-)$ and the two handed states ψ_L and ψ_R of mixed parity, respectively. However, the CP eigenstates $|K_1\rangle$ and $|K_2\rangle$ are not pure since $|K_2\rangle$, which is *odd* with respect to CP, is occasionally observed to decay into products which are *even* with respect to CP. This implies that the Hamiltonian contains a small CP-violating term that mixes $|K_1\rangle$ and $|K_2\rangle$, analogous to the P-violating term that mixes the definite parity states of a chiral molecule. (The long-lived neutral K-meson K_L mentioned in Section 1.9.6 is the same as $|K_2\rangle$, and its decay rate asymmetry is another manifestation of CP violation.)

There is, however, a subtle but fundamental difference between P violation in a chiral molecule and CP violation in the neutral K-meson system: P violation lifts the degeneracy of the P-enantiomers of a chiral molecule (the left- and right-handed states), but CP violation does not lift the degeneracy of the CP-enantiomers of the neutral K-meson (the particle and antiparticle states) because, as already mentioned in Section 1.9.6, CPT invariance guarantees that the rest mass of a particle and its antiparticle are equal. Similarly, CP violation does not lift the degeneracy of the CP-enantiomers of a chiral molecule (a molecule and its mirror image composed of antiparticles, as invoked in Figure 1.23) (Barron, 1994). But it should not be thought that, if antimolecules were accessible, the type of CP

violation observed in the neutral K-meson system might be observed in molecular systems, with molecule–antimolecule superposition states analogous to $|K_1\rangle$ and $|K_2\rangle$ as intermediates bridging the worlds of matter and antimatter. Among other things, such molecule–antimolecule transformations would require a gross violation of the law of baryon conservation, which does not arise in the neutral K-meson system because mesons have baryon numbers zero.

4.4 The symmetry classification of molecular property tensors

In this section point group symmetry arguments are combined with time reversal arguments to establish criteria for the nonvanishing of components of property or transition tensors in a molecule with a given spatial symmetry and in a given quantum state. The example of permanent electric and magnetic dipole moments in Section 4.3.2 gives a preliminary idea of the considerations involved.

4.4.1 Polar and axial, time-even and time-odd tensors

We saw in Section 1.9.2 that it is possible to classify scalar and vector physical quantities with respect to their behaviour under space inversion and time reversal. This classification can be extended to general molecular property tensors by considering relationships such as

$$\mu_\alpha = \alpha_{\alpha\beta} E_\beta$$

in which two measurable quantities are related by means of a property tensor. So if the behaviour under space inversion and time reversal of the two measurable quantities is known, the property tensor can be classified immediately. In this particular example, since μ and E are both polar time-even vectors, $\alpha_{\alpha\beta}$ is a second-rank polar time-even tensor. By applying these considerations to the general expressions (2.6.26) for the induced electric and magnetic multipole moments, the characteristics listed in Table 4.1 are deduced (Buckingham, Graham and Raab, 1971).

4.4.2 Neumann's principle

Neumann's principle (Neumann, 1885) states that any type of symmetry exhibited by the point group of a system is possessed by every physical property of the system. A physical property of a system relates associated measurable quantities: for example, density relates the mass and the volume; and electric polarizability relates the induced electric dipole moment and the applied uniform electric field. Since a point group symmetry operation can be defined as one that leaves the system indistinguishable from its original condition, the same relation must hold between the

Table 4.1 *The behaviour of molecular property tensors*
under space inversion and time reversal

Molecular property tensor	Space inversion	Time reversal
μ_α	polar	even
m_α	axial	odd
$\alpha_{\alpha\beta}$	polar	even
$\alpha'_{\alpha\beta}$	polar	odd
$A_{\alpha,\beta\gamma}$	polar	even
$A'_{\alpha,\beta\gamma}$	polar	odd
$G_{\alpha\beta}$	axial	odd
$G'_{\alpha\beta}$	axial	even
$C_{\alpha\beta,\gamma\delta}$	polar	even
$C'_{\alpha\beta,\gamma\delta}$	polar	odd
$D_{\alpha,\beta\gamma}$	axial	odd
$D'_{\alpha,\beta\gamma}$	axial	even
$\chi_{\alpha\beta}$	polar	even
$\chi'_{\alpha\beta}$	polar	odd

measurable quantities before and after the symmetry operation, and the physical property in question must therefore transform into +1 times itself under all the symmetry operations of the system. Thus, re-expressed in group theoretical terms, Neumann's principle states that any tensor components representing a physical property of a system must transform as the totally symmetric irreducible representation of the system's symmetry group. Curie (1908) provided the following penetrating formulation of Neumann's principle in terms of asymmetry rather than symmetry: 'C'est la dissymmetrie, qui crée le phenomène'. Thus no asymmetry can manifest itself in a property tensor which does not already exist in the system. Birss (1966) and Shubnikov and Koptsik (1974) have discussed Neumann's principle at length. See also Zocher and Török (1953) and Altmann (1992).

Neumann's principle also embraces time reversal symmetry provided the physical property under consideration is static, but it does not apply to transport properties; in other words, it does not apply to phenomena where the entropy of the system is changing. The group theoretical approach is based on the *nonmagnetic* and *magnetic* symmetry groups which are generated from the classical groups by adding new operations generated by combining spatial transformations with time reversal (Birss, 1966, Joshua, 1991). This approach, which is not elaborated here, is most appropriate when considering the magnetic properties of crystals.

Since we are interested mainly in the quantum mechanical properties of individual atoms and molecules in this book, we incorporate time reversal into our symmetry arguments using an alternative approach based on the generalized symmetry selection rule (4.3.37). This takes account of the time reversal characteristics of a physical property by specifying a corresponding time-even or time-odd operator, and takes account of whether the molecule has an even or an odd number of electrons by using a single or a double point group. The diagonal matrix elements give the corresponding property tensor component in particular quantum states, and the off-diagonal matrix elements give corresponding transition tensors. Thus an atom or molecule in a degenerate quantum state that can, according to (4.3.37), support, for example, a magnetic moment, would not have time reversal symmetry; but in the absence of a time-odd external influence, such as a magnetic field that lifts the degeneracy, each atom or molecule will exist in a time-even superposition of states in which the magnetic moments associated with each component state cancel.

4.4.3 Time reversal and the permutation symmetry of molecular property and transition tensors

It has been said that time reversal symmetry is responsible for the intrinsic symmetry of matter tensors (Fumi, 1952). Here we show how time reversal arguments in a quantum mechanical context can be used to glean more detailed information about molecular property and transition tensors than is given by the classical method of Section 4.4.1, particularly when the molecules are in degenerate electronic states. In the case of the polarizability, powerful statements concerning the tensor permutation symmetry emerge. Although analogous statements are not possible for the optical activity tensors, other useful results are obtained.

It is easy to prove the equality (within a phase factor) of the probability amplitudes for the transitions $|1\rangle \rightarrow |2\rangle$ and $|\Theta 2\rangle \rightarrow |\Theta 1\rangle$, where $|1\rangle$ and $|2\rangle$ are any pair of quantum states and $|\Theta 1\rangle$ and $|\Theta 2\rangle$ are the corresponding time-reversed states. Thus using the methods of Section 4.3.2 we can write

$$\langle \Theta 1 | A(\pm) | \Theta 2 \rangle = \langle \Theta 1 | A(\pm)\Theta 2 \rangle$$
$$= \langle \Theta A(\pm)\Theta 2 | \Theta^2 1 \rangle = \langle \Theta A(\pm)\Theta^{-1}\Theta^2 2 | \Theta^2 1 \rangle$$
$$= \pm \langle A(\pm)^\dagger 2 | 1 \rangle = \pm \langle 2 | A(\pm) | 1 \rangle. \tag{4.4.1}$$

This result is independent of whether $A(\pm)$ is Hermitian, antiHermitian or non-Hermitian.

In order to apply (4.4.1) to light scattering, it is necessary to specify a scattering operator with well defined behaviour under time reversal. As shown in Section 4.3.3, the effective polarizability operator $\hat{\alpha}_{\alpha\beta}$ defined in (2.8.14) has a part $\hat{\alpha}^s_{\alpha\beta}$ that is Hermitian and time even and a part $\hat{\alpha}^a_{\alpha\beta}$ that is antiHermitian and time odd. Putting $\hat{\alpha}_{\alpha\beta}$ into (4.4.1) and recalling that a Hermitian operator satisfies $\langle m|V|n\rangle = \langle n|V|m\rangle^*$ and an antiHermitian operator satisfies $\langle m|V|n\rangle = -\langle n|V|m\rangle^*$, we obtain the following fundamental property of the complex transition polarizability (Barron and Nørby Svendsen, 1981; Liu, 1991)):

$$(\tilde{\alpha}_{\alpha\beta})_{mn} = (\tilde{\alpha}_{\beta\alpha})_{\Theta n\Theta m} = (\tilde{\alpha}_{\alpha\beta})^*_{\Theta m\Theta n}. \tag{4.4.2}$$

Despite the approximations used in the derivation of (4.4.2), the result may be shown to be valid for all Raman processes, transparent and resonant (Hecht and Barron, 1993c).

Within the present formalism, the generalization to absorbing frequencies is accomplished by taking account of the lifetimes of the excited intermediate states $|j\rangle$, as discussed in Section 2.6.3. This leads to the introduction of the real dispersion and absorption lineshape functions f and g, and enables us to decompose the (already) complex transition polarizability into dispersive and absorptive parts:

$$(\tilde{\alpha}_{\alpha\beta})_{mn} = (\tilde{\alpha}_{\alpha\beta}(f))_{mn} + i(\tilde{\alpha}_{\alpha\beta}(g))_{mn}. \tag{4.4.3}$$

The fundamental relationship (4.4.2) can now be extended to the case of resonance scattering by means of separate relationships between the dispersive and absorptive parts of the complex transition polarizability:

$$(\tilde{\alpha}_{\alpha\beta}(f))_{mn} = (\tilde{\alpha}_{\alpha\beta}(f))^*_{\Theta m\Theta n}. \tag{4.4.4a}$$

$$(\tilde{\alpha}_{\alpha\beta}(g))_{mn} = (\tilde{\alpha}_{\alpha\beta}(g))^*_{\Theta m\Theta n}. \tag{4.4.4b}$$

Consider first the application of (4.4.2) to an even electron system (integral J). The initial and final states can now be chosen to be either even or odd with respect to time reversal; that is, states of the form (4.3.24) or (4.3.27). If we choose even states (which we always can for integral J), $|\Theta n\rangle = |n\rangle$ and $|\Theta m\rangle = |m\rangle$ so that

$$(\tilde{\alpha}_{\alpha\beta})_{mn} = (\tilde{\alpha}_{\alpha\beta})^*_{mn}. \tag{4.4.5}$$

This result shows that the transition polarizability is pure real, that is $(\tilde{\alpha}_{\alpha\beta})_{mn} = (\alpha_{\alpha\beta})_{mn}$, but says nothing about its permutation symmetry, which implies that both symmetric and antisymmetric parts are allowed by time reversal (unless $m = n$ when only the symmetric part survives).

The application of (4.4.2) to an odd electron system (half odd-integral J) reveals additional richness. As discussed in Section 4.3.2, it is not now possible to construct states that are even or odd with respect to time reversal since a single application of the time reversal operator always generates a state orthogonal to the original one, as demonstrated in (4.3.31). We consider explicitly the most common situation, when the initial and final states are components of a twofold Kramers degenerate electronic level. The conclusions therefore apply immediately to atoms; for molecules we must take the purely electronic part of the transition polarizability that results when the zeroth order Born–Oppenheimer approximation is invoked and so, as discussed later (Section 8.3), the conclusions apply only to Rayleigh scattering and to resonance Raman scattering in totally symmetric modes of vibration. Denoting the two Kramers components by e_n and e'_n, there are four scattering transitions possible: $e_n \leftarrow e_n, e'_n \leftarrow e'_n, e_n \leftarrow e'_n$ and $e'_n \leftarrow e_n$. From (4.3.22) we can write $|\Theta e_n\rangle = |e'_n\rangle$ and $|\Theta e'_n\rangle = -|e_n\rangle$, so from (4.4.2) we have

$$(\tilde{\alpha}_{\alpha\beta})_{e_n e_n} = (\tilde{\alpha}_{\beta\alpha})_{e'_n e'_n} = (\tilde{\alpha}_{\alpha\beta})^*_{e'_n e'_n}, \tag{4.4.6a}$$

$$(\tilde{\alpha}_{\alpha\beta})_{e'_n e_n} = -(\tilde{\alpha}_{\beta\alpha})_{e'_n e_n} = -(\tilde{\alpha}_{\alpha\beta})^*_{e_n e'_n}. \tag{4.4.6b}$$

We deduce from (4.4.6a) that diagonal transitions can generate a complex transition polarizability with a real symmetric and an imaginary antisymmetric part, that is

$$(\alpha_{\alpha\beta})_{e_n e_n} = (\alpha_{\alpha\beta})_{e'_n e'_n} = (\alpha_{\beta\alpha})_{e_n e_n} = (\alpha_{\beta\alpha})_{e'_n e'_n}, \tag{4.4.6c}$$

$$(\alpha'_{\alpha\beta})_{e_n e_n} = -(\alpha'_{\alpha\beta})_{e'_n e'_n} = -(\alpha'_{\beta\alpha})_{e_n e_n} = (\alpha'_{\beta\alpha})_{e'_n e'_n}; \tag{4.4.6d}$$

and from (4.4.6b) that the off-diagonal matrix elements can only generate an antisymmetric transition polarizability, but this can have both real and imaginary parts:

$$(\alpha_{\alpha\beta})_{e'_n e_n} = -(\alpha_{\beta\alpha})_{e'_n e_n} = -(\alpha_{\alpha\beta})_{e_n e'_n}, \tag{4.4.6e}$$

$$(\alpha'_{\alpha\beta})_{e'_n e_n} = -(\alpha'_{\beta\alpha})_{e'_n e_n} = (\alpha'_{\alpha\beta})_{e_n e'_n}. \tag{4.4.6f}$$

In Section 2.8.1 it was shown that antisymmetric Rayleigh scattering is only possible from systems in degenerate states. We are now in a position to offer a better proof: having found that $\hat{\alpha}^a_{\alpha\beta}$ is time odd, we deduce this result immediately from the theorem (Section 4.3.2) that the expectation value of a time-odd operator vanishes for states invariant under time reversal, which can always be constructed for an even electron system and hence for any nondegenerate state. For even electron systems, (4.4.5) tells us that the degeneracy must be such as to support transitions that generate a real antisymmetric tensor, whereas for odd electron systems (4.4.6) tell us that the degeneracy can be such as to support transitions that generate either a real or an imaginary antisymmetric tensor. We now develop a general relationship that embraces all these possibilities for the case of Rayleigh scattering from atoms.

We first use in (4.4.2) the result (4.3.39) for the effect of the time reversal operator on a general atomic state of the form $|JM\rangle$ to write

$$(\tilde{\alpha}_{\alpha\beta})_{J'M', JM} = (-1)^{J+J'-M-M'+p+p'} (\tilde{\alpha}_{\beta\alpha})_{J-M, J'-M'}. \tag{4.4.7}$$

Since we are considering only scattering transitions between components of a degenerate level, we can take $J = J'$ and $p = p'$, in which case (4.4.7) becomes

$$(\tilde{\alpha}_{\alpha\beta})_{JM', JM} = (-1)^{2J-M-M'} (\tilde{\alpha}_{\beta\alpha})_{J-M, J-M'}. \tag{4.4.8}$$

For the special type of off-diagonal transitions where $M' = -M$,

$$(\tilde{\alpha}_{\alpha\beta})_{J-M,JM} = (-1)^{2J}(\tilde{\alpha}_{\beta\alpha})_{J-M,JM} = (-1)^{2J}(\tilde{\alpha}_{\alpha\beta})^*_{JM,J-M}, \quad (4.4.9)$$

so the complex transition polarizability is symmetric if J is integral and antisymmetric if J is half odd-integral, and both real and imaginary parts are allowed in both cases. For diagonal transitions,

$$(\tilde{\alpha}_{\alpha\beta})_{JM,JM} = (-1)^{2(J-M)}(\tilde{\alpha}_{\beta\alpha})_{J-M,J-M} = (-1)^{2(J-M)}(\tilde{\alpha}_{\alpha\beta})^*_{J-M,J-M} \quad (4.4.10)$$

which, for both integral and half odd-integral J, and $M \neq 0$, allows the complex transition polarizability to have a real symmetric and an imaginary antisymmetric part. Notice that (4.4.9) and (4.4.10) accord with (4.4.5). If $M = 0$, which is only possible for integral J,

$$(\tilde{\alpha}_{\alpha\beta})_{J0,J0} = (-1)^{2J}(\tilde{\alpha}_{\alpha\beta})^*_{J0,J0}, \quad (4.4.11)$$

so the complex transition polarizability is pure real and, since it is diagonal, symmetric.

The conclusions in the previous paragraph were reached by considering a complex atomic wavefunction which is neither even nor odd under time reversal. If J is half odd-integral, the wave function cannot be transformed into a time-even or time-odd form, and the conclusions in the previous paragraph stand. But if J is integral, we can always transform the wavefunction into a time-even form, and must therefore take account of the result (4.4.5), which stipulates that all components of the complex transition polarizability must be pure real. By combining this with the conclusions in the previous paragraph, we deduce that if J is integral, the complex transition polarizability is always real and symmetric both for diagonal transitions, and for off-diagonal transitions where $M' = -M$. Notice that, since atoms are spherically symmetric, the symmetric transition polarizability will always be diagonal with respect to its spatial components.

Finally, we note that for off-diagonal transitions where $M' \neq M$, there are additional possibilities. For example, if J is integral, for transitions where $M + M'$ is odd we deduce from (4.4.5) that the complex transition polarizability is pure real, and from (4.4.8) that an antisymmetric part is allowed. In these more general situations, time reversal selection rules are not as restrictive as when $M' = \pm M$ because the initial and final states on each side of (4.4.8) cannot be made equivalent. The least restrictive situation is when $J \neq J'$ and $M \neq M'$.

These general results for the intrinsic symmetry properties of the transition polarizability are developed in more detail in Chapter 8 in the context of antisymmetric scattering.

Relationships analogous to (4.4.2) can be written for the transition optical activity tensors but without the first equality since the real and imaginary parts no longer have well defined permutation symmetry. Using the Hermiticity and reversality characteristics of the corresponding operators deduced in Section 4.3.3, we obtain

$$(\tilde{G}_{\alpha\beta})_{mn} = -(\tilde{G}_{\alpha\beta})^*_{\Theta m \Theta n}, \tag{4.4.12a}$$

$$(\tilde{A}_{\alpha,\beta\gamma})_{mn} = (\tilde{A}_{\alpha,\beta\gamma})^*_{\Theta m \Theta n}. \tag{4.4.12b}$$

For an even electron system (4.4.12) become

$$(\tilde{G}_{\alpha\beta})_{mn} = -(\tilde{G}_{\alpha\beta})^*_{mn}, \tag{4.4.13a}$$

$$(\tilde{A}_{\alpha,\beta\gamma})_{mn} = (\tilde{A}_{\alpha,\beta\gamma})^*_{mn}, \tag{4.4.13b}$$

which shows that $(\tilde{G}_{\alpha\beta})_{mn}$ is pure imaginary and $(\tilde{A}_{\alpha,\beta\gamma})_{mn}$ is pure real, that is $(\tilde{G}_{\alpha\beta})_{mn} = -\mathrm{i}(G'_{\alpha\beta})_{mn}$ and $(\tilde{A}_{\alpha,\beta\gamma})_{mn} = (A_{\alpha,\beta\gamma})_{mn}$.

For an odd-electron system where the initial and final states are components of a twofold Kramers degenerate electronic level we can write from (4.4.12)

$$(\tilde{G}_{\alpha\beta})_{e_n e_n} = -(\tilde{G}_{\alpha\beta})^*_{e'_n e'_n}, \tag{4.4.14a}$$

$$(\tilde{G}_{\alpha\beta})_{e'_n e_n} = (\tilde{G}_{\alpha\beta})^*_{e_n e'_n}, \tag{4.4.14b}$$

$$(\tilde{A}_{\alpha,\beta\gamma})_{e_n e_n} = (\tilde{A}_{\alpha,\beta\gamma})^*_{e'_n e'_n}, \tag{4.4.14c}$$

$$(\tilde{A}_{\alpha,\beta\gamma})_{e'_n e_n} = -(\tilde{A}_{\alpha,\beta\gamma})^*_{e_n e'_n}. \tag{4.4.14d}$$

The reality properties of $(\tilde{A}_{\alpha,\beta\gamma})_{e_n e_n}$ and $(\tilde{A}_{\alpha,\beta\gamma})_{e'_n e_n}$ parallel those of $(\tilde{\alpha}_{\alpha\beta})_{e_n e_n}$ and $(\tilde{\alpha}_{\alpha\beta})_{e'_n e_n}$ and are not discussed further. The other optical activity tensor is more interesting: we deduce from (4.4.14a) that the diagonal matrix elements can generate both real and imaginary parts, that is

$$(G_{\alpha\beta})_{e_n e_n} = -(G_{\alpha\beta})_{e'_n e'_n}, \tag{4.4.15a}$$

$$(G'_{\alpha\beta})_{e_n e_n} = (G'_{\alpha\beta})_{e'_n e'_n}; \tag{4.4.15b}$$

and similarly from (4.4.14b) for the off-diagonal matrix elements:

$$(G_{\alpha\beta})_{e'_n e_n} = (G_{\alpha\beta})_{e_n e'_n}, \tag{4.4.15c}$$

$$(G'_{\alpha\beta})_{e'_n e_n} = -(G'_{\alpha\beta})_{e_n e'_n}. \tag{4.4.15d}$$

In discussing natural and magnetic optical rotation (and indeed any birefringence phenomenon) from systems in degenerate states, it must be remembered that only diagonal transitions can contribute because the phases of the initial and final states must be the same; although they do not need to be the same in Rayleigh and Raman scattering. We see from (4.4.6d) that, although an odd electron atom or molecule in a Kramers degenerate state $|e\rangle$ can support, say, $(\alpha'_{xy})_{e_n e_n}$ and therefore generate Faraday rotation in a light beam travelling along z, this is cancelled by the contribution $(\alpha'_{xy})_{e'_n e'_n}$ from the conjugate state $|e'\rangle$: in order to observe Faraday rotation, an external time-odd influence such as a magnetic field along z is required to lift the degeneracy and prevent exact cancellation. On the other hand, (4.4.15b) shows that natural optical rotation generated by an odd electron chiral molecule in a Kramers degenerate state $|e\rangle$ is equal in sign and magnitude to that generated by the state $|e'\rangle$.

The real optical activity $G_{\alpha\beta}$ has interesting properties because it is generated by an odd-parity time-odd operator $\hat{G}^s_{\alpha\beta}$, and it follows from the foregoing that it can only be supported by a system in a degenerate state. It features in discussions of magnetochiral birefringence (Section 3.4.8), gyrotropic birefringence (Section 3.4.9) and the Jones birefringence (Section 3.4.10). It can be seen immediately from (4.4.15a) that a magnetic field (or some other external time-odd influence) is required to observe any coherent phenomenon from this tensor because Kramers conjugate states generate equal and opposite contributions. On the other hand $\hat{G}^s_{\alpha\beta}$ can, like its polarizability counterpart $\hat{\alpha}^a_{\alpha\beta}$, generate incoherent phenomena such as Rayleigh and Raman scattering, and dispersional intermolecular forces, involving both diagonal and off-diagonal transitions between components of degenerate sets of states. But unlike tensor components generated by $\hat{\alpha}^a_{\alpha\beta}$, which vanish at zero frequency because of (2.8.14e), those generated by $\hat{G}^s_{\alpha\beta}$ appear to describe both *static* and *dynamic* properties because of (2.8.14d). Buckingham and Joslin (1981) have discussed spin-dependent dispersional intermolecular forces generated by $\hat{\alpha}^a_{\alpha\beta}$, and analogous contributions generated by $\hat{G}^s_{\alpha\beta}$ could provide significant discriminating contributions to intermolecular forces between odd electron chiral molecules (Barron and Johnston, 1987). In the examples discussed in Chapter 8 it emerges that, in the absence of vibronic coupling, spin–orbit coupling is an essential ingredient in systems that can support tensor components generated by $\hat{\alpha}^a_{\alpha\beta}$, and the same requirement is anticipated for tensor components generated by $\hat{G}^s_{\alpha\beta}$. Thus crystals and fluids composed of odd electron chiral molecules with large spin–orbit coupling could well show curious new properties.

Barron and Buckingham (2001) have reviewed the application of time reversal symmetry to molecular properties that depend on motion such as those described by $\alpha'_{\alpha\beta}$, $G_{\alpha\beta}$, and $A'_{\alpha,\beta\gamma}$.

4.4.4 The spatial symmetry of molecular property tensors

We now consider the application of Neumann's principle, in conjunction with explicit group theoretical arguments, to reduce a given property tensor to its simplest form in a particular point group. This entails the specification of which tensor components are zero, and of any relationships between the nonzero components. This section is based on a treatment by Birss (1966), which itself follows Fumi (1952) and Fieschi (1957).

We saw in Section 4.2.3 that the components of a polar tensor transform according to

$$P_{\lambda'\mu'\nu'} \ldots = l_{\lambda'\alpha} l_{\mu'\beta} l_{\nu'\gamma} \ldots P_{\alpha\beta\gamma} \ldots \qquad (4.2.32a)$$

and the components of an axial tensor transform according to

$$A_{\lambda'\mu'\nu'} \ldots = (\pm) l_{\lambda'\alpha} l_{\mu'\beta} l_{\nu'\gamma} \ldots A_{\alpha\beta\gamma} \ldots . \qquad (4.2.32b)$$

It follows from Neumann's principle that, if the coordinate transformation corresponds to one of the symmetry operations of the molecule's point group, the corresponding property tensor components are invariant. Since free space is isotropic, a property tensor can depend only on the relative orientation of the molecule and the coordinate axes, and not on their absolute orientation in space. This means that the components of a polar property tensor must satisfy the set of equations

$$P_{\lambda'\mu'\nu'} \ldots = P_{\lambda\mu\nu} \ldots = \sigma_{\lambda\alpha} \sigma_{\mu\beta} \sigma_{\nu\gamma} \ldots P_{\alpha\beta\gamma} \ldots \qquad (4.4.16a)$$

and the components of an axial property tensor must satisfy

$$A_{\lambda'\mu'\nu'} \ldots = A_{\lambda\mu\nu} \ldots = (\pm)\sigma_{\lambda\alpha} \sigma_{\mu\beta} \sigma_{\nu\gamma} \ldots A_{\alpha\beta\gamma} \ldots , \qquad (4.4.16b)$$

where $\sigma_{\lambda\alpha}$ is an element of a matrix corresponding to a particular symmetry operation, and the suffixes $\lambda\mu\nu \ldots$ now refer to the same axis system as $\alpha\beta\gamma \ldots$.

In Section 4.2.2 we considered two sets of axes x, y, z and x', y', z' with a common origin O, and specified the relative orientation of the two sets by a set of nine direction cosines $l_{\lambda'\alpha}$. The set x', y', z' can be generated from x, y, z by some general rotation. The matrix giving the set of direction cosines for a right-handed proper rotation through an angle θ about an axis defined relative to x, y, z by direction cosines l, m, n is (Jeffreys and Jeffreys, 1950)

$$[l_{\lambda'\alpha}] = \begin{pmatrix} \cos\theta + l^2(1 - \cos\theta) & lm(1 - \cos\theta) + n\sin\theta & ln(1 - \cos\theta) - m\sin\theta \\ ml(1 - \cos\theta) - n\sin\theta & \cos\theta + m^2(1 - \cos\theta) & mn(1 - \cos\theta) + l\sin\theta \\ nl(1 - \cos\theta) + m\sin\theta & nm(1 - \cos\theta) - l\sin\theta & \cos\theta + n^2(1 - \cos\theta) \end{pmatrix} .$$

$$(4.4.17)$$

For an improper rotation, which can be considered as a combination of a rotation and an inversion, each element of the matrix (4.4.17) must be multiplied by -1. Thus, for example, the operation C_3 corresponding to a right-handed rotation through

$\theta = 120°$ about the z axis is represented by the set of direction cosines

$$[l_{\lambda'\alpha}] = \begin{pmatrix} \cos 120° & \sin 120° & 0 \\ -\sin 120° & \cos 120° & 0 \\ 0 & 0 & 1 \end{pmatrix} = \begin{pmatrix} -\tfrac{1}{2} & \tfrac{1}{2}\sqrt{3} & 0 \\ -\tfrac{1}{2}\sqrt{3} & -\tfrac{1}{2} & 0 \\ 0 & 0 & 1 \end{pmatrix}. \qquad (4.4.18)$$

As another example, the operation σ_h, a reflection across the xy plane, can be regarded as a rotation through 180° followed by inversion through the origin, and so is represented by

$$[l_{\lambda'\alpha}] = \begin{pmatrix} 1 & 0 & 0 \\ 0 & 1 & 0 \\ 0 & 0 & -1 \end{pmatrix}. \qquad (4.4.19)$$

It is therefore a simple matter to construct a set of *symmetry matrices* $[\sigma_{\lambda\alpha}]$ representing the set of operations of any point group.

One conclusion we can draw immediately is that polar tensors of odd rank and axial tensors of even rank vanish for point groups containing the inversion operation. Thus using the symmetry matrix

$$[\sigma_{\lambda\alpha}] = \begin{pmatrix} -1 & 0 & 0 \\ 0 & -1 & 0 \\ 0 & 0 & -1 \end{pmatrix} \qquad (4.4.20)$$

in (4.4.16) gives

$$P_{\alpha\beta\gamma}\ldots = -P_{\alpha\beta\gamma}\ldots = 0$$

for a polar tensor of odd rank, and

$$A_{\alpha\beta\gamma}\ldots = -A_{\alpha\beta\gamma}\ldots = 0$$

for an axial tensor of even rank.

Another simple example is the polarizability tensor of a molecule with a threefold proper rotation axis. Thus, from (4.4.16a) and (4.4.18),

$$\alpha_{xz} = \sigma_{x\alpha}\sigma_{z\beta}\alpha_{\alpha\beta} = -\tfrac{1}{2}\alpha_{xz} + \tfrac{1}{2}\sqrt{3}\alpha_{yz},$$

$$\alpha_{yz} = \sigma_{y\alpha}\sigma_{z\beta}\alpha_{\alpha\beta} = -\tfrac{1}{2}\sqrt{3}\alpha_{xz} - \tfrac{1}{2}\alpha_{yz},$$

and these two equations can only be satisfied simultaneously if $\alpha_{xz} = \alpha_{yz} = 0$.

In general, by applying the appropriate set of symmetry matrices to (4.4.16), it is possible to achieve the maximum simplification of a polar or axial tensor of any rank for a molecule belonging to a particular point group. In fact it is often not necessary to apply a symmetry matrix for every operation of a point group since there is usually a smaller set of generating operations from which, by taking suitable combinations, the complete set of symmetry operations can be obtained. So it is only necessary to take the set of generating matrices in order to achieve the maximum simplification of a tensor.

The forms of polar and axial tensors up to the fourth rank in the important molecular point groups are displayed in Tables 4.2, adapted from tables given by Birss (1966) which were derived using the methods outlined above. The equalities between property tensors in the important point groups are given in Table 4.2a. The actual form of the tensor represented by a given symbol may be obtained from Tables 4.2b to f for tensors of rank zero to four, respectively. Each column displays the components to which the tensor component at the top of the column reduces in the various point groups; so each row is a list of equalities between pairs of components, and of identities of components to zero. Notations such as $xz(2)$ and $xxy(3)$ indicate the equalities that exist between the two and three tensor components, respectively, which may be obtained by unrestricted permutation of the indices. Notations of the type $yxxx(x.3)$ denote the three distinct components which may be obtained from $yxxx$ by keeping its last index fixed and permuting the others, and notations of the type $xxyy(x:3)$ denote the three distinct components which may be obtained from $xxyy$ by keeping its first index fixed and permuting the others. Notations of the type $xxyz(c4)$ denote the four distinct cyclic permutations. Notations of the type $zzxy(xy: 6)$ denote the six components which can be obtained from $zzxy$ by permuting its indices subject to the restriction that the order of the indices x and y remains unchanged (although x and y need not remain adjacent).

The molecular point groups able to support the appropriate components of the property tensors $G'_{\alpha\beta}$, a second-rank axial tensor, and $A_{\alpha,\beta\gamma}$, a third-rank polar tensor, that are responsible for natural optical rotation as specified in (3.4.42) and (3.4.43) are readily determined from these tables. Thus from Tables 4.2a and 4.2d it is found that $G'_{\alpha\alpha} = G'_{xx} + G'_{yy} + G'_{zz}$, which is responsible for natural optical rotation in isotropic samples, is only supported by molecules belonging to the point groups C_n, D_n, O and T (and also I from the icosahedral system which is not included in these tables) which lack a centre of inversion, reflection planes and rotation-reflection axes. Similarly for $(G'_{xx} + G'_{yy})$ and $(A_{x,yz} - A_{y,xz})$ which contribute to optical rotation in an oriented sample for light propagating along z. Hence natural optical rotation in isotropic samples, and in oriented samples for light propagating along the principal molecular symmetry axis, is supported only by chiral molecules. However, as mentioned in Section 1.9.1, natural optical rotation is possible in some oriented achiral molecules lacking a centre of inversion for light propagating along other directions.

These tables give the simplification of molecular property tensors imposed by point group symmetry considerations *alone*. But additional physical considerations may bring about further simplification. Time reversal arguments are particularly

Table 4.2*a*

System	Schönflies (International) symbol of point group	Orientation of symmetry elements	Polar tensor of even rank m	Axial tensor of even rank m	Polar tensor of odd rank n	Axial tensor of odd rank n
Triclinic	$C_1(1)$	any	A_m	A_m	A_n	A_n
	$C_i(\bar{1})$	any	A_m	–	–	A_n
Monoclinic	$C_2(2)$	$C_2\|\|z$	B_m	B_m	B_n	B_n
	$C_s(m)$	$\sigma_h\|\|z$	B_m	C_m	C_n	B_n
	$C_{2h}(2/m)$	$C_2\|\|z$	B_m	–	–	B_n
Orthorhombic	$D_2(222)$	$C_2\|\|x, C_2\|\|y$	D_m	D_m	D_n	D_n
	$C_{2v}(2mm)$	$\sigma_v\perp x, \sigma_v\perp y$	D_m	E_m	E_n	D_n
	$D_{2h}(mmm)$	$C_2\|\|x, C_2\|\|y$	D_m	–	–	D_n
Tetragonal	$C_4(4)$	$C_4\|\|z$	F_m	F_m	F_n	F_n
	$S_4(\bar{4})$	$S_4\|\|z$	F_m	G_m	G_n	F_n
	$C_{4h}(4/m)$	$C_4\|\|z$	F_m	–	–	F_n
	$D_4(422)$	$C_4\|\|z, C_2\|\|y$	H_m	H_m	H_n	H_n
	$C_{4v}(4mm)$	$C_4\|\|z, \sigma_v\perp y$	H_m	I_m	I_n	H_n
	$D_{2d}(\bar{4}2m)$	$S_4\|\|z, C_2\|\|y$	H_m	J_m	J_n	H_n
	$D_{4h}(4/mmm)$	$C_4\|\|z, C_2\|\|y$	H_m	–	–	H_n
Trigonal	$C_3(3)$	$C_3\|\|z$	K_m	K_m	K_n	K_n
	$S_6(\bar{3})$	$S_6\|\|z$	K_m	–	–	K_n
	$D_3(32)$	$C_3\|\|z, C_2\|\|y$	L_m	L_m	L_n	L_n
	$C_{3v}(3m)$	$C_3\|\|z, \sigma_v\perp y$	L_m	M_m	M_n	L_n
	$D_{3d}(\bar{3}m)$	$C_3\|\|z, C_2\|\|y$	L_m	–	–	L_n
Hexagonal	$C_6(6)$	$C_6\|\|z$	N_m	N_m	N_n	N_n
	$C_{3h}(\bar{6})$	$C_3\|\|z$	N_m	O_m	O_n	N_n
	$C_{6h}(6/m)$	$C_6\|\|z$	N_m	–	–	N_n
	$D_6(622)$	$C_6\|\|z, C_2\|\|y$	P_m	P_m	P_n	P_n
	$C_{6v}(6mm)$	$C_6\|\|z, \sigma_v\perp y$	P_m	Q_m	Q_n	P_n
	$D_{3h}(\bar{6}m2)$	$C_3\|\|z, \sigma_v\perp y$	P_m	R_m	R_n	P_n
	$D_{6h}(6/mmm)$	$C_6\|\|z, C_2\|\|y$	P_m	–	–	P_n
Cubic	$T(23)$	$C_2\|\|x, C_2\|\|y$	S_m	S_m	S_n	S_n
	$T_h(m3)$	$C_2\|\|x, C_2\|\|y$	S_m	–	–	S_n
	$O(432)$	$C_4\|\|x, C_4\|\|y$	T_m	T_m	T_n	T_n
	$T_d(\bar{4}3m)$	$S_4\|\|x, S_4\|\|y$	T_m	U_m	U_n	T_n
	$O_h(m3m)$	$C_4\|\|x, C_4\|\|y$	T_m	–	–	T_n

Table 4.2b

$m = 0$	x
A_0	x
B_0	x
C_0	0
D_0	x
E_0	0
F_0	x
G_0	0
H_0	x
I_0	0
J_0	0
K_0	x
L_0	x
M_0	0
N_0	x
O_0	0
P_0	x
Q_0	0
R_0	0
S_0	x
T_0	x
U_0	0

Table 4.2c

$n = 1$	x	y	z
A_1	x	y	z
B_1	0	0	z
C_1	x	y	0
D_1	0	0	0
E_1	0	0	z
F_1	0	0	z
G_1	0	0	0
H_1	0	0	0
I_1	0	0	z
J_1	0	0	0
K_1	0	0	z
L_1	0	0	0
M_1	0	0	z
N_1	0	0	z
O_1	0	0	0
P_1	0	0	0
Q_1	0	0	z
R_1	0	0	0
S_1	0	0	0
T_1	0	0	0
U_1	0	0	0

Table 4.2d

$m = 2$	xx	yy	zz	xy	yx	$xz(2)$	$yz(2)$
A_2	xx	yy	zz	xy	yx	xz	yz
B_2	xx	yy	zz	xy	yx	0	0
C_2	0	0	0	0	0	xz	yz
D_2	xx	yy	zz	0	0	0	0
E_2	0	0	0	xy	yx	0	0
F_2	xx	xx	zz	xy	$-xy$	0	0
G_2	xx	$-xx$	0	xy	xy	0	0
H_2	xx	xx	zz	0	0	0	0
I_2	0	0	0	xy	$-xy$	0	0
J_2	xx	$-xx$	0	0	0	0	0
K_2	xx	xx	zz	xy	$-xy$	0	0
L_2	xx	xx	zz	0	0	0	0
M_2	0	0	0	xy	$-xy$	0	0
N_2	xx	xx	zz	xy	$-xy$	0	0
O_2	0	0	0	0	0	0	0
P_2	xx	xx	zz	0	0	0	0
Q_2	0	0	0	xy	$-xy$	0	0
R_2	0	0	0	0	0	0	0
S_2	xx	xx	xx	0	0	0	0
T_2	xx	xx	xx	0	0	0	0
U_2	0	0	0	0	0	0	0

Table 4.2e

n = 3	xxx	yyy	zzz	xxy(3)	yyx(3)	xxz(3)	yyz(3)	zzx(3)	zzy(3)	xyz	zzx	xzx	xzy	yxx	yzz	zyx
A_3	xxx	yyy	zzz	xxy	yyx	xxz	yyz	zzx	zzy	xyz	xzy	xzy	xyz	yxy	yzx	−xyz
B_3	0	0	zzz	0	0	xxz	yyz	zzx	0	xyz	xzy	xzy	xyz	yxy	yzx	−xyz
C_3	xxx	yyy	0	xxy	yyx	0	0	0	zzy	0	0	0	0	0	0	xzy
D_3	0	0	0	0	0	0	0	0	0	xyz	xzy	xzy	xyz	yxy	yzx	−xyz
E_3	0	0	zzz	0	0	xxz	xxz	zzx	0	0	0	0	0	0	0	xzy
F_3	0	0	zzz	0	0	xxz	xxz	zzx	0	xyz	xzy	xzy	xyz	−xyz	−xzy	xzy
G_3	0	0	0	0	0	0	−xxz	0	0	xyz	xzy	xzy	xyz	−xyz	−xzy	xzy
H_3	0	0	0	0	0	xxz	xxz	zzx	0	xyz	xzy	xzy	xyz	−xyz	−xzy	xzy
I_3	0	0	zzz	0	0	0	0	0	0	0	0	0	0	0	0	0
J_3	0	0	0	0	0	xxz	xxz	zzx	0	xyz	xzy	xzy	xyz	−xyz	−xzy	xzy
K_3	xxx	yyy	zzz	−yyy	−xxx	xxz	xxz	zzx	0	xyz	xzy	xzy	xyz	−xyz	−xzy	xzy
L_3	0	yyy	0	−yyy	0	xxz	xxz	zzx	0	xyz	xzy	xzy	xyz	−xyz	−xzy	xzy
M_3	xxx	0	zzz	0	−xxx	0	0	0	0	xyz	xzy	xzy	xyz	0	0	xzy
N_3	0	0	0	0	0	xxz	0	zzx	0	xyz	xzy	xzy	xyz	−xyz	−xzy	xzy
O_3	xxx	yyy	zzz	−yyy	−xxx	xxz	xxz	zzx	0	0	0	0	0	−xxx	0	xzy
P_3	0	0	0	0	0	0	0	0	0	0	0	0	0	0	0	0
Q_3	0	0	0	0	0	0	0	0	0	xyz	xzy	xzy	xyz	0	0	xzy
R_3	xxx	0	zzz	0	−xxx	0	0	0	0	0	0	0	0	−xxx	0	xzy
S_3	0	0	0	0	0	0	0	0	0	xyz	xzy	xzy	xyz	xzy	xyz	xyz
T_3	0	0	0	0	0	0	0	0	0	xyz	−xyz	−xyz	xyz	−xyz	xyz	xyz
U_3	0	0	0	0	0	0	0	0	0	xyz	xyz	xyz	xyz	xyz	xyz	xyz

Table 4.2f

$m = 4$	xxxx	yyyy	zzzz	xxxy	yxxx(x.3)	yyyx	xyyy(y.3)	xxxz(4)
A_4	xxxx	yyyy	zzzz	xxxy	yxxx	yyyx	xyyy	xxxz
B_4	xxxx	yyyy	zzzz	xxxy	yxxx	yyyx	xyyy	0
C_4	0	0	0	0	0	0	0	xxxz
D_4	xxxx	yyyy	zzzz	0	0	0	0	0
E_4	0	0	0	xxxy	yxxx	yyyx	xyyy	0
F_4	xxxx	xxxx	zzzz	xxxy	yxxx	−xxxy	−yxxx	0
G_4	xxxx	−xxxx	0	xxxy	yxxx	xxxy	yxxx	0
H_4	xxxx	xxxx	zzzz	0	0	0	0	0
I_4	0	0	0	xxxy	yxxx	−xxxy	−yxxx	0
J_4	xxxx	−xxxx	0	0	0	0	0	0
K_4	yyxx + xyyx + yxyx	xxxx	zzzz	yyxy + xyyy + yxyy	yxxx	−xxxy	−yxxx	xxxz
L_4	yyxx + xyyx + yxyx	xxxx	zzzz	yyxy + xyyy + yxyy	yxxx	−xxxy	−yxxx	xxxz
M_4	0	0	0	0	0	0	0	0
N_4	yyxx + xyyx + yxyx	xxxx	zzzz	yyxy + xyyy + yxyy	yxxx	−xxxy	−yxxx	0
O_4	0	0	0	0	0	0	0	xxxz
P_4	yyxx + xyyx + yxyx	xxxx	0	yyxy + xyyy + yxyy	yxxx	0	0	0
Q_4	0	0	0	0	0	−xxxy	−yxxx	0
R_4	0	0	0	0	0	0	0	0
S_4	xxxx	xxxx	xxxx	0	0	0	0	0
T_4	xxxx	xxxx	xxxx	0	0	0	0	0
U_4	0	0	0	0	0	0	0	0

(Continued)

Table 4.2f (*Continued*)

m = 4	yyyz(4)	zzzx(4)	zzzy(4)	xxyy(x:3)	yyxx(y:3)	xxzz(x:3)	xxzz(z:3)	yyzz(z:3)	yyzz(y:3)(z:3)
A_4	yyyz	zzzx	zzzy	xxyy	yyxx	xxzz	xxzz	yyzz	yyzz
B_4	0	0	0	xxyy	yyxx	xxzz	xxzz	yyzz	yyzz
C_4	yyyz	zzzx	zzzy	0	0	0	0	0	0
D_4	0	0	0	xxyy	yyxx	xxzz	xxzz	yyzz	yyzz
E_4	0	0	0	0	0	0	0	0	0
F_4	0	0	0	xxyy	xxyy	zzxx	zzxx	zzxx	xxzz
G_4	0	0	0	xxyy	$-$xxyy	zzxx	zzxx	zzxx	$-$xxzz
H_4	0	0	0	xxyy	xxyy	zzxx	zzxx	zzxx	xxzz
I_4	0	0	0	0	0	0	0	0	0
J_4	0	0	0	xxyy	$-$xxyy	zzxx	zzxx	zzxx	$-$xxzz
K_4	yyyz	0	0	xxyy	xxyy	zzxx	zzxx	zzxx	xxzz
L_4	0	0	0	xxyy	xxyy	zzxx	zzxx	zzxx	xxzz
M_4	yyyz	0	0	0	0	0	0	0	0
N_4	0	0	0	xxyy	xxyy	zzxx	zzxx	zzxx	xxzz
O_4	yyyz	0	0	0	0	0	0	0	0
P_4	0	0	0	xxyy	xxyy	xxzz	xxzz	xxzz	xxzz
Q_4	0	0	0	0	0	0	0	0	0
R_4	yyyz	0	0	0	0	0	0	0	0
S_4	0	0	0	xxyy	yyxx	yyxx	yyxx	xxyy	yyxx
T_4	0	0	0	xxyy	xxyy	xxyy	xxyy	xxyy	xxyy
U_4	0	0	0	xxyy	$-$xxyy	$-$xxyy	$-$xxyy	xxyy	$-$xxyy

(*Continued*)

Table 4.2f *(Continued)*

m = 4	xxyyz(c4)	xyxyz(c4)	yxxyz(c4)	yyxxz(c4)	yxyyz(c4)	xyyyz(c4)	zzxxy(xy:6)	zzyyx(yx:6)
A_4	xxyz	xyxz	yxxz	yyxz	yxyz	xyyz	zzxy	zzyx
B_4	0	0	0	0	0	0	zzxy	zzyx
C_4	xxyz	xyxz	yxxz	yyxz	yxyz	xyyz	0	0
D_4	0	0	0	0	0	0	0	0
E_4	0	0	0	0	0	0	zzxy	zzyx
F_4	0	0	0	0	0	0	−zzxy	−zzyx
G_4	0	0	0	0	0	0	zzxy	zzyx
H_4	0	0	0	0	0	0	0	0
I_4	0	0	0	0	0	0	−zzxy	−zzyx
J_4	0	0	0	0	0	0	0	0
K_4	−yyyz	−yyyz	−yyyz	−xxxz	−xxxz	−xxxz	−zzxy	−zzyx
L_4	0	0	0	−xxxz	−xxxz	−xxxz	0	0
M_4	−yyyz	−yyyz	−yyyz	0	0	0	−zzxy	−zzyx
N_4	0	0	0	−xxxz	−xxxz	−xxxz	−zzxy	−zzyx
O_4	−yyyz	−yyyz	−yyyz	0	0	0	0	0
P_4	0	0	0	0	0	0	0	0
Q_4	0	0	0	0	0	0	zzxy	zzyx
R_4	−yyyz	−yyyz	−yyyz	0	0	0	0	0
S_4	0	0	0	0	0	0	0	0
T_4	0	0	0	0	0	0	0	0
U_4	0	0	0	0	0	0	0	0

important in this respect because, as we saw in the previous section, they lead to powerful statements about the symmetry or antisymmetry of a tensor with respect to the permutation of its subscripts. For example, it is wrong to conclude on the basis of Tables 4.2a and 4.2d alone that the xy component of an antisymmetric polarizability is supported by molecules belonging to the point groups $C_4, S_4, C_{4h}, C_3, S_6, C_6, C_{3h}$ and C_{6h}. All we can conclude is that $xy - yx$ spans the totally symmetric irreducible representation, but since the antisymmetric part of the effective polarizability operator (2.8.14) is time odd, further considerations involving the generalized selection rule (4.3.37) are required. In any event we know from (2.8.14e) that any antisymmetric polarizability must be dynamic, and further information is provided by (4.4.5) for an even-electron system and by (4.4.6) for an odd-electron system.

4.4.5 *Irreducible cartesian tensors*

The procedure outlined in the previous section for the simplification of molecular property tensors from a consideration of the symmetry operations of the molecule's point group in effect determines the tensor components spanning the totally symmetric irreducible representation. It is desirable to extend this classification to all the irreducible representations of all the point groups. However, this is a formidable task: it has been partially carried out by McClain (1971) and by Mortensen and Hassing (1979), who considered just the components of a second-rank polar tensor in order to discuss conventional Raman scattering, and we refer to these authors for the results. It should be mentioned, however, that it is sometimes possible to obtain this information for certain tensor components very simply: for example, since a second-rank antisymmetric polar tensor transforms the same as an axial vector, the transformation properties of its components can be deduced by consulting standard point group character tables to see which irreducible representations are spanned by components of rotations. But again it must be emphasized that generalized selection rules like (4.3.37) must be used to deduce whether or not a particular property tensor is observable, depending on whether the corresponding operator is time even or time odd and whether the molecule has an even or an odd number of electrons: of course the conventional selection rules can still be used when considering *transition* tensors between initial and final states from different levels.

In this section we content ourselves with a classification with respect to the irreducible representations of the full rotation group R_3 (that is, all the symmetry operations of the sphere, including improper as well as proper rotations). In fact we use the proper rotation group R_3^+ and add subscripts g or u later to distinguish irreducible representations that are even or odd with respect to inversion.

The importance of reducing sets of tensor components is summarized by the following statement from Fano and Racah (1959):

Because the laws of physics are independent of the choice of a coordinate system, the two sides of any equation representing a physical law must transform in the same way under coordinate rotations. It is, of course, convenient to cast both sides of the equations in the form of tensorial sets, so that their transformations will be linear. By resolving these sets into irreducible subsets one pushes the process of simplification to its limit, because one disentangles the physical equations into a maximum number of separate, independent equations.

We denote the irreducible representations of R_3^+ by $D^{(j)}$ where j takes integral values $0, 1, 2, \ldots \infty$. The direct product of two irreducible representations $D^{(j_1)}$ and $D^{(j_2)}$ gives

$$D^{(j_1)} \times D^{(j_2)} = D^{(j_1+j_2)} + D^{(j_1+j_2-1)} + \cdots D^{|j_1-j_2|}. \tag{4.4.21a}$$

In terms of symmetrized (square brackets) and antisymmetrized (curly brackets) direct products, for use with basis sets constructed from products of components of the same set of functions,

$$D^{(j)} \times D^{(j)} = \left[D^{(2j)} + D^{(2j-2)} + \cdots D^{(0)} \right]$$
$$+ \left\{ D^{(2j-1)} + \cdots D^{(1)} \right\}. \tag{4.4.21b}$$

For the double rotation group, the same formulae apply, but now j can take values $0, \frac{1}{2}, 1, \frac{3}{2}, \ldots \infty$.

A scalar transforms as $D^{(0)}$ and a first-rank tensor as $D^{(1)}$. The components of a general second-rank tensor transform like the nine products $x_1x_2, x_1y_2, x_1z_2 \ldots$ according to

$$D^{(1)} \times D^{(1)} = D^{(2)} + D^{(1)} + D^{(0)};$$

but if 1 and 2 refer to the same basis set, only the symmetric irreducible representations $D^{(2)} + D^{(0)}$ survive. The results for tensors up to rank six are given in Table 4.3.

It is well known that a general second-rank polar tensor can be decomposed into a scalar, an antisymmetric second-rank tensor and a symmetric traceless second-rank tensor:

$$P_{\alpha\beta} = P\delta_{\alpha\beta} + P_{\alpha\beta}^a + P_{\alpha\beta}^s; \tag{4.4.22a}$$

$$P = \tfrac{1}{3} P_{\gamma\gamma}, \tag{4.4.22b}$$

$$P_{\alpha\beta}^a = \tfrac{1}{2}(P_{\alpha\beta} - P_{\beta\alpha}), \tag{4.4.22c}$$

$$P_{\alpha\beta}^s = \tfrac{1}{2}(P_{\alpha\beta} + P_{\beta\alpha}) - P\delta_{\alpha\beta}. \tag{4.4.22d}$$

Table 4.3 *Enumeration of the decomposition of general tensors into irreducible parts.*

	$D^{(0)}$	$D^{(1)}$	$D^{(2)}$	$D^{(3)}$	$D^{(4)}$	$D^{(5)}$	$D^{(6)}$
$D^{(1)}$	0	1	0	0	0	0	0
$D^{(1)^2}$	1	1	1	0	0	0	0
$D^{(1)^3}$	1	3	2	1	0	0	0
$D^{(1)^4}$	3	6	6	3	1	0	0
$D^{(1)^5}$	6	15	15	10	4	1	0
$D^{(1)^6}$	15	36	40	29	15	5	1

Clearly $P\delta_{\alpha\beta}$, $P^a_{\alpha\beta}$ and $P^s_{\alpha\beta}$ are irreducible tensors with respect to $D^{(0)}$, $D^{(1)}$ and $D^{(2)}$, and we can rewrite (4.4.22a) as

$$P_{\alpha\beta} = P^{(0)}_{\alpha\beta} + P^{(1)}_{\alpha\beta} + P^{(2)}_{\alpha\beta}. \qquad (4.4.22e)$$

Recalling the dyadic form (4.2.7) of a second-rank tensor, it is instructive to write out $P_{\alpha\beta}$ in terms of irreducible base tensors made up from dyadic products of unit vectors (Fano and Racah, 1959):

$$P\delta_{\alpha\beta} = \tfrac{1}{3}(i_\alpha i_\beta + j_\alpha j_\beta + k_\alpha k_\beta)(P_{xx} + P_{yy} + P_{zz}), \qquad (4.4.22f)$$

$$P^a_{\alpha\beta} = \tfrac{1}{2}[(j_\alpha k_\beta - k_\alpha j_\beta)(P_{yz} - P_{zy}) + (k_\alpha i_\beta - i_\alpha k_\beta)(P_{zx} - P_{xz})$$
$$+ (i_\alpha j_\beta - j_\alpha i_\beta)(P_{xy} - P_{yx})], \qquad (4.4.22g)$$

$$P^s_{\alpha\beta} = \tfrac{1}{2}\big[\tfrac{1}{3}(2k_\alpha k_\beta - i_\alpha i_\beta - j_\alpha j_\beta)(2P_{zz} - P_{xx} - P_{yy})$$
$$+ (i_\alpha i_\beta - j_\alpha j_\beta)(P_{xx} - P_{yy}) + (j_\alpha k_\beta + k_\alpha j_\beta)(P_{yz} + P_{zy})$$
$$+ (k_\alpha i_\beta + i_\alpha k_\beta)(P_{zx} + P_{xz}) + (i_\alpha j_\beta + j_\alpha i_\beta)(P_{xy} + P_{yx}). \qquad (4.4.22h)$$

We can now appreciate the reason behind the choice of the traceless definition (2.4.5) for the electric quadrupole tensor, for it is equivalent to (4.4.22d) and is therefore in irreducible form.

A simple but important application of the decomposition (4.4.22) of a general second-rank polar tensor is to the derivation of angular momentum selection rules in Raman scattering. The polarizability tensor reduces to three parts spanning $D^{(0)}$, $D^{(1)}$ and $D^{(2)}$: if the initial state of the molecule has a total angular momentum quantum number j, it spans $D^{(j)}$, so the final state of the molecule must transform as one of the representations $D^{(0)} \times D^{(j)} = D^{(j)}$, $D^{(1)} \times D^{(j)} = D^{(j+1)} + D^{(j)} + D^{(j-1)}$ or $D^{(2)} \times D^{(j)} = D^{(j+2)} + D^{(j+1)} + D^{(j)} + D^{(j-1)} + D^{(j-2)}$. It therefore follows that the total angular momentum quantum number of the molecule after the Raman scattering process can take only the values j, $j \pm 1$ or $j \pm 2$. Notice that, since $\hat{\alpha}^a_{\alpha\beta}$ spans $D^{(1)}$, these spatial symmetry arguments impose the restriction $\Delta j = 0, \pm 1$ on antisymmetric scattering in addition to the restrictions imposed by time reversal discussed in Section 4.4.3.

The standard general method for reducing an arbitrary cartesian tensor uses Young tableaux (Hamermesh, 1962) and is not elaborated here. But it is instructive to see the irreducible third-rank cartesian tensors written out explicitly. Fortunately, these have been worked out by Andrews and Thirunamachandran (1978), and we simply quote their results. It can be seen from the third row of Table 4.3 that there are three sets spanning $D^{(1)}$ and two sets spanning $D^{(2)}$: for these, only the sums of the sets are determined uniquely; the decomposition into independent tensors is arbitrary and some additional constraint is required. Thus

$$P_{\alpha\beta\gamma} = P^{(0)}_{\alpha\beta\gamma} + \sum_{n=a,b,c} P^{(1n)}_{\alpha\beta\gamma} + \sum_{n=a,b} P^{(2n)}_{\alpha\beta\gamma} + P^{(3)}_{\alpha\beta\gamma}, \tag{4.4.23a}$$

$$P^{(0)}_{\alpha\beta\gamma} = \tfrac{1}{6}\varepsilon_{\alpha\beta\gamma}\varepsilon_{\delta\lambda\mu}P_{\delta\lambda\mu}, \tag{4.4.23b}$$

$$\sum_{n=a,b,c} P^{(1n)}_{\alpha\beta\gamma} = \tfrac{1}{10}[\delta_{\alpha\beta}(4P_{\delta\delta\gamma} - P_{\delta\gamma\delta} - P_{\gamma\delta\delta})$$

$$+ \delta_{\alpha\gamma}(-P_{\delta\delta\beta} + 4P_{\delta\beta\delta} - P_{\beta\delta\delta})$$

$$+ \delta_{\beta\gamma}(-P_{\delta\delta\alpha} - P_{\delta\alpha\delta} + 4P_{\alpha\delta\delta})], \tag{4.4.23c}$$

$$\sum_{n=a,b} P^{(2n)}_{\alpha\beta\gamma} = \tfrac{1}{6}\varepsilon_{\alpha\beta\delta}(2\varepsilon_{\lambda\mu\delta}P_{\lambda\mu\gamma} + 2\varepsilon_{\lambda\mu\gamma}P_{\lambda\mu\delta}$$

$$+ \varepsilon_{\lambda\mu\delta}P_{\gamma\lambda\mu} + \varepsilon_{\lambda\mu\gamma}P_{\delta\lambda\mu} - 2\delta_{\gamma\delta}\varepsilon_{\nu\lambda\mu}P_{\nu\lambda\mu})$$

$$+ \tfrac{1}{6}\varepsilon_{\beta\gamma\delta}(2\varepsilon_{\lambda\mu\delta}P_{\alpha\lambda\mu} + 2\varepsilon_{\lambda\mu\alpha}P_{\delta\lambda\mu} + \varepsilon_{\lambda\mu\delta}P_{\lambda\mu\alpha}$$

$$+ \varepsilon_{\lambda\mu\delta}P_{\lambda\mu\delta} - 2\delta_{\alpha\delta}\varepsilon_{\nu\lambda\mu}P_{\nu\lambda\mu}), \tag{4.4.23d}$$

$$P^{(3)}_{\alpha\beta\gamma} = \tfrac{1}{6}(P_{\alpha\beta\gamma} + P_{\alpha\gamma\beta} + P_{\beta\alpha\gamma} + P_{\beta\gamma\alpha} + P_{\gamma\alpha\beta} + P_{\gamma\beta\alpha})$$

$$- \tfrac{1}{15}[\delta_{\alpha\beta}(P_{\delta\delta\gamma} + P_{\delta\gamma\delta} + P_{\gamma\delta\delta}) + \delta_{\alpha\gamma}(P_{\delta\delta\beta} + P_{\delta\beta\delta} + P_{\beta\delta\delta})$$

$$+ \delta_{\beta\gamma}(P_{\delta\delta\alpha} + P_{\delta\alpha\delta} + P_{\alpha\delta\delta})]. \tag{4.4.23e}$$

The three sets of terms in (4.4.23c) can be regarded as the three linearly independent sets, each spanning $D^{(1)}$. Similarly for the two sets of terms in (4.4.23d).

Notice that, as expected, the isotropic tensors spanning the totally symmetric irreducible representation $D^{(0)}$ of the proper rotation group R_3^+ give the isotropic averages of tensor components discussed in Section 4.2.5. However, in the full rotation group R_3, which includes the inversion, only the scalar $P_{\alpha\beta}^{(0)}$ spans the totally symmetric irreducible representation $D_g^{(0)}$; the pseudoscalar $P_{\alpha\beta\gamma}^{(0)}$ now spans $D_u^{(0)}$ and is no longer 'observable'.

The reduction of a general second-rank axial tensor $A_{\alpha\beta}$ into irreducible parts gives expressions equivalent to (4.4.22) except that, whereas $P_{\alpha\beta}^{(0)}$, $P_{\alpha\beta}^{(1)}$ and $P_{\alpha\beta}^{(2)}$ span $D_g^{(0)}$, $D_g^{(1)}$ and $D_g^{(2)}$ in R_3, $A_{\alpha\beta}^{(0)}$, $A_{\alpha\beta}^{(1)}$, and $A_{\alpha\beta}^{(2)}$ span $D_u^{(0)}$, $D_u^{(1)}$ and $D_u^{(2)}$. This emphasizes the equivalence of an axial tensor and a polar tensor of the next higher rank, for $P_{\alpha\beta\gamma}^{(0)}$, $P_{\alpha\beta\gamma}^{(1)}$ and $P_{\alpha\beta\gamma}^{(2)}$ also span $D_u^{(0)}$, $D_u^{(1)}$ and $D_u^{(2)}$. If $P_{\alpha\beta\gamma}$ is symmetric with respect to permutation of any pair of tensor subscripts, some of its irreducible parts vanish; in particular $D^{(0)}$, which explains why the electric dipole–electric quadrupole tensor (2.6.27c) cannot contribute to optical rotation in an isotropic sample. Notice that the tensor $\zeta'_{\alpha\beta\gamma}$, given by (3.4.13$b$), that combines the electric dipole–electric quadrupole and electric dipole–magnetic dipole contributions to natural optical activity transforms the same as $P_{\alpha\beta\gamma}^{(0)}$.

4.4.6 *Matrix elements of irreducible spherical tensor operators*

Degeneracy in molecular quantum states is an important source of both natural and magnetic optical activity. In order to calculate matrix elements of operators between component states of a degenerate level, it is necessary to classify the wavefunctions and operators with respect to the irreducible representations of the symmetry group of the system, and to employ the celebrated Wigner–Eckart theorem.

The concept of irreducible tensor operators and the development of a formalism for making practical use of them in spherical systems such as atoms are due mainly to Racah. This work was partially based on, and developed concurrently with, advances in the theory of angular momentum made by Wigner. The two authoritative texts by Fano and Racah (1959), and Wigner (1959), summarize this work. The subsequent extension of the theory to the molecular point groups has been summarized by Griffith (1962). We shall not give an account of this work here, but will simply state the formulae required in subsequent chapters and refer the reader to Silver (1976) and Piepho and Schatz (1983) for an introduction to most of the aspects required in this book. In writing down different versions of the Wigner–Eckart theorem for use in different situations, we adhere to the notations of the various authors so that their tables of coupling coefficients can be used directly.

For wavefunctions and operators classified with respect to the proper rotation group R_3^+ we use the following version of the *Wigner–Eckart theorem:*

$$\langle \alpha' j' m' | T_q^k | \alpha j m \rangle = (-1)^{j'-m'} \langle \alpha' j' \| T^k \| \alpha j \rangle \begin{pmatrix} j' & k & j \\ -m' & q & m \end{pmatrix}, \quad (4.4.24)$$

where j and m are the usual angular momentum and magnetic quantum numbers, and α denotes any additional quantum numbers needed to specify the state. T_q^k is the operator written in irreducible spherical tensor form: k denotes the corresponding irreducible representation, and q the component. The $3j$ symbol

$$\begin{pmatrix} j' & k & j \\ -m' & q & m \end{pmatrix}$$

expresses the vector coupling coefficient in a form with special symmetry properties, and $\langle \alpha' j' \| T^k \| \alpha j \rangle$ is the reduced matrix element. In effect, the Wigner–Eckart theorem separates the *physical* part of the problem (the reduced matrix element) from the *geometrical* aspect (the $3j$ symbol). We use the numerical values for $3j$ symbols tabulated by Rotenberg, Bivens, Metropolis and Wooten (1959). The reduced matrix elements can be calculated in some situations, but in many of the applications in this book explicit values are not required because dimensionless expressions for optical activity observables are used and the reduced matrix elements cancel.

Much use is made of matrix elements of cartesian components of the electric dipole moment operator, so we now write them out explicitly in terms of $3j$ symbols and reduced matrix elements. The cartesian components are first written in spherical form:

$$\mu_x = -\frac{1}{\sqrt{2}}(\mu_1^1 - \mu_{-1}^1), \quad \mu_y = \frac{i}{\sqrt{2}}(\mu_1^1 + \mu_{-1}^1), \quad \mu_z = \mu_0^1. \quad (4.4.25a)$$

These follow from the definition of the spherical components using a phase convention consistent with that of Condon and Shortley (1935) for the spherical harmonics:

$$\mu_1^1 = -\frac{1}{\sqrt{2}}(\mu_x + i\mu_y), \quad \mu_0^1 = \mu_z, \quad \mu_{-1}^1 = \frac{1}{\sqrt{2}}(\mu_x - i\mu_y). \quad (4.4.25b)$$

Using (4.4.25a) in (4.4.24), the required matrix elements are

$$\langle \alpha' j'm' | \mu_x | \alpha j m \rangle = (-1)^{j'-m'+1} \frac{1}{\sqrt{2}} \langle \alpha' j' \| \mu \| \alpha j \rangle$$

$$\times \left[\begin{pmatrix} j' & 1 & j \\ -m' & 1 & m \end{pmatrix} - \begin{pmatrix} j' & 1 & j \\ -m' & -1 & m \end{pmatrix} \right], \quad (4.4.26a)$$

$$\langle \alpha' j'm' | \mu_y | \alpha j m \rangle = (-1)^{j'-m'} \frac{i}{\sqrt{2}} \langle \alpha' j' \| \mu \| \alpha j \rangle$$

$$\times \left[\begin{pmatrix} j' & 1 & j \\ -m' & 1 & m \end{pmatrix} + \begin{pmatrix} j' & 1 & j \\ -m' & -1 & m \end{pmatrix} \right], \quad (4.4.26b)$$

$$\langle \alpha' j'm' | \mu_z | \alpha j m \rangle = (-1)^{j'-m'} \langle \alpha' j' \| \mu \| \alpha j \rangle \begin{pmatrix} j' & 1 & j \\ -m' & 0 & m \end{pmatrix}. \quad (4.4.26c)$$

From the properties of the $3j$ symbol, the well known selection rules for electric dipole transitions follow: for the z component, $\Delta j = 0, \pm 1 (0 \longleftrightarrow 0)$, $\Delta m = 0$; and for the x and y components, $\Delta j = 0, \pm 1 (0 \longleftrightarrow 0)$, $\Delta m = \pm 1$ (although if j is purely orbital, parity arguments forbid $\Delta j = 0$). It follows directly from (4.4.26a, b) that

$$\langle jm | \mu_x | j+1\, m\, \pm 1 \rangle = \mp i \langle jm | \mu_y | j+1\, m\, \pm 1 \rangle, \quad (4.4.27)$$

a result that is useful in the discussion of magnetic circular dichroism in atoms.

For analogous calculations on systems belonging to finite molecular point groups, we must use an alternative version of the Wigner–Eckart theorem (Griffith, 1962; Silver, 1976). Thus when it is appropriate to use real basis sets, as in the absence of external magnetic fields, the appropriate version is

$$\langle a\alpha | g_\beta^b | a'\alpha' \rangle = \langle a \| g^b \| a' \rangle V \begin{pmatrix} a & a' & b \\ \alpha & \alpha' & \beta \end{pmatrix}. \quad (4.4.28)$$

When complex basis sets are used, the appropriate version is

$$\langle a\alpha | g_\beta^b | a'\alpha' \rangle = [-1]^{a+\alpha} \langle a \| g^b \| a' \rangle V \begin{pmatrix} a & a' & b \\ -\alpha & \alpha' & \beta \end{pmatrix}. \quad (4.4.29)$$

The state $|a\alpha\rangle$ transforms according to the α component of the irreducible representation a. Care must be taken to use the appropriate sets of real or complex operators and V coefficients depending on which version is employed. We refer to Griffith (1962) or Silver (1976) for the definition of the factor $[-1]^{a+\alpha}$ and the properties of the V coefficients. In order to use Griffith's tables of complex V coefficients, we must write the operator g_β^b in complex form, taking care to use his phase convention, which is actually that of Fano and Racah (1959), rather than the Condon and Shortley phase convention used in (4.4.24). In fact, spherical harmonics in the Fano and Racah phase convention are obtained by multiplying those in the Condon and Shortley phase convention by the factor i^l. Thus in place of the cartesian components (4.4.25*a*) of the electric dipole moment operator we must use, in general,

$$\mu_X = \frac{i}{\sqrt{2}}(\mu_1^1 - \mu_{-1}^1), \quad \mu_Y = \frac{1}{\sqrt{2}}(\mu_1^1 + \mu_{-1}^1), \quad \mu_Z = -i\mu_0^1, \quad (4.4.30)$$

and the matrix elements are now, in place of (4.4.26),

$$\langle a\alpha|\mu_X|a'\alpha'\rangle = [-1]^{a+\alpha} \frac{i}{\sqrt{2}} \langle a||\mu||a'\rangle$$

$$\times \left[V\begin{pmatrix} a & a' & b \\ -\alpha & \alpha' & 1 \end{pmatrix} - V\begin{pmatrix} a & a' & b \\ -\alpha & \alpha' & -1 \end{pmatrix} \right], \quad (4.4.31a)$$

$$\langle a\alpha|\mu_Y|a'\alpha'\rangle = [-1]^{a+\alpha} \frac{1}{\sqrt{2}} \langle a||\mu||a'\rangle$$

$$\times \left[V\begin{pmatrix} a & a' & b \\ -\alpha & \alpha' & 1 \end{pmatrix} + V\begin{pmatrix} a & a' & b \\ -\alpha & \alpha' & -1 \end{pmatrix} \right], \quad (4.4.31b)$$

$$\langle a\alpha|\mu_Z|a'\alpha'\rangle = [-1]^{a+\alpha}(-i) \langle a||\mu||a'\rangle \, V\begin{pmatrix} a & a' & b \\ -\alpha & \alpha' & 0 \end{pmatrix}. \quad (4.4.31c)$$

However, in the dihedral groups $D_n(n > 2)$, Griffith (1962) uses real functions for A_1, A_2, B_1 and B_2 representations and complex functions for E: in other words, his corresponding tables of complex V coefficients are to be used with the complex operators

$$\mu_X = \frac{i}{\sqrt{2}} (\mu_1^1 - \mu_{-1}^1), \quad \mu_Y = \frac{1}{\sqrt{2}}(\mu_1^1 + \mu_{-1}^1)$$

for E, but used with μ_Z left unchanged for A_2. So for $D_n(n > 2)$, (4.4.31*a*, *b*) still apply, but (4.4.31*c*) is replaced by

$$\langle a\alpha|\mu_Z|a'\alpha'\rangle = [-1]^{a+\alpha} \langle a||\mu||\alpha\rangle \, V\begin{pmatrix} a & a' & b \\ -\alpha & \alpha' & 0 \end{pmatrix}. \quad (4.4.31d)$$

The entries b in the V coefficients depend on the irreducible representations spanned by components of μ in the particular point group. Thus in O, (μ_X, μ_Y, μ_Z) span T_1 so $b = T_1$; whereas in D_4, (μ_X, μ_Y) span E and μ_Z spans A_2, so $b = E$ in (4.4.31*a* and *b*) and $b = A_2$ in (4.4.31*d*). In applying (4.4.31), Table C2.3 of Griffith (1962) is used for the V coefficients in O, whereas Table D3.2 (complex) is used for D_4.

Finally, for certain calculations on molecules with odd numbers of electrons, we need an extension of Griffith's methods to the double point groups. Harnung (1973) has provided a suitable extension (see also Dobosh, 1972; and Piepho and Schatz, 1983) and gives the following version of the Wigner–Eckart theorem for the octahedral double group O^*:

$$\langle \Gamma\gamma | \mathcal{D}^K_\kappa | \Gamma'\gamma' \rangle = \sum_\epsilon (-1)^{u(\Gamma-\gamma)} \langle \Gamma \| \epsilon \mathcal{D}^K \| \Gamma' \rangle \begin{pmatrix} \Gamma & K & \Gamma' \\ -\gamma & \kappa & \gamma' \end{pmatrix}_\epsilon . \qquad (4.4.32)$$

The sum over the parameter ϵ arises because O^* is not simply reducible; that is, the direct products of some of the irreducible representations contain repeated representations. We refer to Harnung (1973) for the definition of the factor $(-1)^{u(\Gamma-\gamma)}$, the properties of the 3Γ symbols and tables of 3Γ symbols. The phase conventions of Fano and Racah are again employed, so using operators of the form (4.4.30) we obtain expressions analogous to (4.4.31) for matrix elements of cartesian components of the electric dipole moment operator:

$$\langle \Gamma\gamma | \mu_X | \Gamma'\gamma' \rangle = (-1)^{u(\Gamma-\gamma)} \frac{i}{\sqrt{2}} \langle \Gamma \| \mu \| \Gamma' \rangle$$

$$\times \left[\begin{pmatrix} \Gamma & T_1 & \Gamma' \\ -\gamma & 1 & \gamma' \end{pmatrix} - \begin{pmatrix} \Gamma & T_1 & \Gamma' \\ -\gamma & -1 & \gamma' \end{pmatrix} \right], \quad (4.4.33a)$$

$$\langle \Gamma\gamma | \mu_Y | \Gamma'\gamma' \rangle = (-1)^{u(\Gamma-\gamma)} \frac{1}{\sqrt{2}} \langle \Gamma \| \mu \| \Gamma' \rangle$$

$$\times \left[\begin{pmatrix} \Gamma & T_1 & \Gamma' \\ -\gamma & 1 & \gamma' \end{pmatrix} + \begin{pmatrix} \Gamma & T_1 & \Gamma' \\ -\gamma & -1 & \gamma' \end{pmatrix} \right], \quad (4.4.33b)$$

$$\langle \Gamma\gamma | \mu_Z | \Gamma'\gamma' \rangle = (-1)^{u(\Gamma-\gamma)}(-i)\langle \Gamma \| \mu \| \Gamma' \rangle \begin{pmatrix} \Gamma & T_1 & \Gamma' \\ -\gamma & 0 & \gamma' \end{pmatrix}. \quad (4.4.33c)$$

These 3Γ symbols apply explicitly to O^*, and are given in Table 5 of Harnung (1973).

4.5 Permutation symmetry and chirality

We now turn to a rather different aspect of symmetry in the discussion of molecular properties, namely an algebraic analysis of chirality based on the permutation of ligands among sites on a molecular skeleton. As well as giving insight into the phenomenon of molecular chirality, it provides rigorous algebraic criteria which can be used to assess (at least in principle) any molecular theory of optical activity. Much of this section is based on reviews by Ruch (1972) and Mead (1974), and we refer to these and a later review by King (1991) for further details.

4.5.1 Chirality functions

A molecule can be pictured as a skeleton providing sites to which ligands have been attached. If the skeleton is achiral, any molecular chirality must arise from differences between the ligands. Taking the case of the methane skeleton consisting of a carbon atom with four tetrahedrally directed bonds, it is well known that chirality is only possible if all four ligands are different. This led Crum Brown (1890) and Guye (1890) to propose that the optical rotatory power might be proportional to a 'product of asymmetry' of the form

$$\alpha = (a - b)(b - c)(c - d)(a - c)(a - d)(b - d), \tag{4.5.1}$$

where a, b, c, d are identified with some property of the ligands (which Crum Brown and Guye took to be the masses). Clearly, if any two of a, b, c, d are equal, $\alpha = 0$; and if any two are interchanged, α changes sign. Thus (4.5.1) has the correct form to represent the pseudoscalar observable α, and is called a *chirality function*. The molecular theory of Boys (1934) contains a factor with the same form as (4.5.1), but with the quantities a, b, c, d identified with the radii of the ligands.

Although the chirality function (4.5.1) has the necessary symmetry properties for describing the pseudoscalar optical rotatory parameter, it is not the only one possible. The systematic group theoretical study of chirality functions for general molecular skeletons was taken up by Ruch, Schönhofer and Ugi (1967) and given a definitive form by Ruch and Schönhofer (1970).

Ruch posed the following important question, which he felt a satisfactory theory of chirality functions ought to be able to answer: is it possible to divide chiral molecules into two subclasses which can be designated as right handed and left handed? He quotes the following analogy (Ruch, 1972):

If asked to put our left shoes into one box and our right shoes into a second box we could accomplish the task without mental difficulty, in spite of the fact that the right shoes belonging to different people may be quite different in colour, shape and size and although, probably, there is not a single pair of shoes which are precise mirror images of each other. If asked to solve the same problem with potatoes, we must capitulate. Of course, it is possible that by chance we find an antipodal pair. It is then clear that we must separate them, but for other potatoes different in shape, we have to make new arbitrary decisions each time. Any classification would be very artificial.

We shall see (Section 4.5.6) that the skeleton of any chiral molecule can be assigned to one of two categories. One of these categories is 'shoe-like' in that it admits a classification into right-and left-handed molecules; the other is 'potato-like' in that it permits no such distinction, any classification being arbitrary. For pairs of different chiral molecules with skeletons in the first category, Ruch coined the term *homochiral* if both were either left-handed or right-handed (like two shoes

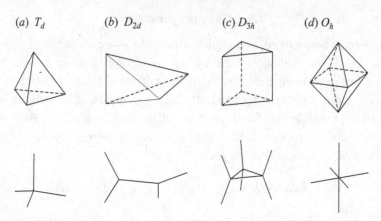

(a) T_d (b) D_{2d} (c) D_{3h} (d) O_h

Fig. 4.5 Typical achiral molecular skeletons: (*a*) methane, (*b*) allene, (*c*) cyclo-propane and (*d*) SF$_6$.

of different make for the same foot) and *heterochiral* if oppositely handed (like two shoes of different make for different feet).

A molecule may often be completely specified by describing a skeleton and the nature (and perhaps the orientation) of the ligand at each site. Thus a particular skeleton can be thought of as defining a class of molecules with individual class members being specified by the ligands at each site. A given molecule can belong to more than one class, depending on which part is taken to be the skeleton and which the ligands: ethane, for example, can be thought of as the six-site ethane skeleton with six hydrogen atoms as ligands; or as the four-site methane skeleton with one methyl and three hydrogen atom ligands.

Here we restrict the discussion to ligands which fulfil the condition that molecules have the symmetry of the bare skeleton if all its ligands are of the same kind. This means that the ligand must possess sufficient symmetry to make all properties invariant under changes of orientation (so the ligand must have a threefold or higher proper rotation axis coincident with the bond linking the ligand to the skeleton); it also excludes intrinsically chiral ligands. If the skeleton is achiral, a molecule containing only ligands of one sort is achiral.

We consider chiral classes which are specified by skeletons such that the molecules are chiral if, at the least, all the ligands are different. Examples are skeletons supporting ligands whose positions are at the corners of regular bodies (Fig. 4.5). The corners of a regular body with T_d symmetry, for example, would correspond to the positions of ligands attached to the methane skeleton (Fig. 4.5*a*). It is assumed that the ligands can be characterized by a physical property associated with a single scalar parameter λ; for example, the radii of spherical ligands.

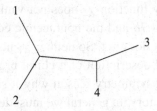

Fig. 4.6 The allene skeleton.

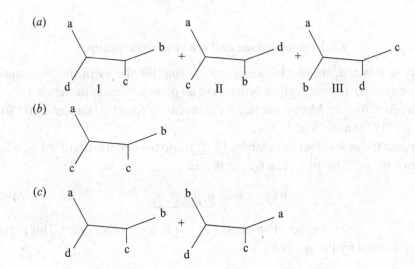

Fig. 4.7 Various isomers of allene.

Take as an example the labelled allene skeleton shown in Fig. 4.6. It is easy to verify that

$$\chi_1 = (\lambda_1 - \lambda_2)(\lambda_3 - \lambda_4), \tag{4.5.2a}$$

$$\chi_2 = (\lambda_1 - \lambda_2)(\lambda_1 - \lambda_3)(\lambda_1 - \lambda_4)(\lambda_2 - \lambda_3)(\lambda_2 - \lambda_4)(\lambda_3 - \lambda_4) \tag{4.5.2b}$$

are both chirality functions for the allene skeleton since they are unchanged under the proper operations and change sign under the improper operations of the D_{2d} skeleton.

However, neither of these chirality functions can be applied without encountering a fundamental difficulty. For example, consider a mixture of the isomers I, II, III of Fig. 4.7a in equal concentrations. The first chirality function χ_1 for this mixture vanishes:

$$\chi_1 = \tfrac{1}{3}[\chi_1(\text{I}) + \chi_1(\text{II}) + \chi_1(\text{III})]$$
$$= \tfrac{1}{3}[(\lambda_a - \lambda_d)(\lambda_b - \lambda_c) + (\lambda_a - \lambda_c)(\lambda_d - \lambda_b) + (\lambda_a - \lambda_b)(\lambda_c - \lambda_d)] = 0;$$

whereas the second chirality function χ_2 does not vanish. On the other hand, for the chiral molecule of Fig. 4.7*b* and the nonracemic equal mixture of isomers in Fig. 4.7*c*, χ_2 vanishes but χ_1 does not. So neither χ_1 nor χ_2 is capable by itself of giving a sufficiently general description of a chiral property, as each vanishes in situations where there is no symmetry reason why it should. In this instance the sum $\chi_1 + \chi_2$ is more satisfactory. In general we must demand that there is no non-racemic mixture of isomers for which the chirality function vanishes. A chirality function of this type is called *qualitatively complete*.

4.5.2 *Permutations and the symmetric group*

In order to proceed, we require some results from the theory of the *permutation group*, or *symmetric group*, which is the set of all permutations of the labels $1, 2, \ldots,$ n and is denoted by $\mathscr{S}_n$. More complete accounts can be found in Hamermesh (1962), Chisholm (1976) and Mead (1974).

Consider the ordered set of numbers $\underline{12\ldots n}$ and the permutation P of $\mathscr{S}_n$ which replaces 1 by p_1, 2 by $p_2, \ldots, n$ by p_n; that is,

$$P\underline{12\ldots n} = \underline{p_1 p_2 \ldots p_n}, \tag{4.5.3}$$

where $p_1 p_2 \ldots p_n$ are the set of numbers $\underline{12\ldots n}$ in some other order. This permutation is denoted by the symbol

$$P = \begin{pmatrix} 1 & 2 & \cdots & n \\ p_1 & p_2 & \cdots & p_n \end{pmatrix}. \tag{4.5.4}$$

A permutation which interchanges m labels cyclically is called an m-cycle and is written as

$$\begin{pmatrix} 1 & 2 & \cdots & m-1 & m \\ 2 & 3 & \cdots & m & 1 \end{pmatrix} \equiv (12\ldots m). \tag{4.5.5}$$

For example, the permutation that changes $\underline{123}$ into $\underline{231}$ is written

$$\begin{pmatrix} 1 & 2 & 3 \\ 2 & 3 & 1 \end{pmatrix} \underline{123} \equiv (123)\underline{123} = \underline{231}.$$

A 2-cycle is called a transposition.

It can be shown that every permutation can be written as a product of cycles which operate on mutually exclusive sets of labels. For example,

$$\begin{pmatrix} 1 & 2 & 3 & 4 & 5 & 6 \\ 2 & 4 & 5 & 1 & 3 & 6 \end{pmatrix} = (124)(35)(6). \tag{4.5.6}$$

Furthermore, every permutation can be expressed as a product of transpositions; for example, $(123) = (13)(12)$. In particular, it can be shown that the $(n-1)$ transpositions $(12), (13), \ldots, (1n)$ constitute a set of generators for the group $\mathscr{S}_n$; that is, every element of $\mathscr{S}_n$ can be written as a suitable product of these transpositions.

It is important to define the effect of a permutation operator on a function of n independent variables $x_1, x_2, \ldots, x_n$ of the form

$$
\begin{aligned}
f(x_1, x_2, \ldots, x_n) &= (x_1 - x_2)(x_1 - x_3) \ldots (x_1 - x_n) \\
&\times (x_2 - x_3) \ldots (x_2 - x_n) \\
&\times (x_{n-1} - x_n).
\end{aligned}
\tag{4.5.7}
$$

If P is an element of $\mathscr{S}_n$, it is clear that

$$
Pf = \varepsilon_p f,
\tag{4.5.8}
$$

where $\varepsilon_p = \pm 1$. If $\varepsilon_p = +1$ the permutation is said to be even, while if $\varepsilon_p = -1$ it is said to be odd. Clearly, even and odd permutations consist of an even and an odd number of transpositions, respectively. Also, the product of two even or two odd permutations is even, whereas the product of an even and an odd permutation is odd.

We now turn to the matter of partitions and conjugate classes. If P and Q are elements of $\mathscr{S}_n$, Q is in the same class as P if there exists an element T for which

$$
Q = TPT^{-1}.
\tag{4.5.9}
$$

Suppose that in the decomposition of P into cycles there occur ν_1 1-cycles, ν_2 2-cycles, $\ldots$, ν_n n-cycles. P is then said to have the cycle structure

$$
(\nu) \equiv (1^{\nu_1} 2^{\nu_2} \ldots n^{\nu_n}).
\tag{4.5.10}
$$

Denoting the cycles by c_i we have

$$
P = c_1 c_2 \ldots c_h,
\tag{4.5.11}
$$

where $h = \nu_1 + \nu_2 + \ldots + \nu_n$. Since there are n numbers in the set, it follows that

$$
\nu_1 + 2\nu_2 + \ldots n\nu_n = n.
\tag{4.5.12}
$$

The conjugate element Q is now given by

$$
Q = (Tc_1 T^{-1})(Tc_2 T^{-1}) \ldots (Tc_n T^{-1}),
\tag{4.5.13}
$$

and it can be shown from this that Q has the same cycle structure as P. Thus all the elements in a given conjugate class have the same cycle structure. It follows that each solution of (4.5.12) in nonnegative integers $\nu_1, \nu_2, \ldots \nu_n$ determines a cycle structure and hence a conjugate class. The number of classes in $\mathscr{S}_n$ is therefore given by the number of solutions of (4.5.13). Writing

$$\nu_1 + \nu_2 + \ldots + \nu_n = \delta_1$$
$$\nu_2 + \ldots + \nu_n = \delta_2 \qquad\qquad (4.5.14a)$$
$$\vdots$$
$$\nu_n = \delta_n,$$

we have

$$\delta_1 + \delta_2 + \ldots + \delta_n = n, \qquad\qquad (4.5.14b)$$

with

$$\delta_1 \geqslant \delta_2 \geqslant \ldots \geqslant \delta_n \geqslant 0. \qquad\qquad (4.5.14c)$$

We say that (4.5.14b) is a partition of n and denote it by $[\delta] \equiv [\delta_1 \delta_2 \ldots \delta_n]$. There is a one to one correspondence between partitions of n and solutions of (4.5.12), since from (4.5.14a) we have

$$\nu_1 = \delta_1 - \delta_2$$
$$\nu_2 = \delta_2 - \delta_3$$
$$\vdots \qquad\qquad\qquad (4.5.14d)$$
$$\nu_n = \delta_n.$$

Consequently, the number of classes in $\mathscr{S}_n$ is given by the number of partitions of n.

It can be shown that the number of elements in the conjugate class with the cycle structure $(1^{\nu_1} 2^{\nu_2} \ldots n^{\nu_n})$ is

$$g = \frac{n!}{(1^{\nu_1} \nu_1!)(2^{\nu_2} \nu_2!) \ldots (n^{\nu_n} \nu_n!)}. \qquad\qquad (4.5.15)$$

Consider $\mathscr{S}_4$ as an example. The partitions of 4 are [4], [3 1], [2 2] $\equiv$ [2^2], [2 1 1] $\equiv$ [2 1^2] and [1 1 1 1] $\equiv$ [1^4]. Thus there are five conjugate classes in $\mathscr{S}_4$. Using (4.5.14d) and (4.5.12) we obtain Table 4.4.

Since the number of conjugate classes in $\mathscr{S}_n$ is given by the number of partitions of n, it follows that the number of irreducible representations of $\mathscr{S}_n$ is also given by the number of partitions of n. Thus, associated with each partition of n, there is an irreducible representation of $\mathscr{S}_n$, which leads to a very convenient method for labelling the irreducible representations. Corresponding to each partition $[\delta] = [\delta_1 \delta_2 \ldots \delta_n]$ we can draw a *Young diagram* $\gamma^{[\delta]}$ consisting of δ_1 cells in the first row, δ_2 cells in the second row and so on, with no row longer than the one above it. If

Table 4.4 *Partitions of four*

Partition	Cycle structure	Number of elements in class	Example
[4]	(1^4)	1	(1)(2)(3)(4)
$[1^4]$	(4^1)	6	(1432)
$[2^2]$	(2^2)	3	(14)(23)
$[21^2]$	$(1^1 3^1)$	8	(132)(4)
[31]	$(1^2 2^1)$	6	(12)(3)(4)

the numbers $1, 2, \ldots, n$ are now inserted into the cells we obtain a *Young tableau*. If the numbers are inserted into the cells in such a way that they increase on going down a column and increase on going along a row from left to right we have a *standard* Young tableau $T^{[\delta]}$. The following theorem (see Hamermesh, 1962) is of fundamental importance: the dimension d of the irreducible representation denoted by the partition $[\delta]$ is given by the number of standard Young tableaux $T_1^{[\delta]}, \ldots, T_d^{[\delta]}$ which can be constructed from the Young diagram $\gamma^{[\delta]}$. The result of applying this theorem to $\mathscr{S}_4$ is shown in Table 4.5. We see that for each irreducible representation $[\delta]$ there exists an irreducible representation $[\tilde{\delta}]$ in which the rows and columns have been interchanged. $[\tilde{\delta}]$ is called the dual of $[\delta]$. We also see that $[2^2]$ is self dual, and that dual irreducible representations have the same dimension.

Just as $\mathscr{S}_4$ has two one-dimensional irreducible representations, [4] and $[1^4]$, so $\mathscr{S}_n$ in general has two one-dimensional irreducible representations $[n]$ and $[1^n]$. Since $[n]$ is totally symmetric, it must be spanned by a basis function $\psi^s(1, 2, \ldots, n)$ that is symmetric to any transposition; for example, under any transposition $(1i)$, $i = 2, 3, \ldots, n$, we must have

$$(1i)\psi^s(1, 2, \ldots, n) = \psi^s(1, 2, \ldots, n).$$

The other one-dimensional irreducible representation $[1^n]$ is symmetric under even permutations but antisymmetric under odd permutations (that is, has characters $+1$ and -1, respectively); so, since the transpositions $(1i)$ are all odd, any function $\psi^a(1, 2, \ldots, n)$ spanning $[1^n]$ must satisfy

$$(1i)\psi^a(1, 2, \ldots, n) = -\psi^a(1, 2, \ldots, n).$$

We now associate with the standard Young tableau

1	2		n

the *symmetrizing operator*

$$S = \sum_P P, \tag{4.5.16}$$

Table 4.5 *The irreducible representations of* $\mathscr{S}_4$

Irreducible representation	Standard Young tableaux	Dimension
[4]	$\young(1234)$	1
[1⁴]	$\young(1,2,3,4)$	1
[2²]	$\young(12,34)\quad\young(13,24)$	2
[31]	$\young(123,4)\quad\young(124,3)\quad\young(134,2)$	3
[21²]	$\young(14,2,3)\quad\young(13,2,4)\quad\young(12,3,4)$	3

where the sum runs over all the permutation operations of $\mathscr{S}_n$. Then if $\psi(1, 2, \ldots, n)$ is an arbitrary function, the function $S\psi$ is a symmetry adapted basis for $[n]$. Similarly we associate with the standard Young tableau

$$\young(1,2,\vdots,n)$$

the *antisymmetrizing operator*

$$A = \sum_{P} \varepsilon_p P. \qquad (4.5.17)$$

Then $A\psi$ is symmetry adapted to $[1^n]$.

These ideas can be generalized to irreducible representations of dimension greater than one. Two types of permutation are defined: horizontal permutations which interchange only symbols in the same row of a standard tableau, and vertical permutations which interchange only symbols in the same column of a standard tableau. The following *Young operator* is now associated with the standard tableau $T_i^{[\delta]}$:

$$Y_i^{[\delta]} = AS, \tag{4.5.18}$$

where S effects only horizontal permutations and A effects only vertical permutations.

Consider, for example, the two-dimensional irreducible representation [21] of $\mathscr{S}_3$ which is associated with the standard tableaux

$$T_1^{[21]} = \begin{array}{|c|c|} \hline 1 & 2 \\ \hline 3 \\ \cline{1-1} \end{array}, \quad T_2^{[21]} = \begin{array}{|c|c|} \hline 1 & 3 \\ \hline 2 \\ \cline{1-1} \end{array}.$$

The corresponding Young operators are

$$Y_1^{[21]} = [I - (13)][1 + (12)], \tag{4.5.19a}$$
$$Y_2^{[21]} = [I - (12)][1 + (13)], \tag{4.5.19b}$$

where $I = (1)(2)(3)$ is the identity operation.

4.5.3 Chirality functions: qualitative completeness

We now use the formalism of the permutation group to give mathematical structure to the concept of qualitative completeness introduced in Section 4.5.1.

The group $\mathscr{S}_n$ generates all the possible isomers of a molecule M belonging to a skeleton with n sites. $\mathscr{S}_n$ possesses a subgroup $\mathscr{G}$ consisting of those ligand permutations which can be interpreted as point group symmetry operations. $\mathscr{G}$ is often, but not always, isomorphic with the point group of the skeleton. For example, all possible permutations of the four ligands on the allene skeleton (Fig. 4.6) make up the permutation group $\mathscr{S}_4$ which contains 24 elements in all: of these, only eight elements are equivalent to the operations of the D_{2d} point group of the allene skeleton, namely, the identity $(1)(2)(3)(4)$; the proper rotations $(12)(34)$, $(13)(24)$ and $(14)(23)$ (equivalent to the three distinct C_2 operations); and the improper rotations $(1)(2)(34)$, $(12)(3)(4)$ (equivalent to the two distinct σ_d operations), (1324) and (1423) (equivalent to the two distinct S_4 operations). A chirality function must by definition belong to the chirality (or pseudoscalar) representation Γ_λ of the subgroup $\mathscr{G}$, which has characters $+1$ for proper rotations and -1 for improper rotations. It is now shown how the transformation properties of chirality functions in the point group of the skeleton are related to their behaviour in the full permutation group of the ligand sites.

Table 4.6 *The character table for* $\mathscr{S}_4$

Irreducible representation Γ_r	Young diagram	(1^4) 1	$(1^2 2^1)$ 6	$(1^1 3^1)$ 8	(2^2) 3	(4^1) 6
Γ_1	[4]	1	1	1	1	1
Γ_2	[31]	3	1	0	-1	-1
Γ_3	$[2^2]$	2	0	-1	2	0
Γ_4	[21]	3	-1	0	-1	1
Γ_5	$[1^4]$	1	-1	1	1	-1

An *ensemble operator* consisting of a linear combination of permutation operators of $\mathscr{S}_n$ is introduced which, when applied to any molecule, generates a mixture of isomers by permuting the n ligands among the n sites on the skeleton:

$$a = \sum_P a(P)P. \qquad (4.5.20)$$

The $a(P)$ are positive real coefficients to be interpreted in terms of concentrations. In the general case that all the ligands are different, an ensemble operator is said to be chiral or achiral depending on whether the resulting mixture of isomers is nonracemic or racemic, respectively.

Given a skeleton with n sites we can form a molecule M by distributing n ligands (in general all different) in an arbitrary way among the sites. A chirality function $\chi(M)$ will have a particular value for this molecule. The corresponding chirality function for the mixture aM is

$$\chi(a\text{M}) = \sum_P a(P)\chi(P\text{M}). \qquad (4.5.21)$$

Qualitative completeness means that if a is not the operator for a racemic mixture, then $\chi(a\text{M})$ does not vanish. We now quote the following theorem: it is necessary and sufficient for qualitative completeness of χ that χ contain z_r independent components transforming according to each irreducible representation Γ_r of $\mathscr{S}_n$, where z_r is the number of times Γ_χ is subduced by Γ_r in $\mathscr{G}$; and that the induction from Γ_χ of $\mathscr{G}$ to $\mathscr{S}_n$ is regular. We refer to Mead (1974) for the proof of this theorem, together with an account of subduced and induced representations.

The meaning of this will become clear by considering as an explicit example the allene skeleton. The character table for $\mathscr{S}_4$ is shown in Table 4.6. The required subgroup of $\mathscr{S}_4$ is D_{2d}, and Table 4.7 shows the classes of D_{2d}, the number of elements in each, the class of $\mathscr{S}_4$ to which each belongs and the character of each for the chirality representation $\Gamma_\chi(\equiv B_1)$ of D_{2d}.

Table 4.7 *Some properties of the classes in* D_{2d}

	I	C_2	$2C_2'$	$2\sigma_d$	$2S_4$
Class in $\mathscr{S}_4$	(1^4)	(2^2)	(2^2)	$(1^2 2^1)$	(4^1)
Character in $\Gamma_\chi (\equiv B_1)$	1	1	1	-1	-1

Table 4.8 *The characters of the irreducible representations for the operations of* $\mathscr{S}_4$ *that are also in* D_{2d}

	I	C_2	$2C_2'$	$2\sigma_d$	$2S_4$
Γ_1	1	1	1	1	1
Γ_2	3	-1	-1	1	-1
Γ_3	2	2	2	0	0
Γ_4	3	-1	-1	-1	1
Γ_5	1	1	1	-1	-1

To find the characters for the representation of D_{2d} subduced by a given representation of $\mathscr{S}_4$ we simply write down the characters of that representation for the elements of $\mathscr{S}_4$ which are also in D_{2d}, and this is done in Table 4.8. Comparing Tables 4.7 and 4.8 and using the standard formula for finding the irreducible parts of a representation by means of the characters, we find that only the representations subduced by Γ_3 and Γ_5 contain Γ_χ, and then only once each. So in this case $z_1 = z_2 = z_4 = 0$, and $z_3 = z_5 = 1$. This also means that the regular induction from Γ_χ of D_{2d} to $\mathscr{S}_4$ gives a representation containing Γ_3 and Γ_5 once each, and the others not at all. Thus a qualitatively complete chirality function for the allene skeleton must have two independent components: one, denoted $\chi^{(\Gamma_3)}$, transforming according to Γ_3 and the other, denoted $\chi^{(\Gamma_5)}$, transforming according to Γ_5 of $\mathscr{S}_4$.

It is left to the reader to verify that a qualitatively complete chirality function for the four-site methane skeleton has just one component transforming according to Γ_5 of $\mathscr{S}_4$, because the regular induction from $\Gamma_\chi (\equiv A_2)$ of T_d to $\mathscr{S}_4$ gives only Γ_5.

4.5.4 Chirality functions: explicit forms

It was indicated in the previous section that a qualitatively complete chirality function χ contains $\sum_r z_r$ components, so the explicit construction of χ reduces to the construction of its components.

The formal procedure for constructing a chirality function belonging to a particular irreducible representation Γ_r of $\mathscr{S}_n$ is as follows. A Young operator $Y^{(\Gamma_r)}$, defined in (4.5.18), is applied to an arbitrary function $\psi(1, 2, \ldots, n)$. If the result is not zero, it will be a function belonging to Γ_r, but not necessarily a chirality function. A projection operator $C^{(\Gamma_\chi)}$ corresponding to the chiral irreducible representation of $\mathscr{G}$ is then applied, and if the result is still not zero it will be a chirality function with the required properties. Thus

$$\chi^{(\Gamma_r)} = C^{(\Gamma_\chi)} Y^{(\Gamma_r)} \psi(1, 2, \ldots, n). \tag{4.5.22}$$

If $z_r > 1$, it is necessary to construct z_r independent functions in this way. In principle the starting function $\psi(1, 2, \ldots, n)$ is arbitrary, and an unlimited number of functions belonging to the same representation is possible. But in practice the functions are chosen to correspond with the model on which a particular theory of optical activity is being constructed. We consider two particularly useful types of functions: the *first procedure* generates polynomials of lowest possible order, and the *second procedure* generates functions of as few ligands as possible.

Consider the first procedure, which generates chirality functions of lowest possible order in one or more ligand parameters. The two functions (4.5.2) provide a simple example. Since we are considering only achiral ligands, they can be characterized by a single scalar parameter λ. The starting function $\psi(1, 2, \ldots, n)$ in (4.5.22) is chosen to be a monomial of the lowest order which is not annihilated by the operations of (4.5.22). The Young operator $Y^{(\Gamma_r)}$ in (4.5.22) antisymmetrizes with respect to permutations of sites in the same column in its tableau. Our monomial cannot therefore be symmetric with respect to any two sites in the same column, that is, it cannot contain the same power of λ for any two such sites. The powers of λ for the sites in a given column must therefore all be different, and the lowest possible choice is $0, 1, 2, \ldots, n$ for a column of length n. The total order is therefore

$$h = \sum_j (j - 1)\delta_j, \tag{4.5.23}$$

where δ_j is the number of sites in the jth row.

This will become clear by considering again the simple example of the allene skeleton. According to the results of Section 4.5.3, a qualitatively complete χ must contain two components, one belonging to Γ_3 and the other to Γ_5. For the irreducible representation Γ_3, the Young diagrams are those of $[2^2]$ in Table 4.5 and so (4.5.23) tells us that $h = 2$. We choose $\psi(1, 2, \ldots, n) = \lambda_2 \lambda_4$ and the tableau

$$T_2^{[2^2]} = \begin{array}{|c|c|} \hline 1 & 3 \\ \hline 2 & 4 \\ \hline \end{array}.$$

Using (4.5.18), the corresponding Young operator is

$$Y_2^{(\Gamma_3)} = [I - (12)][I - (34)][I + (13)][I + (24)]. \qquad (4.5.24)$$

Applying this to $\psi(1, 2, \ldots, n)$ we find

$$Y_2^{(\Gamma_3)} \lambda_2 \lambda_4 = 2(\lambda_2 - \lambda_1)(\lambda_4 - \lambda_3). \qquad (4.5.25)$$

For the projection operator $C^{(\Gamma_x)}$ we have

$$\begin{aligned} C^{(\Gamma_x)} = {} & I + (12)(34) + (13)(24) + (14)(23) - (1)(2)(34) \\ & - (12)(3)(4) - (1324) - (1423). \end{aligned} \qquad (4.5.26)$$

It is easily verified that $C^{(\Gamma_x)}$ applied to $Y_2^{(\Gamma_3)} \lambda_2 \lambda_4$ simply multiplies it by a constant. Dropping the multiplicative constant, the required function is

$$\chi^{(\Gamma_3)} = (\lambda_2 - \lambda_1)(\lambda_4 - \lambda_3), \qquad (4.5.27)$$

which is identical with (4.5.2a). In a similar way it is found that $\chi^{(\Gamma_5)}$ is identical with (4.5.2b).

Now consider the second procedure, which generates chirality functions that depend on as few ligands as possible. The starting function $\psi(1, 2, \ldots, n)$ is not now required to be a monomial, but is simply required to depend on as few ligands as possible, otherwise being arbitrary. If $\psi(1, 2, \ldots, n)$ depends on only b ligands, it must be totally symmetric under both permutations and reflections of the other $(n-b)$ ligands. Hence no two of these $(n-b)$ sites may be in the same column of the tableau of the Young operator $Y^{(\Gamma_r)}$. Thus $(n-b)$ cannot be greater than the number of columns, which is the same as the length of the first row, of the Young diagram. The lowest value of b is therefore $(n-\delta_1)$.

A simple example is again the component spanning Γ_3 of the D_{2d} allene skeleton. We choose $\psi = f(2,4)$, where f is an arbitrary function, and again apply the operators (4.5.24) and (4.5.26). The result is

$$\chi^{(\Gamma_3)'} = g(2,4) - g(1,4) - g(2,3) + g(1,3), \qquad (4.5.28)$$

where $g(i,j) = f(i,j) + f(j,i)$ is a totally symmetric function with respect to interchange of ligands i and j.

The two procedures outlined above for generating explicit chirality functions can be applied, with varying degrees of difficulty, to any class of skeleton. Mead (1974) has given an extensive list of both types of chirality function for a number of important skeletons.

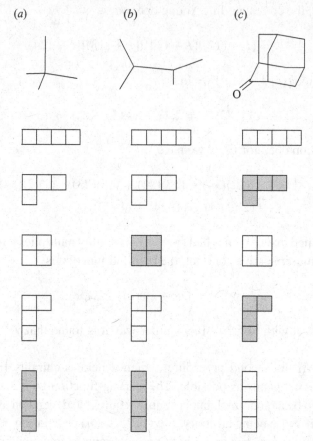

Fig. 4.8 Partition diagrams for (*a*) the T_d methane skeleton, (*b*) the D_{2d} allene skeleton and (*c*) the C_{2v} adamantanone skeleton. The shaded partitions indicate the irreducible representations for which $z_r \neq 0$.

4.5.5 Active and inactive ligand partitions: chirality numbers

We now turn to the following question. Given a set of n ligands to be distributed among the sites of a particular skeleton, which components (if any) of the qualitatively complete chirality function will vanish if any of the ligands happen to be identical? The answer for the case of the allene skeleton can be deduced immediately, since $\chi^{(\Gamma_5)}$, given by (4.5.2*b*), vanishes if any two ligands are identical; whereas $\chi^{(\Gamma_3)}$, given by (4.5.2*a*), can be different from zero with two ligands identical and the other two different, or with two pairs of identical ligands. We now formulate the problem more generally.

It is possible to associate an assortment of ligands with a Young diagram. A *ligand partition* is defined as the list of numbers δ_1, δ_2, etc. of identical ligands. So a partition corresponds to a set of δ_1 identical ligands, δ_2 identical ligands different from the δ_1 previously listed, etc.; and the sum of the δs must equal n. The *partition diagram* $\gamma^{[\delta]}$ is just the Young diagram whose row lengths are δ_1, δ_2, etc. Figure 4.8 shows the partition diagrams of order four for three different four-site skeletons: the T_d methane skeleton, the D_{2d} allene skeleton and the C_{2v} adamantanone skeleton. The shaded diagrams belong to irreducible representations for which $z_r \neq 0$. A ligand partition is called *active* if a chiral molecule can be constructed by *properly* distributing the ligands on the skeleton sites (not every permutation of ligands specified by an active partition leads to a chiral molecule).

It can be proved that all ligand partitions represented by shaded diagrams are active, but we infer from the allene example that further active partitions can exist because, in addition to Γ_3 and Γ_5, it is easy to see that Γ_4 is also active because an allene skeleton dressed with two identical ligands and two different ones is chiral. A method for finding all the active partitions uses the following definition (see Ruch, 1972): a Young diagram γ is *smaller* than another one $\gamma'(\gamma \subset \gamma')$ if γ can be constructed from γ' by pulling boxes from upper lines to lower ones without at any point producing an array of boxes which is not a Young diagram. This definition is supplemented by saying that $\gamma \subset \gamma$ for each diagram γ. This definition can be given another form by using partial sums o_i of the lengths δ_i of the first i horizontal rows: $o_1 = \delta_1, o_2 = \delta_1 + \delta_2, o_3 = \delta_1 + \delta_2 + \delta_3$, etc. Then $\gamma \subset \gamma'$ if, and only if, $o_1 \leqslant o_i'$ for all i. It can be proved that all partitions smaller than any shaded one, and only those, are active for molecules; and furthermore that all partitions smaller than a given shaded one, and only those, are active for corresponding chiral ensemble operators. Active partitions are said to be represented by chiral Young diagrams (or chiral irreducible representations of $\mathscr{S}_n$).

Active partitions are now defined a little more precisely. Given a representation Γ_r of $\mathscr{S}_n$ with $z_r \neq 0$, partition $[\delta]$ is said to be active with respect to Γ_r if there is some molecule belonging to $[\delta]$ for which at least one component $\chi_j^{(\Gamma_r)}$ does not vanish. A partition is simply called active if it is Γ_r active for any Γ_r with $z_r \neq 0$. The question posed at the start of this section can now be formulated precisely: given $[\delta]$ and Γ_r, how do we determine whether $[\delta]$ is Γ_r active? It can be shown (see, for example, Mead, 1974) that a necessary condition for Γ_r activity is that

$$\gamma^{(\Gamma_r)} \supset \gamma^{[\delta]}; \tag{4.5.29}$$

in other words, the ligand partition $[\delta]$ must have a Young diagram smaller than the shaded one corresponding to the Γ_r of interest. Thus looking again at allene, from Fig. 4.8 we see that (4.5.29) leads to the same result as was deduced at the start of this section.

The set of active partitions (or chiral Young diagrams) for a given skeleton generate a set of numbers which characterize the chiral properties of the skeleton. It follows from the results of this section that, in relation to a given diagram, there exists no smaller diagram the first line of which is longer or the first row of which is shorter. This means that within the set of all shaded diagrams we can specify four numbers which characterize the chirality properties of a given class of skeleton: these are the longest and the shortest first line and column of all shaded diagrams. The two most important are the *chirality order o*, defined as the longest first line of all shaded diagrams; and the *chirality index u*, defined as the shortest first column of all shaded diagrams. It follows that the chirality order defines the maximum number of equal ligands, and the chirality index defines the minimum number of different ligands, which can be present in a chiral molecule belonging to a particular class of skeleton.

In Section 4.5.1 a chiral class of skeleton was introduced as one which permits of chiral molecules with exclusively achiral ligands. For such classes, Ruch and Schönhofer (1970) proved that $n - 3 \leqslant o \leqslant n$ and $1 \leqslant u \leqslant 4$. Five cases can be distinguished, each requiring a distinct type of theory to describe the generation of optical activity:

$o = n$. This defines skeletons that can support chiral molecules with ligands all of the same type. The skeleton must therefore be intrinsically chiral, so any theory of optical activity must be concerned with skeletal chirality.

$o = (n - 1)$. This defines skeletons that can support chiral molecules if just one ligand is different, all the others being the same. Thus optical activity is generated by perturbations from a single ligand, which gives rise to sector rules of the quadrant and octant type, as in adamantanone derivatives.

$o = (n - 2)$. For this type of skeleton, two ligands must be different, and optical activity is generated by simultaneous perturbations from two ligands, as in allene derivatives.

$o = (n - 3)$. This type of skeleton requires three different ligands to support a chiral molecule, and three ligand interactions are required to generate optical activity, as in methane derivatives.

$o = 0$. This class of skeleton is achiral by definition since it cannot support a chiral molecule even if all the ligands are different. An example is the benzene skeleton.

4.5.6 Homochirality

An important property required of satisfactory chirality functions is that, for shoe-like skeletons, they accommodate the concepts of homochirality and heterochirality introduced in Section 4.5.1. Chiral relatedness, that is chiral similarity of molecules belonging to a particular class of skeleton, must be based on similarity of ligands. Since we are specifying an achiral ligand by means of a single scalar parameter λ, the molecule is specified by the value of λ at each of the n sites in the skeleton class; that is, a particular molecule corresponds to a point in an n-dimensional

λ space. By continuously varying the λs we can transform any molecule of the class continuously into any other without leaving the class. Shoe-like molecules must therefore be described by continuous pseudoscalar functions which have the same sign for homochiral pairs, have opposite signs for heterochiral pairs, and have only *achiral zeros*, that is they vanish only for achiral molecules. Potato-like molecules are described by less well defined chirality functions: one characteristic which distinguishes them from chirality functions for shoe-like molecules is that they possess *chiral zeros*, that is they vanish for some chiral molecules.

An acceptable division of chiral molecules into right and left therefore means a division of the λ space into two regions, say R and L, such that (i) every chiral molecule is in either R or L and not on the boundary between them; (ii) if a given chiral molecule is in R, its mirror image is in L, and *vice versa*; (iii) achiral molecules are in neither R nor L, but on the boundary between them. Thus the boundary between the regions R and L must be the subspace of the achiral molecules; and since the boundary between two regions of an n-dimensional space must have $(n-1)$-dimensions, the subset of achiral molecules corresponds to a set of $(n-1)$-dimensional hypersurfaces.

An achiral molecule is left invariant by an improper rotation of the point group of the skeleton. It was mentioned in Section 4.5.2 that every permutation can be written as a product of cycles which operate on mutually exclusive labels, as in (4.5.6). So by writing the permutation P corresponding to a particular improper rotation in cyclic form,

$$P = (1, 2, \ldots, s)(s+1, s+2, \ldots, s+t)(s+t, \ldots)\ldots, \qquad (4.5.30)$$

we see that a molecule will be left invariant by P only if sites in the same cycle are occupied by identical ligands, that is if

$$\lambda_1 = \lambda_2 = \ldots \lambda_s,$$
$$\lambda_{s+1} = \lambda_{s+2} = \ldots \lambda_{s+1}, \text{ etc.} \qquad (4.5.31)$$

If P consists of h cycles, the subspace in which (4.5.31) is satisfied is h dimensional. The dimension h is equal to $(n-1)$ only if h consists of a single 2-cycle and $(n-2)$ 1-cycles; that is, P must be a single transposition.

Let $\mathscr{H}$ denote the set of all pairs i, j of sites such that the transposition (ij) corresponds to an improper rotation. The set of $(n-1)$-dimensional hypersurfaces determined by

$$\lambda_j = \lambda_k \qquad (4.5.32)$$

for each pair (jk) contained in $\mathscr{H}$ are subspaces corresponding to achiral molecules. If the hypersurfaces determined by (4.5.32) contain *all* achiral molecules, the subset of the achiral molecules will indeed be a set of $(n-1)$-dimensional hypersurfaces. This will be true only if the subspaces determined by (4.5.31) are all subspaces of those determined by (4.5.32); that is, if the following is satisfied: every cycle of every permutation P corresponding to an improper rotation of the point group of the skeleton must contain at least two sites j,k such that (jk) is contained in $\mathscr{H}$; or equivalently, every achiral molecule of the class must have at least one symmetry operation which in permutation form corresponds to a transposition.

If this condition is satisfied we can choose the surfaces (4.5.32) as the boundary between R and L, so that criteria (i) and (iii) are satisfied. In fact (ii) is also satisfied for, if (ij) is contained in $\mathscr{H}$, then mirror-image molecules correspond to interchanged values of λ_j and λ_k. Thus shoe-like skeletons are those for which this condition holds and an acceptable classification into R and L is possible; whereas potato-like skeletons are those for which this condition does not hold so no direct classification into R and L is possible.

It is emphasized that for shoe-like skeletons the designation of R and L for the two regions of opposite chirality is arbitrary. Also by changing the definition of the ligand parameter λ a molecule originally assigned to R, say, might find itself in L.

Allene (Fig. 4.6) provides a convenient example of a shoe-like skeleton. The surfaces determined by the improper rotations (12) and (34) are $\lambda_1 = \lambda_2$ and $\lambda_3 = \lambda_4$. The improper rotations (1324) and (1423) both determine the one-dimensional space $\lambda_1 = \lambda_2 = \lambda_3 = \lambda_4$, which is a subspace of the above.

A simple example of a potato-like skeleton is the four-site skeleton of symmetry C_{4v} shown in Fig. 4.9. The improper rotations (24) and (13) determine the $(n-1) =$ three-dimensional hypersurfaces $\lambda_1 = \lambda_3$ and $\lambda_2 = \lambda_4$. On the other hand, the improper rotations (12) (34) and (14) (23) determine the two-dimensional hypersurfaces $\lambda_1 = \lambda_2, \lambda_4 = \lambda_3$, and $\lambda_1 = \lambda_4, \lambda_2 = \lambda_3$, which are not subspaces of the above.

The condition developed above for a precise homochirality and heterochirality classification for shoe-like skeletons can be expressed in a form more readily applicable to a given skeleton. A skeleton is shoe-like if, and only if, either the skeleton has only two sites for ligands, or the number of sites, n, is larger, but the symmetry of the skeleton contains mirror planes and each mirror plane contains $(n-2)$

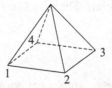

Fig. 4.9 A 'potato-like' skeleton with C_{4v} symmetry.

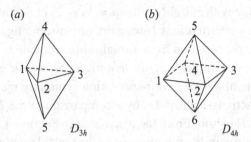

Fig. 4.10 (a) A 'shoe-like' skeleton and (b) a 'potato-like' skeleton.

sites. All other skeletons are potato-like. In addition to the allene and four-site C_{4v} skeletons discussed above, another instructive pair of examples are the shoe-like trigonal bipyramid and the potato-like tetragonal bipyramid in Fig. 4.10.

The tetragonal bipyramid can be used to illustrate the lack of a homochirality concept for a potato-like skeleton. Referring to Fig. 4.10b, we can envisage varying the ligands at positions 1 and 2 continuously so as to finish up with the original positions interchanged. Since an achiral situation is not encountered at any time, we may assign any pair of neighbouring molecules encountered on the path to the same enantiomeric subclass, and so all molecules generated on this path belong to the same subclass. Subsequently, we perform the same variation with the ligands at positions 3 and 4, and the same argument applies. But the end result, with ligands 1 and 2 interchanged and ligands 3 and 4 interchanged, is the enantiomer of the original molecule. Thus chiral relatedness between neighbouring molecules must be interrupted somewhere on the path from the original molecule to its enantiomer; but since no privileged point can be found we conclude that a homochirality concept does not exist.

4.5.7 Chirality functions: concluding remarks

We have presented Ruch's ideas on permutation symmetry and chirality in some detail because it seems to be of fundamental significance in the theory of optical activity, even though at present its applications in stereochemistry have been limited: indeed we make little use of it in subsequent chapters. One reason for its limited applicability in its present form is the restriction to effectively spherical ligands characterized by a single *scalar* parameter, although Mead (1974) has extended the theory to include chiral ligands (still spherical); whereas in the molecular theories of optical activity developed in later chapters it is found that the introduction of *vector* and *tensor* properties of anisotropic ligands usually leads to more tractable expressions. However, the above results based on scalar ligand properties can sometimes be applied to anisotropic ligands: for example, the generation of optical rotation in the case of a molecule with chirality order $o = (n - 2)$ in Section 4.5.5 was said to require simultaneous contributions from pairs of isotropic ligands; but this could be reinterpreted as a contribution from a single anisotropic ligand because a pair of interacting atoms, for example, is equivalent to an anisotropic group.

A simple example of the value of permutation symmetry methods in criticizing theories of optical activity is provided by a comparison of the D_{2d} allene skeleton and the T_d methane skeleton. From the discussion in Section 4.5.3 we know that a qualitatively complete chirality function giving the optical rotation, say, for an allene derivative will contain two independent contributions $\chi^{(\Gamma_3)}$ and $\chi^{(\Gamma_5)}$, perhaps of the form (4.5.2a and b), which transform according to Γ_3 and Γ_5 of $\mathcal{S}_4$; whereas that for a methane derivative contains just one term $\chi^{(\Gamma_5)}$, perhaps of the form (4.5.2b), which transforms according to Γ_5. The relative weights of $\chi^{(\Gamma_1)}$ and $\chi^{(\Gamma_5)}$, which can be determined by measurements of the optical rotation of various nonracemic mixtures of chiral allene isomers, tells us to what extent we are justified in taking a chiral allene derivative as having the approximate symmetry of a regular tetrahedron. Also, if a chiral methane derivative has a skeleton distorted from T_d symmetry we must include a second contribution of the form $\chi^{(\Gamma_3)}$.

King (1991) has reviewed experimental tests, based on optical rotation measurements, of chirality functions determined by chirality algebra. It appears that their success depends greatly on the complexity of the skeleton, particularly with respect to the numbers of sites and chiral ligand partitions. Chirality functions provide fair to good approximations of optical rotation data for chiral derivatives of shoe-like skeletons such as those of methane and allene. However, the approximations deteriorate rapidly for chiral derivatives of more complicated shoe-like skeletons having several chiral ligand partitions or of potato-like skeletons.

The formal development of chirality functions is based on just one pseudoscalar observable, the optical rotatory parameter at transparent wavelengths. It is only in terms of this observable that the arguments about qualitative completeness in Section 4.5.1 appear to be valid, because each component of a nonracemic mixture of isomers contributes *coherently* to the net observed optical rotation. It therefore appears that the formalism is not immediately applicable to pseudoscalar observables such as Rayleigh and Raman optical activity which are generated by *incoherent* scattering processes. This can be seen straight away in the Raman case because nonenantiomeric isomers generate Raman lines at different sets of frequencies. In the Rayleigh case the scattered frequencies are the same from all the isomers, but there is the complication that the isotropic and anisotropic scattering contributions depend differently on sample density, and the relative amounts of isotropic and anisotropic scattering will be different for the different isomers. However, it may be that in the limit of an ideal transparent gas the concept of qualitative completeness is applicable to the Rayleigh case because the total observed circular intensity difference $I^R - I^L$ will then be a simple sum of separate contributions from each molecule in the sample. On the other hand, it may be that it is only the *interpretation* of qualitative completeness in terms of the nonvanishing of the chirality function for any nonracemic mixture of isomers that is suspect in the case of incoherent processes and that the group theoretical analysis of qualitative completeness in Section 4.5.3 remains generally valid. In any event, the entire chirality function formalism will need to be carefully re-evaluated, and perhaps reformulated, before any attempt is made to apply it to Rayleigh and Raman optical activity.

5

Natural electronic optical activity

A theory has only the alternative of being right or wrong. A model has a
third possibility: it may be right, but irrelevant.

(Manfred Eigen)

5.1 Introduction

This chapter is concerned with optical rotation and circular dichroism of visible
and near ultraviolet light in the absence of an external influence such as a static
magnetic field; in other words, natural optical activity in the electronic spectrum.
Natural optical activity is generated by appropriate components of the molecular
property tensors $G'_{\alpha\beta}$ and $A_{\alpha,\beta\gamma}$ which involve interference between an electric
dipole transition moment and either a magnetic dipole or an electric quadrupole
transition moment, respectively. Optical activity for light propagation along an arbitrary direction in a general anisotropic medium is complicated and is not considered
here. We discuss only the most important situations in molecular optics; namely
complete isotropy, as in a liquid or solution, and isotropy in the plane perpendicular
to the direction of propagation. In the language of crystal optics, the latter situation is specified as light propagation along the optic axis of a uniaxial medium: it
also corresponds to light propagation in the direction of a static field applied to an
isotropic medium. As discussed in Section 4.4.4, in these situations the appropriate
components of $G'_{\alpha\beta}$ and $A_{\alpha,\beta\gamma}$ are supported only by chiral molecules.

5.2 General aspects of natural optical rotation and circular dichroism

5.2.1 The basic equations

In Chapter 3, expressions for natural optical rotation and circular dichroism were derived using the refringent scattering approach. Thus from Section 3.4.6 we can write
the following expressions for the optical rotation and circular dichroism generated

in a light beam propagating along the z direction in an oriented sample:

$$\Delta\theta \approx -\tfrac{1}{2}\omega\mu_0 l N\left[\tfrac{1}{3}\omega(A_{x,yz}(f) - A_{y,xz}(f)) + G'_{xx}(f) + G'_{yy}(f)\right], \quad (5.2.1a)$$

$$\eta \approx -\tfrac{1}{2}\omega\mu_0 l N\left[\tfrac{1}{3}\omega(A_{x,yz}(g) - A_{y,xz}(g)) + G'_{xx}(g) + G'_{yy}(g)\right]. \quad (5.2.1b)$$

Optical rotation depends on the dispersion lineshape function f, and circular dichroism on the absorption lineshape function g. Equation (5.2.1b) applies to small ellipticities developed in a light beam that is initially linearly polarized; expressions for more general observations are given in Section 3.4.6.

In an isotropic sample, the average over all orientations yields

$$\Delta\theta \approx -\tfrac{1}{3}\omega\mu_0 l N G'_{\alpha\alpha}(f), \quad (5.2.2a)$$

$$\eta \approx -\tfrac{1}{3}\omega\mu_0 l N G'_{\alpha\alpha}(g), \quad (5.2.2b)$$

the first being the Rosenfeld equation. The electric dipole–electric quadrupole contributions average to zero.

According to (2.6.35), general components of $G'_{\alpha\beta}$ and $A_{\alpha,\beta\gamma}$ are origin dependent. However, it is easily shown that the combinations of components specified in (5.2.1) are independent of the choice of molecular origin, as befits an expression for an observable quantity (Buckingham and Dunn, 1971). It is emphasized that the separate electric dipole–magnetic dipole and electric dipole–electric quadrupole contributions are origin dependent in an oriented sample: the change in one contribution on moving the origin is cancelled by the change in the other. Consequently, the analysis of optical rotation or circular dichroism data on oriented systems can be quite wrong if only the electric dipole–magnetic dipole contribution is considered.

5.2.2 Optical rotation and circular dichroism through circular differential refraction

Although the refringent scattering approach provides the most fundamental and complete description of optical rotation and circular dichroism, it is less familiar than the description in terms of circular differential refraction. For comparison, the basic equations are now derived using the more conventional approach. The derivation is based on that given by Buckingham and Dunn (1971).

It is shown in Section 1.2 that the optical rotation and circular dichroism can be formulated in terms of the refractive indices n^R and n^L and absorption indices n'^R and n'^L for right- and left-circularly polarized light:

$$\Delta\theta = \frac{\omega l}{2c}(n^L - n^R), \quad (5.2.3a)$$

$$\eta = \frac{\omega l}{2c}(n'^L - n'^R). \quad (5.2.3b)$$

The refractive index and absorption index are best introduced through the exponent of the complex electric vector (2.2.11) of the plane wave light beam in the medium:

$$\tilde{E}_\alpha = \tilde{E}_\alpha^{(0)} \exp\left[i\left(\frac{\omega}{c}\tilde{n}_\beta r_\beta - \omega t\right)\right],$$

(5.2.4)

where $\tilde{\mathbf{n}}$ is a complex propagation vector,

$$\tilde{\mathbf{n}} = \mathbf{n} + i\mathbf{n}'.$$

(5.2.5)

$n = |\mathbf{n}|$ is the refractive index, and $n' = |\mathbf{n}'|$ is the absorption index. Clearly the presence of n' leads to an attenuation of the wave. If the medium is nonconducting, the Maxwell equations (2.2.3c and d) for this plane wave become

$$\frac{1}{c}\tilde{n}_\beta \varepsilon_{\alpha\beta\gamma} \tilde{E}_\gamma = \tilde{B}_\alpha,$$

(5.2.6a)

$$\frac{1}{c}\tilde{n}_\beta \varepsilon_{\alpha\beta\gamma} \tilde{H}_\gamma = -\tilde{D}_\alpha.$$

(5.2.6b)

In the theory of crystal optics, the material connections (2.2.2) are generalized to

$$\tilde{D}_\alpha = \tilde{\epsilon}_{\alpha\beta} \epsilon_0 \tilde{E}_\beta,$$

(5.2.7a)

$$\tilde{B}_\alpha = \tilde{\mu}_{\alpha\beta} \mu_0 \tilde{H}_\beta,$$

(5.2.7b)

where the dielectric constant and magnetic permeability are now complex tensors. $\tilde{\mathbf{D}}$ and $\tilde{\mathbf{H}}$ can also be written in terms of a *bulk polarization* $\tilde{\mathbf{P}}$, a *quadrupole polarization* $\tilde{\mathbf{Q}}$ and a *magnetization* $\tilde{\mathbf{M}}$ developed in the medium (Rosenfeld, 1951):

(5.2.8a)

$$\tilde{D}_\alpha = \epsilon_0 \tilde{E}_\alpha + \tilde{P}_\alpha - \tfrac{1}{3}\nabla_\beta \tilde{Q}_{\alpha\beta},$$

(5.2.8b)

$$\tilde{H}_\alpha = \frac{1}{\mu_0}\tilde{B}_\alpha - \tilde{M}_\alpha.$$

(5.2.8c)

This definition of $\tilde{\mathbf{D}}$ differs from that generally used in the macroscopic Maxwell theory by the addition of the quadrupole term; there are further contributions to $\tilde{\mathbf{D}}$ and $\tilde{\mathbf{H}}$ arising from higher multipole polarizations which we do not consider. The bulk multipole polarizations can be related to the multipole moments of the constituent molecules through

$$\tilde{P}_\alpha = N\bar{\tilde{\mu}}_\alpha,$$

(5.2.9a)

$$\tilde{Q}_{\alpha\beta} = N\bar{\tilde{\Theta}}_{\alpha\beta},$$

(5.2.9b)

$$\tilde{M}_\alpha = N\bar{\tilde{m}}_\alpha,$$

(5.2.9c)

where N is the number density of the molecules and the bar denotes a statistical average appropriate to the particular medium.

If we write

$$\tilde{H}_\gamma = \frac{1}{\mu_0}\tilde{B}_\gamma - \frac{1}{\mu_0}\tilde{B}_\gamma + \tilde{H}_\gamma,$$

the Maxwell equations (5.2.6) can be combined into one equation,

$$\tilde{n}_\alpha \tilde{n}_\beta \tilde{E}_\beta - \tilde{n}^2 \tilde{E}_\alpha + \mu_0 c^2 \tilde{D}'_\alpha = 0, \tag{5.2.10}$$

where

$$\tilde{D}'_\alpha = \tilde{D}_\alpha - \frac{1}{c}\tilde{n}_\beta \varepsilon_{\alpha\beta\gamma}\left(\frac{1}{\mu_0}\tilde{B}_\gamma - \tilde{H}_\gamma\right). \tag{5.2.11}$$

If the magnetic permeability is isotropic and unity so that $\tilde{\mathbf{B}} = \mu_0 \tilde{\mathbf{H}}$, $\tilde{\mathbf{D}}'$ reduces to $\tilde{\mathbf{D}}$ and (5.2.10) becomes the fundamental equation used in the theory of light propagation in dielectric crystals. Now introduce (5.2.8) and (5.2.9) into (5.2.11):

$$\begin{aligned}\tilde{D}'_\alpha &= \epsilon_0 \tilde{E}_\alpha + \tilde{P}_\alpha - \tfrac{1}{3}\nabla_\beta \tilde{Q}_{\alpha\beta} - \frac{1}{c}\tilde{n}_\beta \varepsilon_{\alpha\beta_\gamma}\tilde{M}_\gamma \\ &= \epsilon_0 \tilde{E}_\alpha + N\left(\tilde{\mu}_\alpha - \tfrac{1}{3}\nabla_\beta \tilde{\Theta}_{\alpha\beta} - \frac{1}{c}\tilde{n}_\beta \varepsilon_{\alpha\beta\gamma}\tilde{m}_\gamma + \cdots\right),\end{aligned} \tag{5.2.12}$$

where, for simplicity, we have omitted the bars denoting statistical averages. If the molecular multipole moments are induced by the external light wave, (2.6.43) can be used to introduce the molecular polarizability tensors:

$$\begin{aligned}\tilde{D}'_\alpha = \epsilon_0 \tilde{E}_\alpha + N\Big[&\tilde{\alpha}_{\alpha\beta} + \frac{i\omega}{3c}\tilde{n}_\gamma(\tilde{A}_{\alpha,\beta\gamma} - \mathscr{A}_{\beta,\alpha\gamma}) \\ &+ \frac{\tilde{n}_\gamma}{c}(\varepsilon_{\delta\gamma\beta}\tilde{G}_{\alpha\delta} + \varepsilon_{\delta\gamma\alpha}\mathscr{G}_{\delta\beta}) + \cdots\Big]\tilde{E}_\beta,\end{aligned} \tag{5.2.13}$$

where we have assumed that the 'effective fields' at the molecule are those of the light wave in free space. Now introduce into (5.2.13) the tensor $\tilde{\zeta}_{\alpha\beta\gamma}$ defined in (3.4.11):

$$\tilde{D}'_\alpha = \epsilon_0 \tilde{E}_\alpha + N(\tilde{\alpha}_{\alpha\beta} + \tilde{n}_\gamma \tilde{\zeta}_{\alpha\beta\gamma} + \cdots)\tilde{E}_\beta. \tag{5.2.14}$$

In the case of complete isotropy, and isotropy in the plane perpendicular to the direction of propagation, $\tilde{\mathbf{n}} \cdot \tilde{\mathbf{E}} = 0$ and the fundamental equation (5.2.10) becomes

$$[(\tilde{n}^2 - 1)\delta_{\alpha\beta} - \mu_0 c^2 N(\tilde{\alpha}_{\alpha\beta} + \tilde{n}_\gamma \tilde{\zeta}_{\alpha\beta\gamma} + \cdots)]\tilde{E}_\beta = 0. \tag{5.2.15}$$

From (2.3.2), the complex electric vector of a right- or left-circularly polarized light beam is

$$\tilde{E}_{\substack{R\\L}\alpha} = \frac{1}{\sqrt{2}}E^{(0)}(i_\alpha \mp ij_\alpha)\exp\left[i\left(\frac{\omega}{c}\tilde{n}_\beta^{\substack{R\\L}}r_\beta - \omega t\right)\right]$$

so for circularly polarized light (5.2.15) provides two equations:

$$(\overset{R}{\tilde{n}}{}^{2}_{L} - 1) - \mu_0 c^2 N[\tilde{\alpha}_{xx} + \tilde{\zeta}_{xxz} \mp i(\tilde{\alpha}_{xy} + \tilde{\zeta}_{xyz}) + \cdots] = 0, \quad (5.2.16a)$$

$$(\overset{R}{\tilde{n}}{}^{2}_{L} - 1) - \mu_0 c^2 N[\tilde{\alpha}_{yy} + \tilde{\zeta}_{yyz} \pm i(\tilde{\alpha}_{yx} + \tilde{\zeta}_{yxz}) + \cdots] = 0, \quad (5.2.16b)$$

where we have taken z to be the direction of propagation. These can be combined into

$$(\overset{R}{\tilde{n}}{}^{2}_{L} - 1) - \tfrac{1}{2}\mu_0 c^2 N[\tilde{\alpha}_{xx} + \tilde{\alpha}_{yy} + \tilde{\zeta}_{xxz} + \tilde{\zeta}_{yyz}$$
$$\mp i(\tilde{\alpha}_{xy} - \tilde{\alpha}_{yx}) \mp i(\tilde{\zeta}_{xyz} - \tilde{\zeta}_{yxz}) + \cdots] = 0. \quad (5.2.16c)$$

If the molecular medium is dilute, the second term is very small so we can write

$$\overset{R}{\tilde{n}}_{L} \approx 1 + \tfrac{1}{4}\mu_0 c^2 N[\tilde{\alpha}_{xx} + \tilde{\alpha}_{yy} + \tilde{\zeta}_{xxz} + \tilde{\zeta}_{yyz}$$
$$\mp i(\tilde{\alpha}_{xy} - \tilde{\alpha}_{yx}) \mp i(\tilde{\zeta}_{xyz} - \tilde{\zeta}_{yxz}) + \cdots]. \quad (5.2.17)$$

The real and imaginary parts of these complex circular refractive indices are

$$\overset{R}{n}_{L} \approx 1 + \tfrac{1}{4}\mu_0 c^2 N[\alpha_{xx}(f) + \alpha_{yy}(f) + \zeta_{xxz}(f) + \zeta_{yyz}(f)$$
$$\mp 2(\alpha'_{xy}(f) + \zeta'_{xyz}(f)) + \cdots], \quad (5.2.18a)$$

$$\overset{R}{n}_{L} \approx \tfrac{1}{4}\mu_0 c^2 N[\alpha_{xx}(g) + \alpha_{yy}(g) + \zeta_{xxz}(g) + \zeta_{yyz}(g)$$
$$\mp 2(\alpha'_{xy}(g) + \zeta'_{xyz}(g)) + \cdots], \quad (5.2.18b)$$

where we have introduced the dispersion and absorption lineshape functions f and g.

Using (5.2.18) in (5.2.3), the optical rotation and circular dichroism are found to be

$$\Delta\theta \approx -\tfrac{1}{2}\omega\mu_0 l N[-c\alpha'_{xy}(f) + \tfrac{1}{3}\omega(A_{x,yz}(f) - A_{y,xz}(f))$$
$$+ G'_{xx}(f) + G'_{yy}(f)], \quad (5.2.19a)$$

$$\eta \approx -\tfrac{1}{2}\omega\mu_0 l N[-c\alpha'_{xy}(g) + \tfrac{1}{3}\omega(A_{x,yz}(g) - A_{y,xz}(g))$$
$$+ G'_{xx}(g) + G'_{yy}(g)]. \quad (5.2.19b)$$

The antisymmetric polarizability $\alpha'_{\alpha\beta}$ is responsible for magnetic optical activity, and is disregarded in the rest of this chapter.

It is instructive to expose the connection between the modified electric displacement vector (5.2.14) and the equivalent expression derived in the theory of crystal optics to account for optical activity. If the medium is transparent and nonmagnetic,

(5.2.14) becomes

$$D'_\alpha = \epsilon_0 E_\alpha + N[\alpha_{\alpha\beta}(f) - \text{i}n_\gamma \zeta'_{\alpha\beta\gamma}(f) + \cdots]E_\beta.$$

If we identify the dielectric tensor with

$$\epsilon_{\alpha\beta} = \epsilon_0 \delta_{\alpha\beta} + N\alpha_{\alpha\beta}(f), \tag{5.2.20}$$

and introduce the *gyration vector* **g** through

$$Nn_\gamma \zeta'_{\alpha\beta\gamma}(f) = \varepsilon_{\alpha\beta\gamma} g_\gamma, \tag{5.2.21}$$

the relation between **D'** and **E** takes the form

$$D'_\alpha \approx \epsilon_{\alpha\beta} E_\beta - \text{i}\varepsilon_{\alpha\beta\gamma} E_\beta g_\gamma. \tag{5.2.22}$$

This parallels equation (82.7a) of Landau and Lifshitz (1960), and equation (IX.2) of Born and Huang (1954). A symmetric *gyration tensor* defined by

$$g_\alpha = g_{\alpha\beta} n_\beta \tag{5.2.23}$$

is also used in crystal optics, particularly in the discussion of the crystal symmetry requirements for natural optical activity (Nye, 1985). It is clear that

$$N\zeta'_{\alpha\beta\gamma}(f) = \varepsilon_{\alpha\beta\delta} g_{\delta\gamma}. \tag{5.2.24}$$

By using (4.2.42) and (4.2.38), we can rewrite this in the form

$$g_{\alpha\beta} = \tfrac{1}{2}N\varepsilon_{\alpha\gamma\delta}\zeta'_{\gamma\delta\beta}(f). \tag{5.2.25}$$

The tensor $g_{\alpha\beta}$ is in general not symmetric, but since only the symmetric part contributes to optical rotation the crystal symmetry requirements for optical rotation can be discussed in terms of just the symmetric part (Landau and Lifshitz, 1960).

Further discussion and developments of the circular differential refraction approach to natural optical rotation and circular dichroism in chiral media may be found in Raab and Cloete (1994), Theron and Cloete (1996) and Kaminsky (2000).

5.2.3 Experimental quantities

The optical rotation (5.2.2a) is in radians, with the path length l in metres since SI units are used. For applications, it should be translated into experimental units. Experimental results are traditionally reported as specific rotations, defined by (see (1.2.9))

$$[\Delta\theta] = \frac{\text{optical rotation in degrees per decimetre}}{\text{density of optically active material in grams per cubic centimetre}}.$$

For an isotropic sample, the optical rotation in degrees per decimetre is given by (5.2.2a) multiplied by $18/\pi l$. Since N is the number of molecules per cubic metre (the unit volume in SI), the number of grams of optically active material per cubic centimetre is $NM/10^6 N_0$, where M is the molecular weight and N_0 is Avogadro's number (6.023×10^{23}). The specific rotation is therefore

$$[\Delta\theta] \approx -6 \times 10^6 \left(\frac{\omega\mu_0 N_0}{\pi M}\right)\left(\frac{n^2+2}{3}\right)G'_{\alpha\alpha}, \qquad (5.2.26)$$

where we have included the Lorentz factor $(n^2+2)/3$ to take approximate account of the influence of the refractive index n of the medium.

The dissymmetry factor, giving the ratio of the circular dichroism to the absorption, is given in terms of molecular property tensors by (3.4.50). Averaging this over all orientations, and using (2.6.42), the dissymmetry factor for an isotropic sample can be written as follows in terms of molecular transition moments:

$$g(j \leftarrow n) = \frac{4R(j \leftarrow n)}{cD(j \leftarrow n)}, \qquad (5.2.27)$$

where

$$R(j \leftarrow n) = \text{Im}(\langle n|\boldsymbol{\mu}|j\rangle \cdot \langle j|\mathbf{m}|n\rangle), \qquad (5.2.28a)$$

$$D(j \leftarrow n) = \text{Re}(\langle n|\boldsymbol{\mu}|j\rangle \cdot \langle j|\boldsymbol{\mu}|n\rangle), \qquad (5.2.28b)$$

are the *rotational strength* and *dipole strength* of the $j \leftarrow n$ transition. The dissymmetry factor (5.2.27) differs from that encountered in earlier literature by a factor of $1/c$ because we are working in SI.

For light propagating along z in an oriented sample, the same dissymmetry factor (5.2.27) can be used if the rotational strength and dipole strength are now generalized to

$$R_z(j \leftarrow n) = -\left\{\tfrac{1}{3}\omega_{jn}[\text{Re}(\langle n|\mu_x|j\rangle\langle j|\Theta_{yz}|n\rangle) - \text{Re}(\langle n|\mu_y|j\rangle\langle j|\Theta_{xz}|n\rangle)]\right.$$
$$\left. - \text{Im}(\langle n|\mu_x|j\rangle\langle j|m_x|n\rangle) - \text{Im}(\langle n|\mu_y|j\rangle\langle j|m_y|n\rangle)\right\}, \quad (5.2.29a)$$

$$D_z(j \leftarrow n) = \text{Re}(\langle n|\mu_x|j\rangle\langle j|\mu_x|n\rangle) + \text{Re}(\langle n|\mu_y|j\rangle\langle j|\mu_y|n\rangle). \qquad (5.2.29b)$$

Using (2.4.3), (2.4.9) and (2.4.14) for the origin dependence of the electric dipole, electric quadrupole and magnetic dipole moment operators, together with the velocity–dipole transformation (2.6.31b), it is easy to show that the generalized rotational strength (5.2.29a) is independent of the choice of the molecular origin.

Since polar and axial vectors only transform the same under proper rotations, the same components of the polar electric dipole vector $\boldsymbol{\mu}$ and the axial magnetic dipole vector $\mathbf{m}$ only span the same irreducible representations and hence are able to connect the same states $|n\rangle$ and $|j\rangle$ in systems lacking a centre of inversion, reflection planes and rotation–reflection axes. Hence the isotropic rotational strength (5.2.28a) is only nonzero for molecules belonging to point groups containing no more than proper rotation axes, namely C_n, D_n, O, T and I. As previously noted in Sections 1.9.1 and 4.4.4, these are the chiral point groups. This type of argument is not as straightforward for the electric dipole–electric quadrupole terms in the generalized rotational strength (5.2.29a), but from considerations similar to those used in Section 4.4.4 the same conclusion obtains.

5.2.4 Sum rules

An important sum rule, first propounded by Condon (1937), exists for the isotropic rotational strength (5.2.28a). Summing over all states j except the initial state n,

$$\sum_{j \neq n} R(j \leftarrow n) = \sum_{j \neq n} \mathrm{Im}(\langle n|\mu_\alpha|j\rangle\langle j|m_\alpha|n\rangle)$$

$$= \mathrm{Im}\langle n|\mu_\alpha m_\alpha|n\rangle - \mathrm{Im}(\langle n|\mu_\alpha|n\rangle\langle n|m_\alpha|n\rangle) = 0, \quad (5.2.30)$$

where we have used the fact that, according to (2.6.67), the same components of the Hermitian electric and magnetic dipole moment operators commute so that $\mu_\alpha m_\alpha$ is also pure Hermitian and so possesses only pure real expectation values (along with μ_α and m_α separately). It is emphasized that the sum is over all the molecular states, not just the electronic states, and so includes vibrational and rotational components.

It can also be shown that a similar sum rule exists for the rotational strength (5.2.29a) of an oriented sample, that is,

$$\sum_{j \neq n} R_z(j \leftarrow n) = 0. \quad (5.2.31)$$

The electric dipole–magnetic dipole terms sum to zero for the reasons given above, and the electric dipole–electric quadrupole terms can be shown to sum to zero by using

$$im\omega_{jn}\langle j|r_\alpha r_\beta|n\rangle = \langle j|r_\alpha p_\beta + r_\beta p_\alpha - i\hbar\delta_{\alpha\beta}|n\rangle, \quad (5.2.32)$$

which follows from the commutation relation (2.5.22).

We can now see that optical rotation in both isotropic and oriented samples tends to zero at very low and very high frequency. This low frequency behaviour follows directly from the forms of (5.2.1) and (5.2.2). The high frequency behaviour follows from the sum rules (5.2.30) and (5.2.31). For example,

$$G'_{\alpha\alpha} = \frac{2}{\hbar\omega} \sum_{j \neq n} \mathrm{Im}(\langle n|\mu_\alpha|j\rangle\langle j|m_\alpha|n\rangle) = 0, \quad (\omega > \omega_{\max}). \quad (5.2.33)$$

Notice that another version of the Condon sum rule (5.2.30) follows from the Kramers–Kronig relations outlined in Section 2.6.4. Thus

$$\int_0^\infty G'_{\alpha\alpha}(g_\xi)\mathrm{d}\xi = 0. \tag{5.2.34}$$

Also, since optical rotation and circular dichroism are determined by the dispersive and absorptive parts, respectively, of the optical activity tensors, with all other factors the same, we can write Kramers–Kronig relations directly for the optical rotation and circular dichroism. Thus from (2.6.61),

$$\Delta\theta(f_\omega) = \frac{2\omega^2}{\pi}\mathscr{P}\int_0^\infty \frac{\Delta\eta(g_\xi)\mathrm{d}\xi}{\xi(\xi^2 - \omega^2)}, \tag{5.2.35a}$$

$$\Delta\eta(g_\omega) = -\frac{2\omega^3}{\pi}\mathscr{P}\int_0^\infty \frac{\Delta\theta(f_\xi)\mathrm{d}\xi}{\xi^2(\xi^2 - \omega^2)}. \tag{5.2.35b}$$

So a knowledge of the complete optical rotation spectrum of a molecule gives straight away the circular dichroism spectrum, and *vice versa*. The application of Kramers–Kronig relations to optical rotation and circular dichroism has been developed in detail by Moscowitz (1962).

5.3 The generation of natural optical activity within molecules

The essential feature of any source of natural optical activity is the stimulation by the light wave of oscillating electric dipole, magnetic dipole and electric quadrupole moments within the molecule which mutually interfere. This is expressed quantum mechanically by the transition moment terms $\mathrm{Im}(\langle n|\mu_\alpha|j\rangle\langle j|m_\beta|n\rangle)$ and $\mathrm{Re}(\langle n|\mu_\alpha|j\rangle\langle j|\Theta_{\beta\gamma}|n\rangle)$ which appear in $G'_{\alpha\beta}$ and $A_{\alpha,\beta\gamma}$ and in the associated rotational strengths. Quantum chemical computations of natural optical activity observables requires a knowledge of the ground and excited state wavefunctions. Accurate determination of the wavefunctions for large chiral molecules is still a difficult problem, and we refer to Koslowski, Sreerama and Woody (2000) for an account of such calculations. One notable success, however, has been the ab initio computation, via (5.6.26), of specific rotations at transparent wavelengths for small chiral molecules based on evaluation of $G'_{\alpha\alpha}$ using the static approximation (2.6.75) for the electric dipole–magnetic dipole optical activity tensor. These calculations, pioneered by Polavarapu (1997), provide a simple and reliable means of assigning absolute configuration (Kondru, Wipf and Beratan, 1998; Stephens *et al.*, 2002; Polavarapu, 2002b). A historical note which highlights the unifying theme of molecular light scattering of this book is that the idea for performing ab initio computations of specific rotations via $G'_{\alpha\alpha}$ originated in the late 1980s when calculations of Raman optical activity, which require computations of general components of $G'_{\alpha\beta}$ and $A_{\alpha,\beta\gamma}$, were initiated (see Section 7.3.1).

We shall not elaborate further on quantum chemical computations, but instead will concentrate on *coupling models* which are in keeping with the semiclassical light scattering formalism of this book and which provide physical insight into how natural electronic optical activity is generated by chiral molecular structures. Coupling models apply when all groups within a molecule are inherently achiral and no electron exchange exists between them. Electrons are thus localized on symmetric groups, and any optical activity is assumed to arise from perturbations of the intrinsic group electronic states by the chiral intramolecular environment. Among other things, such models serve as a framework for point group symmetry arguments, thereby providing rules which relate the signs and magnitudes of rotatory dispersion and circular dichroism bands to stereochemical and structural features. The opposite case, known as the *inherently chiral chromophore model* (Caldwell and Eyring, 1971; Charney, 1979) applies when electronic states are significantly delocalized over a chiral nuclear framework and will not be considered here. Models sometimes produce useful quantitative results, an example being a coupling treatment of hexa-helicene which provides the correct sign for the specific rotation and a magnitude comparable in accuracy with current ab initio computations (Section 5.4.3).

Two types of coupling model can be distinguished. The *static coupling*, or *one electron*, theory of Condon, Altar and Eyring (1937) emphasizes perturbations due to the electrostatic fields of other groups. The *dynamic coupling*, or *coupled oscillator*, model, put forward independently by Born (1915) and Oseen (1915) and later developed by Kuhn (1930), Boys (1934) and Kirkwood (1937), emphasizes perturbations due to the electrodynamic fields radiated by other groups under the influence of the light wave. The general hypothesis of the dynamic coupling model was well expressed by Lowry (1935):

A molecule is regarded as a system of discrete units, which are fixed more or less rigidly relative to one another. Each of these units possesses the property of assuming an induced polarization under the action of an applied electric field. When a beam of plane polarized light is incident upon such a molecule, the components become polarized under the action of the electric vector of the light wave. Each of these polarized units then produces a field of force which in turn acts upon each of the other units. The resultant polarization of each unit is determined by the combined influence of the applied external field and of the fields created by all the other units of the molecule. The phenomenon by which the state of one of the units of a molecule is thus influenced by the state of other units of the same molecule is described as coupling.

The static and dynamic coupling models can make comparable contributions in the same molecule, and there can be higher order terms involving simultaneous static and dynamic perturbations. When two or more dynamically coupled groups are identical, an *exciton* or *degenerate coupled oscillator* treatment is required in which the electronic excitations are 'shared' between the groups.

These models are usually used to account for optical rotation and circular dichroism in isotropic collections of molecules, in other words to generate $G'_{\alpha\alpha}$ (the trace of the optical activity tensors) and the isotropic rotational strength (5.2.28a). We extend these models to other components of the optical activity tensors $G'_{\alpha\beta}$ and $A_{\alpha,\beta\gamma}$ in order to generate the rotational strength (5.2.29a) of oriented molecules, and to deal later with circular differential scattering.

Although coupling models can be applied to optical rotation at transparent frequencies and to Cotton effects at absorbing frequencies in any chiral molecule, they have been most successful in situations where Cotton effects are induced in electronic transitions of a single intrinsically achiral chromophore (such as the carbonyl group) by chiral intramolecular perturbations. In such a situation, the dominant static and dynamic chiral perturbing fields at the chromophore often originate in just one of the several other groups in the molecule, and so the problem can be reduced to considerations of a simple chiral two-group structure comprising a chromophore and a perturbing group. Such *two-group* models are emphasized in much of the rest of this chapter: these models can be generalized by summing over all groups in a molecule that constitute chiral pairs, although the selection of such pairs is often rather arbitrary.

In applying coupling models explicitly to a particular structure, it is often necessary to know the distribution of the components of the local group tensors $\alpha_{\alpha\beta}$, $G'_{\alpha\beta}$ and $A_{\alpha,\beta\gamma}$ among the irreducible representations of the point group of the unperturbed group. The general methods outlined in Section 4.4.4 can be used for this. But in the case of optical activity induced in a particular transition of a single chromophore, one can simply use a classification of the irreducible representations spanned by components of the electric dipole, magnetic dipole and electric quadrupole moment operators, which can be read directly from character tables.

In the case of the isotropic rotational strength (5.2.28a), for example, electronic transitions on a chromophore will always fall into one of the following categories:

1. Electric dipole allowed, magnetic dipole forbidden; or *vice versa*. Point groups C_i, C_{nh}, D_{nh}, D_{nd} ($n \neq 2$), S_{2n} (n odd), O_h, T_d, I_h.
2. Electric dipole and magnetic dipole allowed, but perpendicular. Point groups C_s, C_{nv}, D_{2d}, S_{2n} (n even).
3. Electric dipole and magnetic dipole allowed and parallel. Point groups C_n, D_n, O, T, I.

The third class contains, of course, the chiral point groups and corresponds to an inherently chiral chromophore. An example of the second class is the $\pi^* \leftarrow n$

transition of the carbonyl chromophore: in this case the static or dynamic chiral perturbation serves to induce an electric dipole transition moment parallel to the fully allowed magnetic one.

We make little use of the algebra of chirality functions in criticizing these models for the reason given in Section 4.5.7, namely that it is restricted in its present form to ligands characterized by a single scalar parameter, whereas most of the coupling theory results developed below specify anisotropic group properties.

The treatment given here has been greatly influenced by the following articles: Moscowitz (1962), Tinoco (1962), Schellman (1968), Höhn and Weigang (1968) and Buckingham and Stiles (1974). See also Rodger and Norden (1997) for a detailed account of the degenerate coupled oscillator model and its application to some typical chiral molecules.

5.3.1 The static coupling model

We consider first the optical activity generated by two groups 1 and 2 that together constitute a chiral structural unit. The two groups are intrinsically achiral so that $G'_{1\alpha\alpha}$ and $G'_{2\alpha\alpha}$ are zero, although each group might be able to support certain components of $G'_{\alpha\beta}$ and $A'_{\alpha,\beta\gamma}$. In the static coupling model, the optical activity is assumed to arise from perturbations of group optical activity tensors by static fields from other groups. The perturbed optical activity tensors of the group i (referred to a local origin on i) in the electrostatic field and field gradient from the group j are analogous to (2.7.1):

$$A_{i\alpha,\beta\gamma}(\mathbf{E}_i, \nabla \mathbf{E}_i) = A_{i\alpha,\beta\gamma} + A^{(\mu)}_{i\alpha,\beta\gamma,\delta} E_{i\delta} + \tfrac{1}{3} A^{(\Theta)}_{i\alpha,\beta\gamma,\delta\epsilon} E_{i\delta\epsilon} + \cdots, \quad (5.3.1a)$$

$$G'_{i\alpha\beta}(\mathbf{E}_i, \nabla \mathbf{E}_i) = G'_{i\alpha\beta} + G'^{(\mu)}_{i\alpha\beta,\gamma} E_{i\gamma} + \tfrac{1}{3} G'^{(\Theta)}_{i\alpha\beta,\gamma\delta} E_{i\gamma\delta} + \cdots, \quad (5.3.1b)$$

where the perturbed tensors have quantum mechanical forms analogous to (2.7.6) at transparent frequencies, and (2.7.8) at absorbing frequencies. The electrostatic field at group i arising from group j is given by (2.4.25) as

$$E_{i\alpha} = \frac{1}{4\pi\epsilon_0}\left(- T_{ij\alpha} q_j + T_{ij\alpha\beta}\mu_{j\beta} - \tfrac{1}{3} T_{ij\alpha\beta\gamma}\Theta_{j\beta\gamma} + \cdots\right), \quad (5.3.2a)$$

and the corresponding field gradient is

$$E_{i\alpha\beta} = \frac{1}{4\pi\epsilon_0}(-T_{ij\alpha\beta} q_j + T_{ij\alpha\beta\gamma}\mu_{j\gamma} + \cdots), \quad (5.3.2b)$$

where q_j, $\mu_{j\alpha}$ and $\Theta_{j\alpha\beta}$ are the permanent charge, electric dipole moment and electric quadrupole moment of j, and the subscript ij on the $\mathbf{T}$ tensors indicates they are functions of the vector $\mathbf{R}_{ij} = \mathbf{R}_i - \mathbf{R}_j$ from the origin on j to that on i.

If group 1, but not 2, is a chromophore at the frequency of the exciting light, the static coupling contribution to the optical activity of the two-group structure is determined by the chromophore transition between appropriate initial and final electronic states of 1 perturbed by the static fields from 2. Thus the isotropic rotational strength of the $j_1 \leftarrow n_1$ chromophore transition follows from (5.2.28a), (5.3.1) and (5.3.2) (or simply by using perturbed eigenstates (2.6.14) in the rotational strength):

$$
\begin{aligned}
R(j \leftarrow n) = & \left\{ \sum_{k_1 \neq n_1} \frac{1}{\hbar \omega_{k_1 n_1}} \text{Im}[\langle k_1|\mu_{1_\beta}|n_1\rangle(\langle n_1|\mu_{1_\alpha}|j_1\rangle\langle j_1|m_{1_\alpha}|k_1\rangle \right. \\
& - \langle n_1|m_{1_\alpha}|j_1\rangle\langle j_1|\mu_{1_\alpha}|k_1\rangle)] \\
& + \sum_{k_1 \neq j_1} \frac{1}{\hbar \omega_{k_1 j_1}} \text{Im}[\langle j_1|\mu_{1_\beta}|k_1\rangle(\langle n_1|\mu_{1_\alpha}|j_1\rangle\langle k_1|m_{1_\alpha}|n_1\rangle \\
& \left. - \langle n_1|m_{1_\alpha}|j_1\rangle\langle k_1|\mu_{1_\alpha}|n_1\rangle)] \right\} \\
& \times \frac{1}{4\pi\epsilon_0}\left(-T_{12_\beta}q_2 + T_{12_{\beta\gamma}}\mu_{2\gamma} - \tfrac{1}{3}T_{12_{\beta\gamma\delta}}\Theta_{2\gamma\delta} + \cdots \right) \\
& + \{\text{same expression with } \Theta_{\beta\gamma} \text{ replacing } \mu_\beta\} \\
& \times \frac{1}{4\pi\epsilon_0}(-T_{12_{\beta\gamma}}q_2 + T_{12_{\beta\gamma\delta}}\mu_{2\delta} + \cdots) + \cdots .
\end{aligned}
\tag{5.3.3}
$$

Since group 1 is intrinsically achiral, there is no term analogous to the first term in (2.7.6b) because $\text{Im}(\langle n_1|\mu_{1_\alpha}|j_1\rangle\langle j_1|m_{1_\alpha}|n_1\rangle)$ is zero. An analogous expression for the generalized rotational strength (5.2.29a) of an oriented molecule can be written down easily if required.

If we allow the perturbing group 2 to have only isotropic properties, only the first terms containing the charge q_2 survive. As discussed in Section 4.5.5, this means that, in order to show optical activity, the skeleton supporting groups 1 and 2 must be defined by a chirality order $o = (n - 1)$. This is realized if, say, the skeleton to which groups 1 and 2 are attached has C_{2v} symmetry, as in adamantanone derivatives.

In molecules containing more than two groups, we sum the interactions of individual groups with the chromophore. In chiral methane derivatives such as CHFClBr, optical activity within the static coupling model is induced in an atom through its simultaneous interaction with the electrostatic fields of at least three other atoms, in accordance with the chirality order $o = (n - 3)$ for the methane skeleton. Since the free atoms are uncharged and nondipolar, any associated fields in the molecule originate in effects such as incomplete shielding of nuclear charges at short distances, and dipole moments induced by other atoms. These effects are usually small, and the optical activity of such molecules is probably determined largely by dynamic coupling, as discussed in the next section.

5.3.2 The dynamic coupling model

The molecule is divided into a convenient, but otherwise arbitrarily selected, set of atoms or groups such as a chiral carbon atom and its four substituent groups. The oscillating multipole moments induced in the molecule are the sums of the moments induced in individual groups referred to local group origins, together with additional contributions from the origin-dependent moments referred to a convenient centre within the molecule. The induced moments can arise both from the direct influence of the radiation field on individual groups and from the secondary fields arising from oscillating multipole moments generated in other groups. Expressions for the optical activity tensors in terms of molecular structural units are obtained from the total induced magnetic dipole and electric quadrupole moments through

$$G'_{\alpha\beta} = -\omega \left[\frac{\partial m'_\beta}{\partial (\dot{E}_\alpha)_0} \right]_{(\dot{E}_\alpha)_0 = 0,} \tag{5.3.4a}$$

$$A_{\alpha,\beta\gamma} = \left[\frac{\partial \Theta_{\beta\gamma}}{\partial (E_\alpha)_0} \right]_{(E_\alpha)_0 = 0,} \tag{5.3.4b}$$

which follow from (2.6.26). The optical activity tensors can also be obtained via the induced electric dipole moment, but the calculation is more complicated.

We consider first the optical activity generated by two neutral groups 1 and 2 that constitute a chiral structural unit. When (5.3.4) are used to calculate the optical activity tensors, the differentiations are performed with respect to fields evaluated at a fixed molecular origin. All the group multipole moments must be referred to this origin, which we choose for convenience to be the local origin on 1. All expressions for observables subsequently obtained are independent of this choice of origin.

The total multipole moments of the two-group structure are then, using (2.4.3), (2.4.9) and (2.4.14),

$$\mu_\alpha = \mu_{1\alpha} + \mu_{2\alpha}, \tag{5.3.5a}$$

$$\Theta_{\alpha\beta} = \Theta_{1\alpha\beta} + \Theta_{2\alpha\beta} - \tfrac{3}{2} R_{12\alpha} \mu_{2\beta} - \tfrac{3}{2} R_{12\beta} \mu_{2\alpha} + R_{12\gamma} \mu_{2\gamma} \delta_{\alpha\beta}, \tag{5.3.5b}$$

$$m_\alpha = m_{1\alpha} + m_{2\alpha} - \tfrac{1}{2} \varepsilon_{\alpha\beta\gamma} R_{12\beta} \dot{\mu}_{2\gamma}. \tag{5.3.5c}$$

The multipole moments of each group i are written in terms of dynamic group property tensors coupled with the dynamic fields $(E_\alpha)_i$, $(B_\alpha)_i$ and $(E_{\alpha\beta})_i$ at the origin of i arising from the light wave, and the dynamic fields $(E'_\alpha)_i$, $(B'_\alpha)_i$ and $(E'_{\alpha\beta})_i$ at i radiated by the oscillating multipole moments induced by the light wave in the other group:

$$\mu_{i\alpha} = \alpha_{i\alpha\beta} [(E_\beta)_i + (E'_\beta)_i] + \tfrac{1}{3} A_{i\alpha,\beta\gamma} [(E_{\beta\gamma})_i + (E'_{\beta\gamma})_i]$$

$$+ \frac{1}{\omega} G'_{i\alpha\beta} [(\dot{B}_\beta)_i + (\dot{B}'_\beta)_i] + \cdots, \tag{5.3.6a}$$

$$\Theta_{i\alpha\beta} = A_{i\gamma,\alpha\beta}[(E_\gamma)_i + (E'_\gamma)_i] + C_{i\alpha\beta,\gamma\delta}[(E_{\gamma\delta})_i + (E'_{\gamma\delta})_i]$$

$$- \frac{1}{\omega} D_{i\gamma,\alpha\beta}[(\dot{B}_\gamma)_i + (\dot{B}'_\gamma)_i] + \cdots, \tag{5.3.6b}$$

$$m'_{i\alpha} = \chi_{i\alpha\beta}[(B_\beta)_i + (B'_\beta)_i] + \frac{1}{3\omega} D'_{i\alpha,\beta\gamma}[(\dot{E}_{\beta\gamma})_i + (\dot{E}'_{\beta\gamma})_i]$$

$$- \frac{1}{\omega} G'_{i\beta\alpha}[(\dot{E}_\beta)_i + (\dot{E}'_\beta)_i] + \cdots. \tag{5.3.6c}$$

By neglecting the tensors $\alpha'_{\alpha\beta}$, $A'_{\alpha,\beta\gamma}$, $G_{\alpha\beta}$, $C'_{\alpha\beta,\gamma\delta}$, $D_{\alpha,\beta\gamma}$ and $\chi'_{\alpha\beta}$, we are assuming that the group has an even number of electrons and no static magnetic fields are present. Three distinct expressions are derived in Section 2.4.5 for the electric and magnetic fields arising from oscillating multipole moments, depending on whether the distance is much smaller than, comparable with, or much larger than the wavelength. We assume that the first case obtains here, so from (2.4.44) the electric field, electric field gradient and magnetic field at the ith group radiated by the jth group are

$$(E'_\alpha)_i = \frac{1}{4\pi\epsilon_0}(T_{ij\alpha\beta}\mu_{j\beta} - \tfrac{1}{3}T_{ij\alpha\beta\gamma}\Theta_{j\beta\gamma} + \cdots), \tag{5.3.7a}$$

$$(E'_{\alpha\beta})_i = \frac{1}{4\pi\epsilon_0}(T_{ij\alpha\beta\gamma}\mu_{j\gamma} + \cdots), \tag{5.3.7b}$$

$$(B'_\alpha)_i = \frac{\mu_0}{4\pi}(T_{ij\alpha\beta}m_{j\beta} + \cdots), \tag{5.3.7c}$$

where the time dependence has been absorbed into the multipole moments. In this approximation there is no contribution to the radiated electric field from the magnetic dipole moment, nor to the radiated magnetic field from the electric dipole moment. One difficulty in applying (5.3.7) to dynamic coupling between groups is that the distance must be much larger than the separation of charges within the radiating group (but still much smaller than the wavelength), which is not true for groups in compact molecules.

Before using these results to write down general dynamic coupling expressions for the optical activity tensors, it is helpful to show the steps in the derivation of *Kirkwood's term*, which is the simplest dynamic coupling contribution to the trace of the optical activity tensors. This is obtained from the term $-\tfrac{1}{2}\epsilon_{\alpha\beta\gamma}R_{12\beta}\mu_{2\gamma}$ in (5.3.5c) if $\mu_{2\gamma}$ is the electric dipole moment induced in 2 by the electric field radiated by 1 when stimulated by the external light wave. From (5.3.6a) and (5.3.7a),

$$\dot{\mu}_{2\gamma} = \alpha_{2\gamma\delta}(\dot{E}'_\delta)_2 = \frac{1}{4\pi\epsilon_0}\alpha_{2\gamma\delta}T_{21\delta\epsilon}\dot{\mu}_{1\epsilon}$$

$$= \frac{1}{4\pi\epsilon_0}\alpha_{2\gamma\delta}T_{21\delta\epsilon}\alpha_{1\epsilon\lambda}(\dot{E}_\lambda)_1. \tag{5.3.8}$$

Since $(E_\lambda)_1$ is the field of the light wave at the origin of group 1, which is also our choice of molecular origin, $(E_\lambda)_1 \equiv (E_\lambda)_0$ and so the isotropic part of the optical activity tensors now follows from (5.3.4a):

$$G'_{\alpha\alpha} = -\omega \frac{\partial}{\partial(\dot{E}_\alpha)_0}\left(-\tfrac{1}{2}\varepsilon_{\alpha\beta\gamma}R_{12\beta}\dot{\mu}_{2\gamma}\right)$$

$$= \frac{\omega}{8\pi\epsilon_0}\varepsilon_{\alpha\beta\gamma}R_{12\beta}\alpha_{2\gamma\delta}T_{21\delta\epsilon}\alpha_{1\epsilon\alpha}. \qquad (5.3.9)$$

Kirkwood (1937) originally derived this term by substituting the operator equivalents of the electric and magnetic dipole moments (5.3.5) into the transition matrix elements of the quantum mechanical expression for $G'_{\alpha\alpha}$. He introduced the dynamic dipole–dipole coupling as a perturbation of the electronic wavefunctions of the two groups, and by a series of transformations was able to express the result in terms of group polarizabilities, as above.

It is important to realize that the Kirkwood term (5.3.9) depends on the choice of local origins within the two groups. This difficulty is removed if we include a further term arising from the intrinsic group magnetic moments in (5.3.5c). From (5.3.6c) and (5.3.7a),

$$m_{1\alpha} + m_{2\alpha} = -\frac{1}{\omega}[G'_{1\beta\alpha}(\dot{E}'_\beta)_1 + G'_{2\beta\alpha}(\dot{E}'_\beta)_2]$$

$$= -\frac{1}{4\pi\epsilon_0\omega}(G'_{1\beta\alpha}T_{12\beta\gamma}\dot{\mu}_{2\gamma} + G'_{2\beta\alpha}T_{21\beta\gamma}\dot{\mu}_{1\gamma})$$

$$= -\frac{1}{4\pi\epsilon_0\omega}(G'_{1\beta\alpha}T_{12\beta\gamma}\alpha_{2\gamma\delta} + G'_{2\beta\alpha}T_{21\beta\gamma}\alpha_{1\gamma\delta})(\dot{E}_\delta)_1, \quad (5.3.10)$$

and we obtain the following additional contribution to $G'_{\alpha\alpha}$:

$$\frac{1}{4\pi\epsilon_0}(G'_{1\beta\alpha}T_{12\beta\gamma}\alpha_{2\gamma\alpha} + G'_{2\beta\alpha}T_{21\beta\gamma}\alpha_{1\gamma\alpha}).$$

The combination

$$G'_{\alpha\alpha} = \frac{1}{4\pi\epsilon_0}\left(\tfrac{1}{2}\omega\varepsilon_{\alpha\beta\gamma}R_{12\beta}\alpha_{2\gamma\delta}T_{21\delta\epsilon}\alpha_{1\epsilon\alpha} + G'_{1\beta\alpha}T_{12\beta\gamma}\alpha_{2\gamma\alpha}\right.$$

$$\left. + G'_{2\beta\alpha}T_{21\beta\gamma}\alpha_{1\gamma\alpha}\right) \qquad (5.3.11)$$

is independent of the choice of local group origins, as may be verified by the replacements

$$R_{i\alpha} \rightarrow R_{i\alpha} + \Delta r_{i\alpha}, \qquad (5.3.12a)$$

$$R_{ij\alpha} \rightarrow R_{ij\alpha} + \Delta r_{i\alpha} - \Delta r_{j\alpha}, \qquad (5.3.12b)$$

$$G'_{i\alpha\beta} \rightarrow G'_{i\alpha\beta} + \tfrac{1}{2}\omega\varepsilon_{\beta\gamma\delta}\Delta r_{i\gamma}\alpha_{i\delta\alpha}, \qquad (5.3.12c)$$

where $\Delta \mathbf{r}_i$ is the shift of the local origin on group i and the last result follows from (2.6.35c).

From (5.3.4–7) the general dynamic coupling contributions to the complete optical activity tensors of the two-group structure are found to be

$$
G'_{\alpha\beta} = G'_{1\alpha\beta} + G'_{2\alpha\beta} + \tfrac{1}{2}\omega\varepsilon_{\beta\gamma\delta}R_{12\gamma}\alpha_{2\delta\alpha}
$$

$$
+ \frac{1}{4\pi\epsilon_0}\big[\tfrac{1}{2}\omega\varepsilon_{\beta\gamma\delta}R_{12\gamma}\alpha_{2\delta\epsilon}T_{12\epsilon\lambda}\alpha_{1\lambda\alpha} + (G'_{1\gamma\beta}T_{12\gamma\delta}\alpha_{2\delta\alpha} + G'_{2\gamma\beta}T_{12\gamma\delta}\alpha_{1\delta\alpha})
$$

$$
- \tfrac{1}{3}(D'_{1\beta,\gamma\delta}T_{12\gamma\delta\epsilon}\alpha_{2\epsilon\alpha} + D'_{2\beta,\gamma\delta}T_{12\gamma\delta\epsilon}\alpha_{1\epsilon\alpha})
$$

$$
+ \tfrac{1}{6}\omega\varepsilon_{\beta\gamma\delta}R_{12\gamma}(\alpha_{1\alpha\mu}T_{12\epsilon\lambda\mu}A_{2\delta,\epsilon\lambda} - \alpha_{2\delta\epsilon}T_{12\epsilon\lambda\mu}A_{1\alpha,\lambda\mu})\big] + \cdots, \quad (5.3.13a)
$$

$$
A_{\alpha,\beta\gamma} = A_{1\alpha,\beta\gamma} + A_{2\alpha,\beta\gamma} - \tfrac{3}{2}R_{12\beta}\alpha_{2\gamma\alpha} - \tfrac{3}{2}R_{12\gamma}\alpha_{2\beta\alpha} + R_{12\delta}\alpha_{2\delta\alpha}\delta_{\beta\gamma}
$$

$$
+ \frac{1}{4\pi\epsilon_0}\big[- (\tfrac{3}{2}R_{12\beta}\alpha_{2\gamma\delta}T_{12\delta\epsilon}\alpha_{1\epsilon\alpha} + \tfrac{3}{2}R_{12\gamma}\alpha_{2\beta\delta}T_{12\delta\epsilon}\alpha_{1\epsilon\alpha}
$$

$$
- R_{12\mu}\alpha_{2\mu\delta}T_{12\delta\epsilon}\alpha_{1\epsilon\alpha}\delta_{\beta\gamma}) + (A_{1\delta,\beta\gamma}T_{12\delta\epsilon}\alpha_{2\epsilon\alpha} + A_{2\delta,\beta\gamma}T_{12\delta\epsilon}\alpha_{1\epsilon\alpha})
$$

$$
+ (C_{1\beta\gamma,\delta\epsilon}T_{12\delta\epsilon\lambda}\alpha_{2\lambda\alpha} - C_{2\beta\gamma,\delta\epsilon}T_{12\delta\epsilon\lambda}\alpha_{1\lambda\alpha})
$$

$$
+ \tfrac{1}{2}R_{12\beta}(\alpha_{2\gamma\delta}T_{12\delta\epsilon\lambda}A_{1\alpha,\epsilon\lambda} - A_{2\gamma,\delta\epsilon}T_{12\delta\epsilon\lambda}\alpha_{1\lambda\alpha})
$$

$$
+ \tfrac{1}{2}R_{12\gamma}(\alpha_{2\beta\delta}T_{12\delta\epsilon\lambda}A_{1\alpha,\epsilon\lambda} - A_{2\beta,\delta\epsilon}T_{12\delta\epsilon\lambda}\alpha_{1\lambda\alpha})
$$

$$
- \tfrac{1}{6}R_{12\mu}(\alpha_{2\mu\delta}T_{12\delta\epsilon\lambda}A_{1\alpha,\epsilon\lambda} - A_{2\mu,\delta\epsilon}T_{12\delta\epsilon\lambda}\alpha_{1\lambda\alpha})\delta_{\beta\gamma}\big] + \cdots. \quad (5.3.13b)
$$

Although these general optical activity tensors depend on the choice of local group origins, when specified components are used in expressions for observables the results are origin invariant. We illustrate this by writing down the Kirkwood contribution to optical rotation for light propagating in the z direction of an oriented medium. The relevant tensor components are

$$
G'_{xx} + G'_{yy} = \frac{1}{4\pi\epsilon_0}\big[\tfrac{1}{2}\omega(R_{12y}\alpha_{2z\alpha}T_{12\alpha\beta}\alpha_{1\beta x} - R_{12z}\alpha_{2y\alpha}T_{12\alpha\beta}\alpha_{1\beta x}
$$

$$
+ R_{12z}\alpha_{2x\alpha}T_{12\alpha\beta}\alpha_{1\beta y} - R_{12x}\alpha_{2z\alpha}T_{12\alpha\beta}\alpha_{1\beta y})
$$

$$
+ G'_{1\alpha x}T_{12\alpha\beta}\alpha_{2\beta x} + G'_{2\alpha x}T_{12\alpha\beta}\alpha_{1\beta x}
$$

$$
+ G'_{1\alpha y}T_{12\alpha\beta}\alpha_{2\beta y} + G'_{2\alpha y}T_{12\alpha\beta}\alpha_{1\beta y}\big], \quad (5.3.14a)
$$

$$
\tfrac{1}{3}\omega(A_{x,yz} - A_{y,xz}) = \frac{1}{4\pi\epsilon_0}\big[\tfrac{1}{2}\omega(-R_{12z}\alpha_{2y\alpha}T_{12\alpha\beta}\alpha_{1\beta x} - R_{12y}\alpha_{2z\alpha}T_{12\alpha\beta}\alpha_{1\beta x}
$$

$$
+ R_{12z}\alpha_{2x\alpha}T_{12\alpha\beta}\alpha_{1\beta y} + R_{12x}\alpha_{2z\alpha}T_{12\alpha\beta}\alpha_{1\beta y})
$$

$$
+ A_{1\alpha,zy}T_{12\alpha\beta}\alpha_{2\beta x} + A_{2\alpha,zy}T_{12\alpha\beta}\alpha_{1\beta x}
$$

$$
- A_{1\alpha,zx}T_{12\alpha\beta}\alpha_{2\beta y} - A_{2\alpha,zx}T_{12\alpha\beta}\alpha_{1\beta y}\big], \quad (5.3.14b)
$$

so from (5.2.1*a*) the optical rotation is (Barron, 1975*b*)

$$\Delta\theta \approx \frac{\omega^2 \mu_0 l N}{8\pi\epsilon_0}[R_{12z}(\alpha_{2y\alpha}T_{12\alpha\beta}\alpha_{1\beta x} - \alpha_{2x\alpha}T_{12\alpha\beta}\alpha_{1\beta y})$$
$$+ G'_{1\alpha x}T_{12\alpha\beta}\alpha_{2\beta x} + G'_{2\alpha x}T_{12\alpha\beta}\alpha_{1\beta x} + G'_{1\alpha y}T_{12\alpha\beta}\alpha_{2\beta y}$$
$$+ G'_{2\alpha y}T_{12\alpha\beta}\alpha_{1\beta y} + A_{1\alpha, zy}T_{12\alpha\beta}\alpha_{2\beta x} + A_{2\alpha, zy}T_{12\alpha\beta}\alpha_{1\beta x}$$
$$- A_{1\alpha, zx}T_{12\alpha\beta}\alpha_{2\beta y} - A_{2\alpha, zx}T_{12\alpha\beta}\alpha_{1\beta y}]. \tag{5.3.15}$$

Thus the Kirkwood contributions to the electric dipole–magnetic dipole and electric dipole–electric quadrupole optical rotation mechanisms have equal and opposite terms that cancel, and identical terms that reinforce. The invariance of this equation to the choice of local group origins may be verified using the replacements (5.3.12*a*–*c*), together with

$$A_{i\alpha, \beta\gamma} \rightarrow A_{i\alpha, \beta\gamma} - \tfrac{3}{2}\Delta r_{i\beta}\alpha_{i\alpha\gamma} - \tfrac{3}{2}\Delta r_{i\gamma}\alpha_{i\alpha\beta} + \Delta r_{i\delta}\alpha_{i\alpha\delta}\delta_{\beta\gamma}. \tag{5.3.12d}$$

Kruchek (1973) has derived a corresponding rotational strength, also starting from the Buckingham–Dunn equation (5.2.1*a*).

If the two dynamically coupled groups have threefold or higher proper rotation axes, the Kirkwood term can be given a tractable form. If unit vectors $\mathbf{s}_i, \mathbf{t}_i, \mathbf{u}_i$ define the principal axes of the *i*th group, with $\mathbf{u}_i$ along the symmetry axis then, from (4.2.58), its polarizability tensor can be written

$$\alpha_{i\alpha\beta} = \alpha_i(1 - \kappa_i)\delta_{\alpha\beta} + 3\alpha_i\kappa_i u_{i\alpha}u_{i\beta}, \tag{5.3.16a}$$

where

$$\alpha_i = \tfrac{1}{3}(2\alpha_{i\perp} + \alpha_{i\parallel}), \tag{5.3.16b}$$

$$\kappa_i = (\alpha_{i\parallel} + \alpha_{i\perp})/3\alpha_i \tag{5.3.16c}$$

are the mean polarizability and dimensionless polarizability anisotropy. The first part of the Kirkwood contribution (5.3.11) to the isotropic part of the optical activity tensors now becomes

$$G'_{\alpha\alpha} = \frac{9\omega}{8\pi\epsilon_0}(\alpha_1\alpha_2\kappa_1\kappa_2)\varepsilon_{\alpha\beta\gamma}R_{12\beta}(u_{2\gamma}u_{2\delta}T_{12\delta\epsilon}u_{1\epsilon}u_{1\alpha}). \tag{5.3.17}$$

The form of the second part of (5.3.11), which is required for general origin invariance, depends on the precise symmetry of the groups. In the case of C_{nv} ($n > 2$), the only nonzero components of $G'_{\alpha\beta}$ are $G'_{xy} = -G'_{yx}$ (taking *z* to be the C_n rotation axis), as may be verified from Tables 4.2 so that, provided we choose the group origins to lie anywhere along the group symmetry axes, terms such as $G'_{i\beta\alpha}T_{ij\beta\gamma}\alpha_{j\gamma\alpha}$ vanish. The optical rotation in isotropic collections of such structures is then simply

$$\Delta\theta \approx -\frac{3\omega^2\mu_0 l N}{8\pi\epsilon_0}(\alpha_1\alpha_2\kappa_1\kappa_2)\varepsilon_{\alpha\beta\gamma}R_{12\beta}(u_{2\gamma}u_{2\delta}T_{12\delta\epsilon}u_{1\epsilon}u_{1\alpha}). \tag{5.3.18}$$

The corresponding Kirkwood contribution to optical rotation in oriented samples is

$$\Delta\theta \approx \frac{3\omega^2\mu_0 l N}{8\pi\epsilon_0}\alpha_1\alpha_2 R_{12_z}[\kappa_1(1-\kappa_2)(u_{1\alpha}u_{1x}T_{12y\alpha} - u_{1\alpha}u_{1y}T_{12x\alpha})$$

$$+ \kappa_2(1-\kappa_1)(u_{2y}u_{2\alpha}T_{12\alpha x} - u_{2x}u_{2\alpha}T_{12\alpha y})$$

$$+ 3\kappa_1\kappa_2(u_{2y}u_{2\alpha}T_{12\alpha\beta}u_{1\beta}u_{1x} - u_{2x}u_{2\alpha}T_{12\alpha\beta}u_{1\beta}u_{1y})]. \qquad (5.3.19)$$

We see from (5.3.18) that if one of the two groups is isotropically polarizable, the Kirkwood term does not contribute to optical rotation in an isotropic sample; likewise if the symmetry axes of the two groups lie in the same plane: both situations correspond, of course, to achiral structures. These dynamic coupling results can be extended to molecules containing more than two groups by summing all pairwise interactions. If a molecule contains three groups, at least one of them must be anisotropically polarizable for the Kirkwood mechanism to contribute to optical rotation in an isotropic sample. In fact the Kirkwood mechanism can only contribute in an isotropic collection of molecules consisting of isotropically polarizable groups if dynamic coupling extends over a chiral arrangement of at least four groups: this is the *Born–Boys model*, which we now consider explicitly.

In the Born–Boys model, the excitation is relayed from the first group encountered by the light wave successively to the other three groups. The induced magnetic dipole moment of the complete system is written as a sum of the moments induced at each group by a wave that has suffered sequential scattering from each of the other three groups. The molecular origin is chosen to be the local origin on group 1, and we retain only the part of a group's magnetic moment that is referred to the origin on group 1, that is $m_{i\alpha} = -\frac{1}{2}\varepsilon_{\alpha\beta\gamma}R_{1i\beta}\dot{\mu}_{i\gamma}$. The total magnetic dipole moment of the molecule is then

$$m_\beta = \sum_{i=1}^{4} m_{i\beta} = -\tfrac{1}{2}\varepsilon_{\beta\gamma\delta}\sum_{i=2}^{4} R_{1i\gamma}\dot{\mu}_{i\delta}$$

$$= -\tfrac{1}{2}\varepsilon_{\beta\gamma\delta}\sum_{i=2}^{4} R_{1i\gamma}\alpha_{i\delta\epsilon}[(\dot{E}_\epsilon)_i + (\dot{E}'_\epsilon)_i]$$

$$= -\tfrac{1}{2}\varepsilon_{\beta\gamma\delta}\sum_{i=2}^{4} R_{1i\gamma}\alpha_{i\delta\epsilon}\left\{(\dot{E}_\epsilon)_i + \frac{1}{4\pi\epsilon_0}\sum_{j\neq i} T_{ij\epsilon\lambda}\alpha_{j\lambda\mu}\right.$$

$$\times\left[(\dot{E}_\mu)_j + \frac{1}{4\pi\epsilon_0}\sum_{k\neq j} T_{jk\mu\nu}\alpha_{k\nu\theta}((\dot{E}_\theta)_k + \frac{1}{4\pi\epsilon_0}\sum_{l\neq k} T_{kl\theta\phi}\alpha_{l\phi\alpha}(\dot{E}_\alpha)_l)\right]\right\}.$$

$$(5.3.20)$$

Retaining just the isotropic part of each group polarizability tensor, $\alpha_{\alpha\beta} = \alpha\delta_{\alpha\beta}$, we find from (5.3.4)

$$G'_{\alpha\beta} = \frac{\omega}{2}\varepsilon_{\beta\gamma\alpha}\sum_{i=2}^{4}\alpha_i R_{1i\gamma} + \frac{\omega}{2}\left(\frac{1}{4\pi\epsilon_0}\right)\varepsilon_{\beta\gamma\delta}\sum_{i=2}^{4}\sum_{j\neq i}\alpha_i\alpha_j R_{1i\gamma}T_{ij\delta\alpha}$$

$$+ \frac{\omega}{2}\left(\frac{1}{4\pi\epsilon_0}\right)^2\varepsilon_{\beta\gamma\delta}\sum_{i=2}^{4}\sum_{j\neq i}\sum_{k\neq j}\alpha_i\alpha_j\alpha_k R_{1i\gamma}T_{ij\delta\epsilon}T_{jk\epsilon\alpha}$$

$$+ \frac{\omega}{2}\left(\frac{1}{4\pi\epsilon_0}\right)^3\varepsilon_{\beta\gamma\delta}\sum_{i=2}^{4}\sum_{j\neq i}\sum_{k\neq j}\sum_{l\neq k}\alpha_i\alpha_j\alpha_k\alpha_l R_{1i\gamma}T_{ij\delta\epsilon}T_{jk\epsilon\lambda}T_{kl\lambda\alpha},$$

$$(5.3.21)$$

with a similar expression for $A_{\alpha,\beta\gamma}$. The relevant tensor components for the Born–Boys contribution to optical rotation for light propagating in an oriented sample can now be written down immediately if required. But we shall be content with just the part which gives the optical rotation in an isotropic sample:

$$G'_{\alpha\alpha} = \frac{\omega}{2}\left(\frac{1}{4\pi\epsilon_0}\right)^3\alpha_1\alpha_2\alpha_3\alpha_4\varepsilon_{\alpha\gamma\delta}\sum_{i=2}^{4}\sum_{\substack{j\neq i}}\sum_{\substack{k\neq j \\ \neq i}}\sum_{\substack{l\neq k \\ \neq j \\ \neq i}}R_{1i\gamma}T_{ij\delta\epsilon}T_{jk\epsilon\lambda}T_{kl\lambda\alpha}.$$

$$(5.3.22)$$

It is easy to see that these are the only nonzero terms: if each dynamically coupled pair of atoms is regarded as an anisotropically polarizable group, all the other terms correspond to dynamic coupling between pairs of 'anisotropically polarizable groups' with their symmetry axes lying in the same plane. Thus (5.3.22) demonstrates explicitly that dynamic coupling must extend over all four atoms for optical rotation to be generated in an isotropic sample.

The Born–Boys model would not, in fact, provide the lowest order contribution to optical rotation in a simple chiral molecule such as CHFClBr because the C–X bonds constitute anisotropically polarizable groups. Thus optical rotation can be generated through dynamic coupling between a bond and a pair of dynamically coupled atoms. We refer to Applequist (1973) for a critical discussion of classical dynamic coupling models of optical rotation, and for details of how the complicated general formulae taking account of all orders of coupling can be handled in numerical calculations.

The dynamic coupling mechanisms are illustrated in Fig. 5.1. Recalling from Section 3.4 that polarization effects in transmitted light originate in interference between forward-scattered waves and unscattered waves, the dynamic coupling mechanisms can be visualized in terms of interference in the intensity measurement at the detector of an unscattered photon and a photon that has sampled the chirality

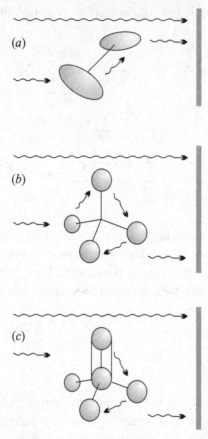

Fig. 5.1 The generation of optical rotation through dynamic coupling between (*a*) two anisotropic groups (the Kirkwood model), (*b*) four isotropic groups (the Born–Boys model) and (*c*) a bond and two isotropic groups.

of a molecule by being deflected from one group to another before emerging in the forward direction. The lowest order dynamic coupling mechanism that can generate optical rotation is the one involving the least number of deflections in generating a 'chiral pathway' for the photon within the molecule. This picture is oversimplified in that the waves scattered from a large number of molecules must first be combined into a net plane wave moving in the forward direction before interfering with the unscattered wave. Such pictures form the basis of a treatment of optical activity using quantum electrodynamics (Atkins and Woolley, 1970).

The dispersive and absorptive parts of the optical activity tensors can be obtained by writing each group polarizability tensor as a function of $(f + ig)$, and equating real and imaginary parts. Thus if we consider a single transition on each group with dispersion and absorption lineshapes f_1, f_2 and g_1, g_2, the dispersive and absorptive parts of the Kirkwood contribution (5.3.11) are

$$
\begin{aligned}
G'_{\alpha\alpha}(f^2-g^2) = \frac{1}{4\pi\epsilon_0}\Big\{ &\tfrac{1}{2}\omega\varepsilon_{\alpha\beta\gamma}R_{12_\beta}[\alpha_{2_{\gamma\delta}}(f_2)T_{21_{\delta\epsilon}}\alpha_{1_{\epsilon\alpha}}(f_1) \\
&-\alpha_{2_{\gamma\delta}}(g_2)T_{21_{\delta\epsilon}}\alpha_{1_{\epsilon\alpha}}(g_1)] \\
&+G'_{1_{\beta\alpha}}(f_1)T_{12_{\beta\gamma}}\alpha_{2_{\gamma\alpha}}(f_2)-G'_{1_{\beta\alpha}}(g_1)T_{12_{\beta\gamma}}\alpha_{2_{\gamma\alpha}}(g_2) \\
&+G'_{2_{\beta\alpha}}(f_2)T_{21_{\beta\gamma}}\alpha_{1_{\gamma\alpha}}(f_1)-G'_{2_{\beta\alpha}}(g_2)T_{21_{\beta\gamma}}\alpha_{1_{\gamma\alpha}}(g_1)\Big\},
\end{aligned}
$$

$$(5.3.23a)$$

$$
\begin{aligned}
G'_{\alpha\alpha}(fg) = \frac{1}{4\pi\epsilon_0}\Big\{ &\tfrac{1}{2}\omega\varepsilon_{\alpha\beta\gamma}R_{12_\beta}[\alpha_{2_{\gamma\delta}}(f_2)T_{21_{\delta\epsilon}}\alpha_{1_{\epsilon\alpha}}(g_1) \\
&+\alpha_{2_{\gamma\delta}}(g_2)T_{21_{\delta\epsilon}}\alpha_{1_{\epsilon\alpha}}(f_1)] \\
&+G'_{1_{\beta\alpha}}(f_1)T_{12_{\beta\gamma}}\alpha_{2_{\gamma\alpha}}(g_2)+G'_{1_{\beta\alpha}}(g_1)T_{12_{\beta\gamma}}\alpha_{2_{\gamma\alpha}}(f_2) \\
&+G'_{2_{\beta\alpha}}(f_2)T_{21_{\beta\gamma}}\alpha_{1_{\gamma\alpha}}(g_1)+G'_{2_{\beta\alpha}}(g_2)T_{21_{\beta\gamma}}\alpha_{1_{\gamma\alpha}}(f_1)\Big\}.
\end{aligned}
$$

$$(5.3.23b)$$

If the two groups are different and we are interested in an electronic absorption of group 1, then $\alpha_{2_{\alpha\beta}}(g_2)$ is zero and (5.3.23) simply describe the perturbation of the group 1 chromophore by a chiral electrodynamic field from group 2. A corresponding rotational strength can be written for the $j_1 \leftarrow n_1$ transition:

$$
\begin{aligned}
R(j_1 \leftarrow n_1) = -\frac{1}{4\pi\epsilon_0}\Big\{ &\tfrac{1}{2}\omega\varepsilon_{\alpha\beta\gamma}\,R_{12_\beta}\alpha_{2_{\gamma\delta}}T_{21_{\delta\epsilon}}\operatorname{Re}(\langle n_1|\mu_{1_\epsilon}|j_1\rangle\langle j_1|\mu_{1_\alpha}|n_1\rangle) \\
&-\alpha_{2_{\gamma\alpha}}T_{12_{\beta\gamma}}\operatorname{Im}(\langle n_1|\mu_{1_\beta}|j_1\rangle\langle j_1|m_{1_\alpha}|n_1\rangle) \\
&+G'_{2_{\beta\alpha}}T_{12_{\beta\gamma}}\operatorname{Re}(\langle n_1|\mu_{1_\gamma}|j_1\rangle\langle j_1|\mu_{1_\alpha}|n_1\rangle) \\
&+\tfrac{1}{3}[\alpha_{2_{\delta\alpha}}T_{12_{\beta\gamma\delta}}\operatorname{Im}(\langle n_1|m_{1_\alpha}|j_1\rangle\langle j_1|\Theta_{1_{\beta\gamma}}|n_1\rangle) \\
&-D'_{2_{\alpha\beta\gamma}}T_{12_{\beta\gamma\delta}}\operatorname{Re}(\langle n_1|\mu_{1_\delta}|j_1\rangle\langle j_1|\mu_{1_\alpha}|n_1\rangle)] \\
&-\tfrac{1}{6}\omega\varepsilon_{\alpha\beta\gamma}R_{12_\beta}[\alpha_{2_{\gamma\delta}}T_{12_{\delta\epsilon\lambda}}\operatorname{Re}(\langle n_1|\mu_{1_\alpha}|j_1\rangle\langle j_1|\Theta_{1_{\epsilon\lambda}}|n_1\rangle) \\
&-A_{2_{\gamma,\delta\epsilon}}T_{12_{\delta\epsilon\lambda}}\operatorname{Re}(\langle n_1|\mu_{1_\alpha}|j_1\rangle\langle j_1|\mu_{1_\lambda}|n_1\rangle)]+\cdots\Big\},
\end{aligned}
$$

$$(5.3.24)$$

where we have included terms of higher order than Kirkwood's since these can dominate when the Kirkwood contribution is symmetry forbidden.

The dynamic coupling mechanism is sometimes called a dispersion mechanism because there is some similarity with the dispersion contribution to intermolecular forces. Thus (5.3.9) giving the isotropic optical activity contains the polarizabilities of the two groups, and these polarizabilities remain finite even at zero frequency (although the optical activity itself becomes zero because it depends on ω). However, at infrared frequencies the polarizability is *dominated* by the static part. Thus at visible and ultraviolet frequencies this contribution is best regarded as a dynamic coupling mechanism, but at infrared frequencies it is best regarded as a dispersion mechanism.

5.3.3 *Exciton coupling (the degenerate coupled oscillator model)*

Until now, we have used wave functions localized on two or more separate groups of a chiral structure. This is acceptable when the groups are different and their energy levels do not coincide, but when the groups are *identical* the wavefunctions must be defined more carefully. The ground state wavefunction of a dimer is written as the direct product of the individual group ground state wavefunctions,

$$|n\rangle = |n_1 n_2\rangle. \tag{5.3.25}$$

Since the wavefunctions of the dimer must reflect the fact that there is an equal probability of finding an excitation induced by the light wave on either group, the dimer wavefunction corresponding to a transition to a particular excited state $|j_i\rangle$ of a group i is

$$|j_\pm\rangle = \frac{1}{\sqrt{2}}(|n_1 j_2\rangle \pm e^{i\alpha}|j_1 n_2\rangle), \tag{5.3.26}$$

where we have used the notation of (4.3.56). This also reflects the fact that the dimer has a C_2 axis, so the true molecular wavefunctions, having a definite energy, must be either symmetric or antisymmetric with respect to the C_2 rotation. From (4.3.55), interaction between the two singly excited local group states results in the following *exciton splitting* of the degeneracy of the states $|j_\pm\rangle$ (Craig and Thirunamachandran, 1984):

$$W_{j+} - W_{j-} = 2|\langle n_1 j_2|V|j_1 n_2\rangle|. \tag{5.3.27}$$

The interaction Hamiltonian is taken to be the operator equivalent of the interaction energy (2.5.15) between two charge distributions. Since the two groups are neutral, dipole–dipole coupling makes the first contribution:

$$V = -\frac{1}{4\pi\epsilon_0}T_{12_{\alpha\beta}}\mu_{1\alpha}\mu_{2\beta}. \tag{5.3.28}$$

So if $T_{12_{\alpha\beta}}\langle n_1|\mu_{1\alpha}|j_1\rangle\langle j_2|\mu_{2\beta}|n_2\rangle$ is real and negative, the interaction energy itself will be real and positive so that $e^{i\alpha} = +1$ and the symmetric state has the higher energy. (It should be remembered in what follows that the subscripts $\pm$ in $|j_\pm\rangle$ and $\omega_\pm$ refer to the higher and lower energy states, not to the symmetric and antisymmetric states.)

The transition frequencies from the ground state to the first two excited states of the dimer are

$$\omega_\pm = \omega_{j\pm} - \omega_n$$
$$= \omega_{j_1 n_1} \pm \frac{1}{4\pi\epsilon_0\hbar}|T_{12_{\alpha\beta}}\langle n_1|\mu_{1\alpha}|j_1\rangle\langle j_2|\mu_{2\beta}|n_2\rangle|. \tag{5.3.29}$$

If the two exciton levels are well resolved and far from other electronic levels of the dimer, each contributes separately to the optical activity so that the isotropic part of the optical activity tensors is

$$G'_{\alpha\alpha} = G'^+_{\alpha\alpha} + G'^-_{\alpha\alpha}$$

$$= -\frac{2}{\hbar} \left\{ \frac{\omega}{\omega_+^2 - \omega^2} \, \text{Im}(\langle n|\mu_\alpha|j_+\rangle\langle j_+|m_\alpha|n\rangle) \right.$$

$$\left. + \frac{\omega}{\omega_-^2 - \omega^2} \text{Im}(\langle n|\mu_\alpha|j_-\rangle\langle j_-|m_\alpha|n\rangle) \right\}. \tag{5.3.30}$$

Writing the electric and magnetic dipole moment operators of the dimer in the form (5.3.5a) and (5.3.5c), and using the wavefunctions (5.3.26), this becomes

$$G'_{\alpha\alpha} = -\frac{e^{i\alpha}}{\hbar} \left\{ \frac{\omega}{\omega_+^2 - \omega^2} \left[-\tfrac{1}{2}\omega_{j_in_i}\varepsilon_{\alpha\beta\gamma} R_{12_\beta} \text{Re}(\langle n_1|\mu_{1\alpha}|j_1\rangle\langle j_2|\mu_{2\gamma}|n_2\rangle) \right.\right.$$

$$\left. + \text{Im}(\langle n_2|\mu_{2\alpha}|j_2\rangle\langle j_1|m_{1\alpha}|n_1\rangle) + \text{Im}(\langle n_1|\mu_{1\alpha}|j_1\rangle\langle j_2|m_{2\alpha}|n_2\rangle) \right]$$

$$+ \frac{\omega}{\omega_-^2 - \omega^2} \left[\tfrac{1}{2}\omega_{j_in_i}\varepsilon_{\alpha\beta\gamma} R_{12_\beta} \text{Re}(\langle n_1|\mu_{1\alpha}|j_1\rangle\langle j_2|\mu_{2\gamma}|n_2\rangle) \right.$$

$$\left.\left. - \text{Im}(\langle n_2|\mu_{2\alpha}|j_2\rangle\langle j_1|m_{1\alpha}|n_1\rangle) - \text{Im}(\langle n_1|\mu_{1\alpha}|j_1\rangle\langle j_2|m_{2\alpha}|n_2\rangle) \right] \right\}. \tag{5.3.31}$$

Notice that terms in $\text{Im}(\langle n_i|\mu_{i\alpha}|j_i\rangle\langle j_j|m_{j\alpha}|n_j\rangle)$ guarantee the invariance of (5.3.31) to shifts in the local group origins. We have dropped the terms corresponding to the intrinsic rotational strengths of the two groups.

The corresponding exciton rotational strengths are

$$R(j_\pm \leftarrow n) = \mp\frac{e^{i\alpha}}{4}\omega_{j_in_i}\varepsilon_{\alpha\beta\gamma} R_{12_\beta}\text{Re}(\langle n_1|\mu_{1\alpha}|j_1\rangle\langle j_2|\mu_{2\gamma}|n_2\rangle)$$

$$\pm \tfrac{1}{2}[\text{Im}(\langle n_2|\mu_{2\alpha}|j_2\rangle\langle j_1|m_{1\alpha}|n_1\rangle)$$

$$+ \text{Im}(\langle n_1|\mu_{1\alpha}|j_1\rangle\langle j_2|m_{2\alpha}|n_2\rangle)]. \tag{5.3.32}$$

Dispersive or absorptive lineshape functions can be introduced into (5.3.31). The contributions from the $j_+ \leftarrow n$ and $j_- \leftarrow n$ transitions in (5.3.31) together generate circular dichroism and optical rotatory dispersion line shapes characteristic of degenerate coupled chromophores. These are drawn in Fig. 5.2 for the case where the exciton splitting is larger than the linewidth. The absolute signs of the high and low frequency bands shown in Fig. 5.2 obtain when the *chirality factor*

$$\varepsilon_{\alpha\beta\gamma} R_{12_\beta}\text{Re}(\langle n_1|\mu_{1\alpha}|j_1\rangle\langle j_2|\mu_{2\gamma}|n_2\rangle)$$

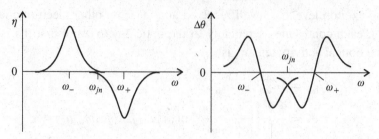

Fig. 5.2 Circular dichroism η and rotatory dispersion $\Delta\theta$ curves for a pair of degenerate coupled chromophores where the exciton splitting is larger than the linewidth. The absolute signs shown obtain when the chirality factor $\varepsilon_{\alpha\beta\gamma} R_{12_\beta} \mathrm{Re}\left(\langle n_1|\mu_{1\alpha}|j_1\rangle\langle j_2|\mu_{2\gamma}|n_2\rangle\right)$ and the coupling factor $T_{12_{\alpha\beta}} \mathrm{Re}\left(\langle n_1|\mu_{1\alpha}|j_1\rangle\langle j_2|\mu_{2\beta}|n_2\rangle\right)$ have opposite signs.

and the *coupling factor*

$$T_{12_{\alpha\beta}} \mathrm{Re}\left(\langle n_1|\mu_{1\alpha}|j_1\rangle\langle j_2|\mu_{2\beta}|n_2\rangle\right)$$

have opposite signs and the additional terms in

$$\mathrm{Im}\left(\langle n_i|\mu_{i\alpha}|j_i\rangle\langle j_j|m_{j\alpha}|n_j\rangle\right)$$

are either zero or can be neglected. The opposite absolute signs obtain when the chirality factor and the coupling factor have the same absolute signs.

The exciton treatment falls within the dynamic coupling model since the exciton splitting originates in an interaction between the electric dipole moments of monomer states excited by the light wave: in the absence of the light wave this interaction does not exist. The exciton treatment is most appropriate in the limiting case of frequency shifts larger than the linewidth. The other limiting case of frequency shifts much smaller than the linewidth is best described by the dynamic coupling expressions (5.3.23); and for the $j \leftarrow n$ transition of two identical monomers, the dispersive and absorptive parts of $G'_{\alpha\alpha}$ depend, respectively, on the functions $f^2 - g^2$ and fg given by (2.7.7). The circular dichroism and optical rotatory dispersion lineshapes now have the forms shown in Fig. 5.3, which are similar to the lineshapes given by the exciton model except that the displacements of the turning points from the band centres are determined by Γ_j rather than by the exciton splitting. The absolute signs of the band structures are determined by the sign of

$$\varepsilon_{\alpha\beta\gamma} R_{12_\beta} \alpha_{2_{\gamma\delta}} T_{12_{\delta\epsilon}} \alpha_{1_{\epsilon\alpha}},$$

which is in effect an 'amalgam' of the chirality factor and the coupling factor.

It is instructive to compare in detail the application of the dynamic coupling results and the exciton results to a simple chiral structure involving two groups

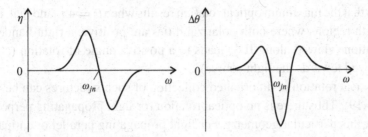

Fig. 5.3 Circular dichroism η and rotatory dispersion $\Delta\theta$ curves for a pair of degenerate coupled chromophores where the exciton splitting is much smaller than the linewidth.

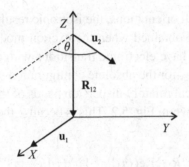

Fig. 5.4 Vectors defining the geometry of a simple chiral two-group structure.

each with $C_{nv}(n > 2)$ symmetry and with their symmetry axes in parallel planes (Fig. 5.4).

The optical rotation of an isotropic collection of such structures can be deduced from (5.3.18). For groups of this particular symmetry, if the local origins are chosen anywhere along the group symmetry axes there is no contribution from terms of the type $G'_{i_{\alpha\beta}} T_{ij_{\beta\gamma}} \alpha_{jy\alpha}$. The choice indicated in Fig. 5.4 is particularly convenient. The only part of $T_{12_{\alpha\beta}}$ that contributes here is $-\delta_{\alpha\beta}/R_{12}^3$, so the net geometrical factor is

$$\varepsilon_{\alpha\beta\gamma} R_{12_{\beta}} (u_{2\gamma} u_{2_\delta} T_{12_{\gamma\epsilon}} u_{1\epsilon} u_{1\alpha}) = -\frac{\sin 2\theta}{2 R_{12}^2}. \tag{5.3.33}$$

Choosing the group origins at other points along the group symmetry axes does not change this result: this can be seen by using the replacement $R_{12_{\beta}} \to R_{12_{\beta}} + \Delta r_1 u_{1_{\beta}} - \Delta r_2 u_{2_{\beta}}$ on the left-hand side of (5.3.33). The optical rotation is then simply

$$\Delta\theta \approx \frac{3\omega^2 \mu_0 \, lN}{16\pi \epsilon_0 R_{12}^2} (\alpha_1 \alpha_2 \kappa_1 \kappa_2) \sin 2\theta. \tag{5.3.34}$$

It follows that the maximum optical rotation results when $\theta = 45°$ and that, in longer wavelength regions where both polarizabilities are positive, a right-handed screw configuration (viewed along $\mathbf{R}_{12}$) leads to a positive angle of rotation (clockwise when viewed towards the light source).

The optical rotation of an oriented collection of such structures can be deduced from (5.3.19). Thus there is no optical rotation for light propagating perpendicular to $\mathbf{R}_{12}$ for this particular geometry. For light propagating parallel or antiparallel to $\mathbf{R}_{12}$ we find the follow optical rotation:

$$\Delta\theta \approx \frac{9\omega^2 \mu_0 \, lN}{16\pi\epsilon_0 R_{12}^2} (\alpha_1 \alpha_2 \kappa_1 \kappa_2) \sin 2\theta. \qquad (5.3.35)$$

If this is averaged over all orientations, the isotropic result (5.3.34) is recovered.

Analogous results are obtained when the exciton model is applied. If the two groups are identical and have electronic transitions with electric dipole transition moments along $\mathbf{u}_1$ and $\mathbf{u}_2$, for the absolute configuration shown in Fig. 5.4 the circular dichroism and optical rotatory dispersion bands of the corresponding exciton levels have the signs shown in Fig. 5.2. This is because the chirality factor reduces to

$$\varepsilon_{\alpha\beta\gamma} R_{12_\beta} \mathrm{Re}(\langle n_1|\mu_{1_\alpha}|j_1\rangle\langle j_2|\mu_{2_\gamma}|n_2\rangle)$$
$$= R_{12}\mathrm{Re}(\langle n_1|\mu_1|j_1\rangle\langle j_2|\mu_2|n_2\rangle) \sin\theta,$$

and the coupling factor reduces to

$$T_{12_{\alpha\beta}} \mathrm{Re}(\langle n_1|\mu_{1_\alpha}|j_1\rangle\langle j_2|\mu_{2_\beta}|n_2\rangle)$$
$$= -\frac{1}{R_{12}^3}\mathrm{Re}(\langle n_1|\mu_1|j_1\rangle\langle j_2|\mu_2|n_2\rangle) \cos\theta.$$

So, for values of θ where $\sin\theta$ and $\cos\theta$ have the same sign ($0 < \theta < \pi/2, \pi < \theta < 3\pi/2$), the chirality factor and the coupling factor have opposite signs. Thus the chirality factor is zero when the two transition moments are parallel and is a maximum when they are perpendicular; the coupling factor is a maximum for the parallel conformation and zero for the perpendicular. Since the amplitudes of the exciton circular dichroism and optical rotatory dispersion curves depend on the magnitude of both the exciton splitting and the intrinsic rotational strengths of each isolated $j_+ \leftarrow n$ and $j_- \leftarrow n$ transition (a large intrinsic rotational strength does no good if there is no splitting), these two conditions lead to an effective overall dependence of $\sin 2\theta$ for the exciton circular dichroism and optical rotatory dispersion amplitudes. This can be seen explicitly for the limiting case when the splitting is much less than the linewidth, which gives the circular dichroism and rotatory dispersion curves shown in Fig. 5.3. The factor

$$\varepsilon_{\alpha\beta\gamma} R_{12_\beta} \alpha_{2_{\gamma\delta}} T_{12_{\delta\epsilon}} \alpha_{1_{\epsilon\alpha}}$$

now automatically provides the sin 2θ dependence.

5.4 Illustrative examples

5.4.1 The carbonyl chromophore and the octant rule

The weak electronic absorption bands in the visible or near ultraviolet spectral regions of organic molecules are often due to the promotion of a lone pair electron on a heteroatom to an antibonding π or σ orbital localized on the chromophoric group. The following are typical chromophores containing a heteroatom:

$$\mathrm{C} = \mathrm{O}, \quad \mathrm{C} = \mathrm{S}, \quad -\mathrm{N} = \mathrm{O}, \quad -\mathrm{NO}_2, \quad -\mathrm{O} - \mathrm{N} = \mathrm{O}.$$

The carbonyl chromophore is of particular importance in electronic optical activity since the accessible electronic absorption band at 250–350 nm is weak, so ample transmitted light is available for measurement, yet the associated Cotton effects can be large. Consequently, a large body of experimental data exists from which symmetry rules have been deduced, and this enables the relative importance of the static and dynamic coupling mechanisms to be assessed.

The relevant localized atomic and molecular orbitals and electronic transitions of the carbonyl chromophore are shown in Fig. 5.5. The symmetry species are assigned on the basis of the local C_{2v} symmetry. The σ and σ^* molecular orbitals result from the in-phase and out-of-phase combinations of carbon and oxygen atomic orbitals; the atomic orbitals are the oxygen $2p_Z$ and the $(2s + \lambda 2p_Z)$ of the carbon sp^2 hybrids. The π and π^* molecular orbitals result from the in-phase and out-of-phase combinations of the oxygen and carbon $2p_X$ orbitals. The nonbonding n orbital is the oxygen $2p_Y$. The ground state has an electronic configuration $\sigma^2\pi^2n^2(^1A_1)$ and is a singlet. The lowest excited state arises from the $\pi^* \leftarrow n$ electron promotion; its configuration $\sigma^2\pi^2n\pi^*(^{1,3}A_2)$ generates both a singlet and a triplet. The weak absorption normally observed in the 250–350 nm region originates in the $\pi^* \leftarrow n$ singlet–singlet transition ($^1A_2 \leftarrow {}^1A_1$). The two strong absorptions normally observed in the 150–250 nm regions originate in the $\sigma^* \leftarrow n$ ($^1B_2 \leftarrow {}^1A_1$) and $\pi^* \leftarrow \pi$ ($^1A_1 \leftarrow {}^1A_1$) transitions.

A cursory inspection of the C_{2v} character table shows that the $\pi^* \leftarrow n$ transition is electric dipole forbidden and magnetic dipole and electric quadrupole allowed. Although relatively weak, the intensity is well above that expected from magnetic dipole and electric quadrupole mechanisms: this intensity originates mainly in $^1B_2 \leftarrow {}^1A_1$ and $^1B_1 \leftarrow {}^1A_1$ electric dipole allowed transitions to vibronic states

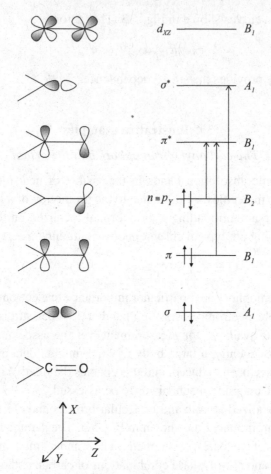

Fig. 5.5 The orbitals and electronic transitions of the carbonyl chromophore (energies not to scale). The origin of the coordinate system is the mid point of the CO bond.

associated with two distinct out-of-plane bending vibrations of symmetry species B_1 and an in-plane bending vibration of species B_2 (see, for example, King, 1964). Another much smaller 'forbidden' electric dipole contribution to the $\pi^* \leftarrow n$ carbonyl transition is present in optically active molecules and originates in the reduction of the C_{2v} symmetry of the chromophore by the chiral intramolecular environment, and is largely responsible for the isotropic optical activity associated with the $\pi^* \leftarrow n$ transition. The $^1B_1, {}^1B_2 \leftarrow {}^1A_1$ vibronic transitions can generate off-diagonal optical activity tensor components (in particular, G'_{XZ} and G'_{YZ}) in the unperturbed chromophore; and can generate isotropic optical activity when the chromophore is perturbed by a chiral intramolecular environment: these mechanisms are responsible for some of the vibronic structure of the $\pi^* \leftarrow n$ Cotton effect curves. It is shown in Section 5.5.3 that the rotational strength associated with a vibronic transition can require a different symmetry rule, and so have a different sign, from the rotational strength associated with the corresponding transition from the ground vibrational state of the ground electronic state to the ground vibrational state of the excited electronic state.

The $\sigma^* \leftarrow n$ transition is electric dipole, magnetic dipole and electric quadrupole allowed, although the components are not such as to generate isotropic optical activity. This again requires a chiral intramolecular environment. Similarly for the $\pi^* \leftarrow \pi$ transition, which is electric dipole and electric quadrupole allowed, and magnetic dipole forbidden.

We now apply (5.3.3) for the static coupling contribution, and (5.3.24) for the dynamic coupling contribution, to the induction of an isotropic rotational strength in the $\pi^* \leftarrow n$ carbonyl transition.

The charge q_2 of the perturber generates static coupling contributions to the rotational strength. Such a charge could originate in an ionic atom or group, or through incomplete shielding of the nuclear charges in a neutral atom or group. For the $\pi^* \leftarrow p_Y$ ($^1A_2 \leftarrow {}^1A_1$) carbonyl transition, n_1 and j_1 correspond to the configurations $\sigma^2 \pi^2 p_Y^2 (^1A_1)$ and $\sigma^2 \pi^2 p_Y \pi^* (^1A_2)$ (we are now using p_Y to denote the oxygen lone pair electron to avoid confusion with the designation n for a general initial state). Then of the possible multipole $j_1 \leftarrow n_1$ transition moments, only $\langle n_1 | m_{1_Z} | j_1 \rangle$ is symmetry allowed, so the static coupling expression (5.3.3) for the isotropic rotational strength reduces to

$$R(j_1 \leftarrow n_1) = \frac{3q_2 R_{12_Z} R_{12_Y}}{2\pi \epsilon_0 R^5}$$

$$\times \left\{ \sum_{k_1 \neq n_1} \frac{1}{\hbar \omega_{k_1 n_1}} \mathrm{Im}(\langle k_1 | \Theta_{1_{XY}} | n_1 \rangle \langle n_1 | m_{1_Z} | j_1 \rangle \langle j_1 | \mu_{1_Z} | k_1 \rangle) \right.$$

$$\left. + \sum_{k_1 \neq j_1} \frac{1}{\hbar \omega_{k_1 j_1}} \mathrm{Im}(\langle j_1 | \Theta_{1_{XY}} | k_1 \rangle \langle n_1 | m_{1_Z} | j_1 \rangle \langle k_1 | \mu_{1_Z} | n_1 \rangle) \right\} + \cdots .$$

$$(5.4.1)$$

In each term, the static perturbation operator must transform as A_2, and Θ_{XY} and $\Theta_{YX} (= \Theta_{XY})$ are the candidates of lowest order. The first excited state k for which $\langle k | \mu_Z | n \rangle$ and $\langle j | \Theta_{XY} | k \rangle$ is symmetry allowed has the configuration $\sigma^2 \pi p_Y^2 \pi^*$, in which case $\langle k | \mu_Z | n \rangle$ and $\langle j | \Theta_{XY} | k \rangle$ effect $\pi^* \leftarrow \pi$ and $\pi \leftarrow p_Y$ single electron transitions, respectively. The first excited state k for which $\langle j | \mu_Z | k \rangle$ and $\langle k | \Theta_{XY} | n \rangle$ are symmetry allowed is $\sigma^2 \pi^2 p_Y d_{XZ}$, where d_{XZ} refers to some combination of carbon and oxygen d_{XZ} orbitals, in which case $\langle j | \mu_Z | k \rangle$ and $\langle k | \Theta_{XY} | n \rangle$ effect $\pi^* \leftarrow d_{XZ}$ and $d_{XZ} \leftarrow p_Y$ orbital transitions, respectively. The corresponding charge induced contribution to the $\pi^* \leftarrow p_Y$ ($^1A_2 \leftarrow {}^1A_1$) isotropic rotational strength now reduces to the following products of orbital transition moments:

$$R(\pi^* \leftarrow p_Y) = \frac{3q\,R_{12_X}R_{12_Y}}{2\pi\,\epsilon_0\,R^5}$$

$$\times \left\{ \frac{1}{\hbar\omega_{d_{xz}p_Y}}\mathrm{Im}(\langle d_{XZ}|\Theta_{XY}|p_Y\rangle\langle p_Y|m_Z|\pi^*\rangle\langle\pi^*|\mu_Z|d_{XZ}\rangle) \right.$$

$$\left. + \frac{1}{\hbar\omega_{p_Y\pi}}\mathrm{Im}(\langle\pi|\Theta_{XY}|p_Y\rangle\langle p_Y|m_Z|\pi^*\rangle\langle\pi^*|\mu_Z|\pi\rangle)\right\}. \quad (5.4.2)$$

Both terms in this expression may be pictured as a helical displacement of charge: $\langle p_Y|m_Z|\pi^*\rangle$ describes a rotation of charge about the Z axis, whereas $\langle\pi^*|\mu_Z|d_{XZ}\rangle\langle d_{XZ}|\Theta_{XY}|p_Y\rangle$ and $\langle\pi^*|\mu_Z|\pi\rangle\langle\pi|\Theta_{XY}|p_Y\rangle$ describe linear displacements of charge with components along the Z axis; the successive rotation and linear displacement is equivalent to a helical path. Many other excited states could contribute to this rotational strength, and it is difficult to assess their relative importance. But the significant feature of all such contributions is their dependence on $R_{12_X}R_{12_Y}$ since this leads to a *quadrant rule*. The two symmetry planes of the carbonyl chromophore divide the surrounding space into quadrants: moving the perturbing charge q from one quadrant into another of the same sign leaves the sign the sign of $R_{12_X}R_{12_Y}$ unchanged, whereas moving q into a quadrant of opposite sign changes the sign of $R_{12_X}R_{12_Y}$ (Fig. 5.6a). An *octant rule*, where an additional nodal plane bisects the $C = 0$ bond, is associated with the higher order interaction of the perturbing charge with an octopole transition moment.

If the perturber is neutral and spherical with no electric multipole moments, such as a ground state hydrogen atom, only dynamic coupling contributes to the rotational strength. For a spherical perturber, $\alpha_{2_{\alpha\beta}} = \alpha_2\delta_{\alpha\beta}$, and the isotropic rotational strength (5.3.24) reduces to

$$R(\pi^* \leftarrow p_Y) = \frac{5\alpha_2 R_{12_X}R_{12_Y}R_{12_Z}}{2\pi\,\epsilon_0\,R^7}\mathrm{Im}(\langle p_Y|m_Z|\pi^*\rangle\langle\pi^*|\Theta_{XY}|p_Y\rangle), \quad (5.4.3)$$

which is the symmetry allowed term of lowest order when $j_1 \leftarrow n_1$ is the $\pi^* \leftarrow p_Y$ ($^1A_2 \leftarrow {}^1A_1$) transition. This generates an octant rule (Fig. 5.6b). In the $\pi^* \leftarrow p_Y$ ($^1B_2 \leftarrow {}^1A_1$) vibronic transition, $\langle n|\mu_Y|j\rangle\langle j|m_x|n\rangle$ is symmetry allowed and leads to a quadrant contribution; and $\langle n|m_X|j\rangle\langle j|\Theta_{YZ}|n\rangle$ is also symmetry allowed and leads to an octant contribution.

If the perturber is axially symmetric and dipolar, both static and dynamic coupling contribute to the rotational strength. Such a perturber could be a bond, a group able to rotate freely about an axis, or a lone pair of electrons on, say, a nitrogen atom. For the $\pi^* \leftarrow p_Y$ ($^1A_2 \leftarrow {}^1A_1$) transition, the first symmetry-allowed contributions to

(a)

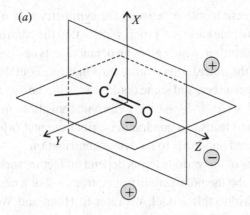

(b)

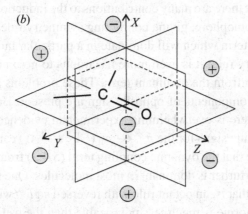

Fig. 5.6 (a) Quadrant and (b) octant rules for the sign of the rotational strength induced in the $\pi^* \leftarrow n$ transition of the carbonyl group by a perturbing group.

the isotropic rotational strengths (5.3.3) and (5.3.24) are

$$
\begin{aligned}
R(\pi^* \leftarrow p_Y) = \frac{3\mu_Z}{2\pi\epsilon_0 R^7} & \left[\frac{1}{\hbar\omega_{d_{XZ}p_Y}} \operatorname{Im}(\langle d_{XZ}|\Theta_{XY}|p_Y\rangle\langle p_Y|m_Z|\pi^*\rangle\langle\pi^*|\mu_Z|d_{XZ}\rangle) \right. \\
& \left. + \frac{1}{\hbar\omega_{p_Y\pi}} \operatorname{Im}(\langle\pi|\Theta_{XY}|p_Y\rangle\langle p_Y|m_Z|\pi^*\rangle\langle\pi^*|\mu_Z|\pi\rangle) \right] \\
& \times [5R_{12_X}R_{12_Y}R_{12_Z}u_{2_Z} + R_{12_X}(5R_{12_Y}R_{12_Y} - R^2)u_{2_Y} \\
& + R_{12_Y}(5R_{12_X}R_{12_X} - R^2)u_{2_X}] \\
& + \frac{1}{2\pi\epsilon_0 R^7} \operatorname{Im}(\langle p_Y|m_Z|\pi^*\rangle\langle\pi^*|\Theta_{XY}|p_Y\rangle) \\
& \times \{5[\alpha_2(1 - \kappa_2) + 3\alpha_2\kappa_2 u_{2_Z}u_{2_Z}]R_{12_X}R_{12_Y}R_{12_Z} \\
& + 3\alpha_2\kappa_2[R_{12_Y}(5R_{12_X}^2 - R^2)u_{2_X}u_{2_Z} \\
& + R_{12_X}(R_{12_Y}^2 - R^2)u_{2_Y}u_{2_Z}]\},
\end{aligned}
\tag{5.4.4}
$$

where $u_{2\alpha}$ is the direction cosine between the symmetry axis $\mathbf{u}_2$ of the perturbing group and the α axis of the carbonyl group. Notice that the contribution (5.4.3) from a neutral spherical perturber, which gives an octant rule, is one part of this more general expression. All the other terms require information about the orientation of the dipolar anisotropic perturber, and some could lead to deviations from simple octant behaviour. In the $\pi^* \leftarrow p_Y$ ($^1B_2 \leftarrow {}^1A_1$) vibronic transition, $\langle n|\mu_Y|j\rangle\langle j|m_X|n\rangle$ is symmetry allowed and leads to a quadrant contribution; and $\langle n|m_X|j\rangle\langle j|\Theta_{YZ}|n\rangle$ is also symmetry allowed and leads to an octant contribution.

The absolute signs of the various terms depend on factors such as the signs of the transition moments and the polarizability anisotropies. For a detailed discussion of the absolute signs within this model, we refer to Höhn and Weigang (1968) and Buckingham and Stiles (1974).

We have seen that there are many contributions to the induction of optical activity in the carbonyl chromophore, giving conflicting symmetry rules. It is usually hazardous to select the term which will dominate in a particular molecule and thereby predict the symmetry rule; it is even more hazardous to go on and deduce the absolute configuration from the dominant term. These problems are compounded if vibronic structure components of opposite sign are present. However, apart from a few anomalies, there is overwhelming experimental evidence for an octant rule with the same absolute signs for the $\pi^* \leftarrow p_Y$ ($^1A_2 \leftarrow {}^1A_1$) carbonyl transition. It is therefore possible that the dynamic coupling term (5.4.3) from a neutral isotropically polarizable perturber is dominant in most molecules. One striking anomaly is the antioctant rule (that is, an octant rule with reversed signs) when the perturber is a fluorine atom. Fluorine is much less polarizable than the usual perturbers (alkyl groups, hydrogen and halogen atoms, etc.), and since a large dipole moment is associated with the C–F bond, the terms in (5.4.4) involving the dipole moment μ_2 of the perturber could dominate and generate the antioctant behaviour. Further details of the octant rule may be found in Rodger and Norden (1997) and Lightner (2000).

The rotational strength expressions deduced above by detailed considerations of transition and perturbation matrix elements for different perturbations accord with the listings in Tables 4.2, which give the allowed components of polar and axial property tensors in the important point groups. For example, Tables 4.2 show that, for a group such as carbonyl with C_{2v} symmetry, all components $G'^{(\mu)}_{\alpha\alpha,\beta}$ of the isotropic part of the optical activity tensor perturbed by a uniform electric field are zero; whereas components $G'^{(\Theta)}_{\alpha\alpha,XY}$ of the corresponding tensor perturbed by an electric field gradient do exist, in agreement with (5.4.1). Similarly, Tables 4.2 show that only components $D'_{Z,XY}$ of the magnetic dipole–electric quadrupole tensor (2.6.27j) are nonzero, in agreement with (5.4.3).

5.4.2 The Co^{3+} chromophore: visible, near ultraviolet and X-ray circular dichroism

In transition metal complexes, electronic optical activity can be induced in the central metal ion by a chiral arrangement of ligands. The classic example is the tris (ethylenediamine) cobalt(III) ion $Co(en)_3^{3+}$ which has D_3 symmetry (Fig. 5.7). Circular dichroism measurements in light propagating along the optic axis (which corresponds to the C_3 axis of the ion) of uniaxial crystals containing this ion have isolated components of the optical activity tensors (McCaffery and Mason, 1963) and provided the first clear example of electric dipole–electric quadrupole optical activity (Barron, 1971). We shall confine the discussion to generalities since the detailed electronic mechanisms are complicated.

The electronic absorption spectrum of $Co(en)_3$ is similar to that of the corresponding O_h complex $Co(NH_3)_6^{3+}$ so the selection rules for electronic transitions in the D_3 complex are assumed to be predominantly those of the parent O_h complex. We consider only the electronic transitions localized on the metal ion. The degeneracy of the metal d orbitals is lifted by the octahedral crystal field in $Co(NH_3)_6^{3+}$: the corresponding irreducible representation $D^{(2)}$ of the full rotation group correlates with the irreducible representations $E_g + T_{2g}$ on descent of symmetry to O_h. The five degenerate d orbitals therefore become two doubly-degenerate e_g orbitals and three triply-degenerate t_{2g} orbitals, and the electronic configuration d^6 of Co^{3+} becomes t_{2g}^6 in the 'strong field' complex $Co(NH_3)_6^{3+}$, which gives rise to a single electronic state $^1A_{1g}$. The promotion of an electron to an e_g orbital generates the configuration $t_{2g}^5 e_g$, which gives rise to the electronic states $^{1,3}T_{1g}$ and $^{1,3}T_{2g}$.

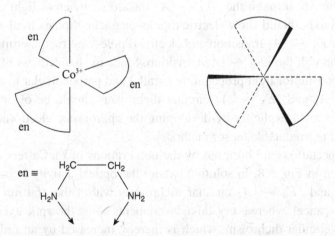

Fig. 5.7 The $(+)$-$Co(en)_3^{3+}$ ion and the view along the C_3 axis.

The weak electronic absorption bands in $Co(NH_3)_6^{3+}$ with maxima at about 476 and 342 nm are ascribed to the transitions $^1T_{1g} \leftarrow {}^1A_{1g}$ and $^1T_{2g} \leftarrow {}^1A_{1g}$, respectively. Both transitions are electric dipole forbidden, the former being magnetic dipole allowed and the latter electric quadrupole allowed, although most of the observed intensity arises through vibronic electric dipole allowed transitions. As in the carbonyl group, such vibronic transitions generate the vibronic structure of the circular dichroism bands when the chromophore is in a chiral environment. The overall rotational strength is determined by the allowed magnetic dipole and electric quadrupole transition moments and the small electric dipole transition moment, induced by the chiral environment, between electronic states in their ground vibrational states. The O_h representations T_{1g} and T_{2g} correlate with $A_2 + E_a$ and $A_1 + E_b$, respectively, on descent of symmetry to D_3, and the weak electronic absorption bands in $Co(en)_3^{3+}$ at 469 and 340 nm are ascribed to the transitions $^1A_2, {}^1E_a \leftarrow {}^1A_1$ and $^1A_1, {}^1E_b \leftarrow {}^1A_1$, respectively.

So we anticipate that the 469 nm absorption band is associated with strong magnetic dipole–weak electric dipole circular dichroism, and the 340 nm band is associated with strong electric quadrupole–weak electric diole circular dichroism. Furthermore, from the irreducible representations of D_3 spanned by particular components of $\boldsymbol{\mu}$, $\mathbf{m}$ and Θ, in conjunction with Tables 4.2, we deduce that the $^1A_1 \leftarrow {}^1A_1$ transition generates no optical activity tensor components; $^1A_2 \leftarrow {}^1A_1$ generates G'_{ZZ}; $^1E_a \leftarrow {}^1A_1$ generates $G'_{XX} = G'_{YY}$; and $^1E_b \leftarrow {}^1A_1$ generates $A_{X,YZ} = -A_{Y,XZ}$. Instead of using Tables 4.2, these conclusions can also be reached by developing the transition matrix elements explicitly using the irreducible tensor methods outlined in Section 4.4.6. So light propagating perpendicular to the C_3 axis (Z) of $Co(en)_3^{3+}$ should show electric dipole–magnetic dipole circular dichroism through the $^1A_2 \leftarrow {}^1A_1$ transition; whereas light propagating along the C_3 axis should show electric dipole–magnetic dipole circular dichroism through the $^1E_a \leftarrow {}^1A_1$ transition and electric dipole–electric quadrupole circular dichroism through the $^1E_b \leftarrow {}^1A_1$ transition. Since the handedness of the helical D_3 ion is opposite for light propagating parallel and perpendicular to the C_3 axis, the $^1A_2 \leftarrow {}^1A_1$ and $^1E_a \leftarrow {}^1A_1$ circular dichroisms should be of opposite sign (this can be shown explicitly by developing the appropriate electronic transition moments using irreducible tensor methods).

These expectations are born out by the observations of McCaffery and Mason (1963) shown in Fig. 5.8. In solution, where the optical activity is isotropic, the $^1A_2 \leftarrow {}^1A_1$ and $^1E_a \leftarrow {}^1A_1$ circular dichroisms within the 469 nm absorption band tend to cancel, whereas crystal measurements along the optic axis isolate the $^1E_a \leftarrow {}^1A_1$ circular dichroism, which is thereby increased by an order of magnitude. The $^1E_b \leftarrow {}^1A_1$ circular dichroism within the 340 nm absorption band is

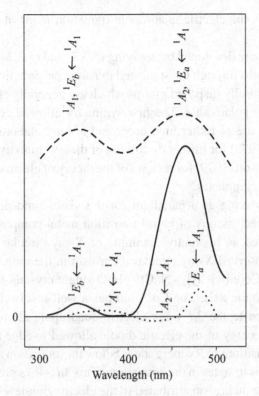

Fig. 5.8 A sketch of the absorption - - - and circular dichroism $\cdots$ of $(+)$-Co(en)$_3^{3+}$ in water, and the circular dichroism —— of the crystal $\{(+)$ - [Co(en)$_3$]Cl$_3\}_2$ · NaCl · 6H$_2$O for light propagating along the optic axis (arbitrary units). Adapted from McCaffery and Mason (1963).

only observed in crystal measurements since electric dipole–electric quadrupole optical activity 'washes out' in isotropic samples. The associated solution circular dichroism is at slightly longer wavelength and could originate in a vibronic contribution to the forbidden $^1A_1 \leftarrow {}^1A_1$ transition which would generate G'_{ZZ} and would therefore not appear in the crystal circular dichroism along the optic axis.

The electric quadrupole optical activity increases with the radius of the central metal ion. Thus the $^1E_a \leftarrow {}^1A_1$ rotational strengths of Rh(Ox)$_3^{3-}$ and Co(Ox)$_3^{3-}$ are similar, whereas the $^1E_b \leftarrow {}^1A_1$ rotational strength is significantly larger in the Rh complex (McCaffery, Mason and Ballard, 1965). Rh and Co occupy equivalent positions in the $4d$ and $3d$ transition series, the main distinction being that the radii of the $4d$ atoms are rather larger than those of the equivalent $3d$ atoms. The magnetic dipole transition moment is independent of the principal quantum number n, whereas the electric quadrupole transition moment increases roughly as n^4/Z_{eff}^2.

Symmetry rules are developed by applying (5.3.3) and (5.3.24) to the generation of rotational strengths through the static and dynamic perturbation of the parent O_h chromophore by chirally disposed groups which are generally charged, multipolar and anisotropically polarizable. The first symmetry allowed contributions to the rotational strength are of rather high order and are not elaborated here. See, for example, Mason (1973) for further discussion of these symmetry rules; and Mason (1979) and Richardson (1979) for reviews of the theory of electronic optical activity in transition metal complexes.

In addition to serving as a paradigm for the visible and near ultraviolet circular dichroism spectroscopy of chiral transition metal complexes, the $Co(en)_3^{3+}$ system has provided an instructive example of X-ray circular dichroism. Stewart *et al.* (1999) observed X-ray circular dichroism in the range $\sim$7690–7770 eV in resolved $\{(+)\text{-}[Co(en)_3]Cl_3\}_2 \cdot NaCl \cdot 6H_2O$ single crystals for radiation propagating along the optic axis. The Co^{3+} ion has a well-resolved pre-edge band at $\sim$7790 eV assigned to the $3d \leftarrow 1s$ electric quadrupole allowed transition that is $\sim$18 eV to low energy of the electric dipole allowed K-edge absorption arising from $np \leftarrow 1s$ transitions to Rydberg states below the ionization threshold and from $\varepsilon p \leftarrow 1s$ transitions to states in the continuum. This $3d \leftarrow 1s$ pre-edge band shows large X-ray circular dichroism attributed to the electric dipole–electric quadrupole mechanism due to a small contribution, induced by the chiral environment, from the nearby $\varepsilon p, np \leftarrow 1s$ electric dipole-allowed transitions. The electric dipole–magnetic dipole mechanism makes a negligible contribution to this X-ray circular dichroism since the $\Delta n = 0$ selection rule on magnetic dipole transitions (arising from the fact that $\mathbf{m}$ is a pure spatial angular momentum operator and so cannot connect states that are radially orthogonal) forbids inter-shell magnetic dipole transitions. Ab initio computations confirm the dominant contribution of the electric dipole–electric quadrupole mechanism to the rotational strength of the $3d \leftarrow 1s$ transition and also support the earlier assignment of the near ultraviolet circular dichroism in the 340 nm absorption band, discussed above, to this mechanism (Peacock and Stewart, 2001).

5.4.3 Finite helices: hexahelicene

The severely overcrowded hydrocarbon hexahelicene (Fig. 5.9) provides an interesting model for discussing the generation of natural optical activity within molecules with a finite helical structure. Steric interference between the regions a, b at one end and c, d at the other end of the molecule force it to take up a right (P)- or left (M)- handed helical conformation. The P- and M-absolute configurations were

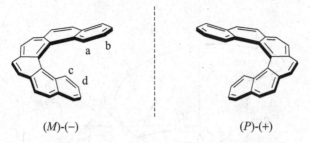

Fig. 5.9 The two enantiomers of hexahelicene.

assigned to enantiomers generating positive (+) and negative (−) specific rotations, respectively, using anomalous X-ray scattering crystallography (Lightner *et al.*, 1972). Since there is complete π electron exchange between the benzenoid rings, the entire molecule should be regarded as a single chromophore with all its transitions fully electric dipole, magnetic dipole and electric quadrupole allowed.

Rather than dwell on the complications of the inherently chiral chromophore aspects of hexahelicene optical activity, we shall calculate the optical rotation at transparent wavelengths by considering dynamic coupling between all 15 pairs of benzenoid rings. Actually, the dynamic coupling theory is not strictly applicable since it assumes that the electronic transitions are localized in the groups. However, a good answer is obtained for the specific rotation, and the correct absolute configuration is deduced. In the event, electron delocalization could probably be incorporated into the dynamic coupling approach by summing contributions from appropriately weighted valence bond structures: this should increase the calculated optical rotation of hexahelicene, and should not affect the sign.

If we assume D_{6h} symmetry for the benzenoid rings, the simplified dynamic coupling contribution (5.3.19) to optical rotation for light propagating in an oriented sample can be applied to calculate the optical rotation components for light travelling along the three nonequivalent directions X, Y, Z (defined in Fig. 5.10) in an oriented hexahelicene molecule (Barron, 1975b). Thus, summing over all pairs of benzenoid rings, for light travelling along Z,

$$\Delta\theta_Z \approx \frac{3\omega^2\mu_0 l N}{8\pi\epsilon_0} \sum_{i>j=1}^{6} \alpha_i\alpha_j R_{ijZ}[\kappa_i(1-\kappa_j)(u_{i\alpha}u_{iX}T_{ijY\alpha} - u_{i\alpha}u_{iY}T_{ijX\alpha})$$
$$+ \kappa_j(1-\kappa_i)(u_{j\alpha}u_{jY}T_{ijX\alpha} - u_{j\alpha}u_{jX}T_{ijY\alpha})$$
$$+ 3\kappa_i\kappa_j(u_{jY}u_{j\alpha}T_{ij\alpha\beta}u_{i\beta}u_{iX} - u_{jX}u_{j\alpha}T_{ij\alpha\beta}u_{i\beta}u_{iY})]. \qquad (5.4.5)$$

The centres of the six benzenoid rings are placed on a right-handed cylindrical helix $X = a\cos\theta, Y = a\sin\theta, Z = b\theta$ at $\theta = 30°, 90°, 150°, 210°, 270°$ and $330°$. The

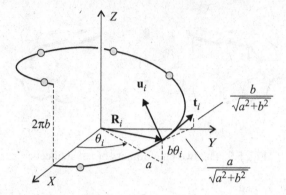

Fig. 5.10 The geometry of right-handed hexahelicene. a is the radius of the helix drawn through the centres of the benzenoid rings and $2\pi b$ is the pitch. $\mathbf{R}_i$ is the radius vector of the centre of the ith benzenoid ring, and $\mathbf{t}_i$ and $\mathbf{u}_i$ are the unit tangent and unit normal.

radius vector of the centre of the ith benzenoid ring is

$$R_{i\alpha} = I_\alpha a \cos\theta_i + J_\alpha a \sin\theta_i + K_\alpha b\theta_i, \tag{5.4.6}$$

where $\mathbf{I}, \mathbf{J}, \mathbf{K}$ are unit vectors along the internal axes X, Y, Z of the hexahelicene molecule. The unit tangent to the helix at $\mathbf{R}_i$ coincides with one of the principal axes of the benzenoid ring (say $\mathbf{t}_i$) and is

$$t_{i\alpha} = (a^2 + b^2)^{-1/2}(-I_\alpha a \sin\theta_i + J_\alpha a \cos\theta_i + K_\alpha b). \tag{5.4.7}$$

The unit vector $\mathbf{u}_i$ along the principal axis of the ith benzenoid ring is normal both to $\mathbf{t}_i$ and to the radius vector

$$a_{i\alpha} = I_\alpha a \cos\theta_i + J_\alpha a \sin\theta_i \tag{5.4.8}$$

perpendicular to the Z axis:

$$u_{i\alpha} = -\frac{1}{a}\varepsilon_{\alpha\beta\gamma} t_{i\beta} a_{i\gamma}$$
$$= (a^2 + b^2)^{-1/2}(I_\alpha b \sin\theta_i - J_\alpha b \cos\theta_i + K_\alpha a). \tag{5.4.9}$$

Using (5.4.9) and (5.4.6) in (5.4.5), it is found that the trigonometric functions combine so that the resulting expression is a function of $\theta_{ij} = \theta_i - \theta_j$ only. The corresponding specific rotation, defined in (5.2.26), is found to be

$$[\Delta\theta_Z] \approx \frac{27 \times 10^6 N_0(n^2 + 2)\alpha^2\kappa^2\gamma}{\epsilon_0^2\lambda^2 Ma^2(1 + \gamma^2)^2} \sum_{i>j=1}^{6} f(\theta_{ij})_Z, \tag{5.4.10a}$$

where $\gamma = b/a$ and

$$f(\theta_{ij})_Z = \frac{\gamma^2\theta_{ij}}{[2(1 - \cos\theta_{ij}) + \gamma^2\theta_{ij}^2]^{3/2}} \left\{ \sin\theta_{ij}(1 + \gamma^2\cos\theta_{ij}) \right.$$

$$\left. + \frac{(\theta_{ij} - \sin\theta_{ij})[2(1 + \gamma^2)(1/\kappa - 1)(1 - \cos\theta_{ij}) - 3\gamma^2\sin\theta_{ij}(\theta_{ij} - \sin\theta_{ij})]}{2(1 - \cos\theta_{ij}) + \gamma^2\theta_{ij}^2} \right\}.$$

Similarly, the specific rotations for light propagating along the two nonequivalent directions X and Y perpendicular to the helix axis are

$$[\Delta\theta_X] = \frac{27 \times 10^6 N_0(n^2 + 2)\alpha^2\kappa^2\gamma}{\epsilon_0^2\lambda^2 Ma^2(1 + \gamma^2)^2} \sum_{i>j=1}^{6} f(\theta_{ij})_X, \qquad (5.4.10b)$$

$$[\Delta\theta_Y] = \frac{27 \times 10^6 N_0(n^2 + 2)\alpha^2\kappa^2\gamma}{\epsilon_0^2\lambda^2 Ma^2(1 + \gamma^2)^2} \sum_{i>j=1}^{6} f(\theta_{ij})_Y, \qquad (5.4.10c)$$

where

$$f(\theta_{ij})_X = \frac{(\cos\theta_i - \cos\theta_j)^2}{[2(1 - \cos\theta_{ij}) + \gamma^2\theta_{ij}^2]^{3/2}} \left\{ (1 + \gamma^2\cos\theta_{ij}) \right.$$

$$\left. - \frac{\gamma^2(\theta_{ij} - \sin\theta_{ij})[(1 + \gamma^2)(1/\kappa - 1)\theta_{ij} + 3(\theta_{ij} - \sin\theta_{ij})]}{2(1 - \cos\theta_{ij}) + \gamma^2\theta_{ij}^2} \right\},$$

$$f(\theta_{ij})_Y = \frac{(\sin\theta_i - \sin\theta_j)^2}{[2(1 - \cos\theta_{ij}) + \gamma^2\theta_{ij}^2]^{3/2}} \left\{ (1 + \gamma^2\cos\theta_{ij}) \right.$$

$$\left. - \frac{\gamma^2(\theta_{ij} - \sin\theta_{ij})[(1 + \gamma^2)(1/\kappa - 1)\theta_{ij} + 3(\theta_{ij} - \sin\theta_{ij})]}{2(1 - \cos\theta_{ij}) + \gamma^2\theta_{ij}^2} \right\}.$$

The specific rotation for an isotropic sample is

$$[\Delta\theta] = \tfrac{1}{3}([\Delta\theta_x] + [\Delta\theta_y] + [\Delta\theta_z])$$

$$= \frac{9 \times 10^6 N_0(n^2 + 2)\alpha^2\kappa^2\gamma}{\epsilon_0^2\lambda^2 Ma^2(1 + \gamma^2)^2} \sum_{i>j=1}^{6} f(\theta_{ij}), \qquad (5.4.11)$$

where

$$f(\theta_{ij}) = \frac{2(1 - \cos\theta_{ij}) + \gamma^2\theta_{ij}\sin\theta_{ij}}{[2(1 - \cos\theta_{ij}) + \gamma^2\theta_{ij}^2]^{3/2}} \left\{ (1 + \gamma^2\cos\theta_{ij}) - \frac{3\gamma^2(\theta_{ij} - \sin\theta_{ij})^2}{2(1 - \cos\theta_{ij}) + \gamma^2\theta_{ij}^2} \right\}.$$

Equation (5.4.11) was first derived by Fitts and Kirkwood (1955).

Using the following SI values,

$a = 2.42 \times 10^{-10}$ m (estimated assuming a bond length of 1.40Å),

$b = 0.486 \times 10^{-10}$ m (from X ray data),

$n = 1.45$ (refractive index of pure chloroform),

$M = 328.4$,

$N_0 = 6.023 \times 10^{23}$,

$\lambda = 5.893 \times 10^{-7}$ m (Na D-line),

$\alpha = 10.4 \times 4\pi\epsilon_0 \times 10^{-30}$m³ and $|\kappa| = 0.18$ (taken from light scattering data on benzene: Bridge and Buckingham, 1966), the calculated specific rotations are, for a right-handed helix,

$$[\Delta\theta_X] = +3880°, [\Delta\theta_Y] = +753°, [\Delta\theta_Z] = +3300°, [\Delta\theta] = +2650°.$$

The observed specific rotation in chloroform solution is +3640° (Newman and Lednicer, 1956) for a right-handed helix, so the calculated isotropic specific rotation is quite good considering the complexity and is comparable in accuracy with recent ab initio computations (Grimme *et al.*, 2002; Autschbach *et al.*, 2002). It is interesting that all three specific rotation components have the same sign: helix optical activity, at least in the form of circular dichroism, is generally found to have opposite signs for light propagating perpendicular and parallel to the helix axis. Unfortunately the three components have not yet been isolated experimentally.

5.5 Vibrational structure in circular dichroism spectra

5.5.1 Introduction

So far, the discussion of the generation of natural electronic optical activity within chiral moleculues has been concerned mainly with 'allowed' contributions to the rotational strength. This depends on the electronic chirality when the nuclei are at their equilibrium positions in the ground electronic state, and reflects the molecular chirality which might be correlated with the sign and magnitude of the rotational strength by a symmetry rule.

But an additional contribution to the rotational strength arises because the electronic chirality changes as the nuclei undergo vibrational motion, and 'forbidden' contributions can be ascribed to each vibrational mode of the molecule. The relationship of the sign and magnitude of these vibronic contributions to the rotational strength is less direct. Vibronic effects are particularly important in circular dichroism because of the low molecular symmetry, and because the transitions which give large dissymmetry factors (such as the carbonyl $\pi^* \leftarrow n$) are often fully magnetic dipole allowed and electric dipole forbidden so that the conventional absorption is usually generated by a vibronic transition.

We shall explore this topic by considering the vibrational perturbation of the ground and excited electronic states in the quantum mechanical expression for the isotropic rotational strength, following a treatment by Weigang (1965).

5.5.2 *The vibronically perturbed rotational strength*

Using the Herzberg–Teller approach outlined in Section 2.8.4, the perturbed electric dipole and magnetic dipole transition moments are written

$$\langle n'|\mu_\alpha|j'\rangle = \langle e_n|\mu_\alpha|e_j\rangle\langle v_n|v_j\rangle + \sum_p C_{\alpha_p}\langle v_n|Q_p|v_j\rangle, \qquad (5.5.1a)$$

$$\langle j'|m_\alpha|n'\rangle = \langle e_j|m_\alpha|e_n\rangle\langle v_j|v_n\rangle + \sum_q B_{\alpha_q}\langle v_j|Q_q|v_n\rangle, \qquad (5.5.1b)$$

where

$$C_{\alpha_p} = \sum_{e_k \neq e_n} \frac{\langle e_n|(\partial H_e/\partial Q_p)_0|e_k\rangle}{\hbar\omega_{e_n e_k}}\langle e_k|\mu_\alpha|e_j\rangle$$

$$+ \sum_{e_k \neq e_j} \frac{\langle e_k|(\partial H_e/\partial Q_p)_0|e_j\rangle}{\hbar\omega_{e_j e_k}}\langle e_n|\mu_\alpha|e_k\rangle, \qquad (5.5.1c)$$

$$B_{\alpha_q} = \sum_{e_k \neq e_n} \frac{\langle e_k|(\partial H_e/\partial Q_q)_0|e_n\rangle}{\hbar\omega_{e_n e_k}}\langle e_j|m_\alpha|e_k\rangle$$

$$+ \sum_{e_k \neq e_j} \frac{\langle e_j|(\partial H_e/\partial Q_q)_0|e_k\rangle}{\hbar\omega_{e_j e_k}}\langle e_k|m_\alpha|e_n\rangle. \qquad (5.5.1d)$$

The rotational strength of vibronic components of a particular electronic transition can now be written

$$R(e_j v_j \leftarrow e_n v_n) = \operatorname{Im}\Bigg[\langle e_n|\mu_\alpha|e_j\rangle\langle e_j|m_\alpha|e_n\rangle\langle v_n|v_j\rangle\langle v_j|v_n\rangle$$

$$+ \Bigg(\sum_q B_{\alpha_q}\langle e_n|\mu_\alpha|e_j\rangle\langle v_n|v_j\rangle\langle v_j|Q_q|v_n\rangle$$

$$+ \sum_p C_{\alpha_p}\langle e_j|m_\alpha|e_n\rangle\langle v_n|Q_p|v_j\rangle\langle v_j|v_n\rangle\Bigg)$$

$$+ \sum_p \sum_q C_{\alpha_p} B_{\alpha_q}\langle v_n|Q_p|v_j\rangle\langle v_j|Q_q|v_n\rangle + \cdots \Bigg]. \qquad (5.5.2)$$

Consider first the approximation that the potential energy surfaces of the ground and excited electronic states are sufficiently similar that the vibrational states in the different electronic manifolds are orthonormal ('vertical' potential surfaces):

$$\langle v_j | v_n \rangle = \delta_{v_j v_n}. \tag{5.5.3}$$

Then for the vibronic band corresponding to the transition from the ground vibrational state of the ground electronic state to the ground vibrational state of the excited electronic state (the 0–0 transition), the rotational strength is determined entirely by the first term of (5.5.2):

$$R(e_j 0_j \leftarrow e_n 0_n) = \mathrm{Im}(\langle e_n | \mu_\alpha | e_j \rangle \langle e_j | m_\alpha | e_n \rangle). \tag{5.5.4}$$

For a vibronic band corresponding to a transition from the ground vibrational state of the ground electronic state to one of the first excited vibrational states of the excited electronic state (a 1–0 transition) the rotational strength is determined entirely by the third term of (5.5.2):

$$R(e_j 1_j \leftarrow e_n 0_n) = \mathrm{Im}(C_{\alpha_p} B_{\alpha_p} \langle v_n | Q_p | v_j \rangle \langle v_j | Q_p | v_n \rangle). \tag{5.5.5}$$

All terms for overtone and combination transitions vanish. The resulting circular dichroism spectrum therefore consists of a 0–0 band followed by a series of single quantum vibronic bands, one for each normal mode, separated from the 0–0 band by their respective fundamental frequencies. Each vibronic rotational strength is determined by the sign and magnitude of $\mathrm{Im}\,(C_{\alpha_p} B_{\alpha_p})$, together with the magnitude of $\langle v_n | Q_p | v_j \rangle \langle v_j | Q_p | v_n \rangle$. If either $\langle e_n | \mu_\alpha | e_j \rangle$ or $\langle e_j | m_\alpha | e_n \rangle$ is forbidden, nontotally symmetric modes are expected to dominate this vibronic coupling mechanism. The last statement applies to transition moments and internal vibrational coordinates localized on an intrinsically achiral chromophore: all transition moments are fully allowed, and all normal modes are totally symmetric, when delocalized over a completely asymmetric structure (but not for a chiral structure that retains a proper rotation axis).

In fact the potential energy surfaces of the ground and excited electronic states are usually different, with equilibrium points no longer vertically disposed ('nonvertical' potential surfaces). The orthonormality condition on vibrational states in different electronic manifolds, (5.5.3), no longer holds: $\langle v_j | v_n \rangle$ are now Franck–Condon overlap integrals and are not necessarily zero when $v_j \neq v_n$. Thus progressions and combination vibronic bands can now arise with (5.5.4) and (5.5.5) describing the first members. But, in addition, the second term of the general vibronic rotational strength (5.5.2) can now contribute. The presence of $\langle v_n | v_j \rangle$ means that the first and second terms can only be nonzero for totally symmetric vibrational states in the excited electronic state: this can arise both from a single totally symmetric mode in a state of excitation associated with both an even or an odd number of quanta, or a single nontotally symmetric mode in a state of excitation associated with only an even number of quanta. Only the third term of (5.5.2) can contribute to the rotational strength in a vibronic transition to a single nontotally symmetric mode in a state of excitation associated with an odd number of quanta. This last term can also contribute to the combination of an odd quanta nontotally symmetric mode with both even and odd quanta totally symmetric modes: such a combination is often observed as a single quantum of a nontotally symmetric mode with a totally symmetric progression.

Applying the closure theorem in the space of the vibrational wavefunctions, we find from (5.5.2) that the rotational strength summed over all vibronic components of a circular dichroism band is

$$
\sum_{v_j} R(e_j v_j \leftarrow e_n v_n) = \text{Im} \Bigg\{ \langle e_n | \mu_\alpha | e_j \rangle \langle e_j | m_\alpha | e_n \rangle + 0
$$

$$
+ \sum_{p,q} C_{\alpha_p} B_{\alpha_p} \langle v_n | Q_p Q_q | v_n \rangle + \cdots \Bigg\}, \quad (5.5.6)
$$

where we have neglected the contribution from vibrationally excited molecules in the ground electronic state. The first term indicates that, within a progression of a circular dichroism band associated with 'allowed' electric and magnetic dipole transitions, the sum of the individual vibronic rotational strengths equals the rotational strength for the 0–0 transition. The disappearance of the second term on performing the summation indicates that it makes no net contribution to the integrated rotational strength, and that its contribution to a particular vibronic rotational strength could equally well be positive or negative, irrespective of the sign of the zero-order rotational strength. The third term, arising from a product of two vibronic mixing factors, does not vanish when summed over all vibrational states, and can produce both positive and negative vibronic circular dichroism bands depending on the sign of $\text{Im} \left(\langle e_n | \mu_\alpha | e_k \rangle \langle e_k | m_\alpha | e_n \rangle \right)$.

5.5.3 The carbonyl chromophore

These concepts are illustrated with some general remarks on the carbonyl chromophore. The $\pi^* \leftarrow n$ transitions of chiral ketones in polar solvents usually show a simple Gaussian circular dichroism band. But in nonpolar solvents, fine structure is often observed, with the spacing of the sublevels corresponding to the spacing of vibrational levels. We consider in particular the application of the third term of (5.5.2) to describe the generation of vibronic circular dichroism induced by nontotally symmetric modes.

Terms involving vibronic mixing of the ground with excited electronic states can usually be neglected in comparison with terms involving mixing of the excited state with other excited states. In that case, the third term of (5.5.2) reduces to

$$R(e_j 1_j \leftarrow e_n 0_n) = \text{Im} \left\{ \langle e_n | \mu_\alpha | e_k \rangle \langle e_k | m_\alpha | e_n \rangle \right.$$

$$\left. \times \frac{1}{\hbar^2 \omega_{e_j e_k}^2} |\langle e_k | (\partial H_e / \partial Q_p)_0 | e_j \rangle|^2 |\langle 0_n | Q_p | 1_j \rangle|^2 \right\}, \quad (5.5.7)$$

where we have assumed that mixing of the resonant electronic state e_j with one particular other excited state e_k dominates. The sign of the corresponding vibronic circular dichroism band is that of the 'allowed' rotational strength $\text{Im}(\langle e_n | \mu_\alpha | e_k \rangle \langle e_k | m_\alpha | e_n \rangle)$. To link up with Section 5.4.1 where symmetry rules were derived for the generation of optical activity in the carbonyl chromophore through coupling with a chiral intramolecular environment, we shall write out (5.5.7) explicitly for the rotational strength generated in particular vibronic transitions through dynamic coupling with a neutral spherical perturber. Thus from (5.3.24) we find, for the $\pi^* \leftarrow p_Y$ ($^1B_1 \leftarrow {}^1A_1$) transition to a vibronic excited state associated with an in-plane bending vibration of symmetry species B_2, that a contribution obeying a quadrant rule is predicted:

$$R[\pi^* \leftarrow p_Y ({}^1B_1 \leftarrow {}^1A_1)]$$

$$= \frac{3\alpha_2 R_X R_Y}{4\pi \epsilon_0 R^5} \text{Im} \left\{ \langle p_Y | \mu_X | d_{XZ} \rangle \langle d_{XZ} | m_Y | p_Y \rangle \right.$$

$$\left. \times \frac{1}{\hbar^2 \omega_{p_Y d_{XZ}}^2} |\langle d_{XZ} | (\partial H_e / \partial Q_{B_2})_0 | \pi^* \rangle|^2 |\langle 0_{p_Y} | Q_{B_2} | 1_{\pi^*} \rangle|^2 \right\}, \quad (5.5.8)$$

where we have taken e_k to be the state generated by the promotion of an electron from the p_Y orbital to the d_{XZ} orbital, which corresponds to the configuration $\sigma^2 \pi^2 p_Y d_{XZ}$, since this is the lowest excited state for which the appropriate vibronic mixing is symmetry allowed. Also $|1_{\pi^*}\rangle$, for example, denotes the vibrational state associated with the excited electronic state e_j that arises from the $\pi^* \leftarrow p_Y$ promotion, with one quantum in the normal mode corresponding to Q_{B_2}. Similarly for the $\pi^* \leftarrow p_Y$ ($^1B_2 \leftarrow {}^1A_1$) transition to a vibronic excited state associated with an -

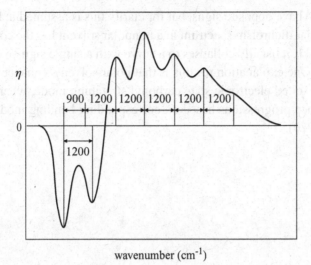

wavenumber (cm^{-1})

Fig. 5.11 Typical vibronic structure for the $\pi^* \leftarrow n$ carbonyl circular dichroism (arbitrary units). Adapted from Weigang (1965).

out-of-plane bending vibration of symmetry species B_1:

$$R[\pi^* \leftarrow p_Y\,({}^1B_2 \leftarrow {}^1A_1)] = \frac{3\alpha_2 R_X R_Y}{4\pi\epsilon_0 R^5}\,\mathrm{Im}\Bigg\{ \langle p_Y|\mu_Y|\sigma^*\rangle\langle\sigma^*|m_X|p_Y\rangle$$

$$\times \frac{1}{\hbar^2\omega^2_{p_Y\sigma^*}}|\langle\sigma^*|\left(\partial H_e/\partial Q_{B_1}\right)_0|\pi^*\rangle|^2|\,\langle 0_{p_Y}|Q_{B_1}|1_{\pi^*}\rangle|^2\Bigg\},$$

$$(5.5.9)$$

except that now the excited state e_k corresponding to the configuration $\sigma^2\pi^2 p_Y\sigma^*$ is the lowest for which the appropriate vibronic mixing is symmetry allowed. Recall that equation (5.4.3) for the corresponding 0–0 rotational strength predicts an octant rule.

The detailed application of such expressions to particular molecules is complicated by the hypersensitivity of the vibronic structure of a circular dichroism band to the solvent medium, and is not attempted here. A generalized vibronic pattern for the circular dichroism band of the $\pi^* \leftarrow n$ transition found in a number of organic carbonyl compounds in nonpolar solvents is shown in Fig. 5.11. A negative 'allowed' progression of the $\sim$1200 cm^{-1} carbonyl stretching mode in the excited electronic state is complemented by a positive 'forbidden' band system based on the same 1200 cm^{-1} totally symmetric progression in combination with a single $\sim$900 cm^{-1} nontotally symmetric mode, either an in-plane or out-of-plane deformation. It is not theoretically necessary for 'allowed' and 'forbidden' circular dichroism progressions to have opposite signs, but for clarity this is assumed in Fig. 5.11. If a carbonyl circular dichroism spectrum in a nonpolar solvent has the couplet appearance of Fig. 5.11, it usually collapses to a curve with a single sign on changing to a polar solvent. One explanation for this is that polar solvents enhance progressions based on the excited electronic state carbonyl stretching mode, whereas nonpolar solvents enhance progressions based on corresponding bending modes (Klingbiel and Eyring, 1970).

6

Magnetic electronic optical activity

I have succeeded in magnetizing and electrifying a ray of light, and in illuminating a magnetic line.

(Michael Faraday)

6.1 Introduction

This chapter is concerned mainly with the visible and near ultraviolet optical rotation and circular dichroism that all molecules show in a static magnetic field. Magnetic optical activity is generated by appropriate components of $\alpha'_{\alpha\beta}$, the imaginary part of the complex dynamic polarizability tensor. As discussed in Chapter 4, $\alpha'_{\alpha\beta}$ is time odd and can only contribute to birefringence phenomena in the presence of some external time-odd influence. This chapter deals mainly with a liquid or solution sample in a static magnetic field, which constitutes a uniaxial medium for light propagating along the field direction.

The formulation of magnetic optical rotation and circular dichroism developed below is based on an article by Buckingham and Stephens (1966), which is itself based on Stephen's dissertation (1964). Although the correct quantum mechanical description had been given much earlier by Serber (1932), it was the Buckingham–Stephens work that initiated a new era in magnetic optical activity, at least in chemistry.

Since magnetochiral birefringence and dichroism are generated by appropriate components of the time-odd molecular property tensors $G_{\alpha\beta}$ and $A'_{\alpha,\beta\gamma}$ in a static magnetic field collinear with the propagation direction of the light beam in a manner analogous to the generation of magnetic optical activity through $\alpha'_{\alpha\beta}$, a quantum mechanical theory of these effects is also developed in this chapter. An important difference, however, is that magnetochiral phenomena are supported only by chiral molecules.

311

6.2 General aspects of magnetic optical rotation and circular dichroism

6.2.1 The basic equations

In Chapter 3, expressions for magnetic optical rotation and circular dichroism were derived using the refringent scattering approach. The same results can be obtained using the more conventional circular differential refraction method, elaborated in Chapter 5 for the case of natural optical rotation and circular dichroism: one simply develops the terms in α'_{xy} in (5.2.19a) and (5.2.19b). Thus from Section 3.4.7 we can write the magnetic optical rotation and circular dichroism for light propagating parallel to a magnetic field along the z direction in an oriented sample as follows:

$$\Delta\theta \approx \tfrac{1}{2}\omega\mu_0\, cl\left(\frac{N}{d_n}\right)B_z\sum_n\left(\alpha''^{(m)}_{xy,z}(f) + m_{n_z}\alpha'_{xy}(f)/kT\right), \qquad (6.2.1a)$$

$$\eta \approx \tfrac{1}{2}\omega\mu_0\, cl\left(\frac{N}{d_n}\right)B_z\sum_n\left(\alpha''^{(m)}_{xy,z}(g) + m_{n_z}\alpha'_{xy}(g)/kT\right). \qquad (6.2.1b)$$

Equation (6.2.1b) applies to small ellipticities developed in a light beam that is initially linearly polarized; expressions for more general magnetic circular dichroism observations can be written down immediately from the results of Section 3.4 if required. Recall that N is the total number of molecules per unit volume in a set of degenerate initial states individually designated ψ_n, d_n is the degeneracy and the sum is over all components of the degenerate set. Also $\alpha''^{(m)}_{\alpha\beta,\gamma}$ is the antisymmetric polarizability perturbed to first order in the magnetic field: the dispersive and absorptive parts are given by (2.7.8e) and (2.7.8f) with μ_γ replaced by m_γ.

The magnetic optical rotation and circular dichroism are conventionally written for the $j \leftarrow n$ transition in the form

$$\Delta\theta \approx -\frac{\mu_0\, clNB_z}{3\hbar}\left[\frac{2\omega_{jn}\omega^2}{\hbar}(f^2-g^2)A + \omega^2 f\left(B+\frac{C}{kT}\right)\right], \qquad (6.2.2a)$$

$$\Delta\eta \approx -\frac{\mu_0\, clN\,B_z}{3\hbar}\left[\frac{4\omega_{jn}\omega^2}{\hbar}(fg)A + \omega^2 g\left(B+\frac{C}{kT}\right)\right], \qquad (6.2.2b)$$

where the *Faraday A-, B- and C-terms*, first introduced by Serber (1932), are

$$A = \frac{3}{d_n}\sum_n(m_{jz}-m_{nz})\,\mathrm{Im}\left(\langle n|\mu_x|j\rangle\langle j|\mu_y|n\rangle\right), \qquad (6.2.2c)$$

$$B = \frac{3}{d_n}\sum_n\mathrm{Im}\left[\sum_{k\neq n}\frac{\langle k|m_z|n\rangle}{\hbar\omega_{kn}}(\langle n|\mu_x|j\rangle\langle j|\mu_y|k\rangle - \langle n|\mu_y|j\rangle\langle j|\mu_x|k\rangle)\right.$$
$$\left. + \sum_{k\neq j}\frac{\langle j|m_z|k\rangle}{\hbar\omega_{kj}}(\langle n|\mu_x|j\rangle\langle k|\mu_y|n\rangle - \langle n|\mu_y|j\rangle\langle k|\mu_x|n\rangle)\right],$$

$$(6.2.2d)$$

$$C = \frac{3}{d_n} \sum_n m_{nz} \, \mathrm{Im}\left(\langle n|\mu_x|j\rangle\langle j|\mu_y|n\rangle\right). \tag{6.2.2e}$$

If the excited state ψ_j is a component of a degenerate set, these expressions must be summed over all transitions $j \leftarrow n$ that are degenerate in the absence of the magnetic field. If, in the absence of the magnetic field, the sample is an isotropic fluid, the magnetic optical rotation and circular dichroism are found using the unit vector average (4.2.49): the basic expressions (6.2.2a) and (6.2.2b) still obtain, but the Faraday A-, B- and C-terms now become

$$A = \tfrac{1}{2}\varepsilon_{\alpha\beta\gamma}\frac{1}{d_n}\sum_n (m_{j\alpha} - m_{n\alpha})\,\mathrm{Im}\left(\langle n|\mu_\beta|j\rangle\langle j|\mu_\gamma|n\rangle\right), \tag{6.2.3a}$$

$$B = \varepsilon_{\alpha\beta\gamma}\frac{1}{d_n}\sum_n \mathrm{Im}\left[\sum_{k\neq n}\frac{\langle k|m_\alpha|n\rangle}{\hbar\omega_{kn}}\langle n|\mu_\beta|j\rangle\langle j|\mu_\gamma|k\rangle\right.$$

$$\left.+\sum_{k\neq j}\frac{\langle j|m_\alpha|k\rangle}{\hbar\omega_{kj}}\langle n|\mu_\beta|j\rangle\langle k|\mu_\gamma|n\rangle\right], \tag{6.2.3b}$$

$$C = \tfrac{1}{2}\varepsilon_{\alpha\beta\gamma}\frac{1}{d_n}\sum_n m_{n\alpha}\,\mathrm{Im}\left(\langle n|\mu_\beta|j\rangle\langle j|\mu_\gamma|n\rangle\right). \tag{6.2.3c}$$

Notice that, parity arguments notwithstanding, only the component of the magnetic field in the direction of the light beam generates nonzero spatial averages. The magnetic optical rotation lineshapes $f^2 - g^2$ (associated with A) and f (associated with B and C), and the magnetic circular dichroism lineshapes fg (associated with A) and g (associated with B and C) have been drawn previously in Figs. 2.4 and 2.5.

The case of degenerate states is of central importance in magnetic optical activity, and is encompassed automatically by the above equations since these specify a sum over transitions from component states of a degenerate set to a particular excited state ψ_j which can itself be a component of a degenerate set. However, the above equations are only valid if the Zeeman components arising from the removal of the degeneracy by the magnetic field are not resolved spectroscopically. The lineshape functions f, g, $f^2 - g^2$ and fg are evaluated using the central frequency ω_{jn} and half width Γ_j of the unresolved $j \leftarrow n$ absorption band. In the other extreme situation where the frequency shifts induced by the magnetic field are larger than the linewidth, so that the Zeeman components are well resolved, the magnetic optical rotation or circular dichroism lineshape is simply the sum of the lineshapes of each component band, given by $\alpha'_{xy}(f)$ or $\alpha'_{xy}(g)$ for each particular Zeeman transition (see Fig. 1.7). A more detailed discussion of the delicate problem of lineshapes in magnetic optical rotation and circular dichroism can be found in Buckingham and Stephens (1966) and Stephens (1970).

We have followed the current convention by giving definitions of A, B and C that are three times larger than those given by Buckingham and Stephens (1966).

It is customary to discuss the ratios A/D, B/D and C/D, where D is the appropriate dipole strength component. This is because the reduced electric dipole moment matrix elements (together with certain correction factors that we have not gone into here) cancel out, leaving simple factors that can be compared with ratios extracted from measured circular dichroism and absorption spectra. For a fluid, the isotropic dipole strength (5.2.28b) is used, except that we now sum over the set of degenerate initial states if necessary:

$$D = \frac{1}{d_n} \sum_n \mathrm{Re}\left(\langle n|\mu_\alpha|j\rangle\langle j|\mu_\alpha|n\rangle\right). \qquad (6.2.3d)$$

6.2.2 Interpretation of the Faraday A-, B- and C-terms

As mentioned in Section 1.3, the Zeeman and Faraday effects are intimately related. In fact the A-term originates in the Zeeman splitting of spectral lines into right- and left-circularly polarized components; the B-term originates in a mixing of energy levels by the magnetic field; and the C-term originates in a change of electronic population of the split ground state. The A-term is nonzero only when either the ground or excited state is a member of a degenerate set, since only then can the matrix elements of the magnetic dipole moment operator be diagonal and a first-order Zeeman effect occur. The C-term is only nonzero when the ground state belongs to a degenerate set. Consequently, only in molecules with a threefold or higher proper rotation axis can A- and C-terms be generated through orbital degeneracy (since orbital degeneracy is only possible in axially or spherically symmetric systems); however, the generation of C-terms by ground state orbital degeneracy is usually complicated by Jahn–Teller effects. In fact ground state degeneracy is most commonly encountered as Kramers degeneracy, so C-terms are important in molecules with an odd number of electrons. The B-term involves only off-diagonal matrix elements of the magnetic dipole moment operator and so is shown by all molecules.

The generation of A- and C-terms is nicely illustrated in the case of atomic $^1S \leftarrow {}^1P$ transitions (Buckingham and Stephens, 1966). The magnetic field lifts the threefold degeneracy of the $M = -1, 0, +1$ states of the ^1P level; the first-order perturbed energies are

$$\begin{aligned} W'\left({}^1P_M\right) &= W\left({}^1P_M\right) - \langle {}^1P_M|m_z|{}^1P_M\rangle B_z \\ &= W\left({}^1P_M\right) + \mu_B M B_z, \end{aligned} \qquad (6.2.4)$$

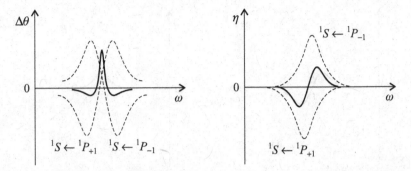

Fig. 6.1 The optical rotation $\Delta\theta$ and circular dichroism η lineshapes generated by the splitting of the component transitions of $^1S \leftarrow {}^1P$ by a magnetic field. These are associated with the Faraday A-term. Adapted from Buckingham and Stephens (1966).

where μ_B is the Bohr magneton. The $^1S \leftarrow {}^1P$ transition therefore generates three electric dipole allowed Zeeman transitions. Using the irreducible tensor methods sketched in Section 4.4.6, in particular the result (4.4.27), we can write

$$\left\langle {}^1S|\mu_x|{}^1P_{\mp1}\right\rangle = \pm i\left\langle {}^1S|\mu_y|{}^1P_{\mp1}\right\rangle. \tag{6.2.5}$$

This means, in effect, that the two mutually orthogonal electric dipole components of a particular $^1S \leftarrow {}^1P_M$ transition in the plane perpendicular to the propagation direction of the light beam are equal in magnitude and $\pi/2$ out of phase, so that together they generate a circular motion of charge. The corresponding contributions to the optical rotation and circular dichroism generated by the transitions $^1S \leftarrow {}^1P_{-1}$ and $^1S \leftarrow {}^1P_{+1}$ in the unperturbed tensor α'_{xy} are equal and opposite, and cancel in the absence of a magnetic field. But the frequency shift induced by the magnetic field results in an incomplete cancellation, thereby generating characteristic optical rotation and circular dichroism lineshapes (Fig. 6.1). For small shifts, these correspond to the lineshape functions $f^2 - g^2$ and fg associated with the A-term in (6.2.2a) and (6.2.2b) that emerge from the perturbation calculation. Furthermore, the populations of the ^1P_M states differ in weak fields in accordance with Boltzmann's law. The $M = -1$ state is of lower energy and more populated than the $M = +1$ state, so the $^1S \leftarrow {}^1P_{-1}$ transition has more intensity than the $^1S \leftarrow {}^1P_{+1}$ transition, again resulting in incomplete cancellation of the corresponding magnetic optical rotation and circular dichroism lineshapes (Fig. 6.2). These correspond to the lineshape functions f and g associated with the C-term that emerge from the perturbation calculation.

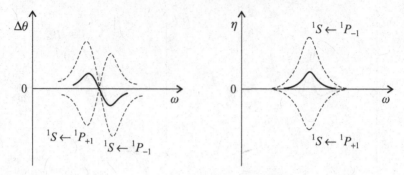

Fig. 6.2 The optical rotation $\Delta\theta$ and circular dichroism η lineshapes generated by population differences in the component transitions of $^1S \leftarrow {}^1P$ split by a magnetic field. These are associated with the Faraday C-term. Adapted from Buckingham and Stephens (1966).

Our example of the $^1S \leftarrow {}^1P$ transition cannot be used to illustrate the generation of the ubiquitous B-term because S and P states are not coupled by the magnetic field. But we can note that, as in the A- and C-terms, mutually orthogonal components of two electric dipole transition moments and one magnetic dipole transition moment are involved. The complex degenerate states required to generate the magnetic moments specified in the A- and C-terms automatically support mutually orthogonal electric dipole transition moments between the same ground and excited states; but in the B-term the mutually orthogonal electric dipole transition moments have just one state in common, the two different states being connected by a magnetic dipole interaction. In molecules of low symmetry with no degenerate states, B is the only term present. The states that are coupled by the magnetic field can often be correlated with components of degenerate sets of states in equivalent molecules of higher symmetry, as we shall see in the case of metal-free porphyrins.

Thus the A-term is responsible for the magnetic optical rotation and circular dichroism curves associated historically with diamagnetic samples (see Section 1.3). The absolute signs predicted for the magnetic optical activity generated by the $^1S \leftarrow {}^1P_M$ transitions are consistent with the generally observed negative magnetic optical rotation at wavelengths outside of electronic absorption bands; the same signs obtain when it is the excited state that is degenerate (e.g. transitions $^1P_M \leftarrow {}^1S$), and are probably quite general. The C-term can only exist in paramagnetic samples since it requires a ground state magnetic moment, and generates the magnetic optical rotation and circular dichroism curves associated historically with paramagnetic samples, including a long wavelength optical rotation opposite in sign to that of diamagnetic samples.

6.3 Illustrative examples

6.3.1 Porphyrins

Porphyrins provide good examples of the generation of Faraday *A*- and *B*-terms. The chromophore is the conjugated ring system; ignoring skeletal distortions and outer substituents (which do not normally produce observable spectral splittings, but can affect band intensities), the effective symmetry of the chromophore is D_{4h} in metal porphyrins I and D_{2h} in free-base porphyrins II. The visible and near ultraviolet absorption spectra are ascribed to transitions, polarized in the molecular plane, among the π electron states of the ring system. A simple treatment involving one electron promotions from the highest filled to the lowest empty molecular orbitals provides a description adequate for our purposes.

I   II

In D_{4h} porphyrin chromophores, the two highest filled orbitals have A_{2u} and A_{1u} symmetry (A_{2u} having the higher energy), and the lowest empty orbitals have E_g symmetry (see Gouterman, 1961, for a review of porphyrin spectra). There are 26 π electrons occupying the 13 molecular orbitals of lowest energy, the highest of which (A_{2u}) is nonbonding. The ground state is therefore of species A_{1g}, and the first and second excited states, arising from the one electron transitions $e_g \leftarrow a_{2u}$ and $e_g \leftarrow a_{1u}$ respectively, are E_u, which we designate E_{u_a} and E_{u_b}. In order to account for the observed relative spectral intensities, it is assumed that the two unperturbed excited states E_{u_a} and E_{u_b} are very close; configuration interaction then separates them (Fig. 6.3).

The higher energy transition $^1E_{u_b} \leftarrow {}^1A_{1g}$ is assigned to the very intense absorption band, called the Soret band, shown by all metal porphyrins at about 400 nm. The lower energy transition $^1E_{u_a} \leftarrow {}^1A_{1g}$ is assigned to a band at about 570 nm, designated Q_0, which is an order of magnitude weaker than the Soret band. A further band, assigned to a vibrational overtone of the $^1E_{u_a} \leftarrow {}^1A_{1g}$ transition and

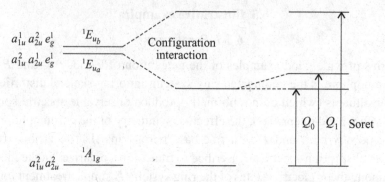

Fig. 6.3 The molecular energy levels involved in the one electron transitions that generate the visible and near ultraviolet absorption spectra of metal porphyrins.

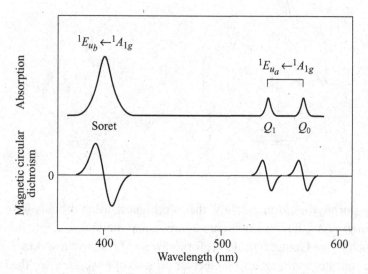

Fig. 6.4 Typical absorption and magnetic circular dichroism spectra, both in arbitrary units, exhibited by D_{4h} metal porphyrin solutions. For simplicity, the vibronic structure often found in Q_1 is not shown.

designated Q_1, occurs at about 540 nm. As illustrated in Fig. 6.4, all three bands show magnetic circular dichroism lineshapes characteristic of A-terms; this provides convincing evidence that the effective chromophore symmetry is indeed D_{4h} and that the three bands are associated with transitions to degenerate excited states of ^1E_u symmetry.

It is instructive to develop the A-term explicitly at this point. Taking the fourfold axis to be the molecule-fixed Z axis, and summing over the two degenerate transitions that constitute each band, (6.2.3a) can be written

$$A = \langle {}^1E|m_Z|{}^1E1\rangle \operatorname{Im}\left(\langle {}^1A_1|\mu_X|{}^1E1\rangle\langle {}^1E1|\mu_Y|{}^1A_1\rangle\right)$$

$$+ \langle {}^1E-1|m_Z|{}^1E-1\rangle \operatorname{Im}\left(\langle {}^1A_1|\mu_X|{}^1E-1\rangle\langle {}^1E-1|\mu_Y|{}^1A_1\rangle\right), \quad (6.3.1)$$

where $|E1\rangle$ and $|E-1\rangle$ are the two components of the degenerate excited state, written in complex form so as to give diagonal matrix elements with m_Z. Using Griffith's formulation of matrix elements of irreducible tensor operators for the point groups, as outlined in Section 4.4.6 (in particular equations 4.4.31), this collapses to a simple product of reduced matrix elements:

$$A = -\frac{i}{2\sqrt{2}} \langle {}^1E\|m\|{}^1E\rangle |\langle {}^1A_1\|\mu\|{}^1E\rangle|^2. \quad (6.3.2)$$

The characteristic A-term circular dichroism lineshape then devolves upon the function fg. The corresponding isotropic dipole strength, summed over the two degenerate transitions, is

$$D = |\langle {}^1A_1 \| \mu \|{}^1E\rangle|^2. \quad (6.3.3)$$

We therefore obtain the ratio

$$\frac{A}{D} = -\frac{i}{2\sqrt{2}} \langle {}^1E\|m\|{}^1E\rangle, \quad (6.3.4)$$

and will not attempt to evaluate the reduced magnetic moment matrix element.

In D_{2h} porphyrin chromophores the X and Y directions are no longer equivalent, removing the degeneracy of the first excited states. The electric dipole transitions responsible for the Soret, Q_0 and Q_1 bands each split into two, along the X and Y directions. Usually this splitting can only be seen in the Q_0 and Q_1 bands and, as shown in Fig. 6.5, magnetic circular dichroism lineshapes characteristic of B-terms are found in the four bands. Although the splitting of the Soret band is not evident in the absorption spectrum, it can be very apparent in the magnetic circular dichroism; in particular, the lineshape characteristic of A-terms is replaced by more complicated structures suggestive of several adjacent B-terms.

Specifically, the ${}^1E_u$ excited states in D_{4h} become ${}^1B_{2u} + {}^1B_{3u}$ in D_{2h}; then the Q_{0x} and Q_{0y} bands are generated by the ${}^1B_{3u_a} \leftarrow {}^1A_g$ and ${}^1B_{2u_a} \leftarrow {}^1A_g$ transitions. Applying (6.2.3b), the only symmetry allowed contributions to B from these transitions are

$$B(Q_{0x}) = \operatorname{Im}\left[\frac{\langle {}^1B_{3u_a}|m_Z|{}^1B_{2u_a}\rangle}{\hbar(\omega_{B_{2u_a}} - \omega_{B_{3u_a}})}\langle {}^1A_g|\mu_X|{}^1B_{3u_a}\rangle\langle {}^1B_{2u_a}|\mu_Y|{}^1A_g\rangle\right], \quad (6.3.5a)$$

$$B(Q_{0y}) = -\operatorname{Im}\left[\frac{\langle {}^1B_{2u_a}|m_Z|{}^1B_{3u_a}\rangle}{\hbar(\omega_{B_{3u_a}} - \omega_{B_{2u_a}})}\langle {}^1A_g|\mu_Y|{}^1B_{2u_a}\rangle\langle {}^1B_{3u_a}|\mu_X|{}^1A_g\rangle\right]. \quad (6.3.5b)$$

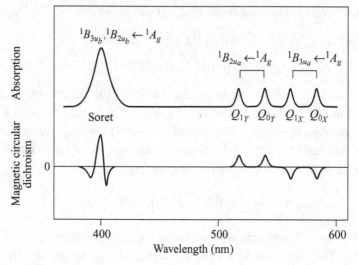

Fig. 6.5 Typical absorption and magnetic circular dichroism spectra exhibited by D_{2h} free base porphyrin solutions.

These equations predict that the B-terms generated by intermixing of states deriving from $^1E_{u_a}$ in the equivalent higher symmetry D_{4h} structure are equal and opposite; this is a general feature associated with the splitting of degenerate states.

We refer to Stephens, Suëtaka and Schatz (1966), and to McHugh, Gouterman and Weiss (1972), for more detailed theoretical discussions of Faraday effects in porphyrins.

6.3.2 Charge transfer transitions in $Fe(CN)_6^{3-}$

Charge transfer transitions in the visible and near ultraviolet spectrum of the low-spin d^5 octahedral complex $Fe(CN)_6^{3-}$ provide good examples of the generation of Faraday C-terms. The account given here is adapted from an article by Schatz *et al.* (1966) and provides a nice illustration of the use of magnetic circular dichroism measurements to obtain definitive electronic absorption band assignments.

Using the well known molecular orbital description of bonding in octahedral complexes (Ballhausen, 1962), appropriate ligand valence orbitals are combined into symmetry adapted linear combinations with respect to the irreducible representations of O_h: interaction of these linear combinations with metal valence orbitals provides the required molecular orbitals. Fig. 6.6 shows a simplified molecular orbital energy level diagram. The sets of orbitals denoted γ and represented by blocks have their major contributions from ligand orbitals. The orbitals denoted e_g and t_{2g} are mostly metal orbitals. In the ground state of low-spin d^5 complexes the γ_u, γ_g levels are fully occupied and the t_{2g} level contains five electrons. Electric

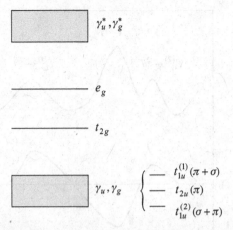

Fig. 6.6 Simplified molecular orbital energy level diagram for an octahedral MX_6 complex. The γs are mainly ligand orbitals, and e_g and t_{2g} are mainly metal. The highest γ_u orbitals when X is cyanide are shown on the right.

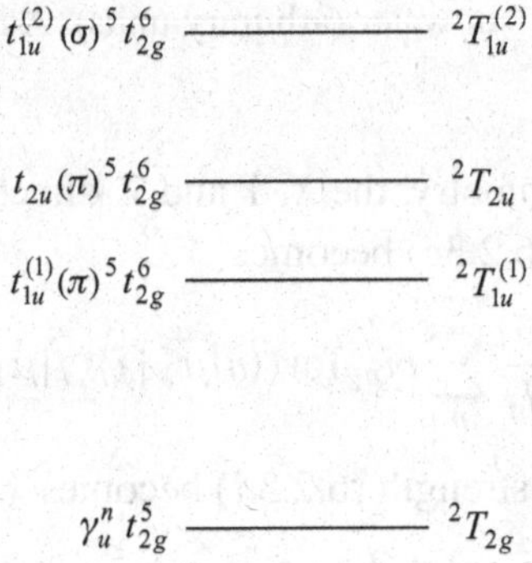

$$t_{1u}^{(2)}(\sigma)^5 t_{2g}^6 \rule{3cm}{0.4pt} {}^2T_{1u}^{(2)}$$

$$t_{2u}(\pi)^5 t_{2g}^6 \rule{3cm}{0.4pt} {}^2T_{2u}$$

$$t_{1u}^{(1)}(\pi)^5 t_{2g}^6 \rule{3cm}{0.4pt} {}^2T_{1u}^{(1)}$$

$$\gamma_u^n t_{2g}^5 \rule{3cm}{0.4pt} {}^2T_{2g}$$

Fig. 6.7 The states arising from the $\gamma_u^n t_{2g}^5$ and $\gamma_u^{n-1} t_{2g}^6$ configurations in $Fe(CN)_6^{3-}$. Spin–orbit splitting is not shown.

dipole allowed transitions in accessible spectral regions are expected to arise from the single electron promotions $t_{2g} \leftarrow \gamma_u$, $e_g \leftarrow \gamma_u$ and $\gamma_u^* \leftarrow t_{2g}$. Since the γs are predominantly ligand orbitals, and e_g and t_{2g} are predominantly metal, these transitions are regarded as *charge transfer*.

We are interested specifically in the $t_{2g} \leftarrow \gamma_u$ transitions. The highest γ_u orbitals expected for cyanide ligands are $t_{1u}^{(1)}(\pi + \sigma)$, $t_{2u}(\pi)$ and $t_{1u}^{(2)}(\sigma + \pi)$. The states arising from the $\gamma_u^n t_{2g}^5$ and $\gamma_u^{n-1} t_{2g}^6$ configurations are shown in Fig. 6.7. Spin–orbit components are not resolved in the absorption and magnetic circular dichroism spectra of $Fe(CN)_6^{3-}$ shown in Fig. 6.8, and so the spin–orbit splitting of the ${}^2T_{2g}, {}^2T_{1u}$ and ${}^2T_{2u}$ states is not invoked.

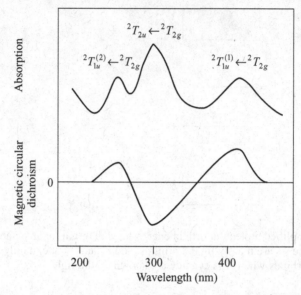

Fig. 6.8 A sketch of the absorption and magnetic circular dichroism spectra of a solution of $K_3Fe(CN)_6^{3-}$ in water (arbitrary units). Adapted from Schatz *et al.* (1966).

In a molecule of O_h symmetry, the X, Y and Z directions are equivalent, so the isotropic Faraday C-term (6.2.3c) becomes

$$C = \frac{3}{d_n} \sum_n m_{n_z} \text{Im} (\langle n|\mu_X|j\rangle \langle j|\mu_Y|n\rangle), \qquad (6.3.6a)$$

and the isotropic oscillator strength (6.2.3d) becomes

$$D = \frac{3}{d_n} \sum_n |\langle n|\mu_X|j\rangle|^2. \qquad (6.3.6b)$$

Following Schatz *et al.* (1966), we neglect spin–orbit coupling in the calculation of C, which results in the simplification that the spin part of the magnetic dipole moment operator does not contribute. This is because the electric dipole moment operator is independent of spin, so states n and j in (6.3.6) must have the same spin quantum numbers: since the sum over the spin magnetic moments of a spin degenerate state is zero, the contribution to the C-term from the ground state spin magnetic moments therefore vanishes. Thus the spin quantum numbers are dropped from the states, and we evaluate just the orbital contribution to the ground state magnetic moment. In effect this is a high temperature approximation valid when the ground state spin–orbit splitting is much less than kT, as in $Fe(CN)_6^{3-}$ at room temperature.

Complex orbital functions are used so as to give diagonal matrix elements with m_Z. The C-term (6.3.6a) can be written explicitly as a sum over the degenerate transitions that constitute a particular band. Thus for a $^2T_{1u} \leftarrow {}^2T_{2g}$ transition we have

$$
\begin{aligned}
C = {} & \langle^2T_{2g}0|m_Z|^2T_{2g}0\rangle \mathrm{Im}\left(\langle^2T_{2g}0|\mu_X|^2T_{1u}^{(1)}0\rangle\langle^2T_{1u}^{(1)}0|\mu_Y|^2T_{2g}0\rangle\right. \\
& + \langle^2T_{2g}0|\mu_X|^2T_{1u}^{(1)}1\rangle\langle^2T_{1u}^{(1)}|\mu_Y|^2T_{2g}0\rangle \\
& + \left.\langle^2T_{2g}0|\mu_X|^2T_{1u}^{(1)}-1\rangle\langle^2T_{1u}^{(1)}-1|\mu_Y|^2T_{2g}0\rangle\right) \\
& + \langle^2T_{2g}1|m_Z|^2T_{2g}1\rangle\mathrm{Im}\left(\langle^2T_{2g}1|\mu_X|^2T_{1u}^{(1)}0\rangle\langle^2T_{1u}^{(1)}0|\mu_Y|^2T_{2g}1\rangle\right. \\
& + \langle^2T_{2g}1|\mu_X|^2T_{1u}^{(1)}1\rangle\langle^2T_{1u}^{(1)}1|\mu_Y|^2T_{2g}1\rangle \\
& + \left.\langle^2T_{2g}1|\mu_X|^2T_{1u}^{(1)}-1\rangle\langle^2T_{1u}^{(1)}-1|\mu_Y|^2T_{2g}1\rangle\right) \\
& + \langle^2T_{2g}-1|m_Z|^2T_{2g}-1\rangle\mathrm{Im}\left(\langle^2T_{2g}-1|\mu_X|^2T_{1u}^{(1)}0\rangle\right. \\
& \times \langle^2T_{1u}^{(1)}0|\mu_Y|^2T_{2g}-1\rangle + \langle^2T_{2g}-1|\mu_X|^2T_{1u}^{(1)}1\rangle\langle^2T_{1u}^{(1)}1|\mu_Y|^2T_{2g}-1\rangle \\
& + \left.\langle^2T_{2g}-1|\mu_X|^2T_{1u}^{(1)}-1\rangle\langle^2T_{1u}^{(1)}-1|\mu_Y|^2T_{2g}-1\rangle\right). \tag{6.3.7}
\end{aligned}
$$

Using equations (4.4.31), this reduces to the following product of reduced matrix elements:

$$
C = \frac{i}{6\sqrt{6}}\langle^2T_{2g}||\mu||^2T_{2g}\rangle|\langle^2T_{2g}||\mu||^2T_{1u}^{(1)}\rangle|^2. \tag{6.3.8}
$$

The corresponding isotropic dipole strength (6.3.6b), summed over the same set of degenerate transitions, is

$$
D = \tfrac{1}{3}|\langle^2T_{2g}||\mu||^2T_{1u}^{(1)}\rangle|^2. \tag{6.3.9}
$$

We finally obtain the ratio

$$
\frac{C(^2T_{1u} \leftarrow {}^2T_{2g})}{D(^2T_{1u} \leftarrow {}^2T_{2g})} = \frac{i}{2\sqrt{6}}\langle^2T_{2g}||\mu||^2T_{2g}\rangle = -\tfrac{1}{2}\mu_B, \tag{6.3.10}
$$

where μ_B is the Bohr magneton. We have used the result of Griffith (1962, p. 23), for the reduced matrix element of the orbital angular momentum operator to obtain the value $i\sqrt{6}\,\mu_B$ for the required magnetic dipole reduced matrix element.

In a similar fashion, the corresponding ratio for a $^2T_{2u} \leftarrow {}^2T_{2g}$ transition is found to be

$$\frac{C\left(^2T_{2u} \leftarrow {}^2T_{2g}\right)}{D\left(^2T_{2u} \leftarrow {}^2T_{2g}\right)} = \tfrac{1}{2}\mu_B. \qquad (6.3.11)$$

The important point to notice is that the sign of the calculated C/D ratio depends on the excited state symmetry, which enables a qualitative property of the observed magnetic circular dichroism to be used in assigning transitions. The observed magnitudes in $Fe(CN)_6^{3-}$ are close to the above calculated values, and the observed signs were in fact used to deduce the band assignments given in Fig. 6.8.

For systems lying close to the opposite low temperature limit in which the ground state spin–orbit splitting is much greater than kT, the approximation used above in which the spin quantum numbers are suppressed must be abandoned, and C/D values calculated for each transition between spin–orbit states. Important examples of such systems are the iridium (IV) hexahalides, and we refer to Henning *et al.* (1968) for an account of their magnetic circular dichroism. See also Dobosh (1974) and Piepho and Schatz (1983).

6.3.3 *The influence of intramolecular perturbations on magnetic optical activity: the carbonyl chromophore*

The theory of the generation of Faraday A- and C-terms is usually straightforward since the effective chromophore symmetries must be sufficiently high to support essential degeneracies. But in general the origin of B-terms is more subtle, which is unfortunate since most organic molecules are made up of structural units with too low a symmetry to support A- and C-terms so that their magnetic optical activity is usually ascribed entirely to B-terms.

Since all groups within a molecule show intrinsic magnetic optical activity, the influence of static and dynamic perturbations from other groups is expected to be much less significant than in natural optical activity, which often depends entirely on such perturbations. Thus we might expect the magnetic optical activity of organic molecules to be dominated by the sum of the B-terms of the individual structural units. There is, however, a notable exception: the magnetic optical activity associated with electronic transitions for which all components are electric dipole forbidden. Again the $\pi^* \leftarrow n$ transition of the carbonyl chromophore provides the classic example. As discussed in Section 5.4.1, all components of this transition are electric dipole forbidden, intensity being gained through vibrational and structural perturbations.

The Faraday B-term involves two perpendicular electric dipole transition moments, and although only one of these involves the $j \leftarrow n$ transition corresponding to the particular absorption, it can be shown quite generally that for electric dipole forbidden transitions, the first non-zero contribution to the B-term is second order in any perturbations that can induce electric dipole strength (Seamans and Moscowitz, 1972). This is born out by the very low magnetic rotational strengths associated with $\pi^* \leftarrow n$ carbonyl transitions, the structures, signs and magnitudes of which are found to be very sensitive to the intramolecular environment.

Since second-order perturbations are required, expressions for the magnetic optical activity induced in the $\pi^* \leftarrow n$ carbonyl transition by static and dynamic coupling are very complicated and are not considered explicitly here. Indeed, the corresponding expressions for natural optical activity, which are only first order in the perturbations, are barely tractable. One further complication in the dynamic coupling mechanism that is not present in natural optical activity is that the magnetic perturbation need not 'reside' on the chromophore: dynamic coupling between a magnetically perturbed group and an unperturbed chromophore can provide a contribution to the magnetic optical activity of the chromophore transitions comparable to that from dynamic coupling between an unperturbed group and a magnetically perturbed chromophore.

It is worth elaborating the least complicated situation where the B-term is generated entirely by vibrational perturbations. This obtains in molecules where the carbonyl symmetry is strictly C_{2v}, as in formaldehyde. Here the X and Y components of the $\pi^* \leftarrow n$ absorption originate in $^1B_1 \leftarrow {}^1A_1$ and $^1B_2 \leftarrow {}^1A_1$ electric dipole allowed transitions to the excited $^1A_2$ electronic state vibronically perturbed by one of two distinct bending vibrations of B_2 symmetry and one of B_1 symmetry, respectively. Applying (6.2.3b), the only allowed contributions to the Faraday B-term from the $\pi^* b_2 \leftarrow p_Y$ and $\pi^* b_1 \leftarrow p_Y$ vibronic transitions are

$$B(\pi^* b_2 \leftarrow p_Y) = \tfrac{1}{3}\mathrm{Im}\left\{ \frac{\langle \pi^* b_2 | m_Z | \pi^* b_1 \rangle}{\hbar(\omega_{\pi^* b_1} - \omega_{\pi^* b_2})} \langle p_Y | \mu_X | \pi^* b_2 \rangle \langle \pi^* b_1 | \mu_Y | p_Y \rangle \right\},$$

$$(6.3.12a)$$

$$B(\pi^* b_1 \leftarrow p_Y) = \tfrac{1}{3}\mathrm{Im}\left\{ \frac{\langle \pi^* b_1 | m_Z | \pi^* b_2 \rangle}{\hbar(\omega_{\pi^* b_2} - \omega_{\pi^* b_1})} \langle p_Y | \mu_Y | \pi^* b_1 \rangle \langle \pi^* b_2 | \mu_X | p_Y \rangle \right\},$$

$$(6.3.12b)$$

where lower case letters refer to vibrational states of that symmetry. Thus two types of vibronic transition provide magnetic circular dichroism contributions (each with g lineshapes) of opposite sign, and since the frequencies differ slightly an s curve is generated. In fact the associated progressions can provide considerable vibronic

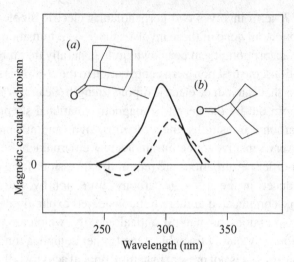

Fig. 6.9 A sketch of the magnetic circular dichroism spectra of (*a*) adamantanone and (*b*) bicyclo-3,3,1-nonan-9-one in cyclohexane solution (arbitrary units) Adapted from Seamans *et al.* (1972).

structure, and we refer to Seamans *et al.* (1972) for the details. Since it is difficult to predict the absolute signs, the signs observed in formaldehyde are taken as a basis for the analysis of the observed magnetic circular dichroism spectra of other carbonyl systems. Thus negative magnetic circular dichroism bands are associated with progressions based on B_1 modes, while positive bands are associated with progressions based on B_2 modes. It should be mentioned, however, that ketones other than formaldehyde possess A_2 modes of vibration: it has been inferred that the associated magnetic circular dichroism bands are negative, although in many instances the affect of A_2 modes on the magnetic circular dichroism spectrum can be neglected.

The magnetic circular dichroism spectra of C_{2v} ketones therefore arise from the superposition of at least three bands, two of which originate in progressions based on B_2 vibronic perturbations and one in a progression based on B_1 vibronic perturbations. Fig. 6.9 shows two spectra which illustrate nicely these vibronic mechanisms. Fig. 6.9*a*, pertaining to adamantanone, shows only a positive band, which indicates that the contributions of B_1 vibrations are suppressed relative to B_2 vibrations (we are neglecting contributions from A_2 vibrations here). This is consistent with the damping of the out-of-plane B_1 bending vibration by the rigidity of the adamantanone skeleton. The spectrum of bicyclo -3,3,1-nonan-9-one (Fig. 6.9*b*), on the other hand, shows two bands of opposite sign with the positive band much more intense than the negative one: since its structure is less rigid than that of adamantanone, the B_1 vibration is not supressed so strongly and generates the weaker negative band.

When the symmetry of the carbonyl chromophore is lower than C_{2v}, the $\pi^* \leftarrow n$ transition becomes electric dipole allowed through coupling with the rest of the molecule. These structural perturbations can be comparable with the vibronic perturbations discussed above, and the analysis becomes very complicated. But it should be remembered that it is only through the analysis of the structural perturbations that stereochemical information is deduced from the observed magnetic circular dichroism spectra, and we refer to Seamans *et al.* (1977) and Linder *et al.* (1977) for further discussion.

6.4 Magnetochiral birefringence and dichroism

This is an appropriate point at which to develop quantum mechanical expressions for magnetochiral birefringence and dichroism. The resulting expressions reveal explicitly how the interplay of chirality and magnetism, alluded to in Section 1.7, generates these subtle phenomena. We start by rewriting the classical expression (3.4.68) for magnetochiral birefringence in a form derived using a quantum statistical average in place of a classical Boltzmann average. Using a development similar to that in Section 3.4.7 leading to (3.4.61) for the Faraday rotation angle we find, for a sample such as a fluid that is isotropic in the absence of the magnetic field,

$$
\begin{aligned}
n^{\uparrow\uparrow} - n^{\uparrow\downarrow} \approx \mu_0 c \left(\frac{N}{d_n} \right) B_Z \sum_n \Big\{ &\tfrac{1}{45}\omega \big[3A'^{(m)}_{\alpha,\alpha\beta,\beta}(f) - A'^{(m)}_{\alpha,\beta\beta,\alpha}(f) \\
&+ (3A'_{\alpha,\alpha\beta}(f)m_{n_\beta} - A'_{\alpha,\beta\beta}(f)m_{n_\alpha})/kT \big] \\
&+ \tfrac{1}{3}\varepsilon_{\alpha\beta\gamma} \big(G^{(m)}_{\alpha\beta,\gamma}(f) + G_{\alpha\beta}(f)m_{n_\gamma}/kT \big) \Big\}.
\end{aligned}
\tag{6.4.1}
$$

As before, N is the total number of molecules per unit volume in a set of degenerate initial states ψ_n, d_n is the degeneracy and the sum is over all components of the degenerate set. A similar expression may be written for the magnetochiral dichroism $n'^{\uparrow\uparrow} - n'^{\uparrow\downarrow}$ in which the dispersion lineshape function f is replaced by the absorption lineshape function g.

The development now proceeds along the lines of the treatment of the Faraday effect given in Section 6.2.1. For the dispersive and absorptive parts of the magnetically perturbed molecular property tensor $G^{(m)}_{\alpha\beta,\gamma}$ we use the expressions (2.7.8b) and (2.7.8c) with μ_β and μ_γ replaced by m_β and m_γ, respectively; and for the dispersive and absorptive parts of $A'^{(m)}_{\alpha,\beta\gamma,\delta}$ we use (2.7.8e) and (2.7.8f) with μ_β and μ_γ replaced by $\Theta_{\beta\gamma}$ and m_δ, respectively. This enables the magnetochiral birefringence and dichroism to be written in the following form (Barron and Vrbancich, 1984):

$$n^{\uparrow\uparrow} - n^{\uparrow\downarrow} \approx \frac{2\mu_0 c N B_z}{3\hbar}\left[\frac{(\omega_{jn}^2 + \omega^2)}{\hbar}(f^2 - g^2)A(G) - \frac{2\omega_{jn}\omega}{\hbar}(f^2 - g^2)A(A')\right.$$

$$\left. + \omega_{jn}f\left(B(G) + \frac{C(G)}{kT}\right) - \omega f\left(B(A') + \frac{C(A')}{kT}\right)\right],$$

$$(6.4.2a)$$

$$n'^{\uparrow\uparrow} - n'^{\uparrow\downarrow} \approx \frac{2\mu_0 c N B_z}{3\hbar}\left[\frac{2(\omega_{jn}^2 + \omega^2)}{\hbar}fgA(G) - \frac{4\omega_{jn}\omega}{\hbar}fgA(A')\right.$$

$$\left. + \omega_{jn}g\left(B(G) + \frac{C(G)}{kT}\right) - \omega g\left(B(A') + \frac{C(A')}{kT}\right)\right], \quad (6.4.2b)$$

where the magnetochiral analogues of the Faraday A-, B- and C-terms (6.2.3) are given by

$$A(G) = \varepsilon_{\alpha\beta\gamma}\frac{1}{d_n}\sum_n(m_{j\gamma} - m_{n\gamma})\mathrm{Re}\left(\langle n|\mu_\alpha|j\rangle\langle j|m_\beta|n\rangle\right), \quad\quad (6.4.2c)$$

$$B(G) = \varepsilon_{\alpha\beta\gamma}\frac{1}{d_n}\sum_n\mathrm{Re}\left[\sum_{k\neq n}\frac{\langle k|m_\gamma|n\rangle}{\hbar\omega_{kn}}(\langle n|\mu_\alpha|j\rangle\langle j|m_\beta|k\rangle + \langle n|m_\beta|j\rangle\langle j|\mu_\alpha|k\rangle)\right.$$

$$\left. + \sum_{k\neq j}\frac{\langle j|m_\gamma|k\rangle}{\hbar\omega_{kj}}(\langle n|\mu_\alpha|j\rangle\langle k|m_\beta|n\rangle + \langle n|m_\beta|j\rangle\langle k|\mu_\alpha|n\rangle)\right], \quad\quad (6.4.2d)$$

$$C(G) = \varepsilon_{\alpha\beta\gamma}\frac{1}{d_n}\sum_n m_{n\gamma}\mathrm{Re}\left(\langle n|\mu_\alpha|j\rangle\langle j|m_\beta|n\rangle\right); \quad\quad (6.4.2e)$$

$$A(A') = \frac{\omega}{15d_n}\sum_n(m_{j\beta} - m_{n\beta})\mathrm{Im}(3\langle n|\mu_\alpha|j\rangle\langle j|\Theta_{\alpha\beta}|n\rangle - \langle n|\mu_\beta|j\rangle\langle j|\Theta_{\alpha\alpha}|n\rangle),$$

$$(6.4.2f)$$

$$B(A') = \frac{\omega}{15d_n}\sum_n\mathrm{Im}\left\{\sum_{k\neq n}\frac{\langle k|m_\beta|n\rangle}{\hbar\omega_{kn}}[3(\langle n|\mu_\alpha|j\rangle\langle j|\Theta_{\alpha\beta}|k\rangle - \langle n|\Theta_{\alpha\beta}|j\rangle\langle j|\mu_\alpha|k\rangle)\right.$$

$$- (\langle n|\mu_\beta|j\rangle\langle j|\Theta_{\alpha\alpha}|k\rangle - \langle n|\Theta_{\alpha\alpha}|j\rangle\langle j|\mu_\beta|k\rangle)]$$

$$+ \sum_{k\neq j}\frac{\langle j|m_\beta|k\rangle}{\hbar\omega_{kj}}[3(\langle n|\mu_\alpha|j\rangle\langle k|\Theta_{\alpha\beta}|n\rangle - \langle n|\Theta_{\alpha\beta}|j\rangle\langle k|\mu_\alpha|n\rangle)$$

$$\left. - (\langle n|\mu_\beta|j\rangle\langle k|\Theta_{\alpha\alpha}|n\rangle - \langle n|\Theta_{\alpha\alpha}|j\rangle\langle k|\mu_\beta|n\rangle)]\right\}, \quad\quad (6.4.2g)$$

$$C(A') = \frac{\omega}{15d_n}\sum_n m_{n\beta}\mathrm{Im}(3\langle n|\mu_\alpha|j\rangle\langle j|\Theta_{\alpha\beta}|n\rangle - \langle n|\mu_\beta|j\rangle\langle j|\Theta_{\alpha\alpha}|n\rangle). \quad (6.4.2h)$$

The application of symmetry arguments to the tensor components specified in (6.4.1) shows that the magnetochiral effect is supported only by chiral molecules. Consider the product $\varepsilon_{\alpha\beta\gamma} G_{\alpha\beta} m_\gamma$. Since $G_{\alpha\beta}$ transforms as a second-rank axial tensor, the antisymmetric combination $\varepsilon_{\alpha\beta\gamma} G_{\alpha\beta}$ transforms as the γ-component of a polar vector, so the specified quantum state ψ_n of the molecule must be able to support the same components of this polar vector and the axial vector m_γ, which is only possible in the chiral point groups C_n, D_n, O, T and I. The same conclusion may be shown to follow from a consideration of the other specified tensor components.

As in the Faraday effect, the magnetochiral A-terms originate in the Zeeman splitting of spectral lines into right- and left-circularly polarized components; the B-terms originate in mixing of the levels by the magnetic field; and the C-terms originate in population differences between components of the split ground level. Thus the magnetochiral A-terms are supported by chiral molecules in which either the ground or excited state is a member of a degenerate set, since only then can the matrix elements of the magnetic dipole moment operator be diagonal and a first-order Zeeman effect occur. Since orbital degeneracy is only possible in axially- or spherically-symmetric systems, the generation of A-terms through excited state orbital degeneracy is only possible in chiral molecules with a threefold or higher proper rotation axis. The B-terms involve only off-diagonal matrix elements of the magnetic dipole moment operator and so are shown by all chiral molecules. The C-terms are only nonzero when the ground level is degenerate, and since ground state degeneracy is most commonly encountered as Kramers degeneracy, C-terms will be important in chiral molecules containing an odd number of electrons.

Few experimental observations of magnetochiral birefringence and dichroism have been reported to date and there are no clear model examples of how the effects are generated. However, considerations such as those given in Chapter 5 for the generation of natural optical activity will be required together with considerations of magnetic structure. Being fully magnetic dipole and electric quadrupole allowed but with only weak electric dipole character induced by the chirally disposed ligands, d–d transitions in chiral transition metal complexes which, as discussed in Section 5.4.2, exhibit large circular dichroism dissymmetry factors, are potentially good subjects for observing magnetochiral A- and C-terms (Barron and Vrbancich, 1984). Magnetochiral dichroism has been observed in d–d transitions in chiral crystals of α-NiSO$_4$.6H$_2$O (Rikken and Raupach, 1998); and the demonstration of magnetochiral enantioselective photochemistry by Rikken and Raupach (2000), mentioned in Section 1.7, was based on magnetochiral dichroism in d–d transitions of a chromium(III)tris oxalato complex. The first measurements of magnetochiral dichroism (in the form of a luminescence anisotropy) by Rikken and Raupach (1997) involved f–f transitions in a chiral europium(III) complex. It had been anticipated (Barron and Vrbancich, 1984) that the largest magnetochiral A- and C-term dissymmetry factors might be found in the $f-f$ absorption bands of certain chiral lanthanide and actinide complexes since these often exhibit even larger natural circular dichroism dissymmetry factors than $d-d$ transitions in chiral transition metal complexes. This is because, like $d-d$ transitions, $f-f$ transitions are fully magnetic dipole and electric quadrupole allowed but the ligand-induced electric dipole character in the chiral complex can be much weaker. Recent ab initio calculations of the magnetochiral birefringence of organic molecules such as carvone, limonene and proline using the expressions derived in this section have been reported recently (Coriani *et al.*, 2002), but for reasons not yet understood the calculated values are several orders of magnitude smaller than the experimental values.

7

Natural vibrational optical activity

What Emanations,
Quick vibrations
And bright stirs are there?

Henry Vaughan (Midnight, *from* Silex Scintillans)

7.1 Introduction

We now turn from the established topic of electronic optical activity to the newer topic of optical activity originating in transitions between the vibrational levels of chiral molecules. Absorption of infrared radiation and Raman scattering of visible radiation provide two distinct methods for obtaining a vibrational spectrum. We shall be concerned equally with manifestations of vibrational optical activity in infrared and Raman spectra. As described in Section 1.5, these take the form of optical rotation and circular dichroism of infrared radiation, and a difference in the intensity of Raman scattering in right- and left-circularly polarized incident light or, equivalently, a circular component in the scattered light using fixed incident polarization.

The fundamental description of natural vibrational optical rotation and circular dichroism parallels the electronic case, being linear in components of the tensors $G'_{\alpha\beta}$ and $A_{\alpha,\beta\gamma}$, except that now the molecule remains in the ground electronic state into which excited electronic states are mixed by vibrational perturbations. On the other hand, the description of natural vibrational Raman optical activity involves cross terms between components of the tensor $\alpha_{\alpha\beta}$ with the tensors $G'_{\alpha\beta}$ and $A_{\alpha,\beta\gamma}$, with excited electronic states providing the pathway for the scattering of visible light. It is the variation of these tensors with the normal coordinates of vibration that brings about the vibrational Raman transitions.

331

An account of the theory of molecular vibrations is not given here since this is covered in numerous texts and needs no reformulation in order to cope with vibrational optical activity, at least within the semiclassical molecular optics approach used here. We refer in particular to Wilson, Decius and Cross (1955) and Califano (1976).

At the time of writing the first edition of this book, theories of both infrared and Raman vibrational optical activity were in a state of flux. Much progress has been made in the intervening years. In particular, the development of a framework for accurate ab initio calculations of vibrational circular dichroism constitutes one of the triumphs of quantum chemistry. General surveys of the theory of vibrational optical activity may be found in the book by Polavarapu (1998) and in reviews by Buckingham (1994) and Nafie and Freedman (2000).

7.2 Natural vibrational optical rotation and circular dichroism

7.2.1 The basic equations

The general aspects of natural vibrational optical rotation and circular dichroism are the same as for the electronic case given in Section 5.2. Thus (5.2.1) apply equally well to optical rotation and circular dichroism generated in an infrared beam propagating along the z direction in an oriented sample; as do (5.2.2) for an isotropic sample. Similarly for experimental quantities such as the specific rotation (5.2.26), the dissymmetry factor (5.2.27), the isotropic rotational strength and dipole strength (5.2.28), and the oriented rotational strength and dipole strength (5.2.29).

In order to describe vibrational optical activity, the isotropic rotational and dipole strengths (5.2.28) are particularized to the $v_j \leftarrow v_n$ vibrational transition for a chiral molecule in the electronic state e_n:

$$R(v_j \leftarrow v_n) = \text{Im}(\langle e_n v_n | \boldsymbol{\mu} | e_n v_j \rangle \cdot \langle e_n v_j | \mathbf{m} | e_n v_n \rangle), \qquad (7.2.1a)$$

$$D(v_j \leftarrow v_n) = \text{Re}(\langle e_n v_n | \boldsymbol{\mu} | e_n v_j \rangle \cdot \langle e_n v_j | \boldsymbol{\mu} | e_n v_n \rangle). \qquad (7.2.1b)$$

These expressions may be developed in several different ways (Polavarapu, 1998; Nafie and Freedman, 2000), of which three will be considered below. In the *fixed partial charge* model, the quantum nature of the electronic states is ignored, thereby leading to a simple and instructive picture of how vibrational optical activity may be generated, but at the expense of computational accuracy. Explicit consideration of the electronic quantum states exposes a subtle problem arising from the fact that, being time odd, the magnetic dipole moment operator has zero expectation values in nondegenerate electronic states (Section 4.3.2). As we shall see, this problem may be overcome by going beyond the adiabatic (Born-Oppenheimer) approximation and considering the dependence of the electronic wavefunction on the nuclear velocities as well as the nuclear positions (Nafie, 1983; Buckingham, Fowler and Galwas, 1987). This provides the foundation for a *bond dipole* model which provides useful physical insight into the generation of vibrational optical activity by particular chiral molecular structures but which does not yield useful quantitative results, and a perturbation theory which provides the foundation for ab initio computations of high accuracy.

The experimental quantity that determines the feasibility of measuring infrared vibrational optical activity is the dissymmetry factor (5.2.27) specialized to a vibrational transition:

$$g(v_j \leftarrow v_n) = \frac{4R(v_j \leftarrow v_n)}{cD(v_j \leftarrow v_n)}. \tag{7.2.2}$$

We shall see that both the fixed partial charge model and the bond dipole model provide expressions for g that are linear in the frequency of the vibrational transition. In the case of electronic optical activity, the expressions for g are linear in the frequency of the optical transition. Since the geometrical factors in the two cases are similar, infrared vibrational optical activity observables can be several orders of magnitude smaller than visible and ultraviolet electronic optical activity observables.

7.2.2 *The fixed partial charge model*

In the fixed partial charge model of infrared vibrational optical activity (Deutsche and Moscowitz, 1968 and 1970; Schellman, 1973), the atoms are taken to be the ultimate particles with residual charges determined by the equilibrium electronic distribution of the molecule. This means that in the vibrational rotational strength (7.2.1a), the quantum aspects of the electronic states are suppressed and we write

$$\mu_\alpha = \sum_i e_i r_{i\alpha}, \tag{7.2.3a}$$

$$m_\alpha = \sum_i \frac{e_i}{2m_i} \varepsilon_{\alpha\beta\gamma} r_{i\beta} p_{i\gamma}, \tag{7.2.3b}$$

with e_i, m_i, $\mathbf{r}_i$ and $\mathbf{p}_i$ interpreted as the charge, position and momentum of the ith atom.

The fixed partial charge model is usually used with normal vibrational coordinates written as sums over a set of atomic displacement coordinates. The components of the atomic position vector $\mathbf{r}_i$ are therefore written

$$r_{i\alpha} = R_{i\alpha} + \Delta r_{i\alpha}, \tag{7.2.4}$$

where $\mathbf{R}_i$ is the equilibrium position of atom i and $\Delta \mathbf{r}_i$ is its instantaneous displacement from equilibrium. The atomic cartesian displacements can be written as a sum over the set of normal coordinates Q_p, and for harmonic oscillations

$$\Delta r_{i\alpha} = \sum_{p=1}^{3N} t_{i\alpha p} Q_p. \tag{7.2.5}$$

Notice that the sum is over $3N$ normal coordinates rather than $3N-6$: this is because the atomic displacement coordinates also encompass rotations and translations of the whole molecule. The **t**-matrix accomplishes both the mass weighting of coordinates and the transformation from normal to cartesian coordinates.

Matrix elements of the normal vibrational coordinate operator have the form (Wilson, Decius and Cross, 1955; Califano, 1976)

$$\langle v|Q|v+1\rangle = \left[\frac{\hbar(v+1)}{2\omega}\right]^{\frac{1}{2}}, \tag{7.2.6a}$$

$$\langle v|Q|v-1\rangle = \left[\frac{\hbar v}{2\omega}\right]^{\frac{1}{2}}, \tag{7.2.6b}$$

$$\langle v|Q|v'\rangle = 0, \quad \text{if} \quad v' \neq v \pm 1. \tag{7.2.6c}$$

Using these together with (7.2.3a), (7.2.4) and (7.2.5), we can write the vibrational matrix element of the electric dipole moment operator for the fundamental transition $1_p \leftarrow 0$ associated with the normal coordinate Q_p as

$$\langle 0|\mu_\alpha|1_p\rangle = \left(\frac{\hbar}{2\omega_p}\right)^{\frac{1}{2}} \sum_i e_i t_{i\alpha p}, \tag{7.2.7}$$

where ω_p is the angular frequency associated with Q_p.

Before developing the vibrational matrix element of the magnetic dipole moment operator, it is as well to express the momentum of the ith atom in terms of normal coordinates. Thus, using (7.2.5), we write

$$p_{i\alpha} = m_i \dot{r}_{i\alpha} = m_i \sum_{p=1}^{3N} t_{i\alpha p} \dot{Q}_p = m_i \sum_{p=1}^{3N} t_{i\alpha p} P_p \tag{7.2.8}$$

since $P_p = \dot{Q}_p$ is the momentum conjugate to the normal coordinate Q_p. Matrix elements of the vibrational momentum operator have the form (Wilson, Decius and

Cross, 1955; Califano, 1976):

$$\langle v|P|v+1\rangle = -\mathrm{i}\left[\frac{\hbar\omega(v+1)}{2}\right]^{\frac{1}{2}}, \tag{7.2.9a}$$

$$\langle v|P|v-1\rangle = \mathrm{i}\left(\frac{\hbar\omega v}{2}\right)^{\frac{1}{2}}, \tag{7.2.9b}$$

$$\langle v|P|v'\rangle = 0 \quad \text{if} \quad v' \neq v \pm 1. \tag{7.2.9c}$$

Comparing these with (7.2.6), we obtain a vibrational version of the velocity–dipole transformation (2.6.31):

$$\langle v|P|v+1\rangle = -\mathrm{i}\omega\langle v|Q|v+1\rangle, \tag{7.2.10a}$$

$$\langle v|P|v-1\rangle = \mathrm{i}\omega\langle v|Q|v-1\rangle. \tag{7.2.10b}$$

Using (7.2.5), the magnetic dipole moment operator can be written

$$
\begin{aligned}
m_\alpha &= \sum_i \frac{e_i}{2}\varepsilon_{\alpha\beta\gamma}\left(R_{i\beta}+\sum_{p=1}^{3N} t_{i\beta p}Q_p\right)\sum_{q=1}^{3N} t_{i\gamma q}P_q \\
&= \sum_i\sum_{q=1}^{3N}\frac{e_i}{2}\varepsilon_{\alpha\beta\gamma}R_{i\beta}t_{i\gamma q}P_q \\
&\quad + \sum_i\sum_{p,q=1}^{3N}\frac{e_i}{2}\varepsilon_{\alpha\beta\gamma}t_{i\beta p}t_{i\gamma q}Q_p P_q.
\end{aligned}
\tag{7.2.11}
$$

According to Faulkner *et al.* (1977), the first term of (7.2.11) may be interpreted as the contribution to the magnetic dipole moment operator arising from the component of the vibrational motion corresponding to rotation of the partially charged atoms about the overall molecular coordinate origin. The second term may be interpreted as the contribution arising from each partially charged atom moving with momentum $t_{i\gamma q}P_q$ on a lever arm $t_{i\beta p}Q_p$ relative to an origin at the equilibrium position of the atom. The second term is usually neglected in calculations of fundamental transitions, but can be important in overtone or combination transitions since the operator QP effects transitions with $\Delta v = 0, \pm 2$. Thus from (7.2.9b) and the first term of (7.2.11), the required magnetic dipole matrix element is

$$\langle 1_p|m_\alpha|0\rangle \approx \frac{\mathrm{i}}{2}\left(\frac{\hbar\omega_p}{2}\right)^{\frac{1}{2}}\sum_i e_i\varepsilon_{\alpha\beta\gamma}R_{i\beta}t_{i\gamma p}. \tag{7.2.12}$$

Combining (7.2.7) and (7.2.12), the isotropic rotational strength for a fundamental vibrational transition associated with the normal coordinate Q_p is found to be

$$R(1_p \leftarrow 0) \approx \frac{\hbar}{4} \sum_{i<j} e_i e_j R_{ji\alpha} \varepsilon_{\alpha\beta\gamma} t_{j\beta p} t_{i\gamma p}, \tag{7.2.13a}$$

where $\mathbf{R}_{ji} = \mathbf{R}_j - \mathbf{R}_i$ is the vector from atom i to atom j at the equilibrium nuclear configuration. The corresponding dipole strength is

$$D(1_p \leftarrow 0) \approx \frac{\hbar}{2\omega_p} \sum_{i,j} e_i e_j t_{i\alpha p} t_{j\alpha p}. \tag{7.2.13b}$$

The application of these fixed partial charge expressions requires a normal coordinate analysis of the molecule. A set of fixed partial charges is also required, which is usually estimated from experimental dipole moment data. We refer to Keiderling and Stephens (1979) for an example of such a calculation. The fixed partial charge model at this level of approximation consistently gives vibrational rotational strengths of about one order of magnitude smaller than actually observed, sometimes of the wrong sign. One refinement allows for charge redistribution during the vibrational excursions away from the equilibrium configuration (Polavarapu, 1998).

7.2.3 The bond dipole model

In the bond dipole model of infrared vibrational optical activity, the molecule is broken down into a convenient arrangement of bonds or groups that can support internal vibrational coordinates s_q such as local bond stretches and angle bendings. These internal vibrational coordinates are written as sums over the set of normal vibrational coordinates (Wilson, Decius and Cross, 1955; Califano, 1976):

$$s_q = \sum_{p=1}^{3N-6} L_{qp} Q_p. \tag{7.2.14}$$

Notice that, unlike (7.2.5), the sum is now over $3N-6$ normal coordinates because the choice of internal vibrational coordinates automatically excludes rotations and translations. The **L**-matrix elements are determined from a normal coordinate analysis.

Within the bond dipole model, infrared intensities are calculated by way of the adiabatic approximation (Section 2.8.2). As in (2.8.33), the adiabatic permanent electric dipole moment in the ground electronic state $\psi_0(r, Q)$ at a fixed nuclear configuration Q is written

$$\mu_\alpha(Q) = \langle \psi_0(r, Q) | \mu_\alpha | \psi_0(r, Q) \rangle, \tag{7.2.15}$$

where the operator μ_α is a function of both electronic and nuclear coordinates. The electric dipole vibrational transition moment is then obtained from a Taylor expansion of $\mu_\alpha(Q)$ in the nuclear displacements around the equilibrium nuclear configuration:

$$\langle v_j|\mu_\alpha(Q)|v_n\rangle = (\mu_\alpha)_0 \delta_{v_j v_n} + \sum_p \left(\frac{\partial \mu_\alpha}{\partial Q_p}\right)_0 \langle v_j|Q_p|v_n\rangle$$

$$+ \frac{1}{2} \sum_{p,q} \left(\frac{\partial^2 \mu_\alpha}{\partial Q_p \partial Q_q}\right)_0 \langle v_j|Q_p Q_q|v_n\rangle + \cdots, \quad (7.2.16)$$

where $(\mu_\alpha)_0$ is the permanent electric dipole moment of the molecule at the equilibrium nuclear configuration of the ground electronic state. Being linear in Q_p, the second term describes fundamental vibrational transitions; whereas the third term, being a function of $Q_p Q_q$, describes first overtone and combination transitions.

A detailed account of the bond dipole theory of infrared intensities can be found in Sverdlov, Kovner and Krainov (1974). Basically, the calculation devolves upon $(\partial \mu_\alpha/\partial Q_p)_0$, which is developed by way of the variation of the molecular dipole moment with local internal coordinates using

$$\left(\frac{\partial \mu_\alpha}{\partial Q_p}\right)_0 = \sum_q \left(\frac{\partial \mu_\alpha}{\partial s_q}\frac{\partial s_q}{\partial Q_p}\right)_0 = \sum_q \left(\frac{\partial \mu_\alpha}{\partial s_q}\right)_0 L_{qp}, \quad (7.2.17)$$

the last step following from (7.2.14). The total electric dipole moment operator of the molecule is written as a sum over the electric dipole moment operators μ_i of all the bonds or groups into which the molecule has been broken down:

$$\mu_\alpha = \sum_i \mu_{i\alpha}. \quad (7.2.18)$$

The components of the bond moments are still functions of Q because they are still specified in the molecular coordinate system and so can change during vibrational excursions. Using the matrix elements (7.2.6), together with the expansion (7.2.16), we find the following expression for the vibrational matrix element of the electric dipole moment operator for the fundamental transition associated with the normal coordinate Q_p:

$$\langle l_p|\mu_\alpha(Q)|0\rangle = \left(\frac{\hbar}{2\omega_p}\right)^{\frac{1}{2}} \sum_i \sum_q \left(\frac{\partial \mu_{i\alpha}}{\partial s_q}\right)_0 L_{qp}. \quad (7.2.19)$$

Since the magnetic dipole moment operator is time odd and therefore has zero expectation values in nondegenerate electronic states, we must be more circumspect in formulating the magnetic dipole vibrational transition moment. It is necessary to go beyond the Born–Oppenheimer approximation by writing the magnetic dipole moment in the ground electronic state as a function of nuclear velocities as well as nuclear positions:

$$m_\alpha(\dot{Q}) = \langle\psi_0(r, Q, \dot{Q})|m_\alpha|\psi_0(r, Q, \dot{Q})\rangle. \qquad (7.2.20)$$

The magnetic dipole vibrational transition moment is then obtained from a Taylor expansion of $m_\alpha(\dot{Q})$ in the nuclear velocities:

$$\langle v_j|m_\alpha(\dot{Q})|v_n\rangle = \sum_p \left(\frac{\partial m_\alpha}{\partial \dot{Q}_p}\right)_0 \langle v_j|\dot{Q}_p|v_n\rangle + \cdots. \qquad (7.2.21)$$

Only terms in odd powers of $\dot{Q}_p$ are nonzero since the associated electronic properties such as $(\partial m_\alpha/\partial \dot{Q}_p)_0$ are time even and therefore may be nonzero in nondegenerate electronic states. The time-odd character of the original magnetic dipole moment operator is now embodied in the operator $\dot{Q}_p$ which can bring about $\Delta v = \pm 1$ transitions between nondegenerate vibrational states even though $\dot{Q}_p$ does not have expectation values.

A crucial step in the development of the bond dipole theory of infrared vibrational optical activity is the inclusion of the origin dependent part of each bond or group magnetic dipole moment. According to (2.4.14), on moving the origin from **O** to **O** + **a** the magnetic dipole moment changes as follows:

$$m_\alpha \rightarrow m_\alpha - \tfrac{1}{2}\varepsilon_{\alpha\beta\gamma}a_\beta\dot{\mu}_\gamma. \qquad (7.2.22)$$

In the present case, however, the vector separating the molecular origin and a local bond or group origin is time dependent on account of the molecular vibrations. Hence on moving the local origin on bond i to the molecular origin, the magnetic dipole moment changes to

$$m_{i\alpha} \rightarrow m_{i\alpha} + \tfrac{1}{2}\varepsilon_{\alpha\beta\gamma}(r_{i\beta}\dot{\mu}_{i\gamma} + \mu_{i\beta}\dot{r}_{i\gamma}), \qquad (7.2.23)$$

where $\boldsymbol{\mu}_i$ and $\mathbf{m}_i$ are the electric and magnetic dipole moments of bond i referred to the local origin on i, $\mathbf{r}_i$ is the vector from the molecular origin to the bond origin, and we have assumed the bond to be neutral. The total magnetic dipole moment of the molecule may therefore be written

$$m_\alpha = \sum_i \left[m_{i\alpha} + \tfrac{1}{2}\varepsilon_{\alpha\beta\gamma}(r_{i\beta}\dot{\mu}_{i\gamma} + \mu_{i\beta}\dot{r}_{i\gamma})\right]. \qquad (7.2.24)$$

Using this expression in (7.2.21), together with $\dot{X} = \left(\partial X/\partial Q_p\right)\dot{Q}_p$ and (7.2.9b), the magnetic dipole transition moment is found to be

$$\langle 1_p|m_\alpha(\dot{Q})|0\rangle = i\left(\frac{\hbar\omega_p}{2}\right)^{\frac{1}{2}} \sum_i \left\{\left(\frac{\partial m_{i\alpha}}{\partial \dot{Q}_p}\right)_0\right.$$

$$\left. + \tfrac{1}{2}\varepsilon_{\alpha\beta\gamma}\left[(r_{i\beta})_0\left(\frac{\partial \mu_{i\gamma}}{\partial Q_p}\right)_0 + (\mu_{i\beta})_0\left(\frac{\partial r_{i\gamma}}{\partial Q_p}\right)_0\right]\right\}. \qquad (7.2.25)$$

Using (7.2.17), this expression may be written in terms of internal coordinates:

$$\langle 1_p|m_\alpha(\dot{Q})|0\rangle = i\left(\frac{\hbar\omega_p}{2}\right)^{\frac{1}{2}}\sum_i\sum_q\left\{\left(\frac{\partial m_{i\alpha}}{\partial \dot{s}_q}\right)_0 L_{qp}\right.$$

$$\left. + \frac{1}{2}\varepsilon_{\alpha\beta\gamma}\left[(r_{i\beta})_0\left(\frac{\partial\mu_{i\gamma}}{\partial s_q}\right)_0 L_{qp} + (\mu_{i\beta})_0\left(\frac{\partial r_{i\gamma}}{\partial s_q}\right)_0 L_{qp}\right]\right\}. \quad (7.2.26)$$

Using (7.2.26) and (7.2.19), the isotropic dipole and rotational strengths for a fundamental vibrational transition associated with the normal coordinate Q_p are found to be

$$D(1_p \leftarrow 0) = \frac{\hbar}{2\omega_p}\left[\sum_i\sum_q\left(\frac{\partial\mu_{i\alpha}}{\partial s_q}\right)_0 L_{qp}\right]\left[\sum_j\sum_r\left(\frac{\partial\mu_{j\alpha}}{\partial s_r}\right)_0 L_{rp}\right],$$

$$(7.2.27a)$$

$$R(1_p \leftarrow 0) = \frac{\hbar}{4}\varepsilon_{\alpha\beta\gamma}\left\{\sum_{i<j}R_{ji\beta}\left[\sum_q\left(\frac{\partial\mu_{i\alpha}}{\partial s_q}\right)_0 L_{qp}\right]\left[\sum_r\left(\frac{\partial\mu_{j\gamma}}{\partial s_r}\right)_0 L_{rp}\right]\right.$$

$$\left. + \left[\sum_i\sum_q\left(\frac{\partial\mu_{i\alpha}}{\partial s_q}\right)_0 L_{qp}\right]\left[\sum_j(\mu_{j\gamma})_0\sum_r\left(\frac{\partial r_{j\beta}}{\partial s_r}\right)_0 L_{rp}\right]\right\}$$

$$+ \frac{\hbar}{2}\text{Im}\left\{\left[\sum_i\sum_q\left(\frac{\partial\mu_{i\alpha}}{\partial s_q}\right)_0 L_{qp}\right]\left[\sum_j\sum_r\left(\frac{\partial m_{j\alpha}}{\partial \dot{s}_r}\right)_0 L_{rp}\right]\right\},$$

$$(7.2.27b)$$

where $\mathbf{R}_{ji} = \mathbf{R}_j - \mathbf{R}_i$ is the vector from the origin on group i to that on group j at the equilibrium nuclear configuration.

The first term in the vibrational rotational strength (7.2.27b) is a sum over all pairs of groups that constitute chiral structures and so represents a generalized two-group term. The second term involves changes in the position vector $\mathbf{r}_j$ of group j relative to the molecule-fixed origin: normal coordinates containing contributions from changes in either the length of $\mathbf{r}_j$, its orientation, or both, will activate this term. We call this the *inertial dipole* term since changes in the orientation of $\mathbf{r}_j$ can generate significant contributions in torsion modes. The last term involves the product of intrinsic group electric and magnetic dipole moment derivatives. Simple examples of the first two contributions are given later.

In applying the vibrational rotational strength (7.2.27b), the question arises of the actual choice of origins within the groups, since it might be thought that this could affect the result. It should first be realized that the complete expression is invariant both to the choice of the molecular origin and to the local group origins: the first two terms are generated by the origin-dependent parts of the total magnetic dipole moment (7.2.24) of the molecule, so any changes in the first two terms caused by changes in the relative disposition of the molecular origin and the local group origins are compensated by a change in the third term. It is natural to choose a group origin to lie along the corresponding dipole axis, which will coincide with the principal proper rotation axis of the group (this can be seen by looking at Table 4.2c which gives the nonzero polar vector components relative to the symmetry elements). The two-group and inertial dipole terms taken together are then invariant to any shift of the group origins along the dipole axes. In fact it is the cross terms ($i \neq j$) in the inertial dipole term that compensate the two-group term; intrinsic group inertial dipole terms ($i = j$) are invariant to origin shifts along the electric dipole axis. It is left to the reader as an exercise to verify explicitly that an origin shift along the dipole axis of a group generates a change in the two-group term that is cancelled by a change in the inertial dipole term.

The value of the bond dipole formulation is that, since it is based on internal vibrational coordinates rather than atomic cartesian displacements, it immediately generates simple geometrical expressions for model structures, and for archetypal structural units in large molecules, allowing group optical activity approximations to be made. A computational version of the bond dipole theory of infrared vibrational optical activity is available (Escribano and Barron, 1988), but has been superseded by ab initio methods.

7.2.4 A perturbation theory of vibrational circular dichroism

Perturbation theories have been particularly successful for accurate calculations of vibrational circular dichroism. Here we develop expressions for the electric and magnetic dipole vibrational transition moments following a treatment due to Buckingham, Fowler and Galwas (1987) in which the electronic contributions to the vibrational transition moments are derived using the vibronic coupling formalism described in Section 2.8.4. We use crude adiabatic vibronic states of the form (2.8.39) represented by $|j\rangle = |e_j v_j\rangle = |e_j\rangle|v_j\rangle$. The electronic parts are taken to be perturbed to first order in $\sum_p (\partial H_e/\partial Q_p)_0 Q_p$ and so are written as in (2.8.43). A general off-diagonal vibrational matrix element of an electronic operator A for an electronic state $|e_j\rangle$ may then be written

$$
\langle e_n v_j | A | e_n v_n \rangle = \sum_{\substack{e_k \neq e_n \\ p}} \left[\frac{\langle e_n|A|e_k\rangle\langle e_k|(\partial H_e/\partial Q_p)_0|e_n\rangle\langle v_j|Q_p|v_n\rangle}{\hbar(\omega_{e_n e_k} + \omega_{v_n v_j})} \right.
$$
$$
\left. + \frac{\langle e_k|A|e_n\rangle\langle e_n|(\partial H_e/\partial Q_p)_0|e_k\rangle\langle v_j|Q_p|v_n\rangle}{\hbar(\omega_{e_n e_k} - \omega_{v_n v_j})} \right], \qquad (7.2.28)
$$

where we have used the orthogonality of the vibrational wavefunctions. Since the electronic energy separations are generally much larger than vibrational quanta, and assuming the operators A and $\sum_p (\partial H_e/\partial Q_p)_0 Q_p$ are Hermitian, this may be rearranged to give

$$\langle e_n v_j | A | e_n v_n \rangle = \sum_{\substack{e_k \neq e_n \\ p}} 2\langle v_j | Q_p | v_n \rangle \left\{ \frac{1}{\hbar \omega_{e_n e_k}} \mathrm{Re}[\langle e_n | A | e_k \rangle \langle e_k | (\partial H_e/\partial Q_p)_0 | e_n \rangle] \right.$$

$$\left. - \mathrm{i} \left(\frac{\omega_{v_n v_j}}{\hbar \omega_{e_n e_k}^2} \right) \mathrm{Im}[\langle e_n | A | e_k \rangle \langle e_k | (\partial H_e/\partial Q_p)_0 | e_n \rangle] \right\}. \quad (7.2.29)$$

We take the wavefunctions to be real, as for nondegenerate electronic states in the absence of a static magnetic field. Then if the operator A is real, such as the electric dipole moment operator, only the first term is nonzero; but if it is imaginary, such as the magnetic dipole moment operator, only the second term is nonzero.

Taking A to be the electronic part of the electric dipole moment operator, we may write from the first term of (7.2.29)

$$\langle e_n v_j | \mu_\alpha^{\mathrm{el}} | e_n v_j \rangle = 2 \sum_p \langle v_j | Q_p | v_n \rangle \mathrm{Re} \left(\langle \psi_n^{(0)} | \mu_\alpha | \partial \psi_n/\partial Q_p \rangle \right), \quad (7.2.30a)$$

where

$$\frac{\partial \psi_n}{\partial Q_p} = \sum_{k \neq n} \frac{1}{\hbar \omega_{nk}} \psi_k^{(0)} \langle \psi_k^{(0)} | (\partial H_e/\partial Q_p)_0 | \psi_n^{(0)} \rangle. \quad (7.2.30b)$$

Comparing with the expansion (7.2.16), we obtain

$$\left(\frac{\partial \mu_\alpha}{\partial Q_p} \right)_0 = 2\mathrm{Re} \left(\langle \psi_n^{(0)} | \mu_\alpha | \partial \psi_n/\partial Q_p \rangle \right). \quad (7.2.31)$$

Taking A to be the electronic part of the magnetic dipole moment operator and using (7.2.10b), we may write from the second term of (7.2.29)

$$\langle e_n v_j | m_\alpha^{\mathrm{el}} | e_n v_n \rangle = -2\hbar \sum_p \langle v_j | \dot{Q}_p | v_n \rangle \mathrm{Im} \left(\langle \partial \psi_n(B_\alpha)/\partial B_\alpha | \partial \psi_n/\partial Q_p \rangle \right),$$

$$(7.2.32a)$$

where $\psi_n(B_\alpha)$ is the electronic wavefunction in the presence of a "fake" static magnetic field so that

$$\frac{\partial \psi_n(B_\alpha)}{\partial B_\alpha} = -\sum_{k \neq n} \frac{1}{\hbar \omega_{nk}} \psi_k^{(0)} \langle \psi_k^{(0)} | m_\alpha | \psi_n^{(0)} \rangle \quad (7.2.32b)$$

is the corresponding first-order correction to the electronic wavefunction for a perturbation by the operator $-m_\alpha B_\alpha$. Comparing with the expansion (7.2.21), we obtain

$$\left(\frac{\partial m_\alpha}{\partial \dot{Q}_p} \right)_0 = -2\hbar \, \mathrm{Im} \left(\langle \partial \psi_n(B_\alpha)/\partial B_\alpha | \partial \psi_n/\partial Q_p \rangle \right). \quad (7.2.33)$$

For computations, it is more convenient to use cartesian displacement coordinates $\Delta r_{i\alpha}$ instead of normal vibrational coordinates Q_p. From (7.2.5) we may write

$$\frac{\partial \psi_n}{\partial Q_p} = \sum_i \frac{\partial \psi_n}{\partial \Delta r_{i\beta}} \frac{\partial \Delta r_{i\beta}}{\partial Q_p} = \sum_i \frac{\partial \psi_n}{\partial \Delta r_{i\beta}} t_{i\alpha p}. \tag{7.2.34}$$

There are also contributions to the vibrational transition moments from nuclear motions, which may be obtained directly from the fixed partial charge results (7.2.7) and (7.2.12) with e_i being taken as the charge of nucleus i rather than the fixed partial charge of atom i. The complete electric and magnetic dipole vibrational transition moments for the $1_p \leftarrow 0$ fundamental transition are then found to be (Buckingham, Fowler and Galwas, 1987; Buckingham, 1994)

$$\langle 1_p | \mu_\alpha | 0 \rangle = \left(\frac{\hbar}{2\omega_p} \right)^{\frac{1}{2}} \sum_i \left[Z_i e \delta_{\alpha\beta} + 2\mathrm{Re}\left(\langle \psi_n^{(0)} | \mu_\alpha | \partial \psi_n / \partial \Delta r_{i\beta} \rangle \right) \right] t_{i\beta p},$$

$$\tag{7.2.35a}$$

$$\langle 1_p | m_\alpha | 0 \rangle = \mathrm{i} \left(\frac{\hbar \omega_p}{2} \right)^{\frac{1}{2}}$$

$$\times \sum_i \left[\tfrac{1}{2} Z_i e \varepsilon_{\alpha\beta\gamma} R_{i\beta} - 2\hbar \, \mathrm{Im}\left(\langle \partial \psi_n(B_\alpha) / \partial B_\alpha | \partial \psi_n / \partial \Delta r_{i\beta} \rangle \right) \right] t_{i\beta p},$$

$$\tag{7.2.35b}$$

where $Z_i e$ is the charge on nucleus i. Similar expressions have been derived by Stephens (1985). The required derivatives may be computed routinely using modern ab initio methods. Testimony to the power of this formalism is the close agreement between observed and calculated vibrational circular dichroism spectra (Stephens and Devlin, 2000; Nafie and Freedman, 2000).

7.3 Natural vibrational Raman optical activity

7.3.1 The basic equations

The observables in Rayleigh and Raman optical activity are a small circularly polarized component in the scattered light, and a small difference in the scattered intensity in right- and left-circularly polarized incident light. General expressions for the optical activity observables in Rayleigh scattering from chiral molecules were derived in Chapter 3 in terms of molecular property tensors. Thus the dimensionless circular intensity difference

$$\Delta = \frac{I^R - I^L}{I^R + I^L} \tag{1.4.1}$$

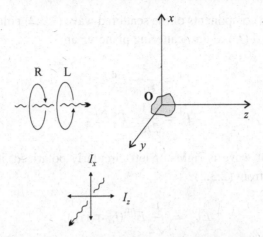

Fig. 7.1 The geometry for polarized light scattering at 90°.

is given by (3.5.34–6). But for the benefit of the reader who wants an uncluttered derivation, we now give a direct calculation of the Rayleigh circular intensity difference for 90° scattering (Barron and Buckingham, 1971).

Consider a molecule at the origin **O** of a coordinate system x, y, z (Fig. 7.1) in an incident plane wave light beam with electric vector

$$\tilde{E}_\alpha = \tilde{E}_\alpha^{(0)} e^{i\omega(z/c-t)} \tag{7.3.1}$$

travelling in the z direction. This wave induces oscillating electric and magnetic multipole moments in the molecule which are the source of the scattered light. For scattering at 90° we require the following expression for the complex electric vector radiated into the y direction at a distance from the origin much greater than the wavelength:

$$\tilde{E}_\alpha^{d} = \frac{\omega^2 \mu_0}{4\pi y} \left[\tilde{\mu}_\alpha^{(0)} - j_\alpha \tilde{\mu}_y^{(0)} - \frac{1}{c} \varepsilon_{\alpha y \beta} \tilde{m}_\beta^{(0)} \right.$$
$$\left. - \frac{i\omega}{3c} (\tilde{\Theta}_{\alpha y}^{(0)} - j_\alpha \tilde{\Theta}_{yy}^{(0)} + \cdots) \right] e^{i\omega(y/c-t)}, \tag{7.3.2}$$

which is a specialization of (3.3.1). The complex multipole moment amplitudes are a specialization of (3.3.2):

$$\tilde{\mu}_\alpha^{(0)} = \left(\tilde{\alpha}_{\alpha\beta} + \frac{i\omega}{3c} \tilde{A}_{\alpha,z\beta} + \frac{1}{c} \varepsilon_{\gamma z\beta} \tilde{G}_{\alpha\gamma} + \cdots \right) \tilde{E}_\beta^{(0)}, \tag{7.3.3a}$$

$$\tilde{\Theta}_{\alpha\beta}^{(0)} = (\mathscr{A}_{\gamma,\alpha\beta} + \cdots) \tilde{E}_\gamma^{(0)}, \tag{7.3.3b}$$

$$\tilde{m}_\alpha^{(0)} = (\mathscr{G}_{\alpha\beta} + \cdots) \tilde{E}_\beta^{(0)}. \tag{7.3.3c}$$

The intensity of the components of the scattered wave (7.3.2) polarized perpendicular (I_x) and parallel (I_z) to the scattering plane yz are

$$I_x^d = \frac{1}{2\mu_0 c}\tilde{E}_x^d \tilde{E}_x^{d*},$$

(7.3.4a)

$$I_z^d = \frac{1}{2\mu_0 c}\tilde{E}_z^d \tilde{E}_z^{d*}.$$

(7.3.4b)

If the incident light wave is right- or left-circularly polarized, its electric vector amplitude follows from (2.3.2) as

$$\tilde{E}_{\substack{R\\L}\alpha}^{(0)} = \frac{1}{\sqrt{2}}E^{(0)}(i_\alpha \mp i j_\alpha).$$

(7.3.5)

Using (7.3.2) to (7.3.5), the required sums and differences of the scattered intensity components in right- and left-circularly polarized incident light are

$$I_x^{dR} + I_x^{dL} = \frac{\omega^4 \mu_0 E^{(0)2}}{32\pi^2 c y^2}(\tilde{\alpha}_{xx}\tilde{\alpha}_{xx}^* + \tilde{\alpha}_{xy}\tilde{\alpha}_{xy}^* + \cdots),$$

(7.3.6a)

$$I_z^{dR} + I_z^{dL} = \frac{\omega^4 \mu_0 E^{(0)2}}{32\pi^2 c y^2}(\tilde{\alpha}_{zx}\tilde{\alpha}_{zx}^* + \tilde{\alpha}_{zy}\tilde{\alpha}_{zy}^* + \cdots),$$

(7.3.6b)

$$I_x^{dR} - I_x^{dL} = \frac{\omega^4 \mu_0 E^{(0)2}}{16\pi^2 c^2 y^2}[\mathrm{Im}(c\tilde{\alpha}_{xy}\tilde{\alpha}_{xx}^* + \tilde{\alpha}_{xy}\tilde{G}_{xy}^*$$

$$+ \tilde{\alpha}_{xx}\tilde{G}_{xx}^* - \tilde{\alpha}_{xy}\tilde{\mathscr{G}}_{zx}^* + \tilde{\alpha}_{xx}\tilde{\mathscr{G}}_{zy}^*$$

$$+ \tfrac{1}{3}\omega\mathrm{Re}\,(\tilde{\alpha}_{xx}\tilde{A}_{x,zy}^* - \tilde{\alpha}_{xy}\tilde{A}_{x,zx}^*$$

$$+ \tilde{\alpha}_{xy}\tilde{\mathscr{A}}_{x,xy}^* - \tilde{\alpha}_{xx}\tilde{\mathscr{A}}_{y,xy}^*) + \cdots],$$

(7.3.6c)

$$I_z^{dR} - I_z^{dL} = \frac{\omega^4 \mu_0 E^{(0)2}}{16\pi^2 c^2 y^2}[\mathrm{Im}(c\tilde{\alpha}_{zy}\tilde{\alpha}_{zx}^* + \tilde{\alpha}_{zy}\tilde{G}_{zy}^*$$

$$+ \tilde{\alpha}_{zx}\tilde{G}_{zx}^* + \tilde{\alpha}_{zy}\tilde{\mathscr{G}}_{xx}^* - \tilde{\alpha}_{zx}\tilde{\mathscr{G}}_{xy}^*$$

$$+ \tfrac{1}{3}\omega\mathrm{Re}\,(\tilde{\alpha}_{zx}\tilde{A}_{z,zy}^* - \tilde{\alpha}_{zy}\tilde{A}_{z,zx}^*$$

$$+ \tilde{\alpha}_{zy}\tilde{\mathscr{A}}_{x,zy}^* - \tilde{\alpha}_{zx}\tilde{\mathscr{A}}_{y,zy}^*) + \cdots].$$

(7.3.6d)

If the molecules are chiral, are in nondegenerate ground states, and no external static magnetic field is present, we need only retain terms in $\alpha G'$ and αA in (7.3.6). For scattering from fluids it is necessary to average these expressions over all orientations of the molecule, making use of the unit vector averages (4.2.53) and (4.2.54). We finally recover the circular intensity difference components obtained in Chapter 3:

$$\Delta_x(90°) = \frac{2(7\alpha_{\alpha\beta}G'^*_{\alpha\beta} + \alpha_{\alpha\alpha}G'^*_{\beta\beta} + \frac{1}{3}\omega\alpha_{\alpha\beta}\varepsilon_{\alpha\gamma\delta}A^*_{\gamma,\delta\beta})}{c(7\alpha_{\lambda\mu}\alpha^*_{\lambda\mu} + \alpha_{\lambda\lambda}\alpha^*_{\mu\mu})}, \qquad (3.5.36a)$$

$$\Delta_z(90°) = \frac{4(3\alpha_{\alpha\beta}G'^*_{\alpha\beta} - \alpha_{\alpha\alpha}G'^*_{\beta\beta} - \frac{1}{3}\omega\alpha_{\alpha\beta}\varepsilon_{\alpha\gamma\delta}A^*_{\gamma,\delta\beta})}{2c(3\alpha_{\lambda\mu}\alpha^*_{\lambda\mu} - \alpha_{\lambda\lambda}\alpha^*_{\mu\mu})}. \qquad (3.5.36b)$$

The same expressions may be used for Raman optical activity if the property tensors are replaced by corresponding vibrational Raman transition tensors. Within Placzek's approximation, discussed in Section 2.8.3, for scattering at transparent frequencies the replacements are

$$\alpha_{\alpha\beta} \to (\alpha_{\alpha\beta})_{v_m v_n} = \langle v_m | \alpha_{\alpha\beta}(Q) | v_n \rangle, \qquad (7.3.7a)$$

$$G'_{\alpha\beta} \to (G'_{\alpha\beta})_{v_m v_n} = \langle v_m | G'_{\alpha\beta}(Q) | v_n \rangle, \qquad (7.3.7b)$$

$$A_{\alpha,\beta\gamma} \to (A_{\alpha,\beta\gamma})_{v_m v_n} = \langle v_m | A_{\alpha,\beta\gamma}(Q) | v_n \rangle. \qquad (7.3.7c)$$

We can now deduce the basic symmetry requirements for natural Rayleigh and Raman optical activity. For natural Rayleigh optical activity, the same components of $\alpha_{\alpha\beta}$ and $G'_{\alpha\beta}$ must span the totally symmetric representation; and for natural vibrational Raman optical activity the same components of $\alpha_{\alpha\beta}$ and $G'_{\alpha\beta}$ must span the irreducible representation of the particular normal coordinate of vibration. This can only happen in the chiral point groups C_n, D_n, O, T, I (which lack improper rotation elements) in which polar and axial tensors of the same rank, such as $\alpha_{\alpha\beta}$ and $G'_{\alpha\beta}$, have identical transformation properties. Furthermore, although $A_{\alpha,\beta\gamma}$ does not transform like $G'_{\alpha\beta}$, the second-rank axial tensor $\varepsilon_{\alpha\gamma\delta}A_{\gamma,\delta\beta}$ that combines with $\alpha_{\alpha\beta}$ in the expressions for optically active scattering has transformation properties identical with $G'_{\alpha\beta}$. Consequently, all the Raman active vibrations in a chiral molecule should show Raman optical activity.

The further development of the Raman optical activity expressions can proceed in several different ways (Polavarapu, 1998; Nafie and Freedman, 2000). There are two models of Raman optical activity which parallel the fixed partial charge and bond dipole models of infrared vibrational optical activity, namely the *atom dipole interaction model* (Prasad and Nafie, 1979) and the *bond polarizability model* (Barron, 1979b; Escribano and Barron, 1988) which break the molecule down into either its constituent atoms or bonds, respectively. We shall emphasize the latter since it is more in keeping with the approach to electronic optical activity used in Chapter 5 in that it is based on bond or group properties and the geometrical disposition of the bonds or groups within the chiral molecule.

Since the property tensors $G'_{\alpha\beta}$ and $A_{\alpha,\beta\gamma}$ responsible for Raman optical activity are time even, there is no fundamental problem in Raman optical activity theory analogous to that arising in the theory of vibrational optical rotation and circular dichroism due to the time-odd nature of the magnetic dipole moment operator. Consequently, ab initio calculations of Raman optical activity appeared several years before the first such vibrational circular dichroism calculations: an ab initio method, based on calculations of $\alpha_{\alpha\beta}$, $G'_{\alpha\beta}$ and $A_{\alpha,\beta\gamma}$ in a static approximation (Section 2.6.5) and how these property tensors vary with the normal vibrational coordinates, was developed by Polavarapu in the late 1980s (Bose, Barron and Polavarapu, 1989; Polavarapu, 1990). Although the first tranche of ab initio calculations of Raman optical activity spectra did not reached the high levels of accuracy that are now routine for calculations of vibrational circular dichroism spectra, they nonetheless proved valuable. For example, the absolute configuration of CHFClBr was reliably assigned as (S)-$(+)$ or (R)-$(-)$ from a comparison of the experimental and ab initio theoretical Raman optical activity spectra (Costante *et al.*, 1997; Polavarapu, 2002*b*). Analogous vibrational circular dichroism studies of this molecule are unfavourable because the frequencies of most of the fundamental normal modes of vibration are too low to be experimentally accessible, whereas all are accessible in the Raman optical activity spectrum. However, by including rarified basis sets containing moderately diffuse *p*-type orbitals on hydrogen atoms, ab initio Raman optical activity calculations of similar high quality to those of infrared circular dichroism have now been achieved (Zuber and Hug, 2004). More generally, the recent work of Hug and his colleagues (Hug, 2001; Hug *et al.*, 2001; Hug, 2002) has provided a fresh approach to Raman optical activity theory that provides a firm foundation not only for accurate calculations of Raman optical activity observables but also for a sound understanding of the underlying mechanisms.

Only natural optical activity in *transparent* Raman scattering is discussed in detail below. Although the basic results in the first part of this section, and also the general Stokes parameters (3.5.3), may be applied to the resonance situation, the subject is still at an early stage of development compared with transparent Raman optical activity at visible exciting wavelengths which, at the time of writing, is a well established and highly informative practical chiroptical technique. Nonetheless a satisfactory theory of natural resonance Raman optical activity has been developed (Nafie, 1996) and the first experimental observations reported (Vargek *et al.*, 1998). Resonance scattering may present a rich variety of new Raman optical activity phenomena that do not arise in transparent scattering, and could be especially valuable in the ultraviolet region for the study of biomolecules.

7.3.2 Experimental quantities

In most light scattering work, absolute intensities are not measured because of problems with the standardization of instrumental factors. Instead, dimensionless quantities such as the depolarization ratio (3.5.9) and, in the case of Rayleigh and

Raman optical activity, the circular intensity difference (1.4.1), are often derived from the intensities measured on the same arbitary scale.

It is useful to rewrite the circular intensity difference expressions (3.5.34–36) in terms of the following tensor invariants (Section 4.2.6):

$$\alpha = \tfrac{1}{3}\alpha_{\alpha\alpha} = \tfrac{1}{3}(\alpha_{XX} + \alpha_{YY} + \alpha_{ZZ}), \tag{7.3.8a}$$

$$\begin{aligned}
\beta(\alpha)^2 &= \tfrac{1}{2}(3\alpha_{\alpha\beta}\alpha_{\alpha\beta} - \alpha_{\alpha\alpha}\alpha_{\beta\beta}) \\
&= \tfrac{1}{2}\big[(\alpha_{XX} - \alpha_{YY})^2 + (\alpha_{XX} - \alpha_{ZZ})^2 + (\alpha_{YY} - \alpha_{ZZ})^2 \\
&\quad + 6(\alpha_{XY}^2 + \alpha_{XZ}^2 + \alpha_{YZ}^2)\big],
\end{aligned} \tag{7.3.8b}$$

$$G' = \tfrac{1}{3}G'_{\alpha\alpha} = \tfrac{1}{3}(G'_{XX} + G'_{YY} + G'_{ZZ}), \tag{7.3.8c}$$

$$\begin{aligned}
\beta(G')^2 &= \tfrac{1}{2}(3\alpha_{\alpha\beta}G'_{\alpha\beta} - \alpha_{\alpha\alpha}G'_{\beta\beta}) \\
&= \tfrac{1}{2}\{(\alpha_{XX} - \alpha_{YY})(G'_{XX} - G'_{YY}) + (\alpha_{XX} - \alpha_{ZZ})(G'_{XX} - G'_{ZZ}) \\
&\quad + (\alpha_{YY} - \alpha_{ZZ})(G'_{YY} - G'_{ZZ}) \\
&\quad + 3[\alpha_{XY}(G'_{XY} + G'_{YX}) + \alpha_{XZ}(G'_{XZ} + G'_{ZX}) + \alpha_{YZ}(G'_{YZ} + G'_{ZY})]\},
\end{aligned} \tag{7.3.8d}$$

$$\begin{aligned}
\beta(A)^2 &= \tfrac{1}{2}\omega\alpha_{\alpha\beta}\varepsilon_{\alpha\gamma\delta}A_{\gamma,\delta\beta} \\
&= \tfrac{1}{2}\omega[(\alpha_{YY} - \alpha_{XX})A_{Z,XY} + (\alpha_{XX} - \alpha_{ZZ})A_{Y,ZX} + (\alpha_{ZZ} - \alpha_{YY})A_{X,YZ} \\
&\quad + \alpha_{XY}(A_{YYZ} - A_{ZYY} + A_{ZXX} - A_{XXZ}) \\
&\quad + \alpha_{XZ}(A_{YZZ} - A_{ZZY} + A_{XXY} - A_{YXX}) \\
&\quad + \alpha_{YZ}(A_{ZZX} - A_{XZZ} + A_{XYY} - A_{YYX})].
\end{aligned} \tag{7.3.8e}$$

Notice that, since $\varepsilon_{\alpha\gamma\delta}A_{\gamma,\delta\beta}$ is traceless, there is no corresponding isotropic tensor analogous to α and G'. In a principal axis system that diagonalizes $\alpha_{\alpha\beta}$, the terms in (7.3.8) involving off-diagonal components of $\alpha_{\alpha\beta}$ vanish. The circular intensity difference expressions (3.5.34–36) are now

$$\Delta(0°) = \frac{4[45\alpha G' + \beta(G')^2 - \beta(A)^2]}{c[45\alpha^2 + 7\beta(\alpha)^2]}, \tag{7.3.9a}$$

$$\Delta(180°) = \frac{24[\beta(G')^2 + \tfrac{1}{3}\beta(A)^2]}{c[45\alpha^2 + 7\beta(\alpha)^2]}, \tag{7.3.9b}$$

$$\Delta_x(90°) = \frac{2[45\alpha G' + 7\beta(G')^2 + \beta(A)^2]}{c[45\alpha^2 + 7\beta(\alpha)^2]}, \tag{7.3.9c}$$

$$\Delta_z(90°) = \frac{12[\beta(G')^2 - \tfrac{1}{3}\beta(A)^2]}{6c\beta(\alpha)^2}. \tag{7.3.9d}$$

Common factors in the numerators and denominators of these circular intensity differences have not been cancelled, so that the relative sum and difference intensities may be compared directly. We refer to Andrews (1980) and Hecht and Barron (1990) for further discussion of the dependence of the circular intensity difference components on the scattering angle and the extraction of the tensor invariants.

There is an additional experimental configuration for circular intensity difference measurements that is of interest. By using the general Stokes parameters (3.5.3) for Rayleigh and Raman scattering within the Mueller matrix formalism, it may be shown that, by setting the transmission axis of the linear polarization analyzer in the light beam scattered at 90° at the *magic angle* of $\pm \sin^{-1} \sqrt{(2/3)} \approx \pm 54.74°$ to the scattering plane yz, the contribution from the electric dipole–electric quadrupole Raman optical activity mechanism vanishes, so that pure electric dipole–magnetic dipole Raman optical activity spectra may be measured (Hecht and Barron, 1989). The associated magic angle circular intensity difference is

$$\Delta_*(90°) = \frac{(20/3)[9\alpha G' + 2\beta(G')^2]}{(10/3)[9\alpha^2 + 2\beta(\alpha)^2]}. \tag{7.3.10}$$

Hug (2001, 2002) found that the invariant combinations of tensor products appearing in the numerator and denominator of (7.3.10) are all that is measured in scattering cross sections integrated over all directions, and pointed out that this is reminiscent of the situation in natural optical rotation and circular dichroism of isotropic samples where the electric dipole–electric quadrupole contributions also average to zero.

It was mentioned in Section 3.5.4 that the degree of circularity of the scattered light gives information equivalent to that from the circular intensity difference. The two experimental strategies are called *scattered circular polarization* and *incident circular polarization* Raman optical activity measurements, respectively. The simultaneous measurement of both incident and scattered circular polarization Raman optical activity, called *dual circular polarization* Raman optical activity, can be advantageous (Nafie and Freedman, 1989; Hecht and Barron, 1990; Nafie and Che, 1994), but we shall not give the detailed analysis here.

In fact incident circular polarization and scattered circular polarization Raman optical activity will only give *identical* information for Rayleigh scattering. For vibrational Raman scattering, this information is approximately the same for scattering at transparent wavelengths; but in the case of resonance scattering at absorbing incident wavelengths, an interesting Stokes–antiStokes asymmetry arises (Barron and Escribano, 1985). This may be shown from the general Stokes parameters (3.5.3) of the scattered light using the general expression (3.3.4) for the scattering tensor and retaining the italic and script forms of the optical activity tensors which are taken as Raman transition tensors between different initial and final states m and n. We require the relationships

$$(\tilde{G}_{\alpha\beta})_{mn} = -(\tilde{\mathscr{G}}_{\beta\alpha})_{nm} = -(\tilde{G}_{\alpha\beta})^*_{mn}, \tag{7.3.11a}$$

$$(\tilde{A}_{\alpha,\beta\gamma})_{mn} = (\tilde{\mathscr{A}}_{\alpha,\beta\gamma})_{nm} = (\tilde{A}_{\alpha,\beta\gamma})^*_{mn}, \tag{7.3.11b}$$

which obtain for vibrational Raman scattering by invoking the Hermiticity of the operators μ, $\mathbf{m}$ and Θ and writing $\omega_{jn} \approx \omega_{jm}$, or from general time reversal arguments (Hecht and Barron, 1993c). It is then found that the degree of circularity $S_3^d(90°)/S_0^d(90°)$ observed in a *Stokes* Raman transition $m \leftarrow n$ in incident light linearly polarized perpendicular to the scattering plane, for example, approximately equals the polarized circular intensity difference (7.3.9c) for an *antiStokes* Raman transition $n \leftarrow m$, and vice versa. The equality is exact for a Stokes/antiStokes reciprocal pair (Hecht and Barron, 1993c), meaning that if the incident frequency is ω for the Stokes process, an incident frequency $\omega - \omega_{mn}$ must be taken for the associated antiStokes process (Hecht and Barron, 1993a). On the other hand, the Stokes and antiStokes degrees of circular polarization will in general be different, as will the Stokes and antiStokes circular intensity differences.

A further optical activity phenomenon, called *linear polarization* Raman optical activity, can occur under resonance conditions in 90° scattering (Hecht and Nafie, 1990; Hecht and Barron, 1993b,c). This involves intensity differences in the Raman scattered light associated with orthogonal linear polarization states at $\pm45°$ to the scattering plane in the incident or scattered radiation or in both simultaneously.

It is clear that vibrational optical activity in Raman scattering may be studied by means of a plethora of different experimental strategies, each having particular advantages (or disadvantages) and revealing different aspects of the phenomenon. Fortunately, however, for most practical applications in chemistry and biochemistry measurement of either the simple incident or scattered circular polarization form of Raman optical activity provides all necessary information. As discussed in Section 7.3.6 below, a bond polarizability model reveals that backscattering is the optimum experimental geometry for most routine applications of Raman optical activity since this provides the optimum signal-to-noise ratio. This discovery was immensely valuable since it was a critical step in the extension of Raman optical activity measurements to biomolecules in aqueous solution.

7.3.3 Optical activity in transmitted and scattered light

Before proceeding with the detailed theoretical development, we shall pause and reflect on the relationship between the fundamental scattering mechanisms responsible for conventional optical rotation and circular dichroism on the one hand, and Rayleigh and Raman optical activity on the other.

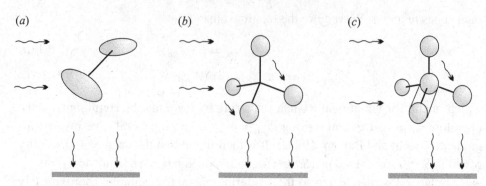

Fig. 7.2 The generation of Rayleigh optical activity by (*a*) two anisotropic groups, (*b*) four isotropic groups and (*c*) a bond and two isotropic groups.

For Rayleigh optical activity, general components of the electronic optical activity tensors $G'_{\alpha\beta}$ and $A_{\alpha,\beta\gamma}$ must be calculated; and for Raman optical activity these tensors must be calculated as functions of the normal vibrational coordinates. We saw in Chapter 5 that useful physical insight into conventional optical rotation and circular dichroism obtains from coupling models. But in extending these models to Rayleigh and Raman optical activity, care must be taken to include the origin-dependent parts of $G'_{\alpha\beta}$ and $A_{\alpha,\beta\gamma}$ since these give rise to a mechanism that has no counterpart in optical rotation and circular dichroism (Barron and Buckingham, 1974). The latter are birefringence phenomena and therefore originate in interference between transmitted and forward-scattered waves. Thus in the Kirkwood model, illustrated in Fig. 5.1*a*, the optical rotation generated by a chiral structure comprising two achiral groups involves dynamic coupling: only forward-scattered waves that have been deflected from one group to the other have sampled the chirality and can generate optical rotation on combining with the transmitted wave. But the transmitted wave is not important in Rayleigh scattering, so interference between two waves scattered independently from the two groups provides chiral information (Fig. 7.2*a*). Dynamic coupling is not required, although it can make higher-order contributions.

This picture can be extended to a chiral tetrahedral structure such as CHFClBr. Since a pair of dynamically coupled spherical atoms constitutes a single anisotropically polarizable group, the Born–Boys model of optical rotation, which considers just the four ligand atoms, requires dynamic coupling over all four atoms (Fig. 5.1*b*); whereas Rayleigh optical activity require interference between waves scattered independently from two pairs of dynamically coupled atoms (Fig. 7.2*b*), or from one atom and three other dynamically coupled atoms. If the central carbon atom is included, the carbon–ligand bonds constitute anisotropic groups, so less dynamic coupling is required (Fig. 7.2*c*). Each diagram represents just one possible scattering sequence: any explicit calculation would sum over all permutations.

The mechanisms just described apply equally to Raman optical activity associated with bond stretching vibrations. It will be seen shortly that, for Raman optical activity in certain modes of vibration involving, in particular, deformations and torsions, rather different mechanisms can dominate.

7.3.4 The two-group model of Rayleigh optical activity

The Rayleigh optical activity generated by a chiral molecule consisting of two neutral achiral groups 1 and 2 is now considered in detail. As well as giving mathematical expression to the scattering mechanism illustrated in Fig. 7.2a, this provides useful background for a consideration of Raman optical activity.

We assume no electron exchange between the groups, and write the polarizability and optical activity tensors of the molecule as sums of the corresponding group tensors. A group tensor must be referred to a fixed origin within the molecule, which we choose to be the local origin on group 1. Using (2.6.35) for the origin-dependences of the tensors, we have

$$\alpha_{\alpha\beta} = \alpha_{1_{\alpha\beta}} + \alpha_{2_{\alpha\beta}} + \text{coupling terms}, \tag{7.3.12a}$$

$$G'_{\alpha\beta} = G'_{1_{\alpha\beta}} + G'_{2_{\alpha\beta}} - \tfrac{1}{2}\omega\varepsilon_{\beta\gamma\delta}R_{21\gamma}\alpha_{2_{\alpha\delta}} + \text{coupling terms}, \tag{7.3.12b}$$

$$A_{\alpha,\beta\gamma} = A_{1_{\alpha,\beta\gamma}} + A_{2_{\alpha,\beta\gamma}} + \tfrac{3}{2}R_{21_\beta}\alpha_{2_{\alpha\gamma}} + \tfrac{3}{2}R_{21_\gamma}\alpha_{2_{\alpha\beta}}$$
$$- R_{21_\delta}\alpha_{2_{\alpha\delta}}\delta_{\beta\gamma} + \text{coupling terms}, \tag{7.3.12c}$$

where $\alpha_{i_{\alpha\beta}}$, $G'_{i_{\alpha\beta}}$ and $A_{i_{\alpha,\beta\gamma}}$ are tensors referred to a local origin on group i, and $\mathbf{R}_{21} = \mathbf{R}_2 - \mathbf{R}_1$ is the vector from the origin on 1 to that on 2. The coupling terms can be developed using the methods of Chapter 5 if required. Even though all components of $G'_{i_{\alpha\beta}}$ and $A_{i_{\alpha,\beta\gamma}}$ may be zero, the origin-dependent parts may not be. Also, although the groups are assumed to be achiral in the usual sense, for certain symmetries such as C_{2v} (see Table 4.2) there are nonzero components of the optical activity tensors that can contribute to Rayleigh optical activity. Using (7.3.12), the relevant polarizability–optical activity products in the circular intensity difference components (3.5.36) can be approximated by

$$\alpha_{\alpha\beta}G'_{\alpha\beta} = -\tfrac{1}{2}\omega\varepsilon_{\beta\gamma\delta}R_{21\gamma}\alpha_{1_{\alpha\beta}}\alpha_{2_{\delta\alpha}} + \alpha_{1_{\alpha\beta}}G'_{2_{\alpha\beta}} + \alpha_{2_{\alpha\beta}}G'_{1_{\alpha\beta}}, \tag{7.3.13a}$$

$$\tfrac{1}{3}\omega\,\alpha_{\alpha\beta}\varepsilon_{\alpha\gamma\delta}A_{\gamma,\delta\beta} = -\tfrac{1}{2}\omega\varepsilon_{\beta\gamma\delta}R_{21\gamma}\alpha_{1_{\alpha\beta}}\alpha_{2_{\delta\alpha}}$$
$$+ \tfrac{1}{3}\omega\alpha_{1_{\alpha\beta}}\varepsilon_{\alpha\gamma\delta}A_{2_{\gamma,\delta\beta}} + \tfrac{1}{3}\omega\,\alpha_{2_{\alpha\beta}}\varepsilon_{\alpha\gamma\delta}A_{1_{\gamma,\delta\beta}}, \tag{7.3.13b}$$

$$\alpha_{\alpha\alpha}G'_{\beta\beta} = 0, \tag{7.3.13c}$$

where the coupling terms have been neglected.

If both groups have threefold or higher proper rotation axes, the equations (7.3.13) can be given a tractable form. If the groups are achiral, they cannot belong to the proper rotation point groups, and for the remaining axially-symmetric point groups it can be deduced from Tables 4.2 that the components of the second-rank axial tensors $G'_{\alpha\beta}$ and $\varepsilon_{\alpha\gamma\delta}A_{\gamma,\delta\beta}$ are either zero or have $G'_{xy} = -G'_{yx}$ and $\varepsilon_{xy\delta}A_{\gamma,\delta y} = -\varepsilon_{yy\delta}A_{\gamma,\delta x}$. The terms $\alpha_{i_{\alpha\beta}}G'_{j_{\alpha\beta}}$ and $\alpha_{i_{\alpha\beta}}\varepsilon_{\alpha\gamma\delta}A_{j_{\gamma,\delta\beta}}$ in (7.3.13) are then zero because $\alpha_{\alpha\beta} = \alpha_{\beta\alpha}$ This conclusion can be reached more simply by invoking (4.2.59). If the unit vectors $\mathbf{s}_i$, $\mathbf{t}_i$, $\mathbf{u}_i$ define the principal axes of group i, with $\mathbf{u}_i$ along the symmetry axis, from (4.2.58) its polarizability tensor can be written

$$\alpha_{i_{\alpha\beta}} = \alpha_i(1 - \kappa_i)\delta_{\alpha\beta} + 3\alpha_i\kappa_i u_{i_\alpha} u_{i_\beta}, \qquad (7.3.14)$$

where α_i and κ_i are the mean polarizability and dimensionless polarizability anisotropy. Then

$$\varepsilon_{\beta\gamma\delta}R_{21\gamma}\alpha_{1_{\alpha\beta}}\alpha_{2_{\delta\alpha}} = 9\varepsilon_{\beta\gamma\delta}R_{21\gamma}\alpha_1\alpha_2\kappa_1\kappa_2 u_{1_\alpha} u_{2_\alpha} u_{1_\beta} u_{2_\delta}. \qquad (7.3.15)$$

For the simplest chiral pair where the principal axes of the two groups are in parallel planes (Fig. 5.4), this becomes

$$\varepsilon_{\beta\gamma\delta}R_{21\gamma}\alpha_{1_{\alpha\beta}}\alpha_{2_{\delta\alpha}} = -\tfrac{9}{2}R_{21}\alpha_1\alpha_2\kappa_1\kappa_2 \sin 2\theta. \qquad (7.3.16)$$

Using (7.3.16) in (7.3.13), the combinations of polarizability–optical activity products required in the circular intensity difference components (3.5.36) are

$$3\alpha_{\alpha\beta}G'_{\alpha\beta} - \alpha_{\alpha\alpha}G'_{\beta\beta} - \tfrac{1}{3}\omega\,\alpha_{\alpha\beta}\varepsilon_{\alpha\gamma\delta}A_{\gamma,\delta\beta} = \tfrac{9}{2}\omega R_{21}\alpha_1\alpha_2\kappa_1\kappa_2 \sin 2\theta, \qquad (7.3.17a)$$

$$7\alpha_{\alpha\beta}G'_{\alpha\beta} + \alpha_{\alpha\alpha}G'_{\beta\beta} + \tfrac{1}{3}\omega\,\alpha_{\alpha\beta}\varepsilon_{\alpha\gamma\delta}A_{\gamma,\delta\beta} = 18\omega R_{21}\alpha_1\alpha_2\kappa_1\kappa_2 \sin 2\theta. \qquad (7.3.17b)$$

We also require combinations of polarizability–polarizability products. First use (7.3.12a) to write

$$\alpha_{\alpha\beta}\alpha_{\alpha\beta} = \alpha_{1_{\alpha\beta}}\alpha_{1_{\alpha\beta}} + \alpha_{2_{\alpha\beta}}\alpha_{2_{\alpha\beta}} + 2\alpha_{1_{\alpha\beta}}\alpha_{2_{\alpha\beta}}. \qquad (7.3.18)$$

From (7.3.14) we can write, for axially-symmetric groups,

$$\alpha_{i_{\alpha\beta}}\alpha_{j_{\alpha\beta}} = 3\alpha_i\alpha_j + 3\alpha_i\alpha_j\kappa_i\kappa_j(3u_{i_\alpha}u_{j_\alpha}u_{i_\beta}u_{j_\beta} - 1) \qquad (7.3.19a)$$

$$= 3\alpha_i\alpha_j + \tfrac{3}{2}\alpha_i\alpha_j\kappa_i\kappa_j(1 + 3\cos 2\theta_{ij}), \qquad (7.3.19b)$$

$$\alpha_{i_{\alpha\alpha}}\alpha_{j_{\beta\beta}} = 9\alpha_i\alpha_j. \qquad (7.3.19c)$$

Consequently,

$$3\alpha_{i_{\alpha\beta}}\alpha_{j_{\alpha\beta}} - \alpha_{i_{\alpha\alpha}}\alpha_{j_{\beta\beta}} = \tfrac{9}{2}\alpha_i\alpha_j\kappa_i\kappa_j(1 + 3\cos 2\theta_{ij}), \qquad (7.3.20a)$$

$$7\alpha_{i_{\alpha\beta}}\alpha_{j_{\alpha\beta}} + \alpha_{i_{\alpha\alpha}}\alpha_{j_{\beta\beta}} = 30\alpha_i\alpha_j + \tfrac{21}{2}\alpha_i\alpha_j\kappa_i\kappa_j(1 + 3\cos 2\theta_{ij}). \qquad (7.3.20b)$$

Using (7.3.20) in (7.3.18), the required expressions are found to be

$$3\alpha_{\alpha\beta}\alpha_{\alpha\beta} - \alpha_{\alpha\alpha}\alpha_{\beta\beta} = 18\left(\alpha_1^2\kappa_1^2 + \alpha_2^2\kappa_2^2\right) + 9\alpha_1\alpha_2\kappa_1\kappa_2(1 + 3\cos 2\theta), \quad (7.3.21a)$$

$$7\alpha_{\alpha\beta}\alpha_{\alpha\beta} + \alpha_{\alpha\alpha}\alpha_{\beta\beta} = 30\left(\alpha_1^2 + \alpha_2^2 + 2\alpha_1\alpha_2\right) + 42\left(\alpha_1^2\kappa_1^2 + \alpha_2^2\kappa_2^2\right)$$
$$+ 21\alpha_1\alpha_2\kappa_1\kappa_2(1 + 3\cos 2\theta). \quad (7.3.21b)$$

If groups 1 and 2 are identical, these results give the following Rayleigh circular intensity difference components (Barron and Buckingham, 1974):

$$\Delta_x(90°) = \frac{24\pi R_{21}\kappa^2 \sin 2\theta}{\lambda[40 + 7\kappa^2(5 + 3\cos 2\theta)]}, \quad (7.3.22a)$$

$$\Delta_z(90°) = \frac{2\pi R_{21} \sin 2\theta}{\lambda(5 + 3\cos 2\theta)}, \quad (7.3.22b)$$

where κ is a group polarizability anisotropy. Positive Δ values obtain for the absolute configuration of Fig. 5.4. Notice that, provided $\lambda \gg R_{21}$, Rayleigh optical activity increases with increasing separation of the two groups. In contrast, the corresponding Kirkwood optical rotation, given by (5.3.34), decreases with increasing separation since it depends on dynamic coupling. No knowledge of the polarizabilities of the two groups is required to calculate $\Delta_z(90°)$.

A twisted biphenyl provides a simple example of a chiral two-group structure with the symmetry axes of the two groups in parallel planes. The symmetry axes $\mathbf{u}_1$ and $\mathbf{u}_2$ are along the sixfold rotation axes of the aromatic rings (for simplicity we disregard the fact that the ring substituents required to constrain the biphenyl to a chiral conformation destroy the axial symmetry of the aromatic rings). With $R_{21} \approx 5 \times 10^{-10}$ m, $\theta = 45°$ and $\lambda = 500$ nm, (7.3.22b) gives $\Delta_z(90°) \approx 1.3 \times 10^{-3}$. Taking $|\kappa| = 0.18$ for benzene (from light scattering data), $\Delta_x(90°) \approx 0.6 \times 10^{-4}$. Thus the depolarized Rayleigh circular intensity difference is at least an order of magnitude larger than the polarized. These estimates apply only to gaseous samples. In liquids, a significant reduction in Rayleigh scattering occurs through interference, the isotropic contribution being suppressed much more than the anisotropic.

We refer to Stone (1975 and 1977) for an extension of the calculation to a two-group structure of more general geometry.

The basic circular intensity difference equations (3.5.36) are valid only when the wavelength of the incident light is much greater than the molecular dimensions, so the two-group results derived above are valid only when $R_{21} \ll \lambda$, which is satisfied for most molecules, except long chain polymers, in visible light. Instead of starting from the basic equations (3.5.36), it is in fact possible to derive the two-group circular intensity difference results in a manner that illustrates directly the physical picture outlined in Section 7.3.3, without invoking the optical activity tensors $G'_{\alpha\beta}$

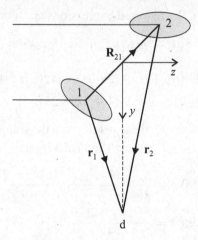

Fig. 7.3 The geometry for scattering at 90° by a chiral two-group molecule. The *x* direction is out of the plane of the paper.

and $A_{\alpha,\beta\gamma}$ at all: at the same time, the restriction to small molecular dimensions is lifted. We simply compute the intensity components at the detector arising from waves radiated independently by the oscillating electric dipole moments induced in the two groups by the incident light wave. So as not to obscure the essential simplicity of the treatment, we neglect contributions from the intrinsic optical activity tensors of the two groups: for reasons given above, the results are then valid if the two groups are intrinsically achiral and have threefold or higher proper rotation axes.

Figure 7.3 shows the chiral two-group structure at some arbitrary orientation to space-fixed axes *x*, *y*, *z* with the origin taken at the mid point of the connecting bond. The complex electric vector of the incident plane wave light beam is given by (7.3.1) Assuming that the incident plane wave front is sufficiently wide that the electric field amplitude (but not the phase) is the same at the two groups, the amplitudes of the complex oscillating electric dipole moments induced in the two groups are, from (7.3.3a),

$$\tilde{\mu}_{1\alpha}^{(0)} = \tilde{\alpha}_{1\alpha\beta} \tilde{E}_{\beta}^{(0)} e^{-i\omega R_{21z}/2c}, \qquad (7.3.23a)$$

$$\tilde{\mu}_{2\alpha}^{(0)} = \tilde{\alpha}_{2\alpha\beta} \tilde{E}_{\beta}^{(0)} e^{i\omega R_{21z}/2c}, \qquad (7.3.23b)$$

where the electric field of the light wave is taken at the appropriate local group origins. From (7.3.2), the corresponding radiated fields at the point d on the *y* axis in the wave zone are

$$\tilde{E}_{1\alpha} = \frac{\omega^2 \mu_0}{4\pi r_1} \tilde{\mu}_{1\alpha}^{(0)} e^{-i\omega(t - r_1/c)}, \qquad (7.3.24a)$$

$$\tilde{E}_{2\alpha} = \frac{\omega^2 \mu_0}{4\pi r_2} \tilde{\mu}_{2\alpha}^{(0)} e^{-i\omega(t - r_2/c)}. \qquad (7.3.24b)$$

Combining (7.3.23) and (7.3.24), the waves radiated from the two groups provide the following contributions to the electric vector of the light scattered at 90° from the complete structure:

$$\tilde{E}_{1\alpha} = \frac{\omega^2 \mu_0}{4\pi \left(y - \frac{1}{2}R_{21y}\right)} \tilde{\alpha}_{1\alpha\beta} \tilde{E}_{\beta}^{(0)} e^{-i\omega[t-(y-\frac{1}{2}R_{21y}-\frac{1}{2}R_{21z})/c]}, \quad (7.3.25a)$$

$$\tilde{E}_{2\alpha} = \frac{\omega^2 \mu_0}{4\pi \left(y + \frac{1}{2}R_{21y}\right)} \tilde{\alpha}_{2\alpha\beta} \tilde{E}_{\beta}^{(0)} e^{-i\omega[t-(y+\frac{1}{2}R_{21y}+\frac{1}{2}R_{21z})/c]} \quad (7.3.25b)$$

where, since $y \gg R_{21}$, we have taken

$$r_1 \approx y - \tfrac{1}{2}R_{21y}, \quad r_2 \approx y + \tfrac{1}{2}R_{21y}.$$

The circular intensity difference components are now obtained from

$$\Delta_x(90°) = \frac{(\tilde{E}_x \tilde{E}_x^*)^R - (\tilde{E}_x \tilde{E}_x^*)^L}{(\tilde{E}_x \tilde{E}_x^*)^R + (\tilde{E}_x \tilde{E}_x^*)^L},$$

$$\Delta_z(90°) = \frac{(\tilde{E}_z \tilde{E}_z^*)^R - (\tilde{E}_z \tilde{E}_z^*)^L}{(\tilde{E}_z \tilde{E}_z^*)^R + (\tilde{E}_z \tilde{E}_z^*)^L},$$

where $E_\alpha = E_{1\alpha} + E_{2\alpha}$ is the total electric field detected at d. Using (7.3.5) for the right- and left-circularly polarized incident electric vectors in (7.3.25) we find, for transparent frequencies in the absence of static magnetic fields,

$$\langle (\tilde{E}_x \tilde{E}_x^*)^R - (\tilde{E}_x \tilde{E}_x^*)^L \rangle = \left(\frac{\omega^2 \mu_0 E^{(0)}}{4\pi y}\right)^2$$
$$\times \left\langle 2(\alpha_{1xx}\alpha_{2xy} - \alpha_{1xy}\alpha_{2xx}) \sin\left[\frac{\omega}{c}(R_{21z} + R_{21y})\right] + \cdots \right\rangle,$$
$$(7.3.26a)$$

$$\langle (\tilde{E}_x \tilde{E}_x^*)^R + (\tilde{E}_x \tilde{E}_x^*)^L \rangle = \left(\frac{\omega^2 \mu_0 E^{(0)}}{4\pi y}\right)^2 \left\langle \alpha_{1xx}^2 + \alpha_{2xx}^2 + \alpha_{1xy}^2 + \alpha_{2xy}^2 \right.$$
$$\left. + 2(\alpha_{1xx}\alpha_{2xx} + \alpha_{1xy}\alpha_{2xy}) \cos\left[\frac{\omega}{c}(R_{21z} + R_{21y})\right] + \cdots \right\rangle,$$
$$(7.3.26b)$$

$$\langle (\tilde{E}_z \tilde{E}_z^*)^R - (\tilde{E}_z \tilde{E}_z^*)^L \rangle = \left(\frac{\omega^2 \mu_0 E^{(0)}}{4\pi y}\right)^2$$
$$\times \left\langle 2(\alpha_{1zx}\alpha_{2zy} - \alpha_{1zy}\alpha_{2zx}) \sin\left[\frac{\omega}{c}(R_{21z} + R_{21y})\right] + \cdots \right\rangle,$$
$$(7.3.26c)$$

$$\langle (\tilde{E}_z \tilde{E}_z^*)^R + (\tilde{E}_z \tilde{E}_z^*)^L \rangle = \left(\frac{\omega^2 \mu_0 E^{(0)}}{4\pi y}\right)^2 \left\langle \alpha_{1zx}^2 + \alpha_{2zx}^2 + \alpha_{1zy}^2 + \alpha_{2zy}^2 \right.$$
$$\left. + 2(\alpha_{1zx}\alpha_{2zx} + \alpha_{1zy}\alpha_{2zy}) \cos\left[\frac{\omega}{c}(R_{21z} + R_{21y})\right] + \cdots \right\rangle,$$
$$(7.3.26d)$$

where we have displayed only terms that contribute to the isotropic averages. It is not obvious that averages of (7.3.26) are obtainable for general values of $R_{21_z} + R_{21_y}$, but if the trigonometric functions are expanded, successive terms in powers of $2\pi(R_{21_z} + R_{21_y})/\lambda$ can be averaged. In the simplest case when $\lambda \gg R_{21_z} + R_{21_y}$, we can use the averages (4.2.53) and (4.2.54) to recover the two-group circular intensity difference components (7.3.22) that were obtained from a consideration of the origin dependence of the optical activity tensors. Without obtaining the isotropic averages for general values of $R_{21_z} + R_{21_y}$, we can still deduce the behaviour of the Rayleigh optical activity at group separations of the order of λ, since this is determined by $\sin[2\pi(R_{21_z} + R_{21_y})/\lambda]$. For a particular orientation, the Rayleigh optical activity is maximized (with alternating sign) when $R_{21_z} + R_{21_y} = \lambda/4,\ 3\lambda/4,\ 5\lambda/4, \ldots$, and is zero when $R_{21_z} + R_{21_y} = 0,\ \lambda/2,\ \lambda, \ldots$.

Notice that no Rayleigh optical activity is generated by the two-group model in the near-forward direction because there is no phase difference between the forward-scattered waves from the two groups. But Rayleigh optical activity is generated in backward scattering since a phase difference of $4\pi R_{21_z}/\lambda$ exists between the two backward-scattered waves. The Rayleigh circular intensity difference in the backward direction is simply twice the polarized two-group circular intensity difference (7.3.22a) for 90° scattering. Dynamic coupling between the groups, as in the Kirkwood model of optical rotation, is required to generate Rayleigh optical activity in the near-forward direction.

Andrews and Thirunamachandran (1977a,b) have discussed the two-group model in detail, including the question of isotropic averages for general group separations. Andrews (1994) has revisited the two-group model and provided a critical assessment from the view point of quantum electrodynamics.

7.3.5 The bond polarizability model of Raman optical activity

The starting point for the bond polarizability model (and indeed for the atom dipole interaction model) of Raman intensity and optical activity is Placzek's approximation, discussed in Section 2.8.3, for the vibrational transition polarizability at transparent frequencies. On expanding the effective polarizability operator $\alpha_{\alpha\beta}(Q)$ in the normal vibrational coordinates, the transition polarizability becomes

$$\langle v_m|\alpha_{\alpha\beta}(Q)|v_n\rangle = (\alpha_{\alpha\beta})_0\delta_{v_m v_n} + \sum_p \left(\frac{\partial\alpha_{\alpha\beta}}{\partial Q_p}\right)_0 \langle v_m|Q_p|v_n\rangle$$

$$+ \frac{1}{2}\sum_{p,q}\left(\frac{\partial^2\alpha_{\alpha\beta}}{\partial Q_p\partial Q_q}\right)_0 \langle v_m|Q_p Q_q|v_n\rangle + \cdots, \quad (7.3.27)$$

where $(\alpha_{\alpha\beta})_0$ is the polarizability of the molecule at the equilibrium nuclear configuration within the ground electronic state. The second term describes fundamental
vibrational Raman transitions and the third term describes first overtone and combination transitions.

The Raman intensity, therefore, is determined by the variation of the molecular
polarizability tensor with a normal coordinate of vibration, and this is calculated by
way of the variation of the tensor with local internal coordinates. We use (7.2.14),

$$\left(\frac{\partial \alpha_{\alpha\beta}}{\partial Q_p}\right)_0 = \sum_q \left(\frac{\partial \alpha_{\alpha\beta}}{\partial s_q}\frac{\partial s_q}{\partial Q_p}\right)_0 = \sum_q \left(\frac{\partial \alpha_{\alpha\beta}}{\partial s_q}\right)_0 L_{qp},\qquad(7.3.28)$$

and write the total molecular polarizability operator as a sum of local bond or group
polarizability operators:

$$\alpha_{\alpha\beta}(Q) = \sum_i \alpha_{i\,\alpha\beta}(Q).\qquad(7.3.29)$$

From the matrix elements (7.2.6), together with (7.3.27) and (7.3.28), we obtain
the following expression for the vibrational matrix element of the polarizability
operator for a fundamental transition associated with the normal coordinate Q_p:

$$\langle 1_p|\alpha_{\alpha\beta}(Q)|0\rangle = \left(\frac{\hbar}{2\omega_p}\right)^{\frac{1}{2}} \sum_i \sum_q \left(\frac{\partial \alpha_{i\alpha\beta}}{\partial s_q}\right)_0 L_{qp}.\qquad(7.3.30)$$

The extension to Raman optical activity involves writing the optical activity
tensors as sums of corresponding bond tensors, taking care to include the origin-
dependent parts. Thus, using a generalization of (7.3.12b), we obtain

$$G'_{\alpha\beta}(Q) = \sum_i G'_{i\alpha\beta}(Q) - \frac{\omega}{2}\varepsilon_{\beta\gamma\delta} \sum_i r_{i\gamma}(Q)\alpha_{i\alpha\delta}(Q),\qquad(7.3.31)$$

where $\mathbf{r}_i(Q)$ is the vector from the molecular origin to the origin of the ith group.
Expanding each term in the normal coordinates, and using (7.2.6) and (7.3.28), we
obtain the following vibrational matrix element:

$$\langle 1_p|G'_{\alpha\beta}(Q)|0\rangle = \left(\frac{\hbar}{2\omega_p}\right)^{\frac{1}{2}} \sum_i \sum_q \left(\frac{\partial G'_{i\alpha\beta}}{\partial s_q}\right)_0 L_{qp}$$

$$- \frac{\omega}{2}\left(\frac{\hbar}{2\omega_p}\right)^{\frac{1}{2}} \varepsilon_{\beta\gamma\delta} \sum_i \left[R_{i\gamma} \sum_q \left(\frac{\partial \alpha_{i\alpha\delta}}{\partial s_q}\right)_0 L_{qp}\right.$$

$$\left.+ (\alpha_{i\alpha\delta})_0 \sum_q \left(\frac{\partial r_{i\gamma}}{\partial s_q}\right)_0 L_{qp}\right].\qquad(7.3.32a)$$

A similar development for $\langle v_m|A_{\alpha,\beta\gamma}(Q)|v_n\rangle$ leads to

$$
\langle 1_p|A_{\alpha,\beta\gamma}(Q)|0\rangle = \left(\frac{\hbar}{2\omega_p}\right)^{\frac{1}{2}} \sum_i \sum_q \left(\frac{\partial A_{i\alpha,\beta\gamma}}{\partial s_q}\right)_0 L_{qp}
$$

$$
+ \left(\frac{\hbar}{2\omega_p}\right)^{\frac{1}{2}} \sum_i \left[\tfrac{3}{2}R_{i\beta} \sum_q \left(\frac{\partial\alpha_{i\alpha\gamma}}{\partial s_q}\right)_0 L_{qp}\right.
$$

$$
+ \tfrac{3}{2}(\alpha_{i\alpha\gamma})_0 \sum_q \left(\frac{\partial r_{i\beta}}{\partial s_q}\right)_0 L_{qp} + \tfrac{3}{2}R_{i\gamma} \sum_q \left(\frac{\partial\alpha_{i\alpha\beta}}{\partial s_q}\right)_0 L_{qp}
$$

$$
+ \tfrac{3}{2}(\alpha_{i\alpha\beta})_0 \sum_q \left(\frac{\partial r_{i\gamma}}{\partial s_q}\right)_0 L_{qp} - R_{i\delta} \sum_q \left(\frac{\partial\alpha_{i\alpha\delta}}{\partial s_q}\right)_0 L_{qp}\delta_{\beta\gamma}
$$

$$
\left.- (\alpha_{i\alpha\gamma})_0 \sum_q \left(\frac{\partial r_{i\delta}}{\partial s_q}\right)_0 L_{qp}\delta_{\beta\gamma}\right]. \tag{7.3.32b}
$$

Finally, using (7.3.30) and (7.3.32), the isotropic Raman intensity and optical activity in a fundamental transition associated with the normal coordinate Q_p are found to be (Barron and Clark, 1982)

$$
\langle 0|\alpha_{\alpha\beta}|1_p\rangle\langle 1_p|\alpha_{\alpha\beta}|0\rangle = \frac{\hbar}{2\omega_p} \left[\sum_i \sum_q \left(\frac{\partial\alpha_{i\alpha\beta}}{\partial s_q}\right)_0 L_{qp}\right]\left[\sum_j \sum_r \left(\frac{\partial\alpha_{j\alpha\beta}}{\partial s_r}\right)_0 L_{rp}\right],
$$
$$\tag{7.3.33a}$$

$$
\langle 0|\alpha_{\alpha\beta}|1_p\rangle\langle 1_p|G'_{\alpha\beta}|0\rangle =
$$

$$
- \frac{\hbar\omega}{4\omega_p}\varepsilon_{\beta\gamma\delta}\left\{\sum_{i<j}R_{ji\gamma}\left[\sum_q \left(\frac{\partial\alpha_{i\alpha\beta}}{\partial s_q}\right)_0 L_{qp}\right]\left[\sum_r \left(\frac{\partial\alpha_{j\delta\alpha}}{\partial s_r}\right)_0 L_{rp}\right]\right.
$$

$$
\left.+ \left[\sum_i \sum_q \left(\frac{\partial\alpha_{i\alpha\beta}}{\partial s_q}\right)_0 L_{qp}\right]\left[\sum_j (\alpha_{j\delta\alpha})_0 \sum_r \left(\frac{\partial r_{j\gamma}}{\partial s_r}\right)_0 L_{rp}\right]\right\}
$$

$$
+ \frac{\hbar}{2\omega_p}\left[\sum_i \sum_q \left(\frac{\partial\alpha_{i\alpha\beta}}{\partial s_q}\right)_0 L_{qp}\right]\left[\sum_j \sum_r \left(\frac{\partial G'_{j\alpha\beta}}{\partial s_r}\right)_0 L_{rp}\right], \tag{7.3.33b}
$$

$$
\tfrac{1}{3}\omega\langle 0|\alpha_{\alpha\beta}|1_p\rangle\langle 1_p|\varepsilon_{\alpha\gamma\delta}A_{\gamma,\delta\beta}|0\rangle =
$$

$$
- \frac{\hbar\omega}{4\omega_p}\varepsilon_{\beta\gamma\delta}\left\{\sum_{i<j}R_{ji\gamma}\left[\sum_q \left(\frac{\partial\alpha_{i\alpha\beta}}{\partial s_q}\right)_0 L_{qp}\right]\left[\sum_r \left(\frac{\partial\alpha_{j\delta\alpha}}{\partial s_r}\right)_0 L_{rp}\right]\right.
$$

$$+ \left[\sum_i \sum_q \left(\frac{\partial \alpha_{i\alpha\beta}}{\partial s_q} \right)_0 L_{qp} \right] \left[\sum_j (\alpha_{j\delta\alpha})_0 \sum_r \left(\frac{\partial r_{j\gamma}}{\partial s_r} \right)_0 L_{rp} \right] \right\}$$

$$+ \frac{\hbar}{6\omega_p} \left[\sum_i \sum_q \left(\frac{\partial \alpha_{i\alpha\beta}}{\partial s_q} \right)_0 L_{qp} \right] \left[\varepsilon_{\alpha\gamma\delta} \sum_j \sum_r \left(\frac{\partial A_{j\gamma,\delta\beta}}{\partial s_r} \right)_0 L_{rp} \right].$$

(7.3.33c)

We also require the products

$$\langle 0|\alpha_{\alpha\alpha}|1_p\rangle\langle 1_p|\alpha_{\beta\beta}|0\rangle = \frac{\hbar}{2\omega_p} \left[\sum_i \sum_q \left(\frac{\partial \alpha_{i\alpha\alpha}}{\partial s_q} \right)_0 L_{qp} \right] \left[\sum_j \sum_r \left(\frac{\partial \alpha_{j\beta\beta}}{\partial s_r} \right)_0 L_{rp} \right],$$

(7.3.33d)

$$\langle 0|\alpha_{\alpha\alpha}|1_p\rangle\langle 1_p|G'_{\beta\beta}|0\rangle = \frac{\hbar}{2\omega_p} \left[\sum_i \sum_q \left(\frac{\partial \alpha_{i\alpha\alpha}}{\partial s_q} \right)_0 L_{qp} \right] \left[\sum_j \sum_r \left(\frac{\partial G'_{j\beta\beta}}{\partial s_r} \right)_0 L_{rp} \right].$$

(7.3.33e)

The terms in these bond polarizability Raman optical activity expressions have interpretations analogous to some of the terms in the bond dipole infrared rotational strength (7.2.27b). Thus the first terms in (7.3.33b,c) are sums over all pairs of groups that constitute chiral structures, in accordance with the two-group mechanism illustrated in Fig. 7.2a, $\mathbf{R}_{ji}$ being the vector from the origin on group i to that on group j at the equilibrium nuclear configuration. The second terms involve changes in the position vector $\mathbf{r}_j$ of a group relative to the molecule-fixed origin: normal coordinates containing contributions from changes in either the length of $\mathbf{r}_j$, its orientation, or both, will activate this term. We call this the *inertial* term. The last terms involve products of intrinsic group polarizability and optical activity tensors. Simple examples of these different contributions are given shortly.

In applying these Raman optical activity expressions the question arises, as it did in the infrared case, of the actual choice of origins within the groups or bonds. Again, the complete expressions are invariant both to the choice of the molecular origin and the local group origins: the first two terms in (7.3.33b,c) are generated by the origin-dependent parts of the total $G'_{\alpha\beta}$ and $A_{\alpha,\beta\gamma}$ tensors of the molecule, so any changes in the first two terms caused by changes in the relative dispositions of the molecular origin and the local group origins are compensated by changes in the third terms.

For achiral axially-symmetric groups with local origins chosen to lie along the symmetry axes, it was shown in Section 7.3.4 that terms involving $\alpha_{i\alpha\beta} G'_{j\alpha\beta}$ and

$\alpha_{i_{\alpha\beta}}\varepsilon_{\alpha\gamma\delta}A_{j_{\gamma,\,\delta\beta}}$ do not contribute to Rayleigh optical activity. By extending the argument to vibrational transition tensors, the corresponding terms in (7.3.33b,c) can also be shown to be zero for all normal modes. The argument runs as follows. The bond polarizability model is based on Placzek's approximation and so the transition polarizability is written in the form

$$\left\langle v_m\left|\sum_i \alpha_{i_{\alpha\beta}}(Q)\right|v_n\right\rangle,$$

the sum being over all groups or bonds into which the molecule has been broken down. The time reversal arguments in Section 4.4.3 tell us that the real polarizability, and hence the effective operator $\Sigma_i\alpha_{i_{\alpha\beta}}(Q)$, is always pure symmetric, and this is also true of the individual bond polarizabilities $\alpha_{i_{\alpha\beta}}(Q)$. Thus the bond polarizability theory automatically implies pure symmetric transition polarizabilities. A different theory must be used to deal with those exotic situations which can lead to antisymmetric transition polarizabilities. Although statements at this level of generality cannot be applied to the transition optical activities because $G'_{\alpha\beta}$ and $A_{\alpha,\beta\gamma}$ do not have well defined permutation symmetry for all the point groups, some useful statements can be made for axial symmetry. For achiral axial symmetry, $G'_{\alpha\beta}$ is either antisymmetric or zero (Tables 4.2) and this enabled us to write in Chapter 4 that

$$G'_{\alpha\beta} = G'\varepsilon_{\alpha\beta\gamma}K_\gamma. \tag{4.2.59}$$

Within Placzek's approximation, the corresponding transition optical activity is written out as in (7.3.32a), and so for a chiral structure made up of achiral axially-symmetric groups, the intrinsic group contribution

$$\left\langle v_m\left|\sum_i G'_{i_{\alpha\beta}}(Q)\right|v_n\right\rangle$$

is pure antisymmetric, provided that the local group origins are chosen to lie along the group symmetry axes and the symmetry of the individual groups is maintained during the normal mode excursion. Of course, the origin-dependent parts of the transition optical activity (7.3.32a) have no particular permutation symmetry (indeed, if they were antisymmetric in α, β the complete structure could not be chiral). So for any pair of axially-symmetric groups i and j, $(\partial\alpha_{i_{\alpha\beta}}/\partial Q)_0$ is pure symmetric, $(\partial G'_{j_{\alpha\beta}}/\partial Q)_0$ is pure antisymmetric, and their product is zero. Similarly for

$$(\partial\alpha_{i_{\alpha\beta}}/\partial Q)_0\varepsilon_{\alpha\gamma\delta}(\partial A_{j_{\gamma,\,\delta\beta}}/\partial Q)_0.$$

Thus for achiral axially symmetric groups, with origins chosen to lie along the symmetry axes, all the Raman optical activity is generated by the two-group and inertial terms in (7.3.33b,c). These two terms, taken together, are invariant to displacements of the local group origins along the symmetry axes; for it is easily verified by invoking (7.3.14) for each group that the change in one term is compensated by an equal and opposite change in the other term. Notice that it is only changes in the inertial terms with $i \neq j$ that compensate changes in the two-group terms: intrinsic group intertial terms, corresponding to $i = j$, are invariant to the position of the origin anywhere along the symmetry axis.

For non axially-symmetric groups, the application of the bond polarizability expressions is much more complicated since the last terms of (7.3.33*b*,*c*), involving intrinsic group optical activity tensors, are now expected to make contributions comparable with the two-group and inertial terms, and all must be evaluated in order to guarantee a result that is invariant both to the choice of molecular origin and local group origins.

7.3.6 The bond polarizability model in forward, backward and 90° scattering

An important feature of the two-group model of Rayleigh optical activity follows from (7.3.13), namely that, for a pair of idealized axially-symmetric achiral groups or bonds, the isotropic contribution vanishes and the magenetic dipole and electric quadrupole mechanisms make equivalent contributions:

$$\alpha G' = 0, \quad \beta(G')^2 = \beta(A)^2. \tag{7.3.34}$$

It follows from (7.3.33) that equivalent results obtain within the bond polarizability model of Raman optical activity for a molecule composed entirely of idealized axially-symmetric achiral groups or bonds. This leads to some valuable simplifications of the general Raman optical activity expressions.

Consider first the circular intensity differences for polarized and depolarized 90° scattering given by (7.3.9*c*,*d*) respectively. Using (7.3.34), these reduce to

$$\Delta_x(90°) = \frac{16\beta(G')^2}{c[45\alpha^2 + 7\beta(\alpha)^2]}, \tag{7.3.35a}$$

$$\Delta_z(90°) = \frac{8\beta(G')^2}{6c\beta(\alpha)^2}, \tag{7.3.35b}$$

so that the ratio of the polarized to the depolarized Raman optical activity becomes (Barron, Escribano and Torrance, 1986)

$$\left(I_x^R - I_x^L\right) / \left(I_z^R - I_z^L\right) = 2. \tag{7.3.36}$$

Deviations from this factor of two provide a measure of the breakdown of the bond polarizability model and may give insight into Raman optical activity mechanisms.

Using (7.3.34), the circular intensity differences for forward and backward scattering given by (7.3.9a) and (7.3.9b), respectively, reduce to

$$\Delta(0°) = 0, \tag{7.3.34a}$$

$$\Delta(180°) = \frac{64\beta(G')^2}{2c[45\alpha^2 + 7\beta(\alpha)^2]}. \tag{7.3.34b}$$

Hence within the bond polarizability model we obtain the remarkable result that the Raman optical activity vanishes in the forward direction, but is maximized in the backward direction. As already mentioned at the end of Section 7.3.4, by considering a simple chiral two-group structure, it is easy to understand why no Rayleigh or Raman optical activity is generated in the forward direction because the two waves scattered independently from the two groups have covered the same optical path distance and so have the same phase. Compared with polarized 90° scattering, the Raman optical activity intensity is four times greater in backscattering with the associated conventional Raman intensity increased twofold. This represents a $2\sqrt{2}$-fold enhancement in the signal-to-noise ratio for the Raman optical activity within the same measurement time, so that a given signal-to-noise ratio is achieved eight times faster (Hecht, Barron and Hug, 1989).

7.4 The bond dipole and bond polarizability models applied to simple chiral structures

The bond dipole model of infrared vibrational optical activity and the bond polarizability model of Raman optical activity are both based on a decomposition of the molecule into bonds or groups supporting local internal vibrational coordinates. In principle, given a normal coordinate analysis and a set of bond dipole and bond polarizability parameters, the infrared and Raman optical activity associated with every normal mode of vibration of a chiral molecule may be calculated from (7.2.27) and (7.3.33) or, preferably, from more refined computational expressions given by Escribano and Barron (1988). However, due to the approximations inherent in these models, such calculations do not reproduce experimental data at all well: as mentioned previously, ab initio computations of vibrational circular dichroism and Raman optical activity are far superior and much easier to implement. Nonetheless, these two models do provide valuable insight into the generation of infrared and Raman vibrational optical activity, which we illustrate in this section by applying them to idealized normal modes, containing just one or two internal coordiantes, of some simple chiral molecular structures.

But first it is instructive to compare the magnitudes of the corresponding Raman and infrared optical activity observables, namely the dimensionless Raman circular intensity difference Δ and the infrared dissymmetry factor g. Both the infrared dipole strength (7.2.27a) and the Raman intensity (7.3.33a) depend on $1/\omega_p$ and so, other things being equal, both increase with decreasing vibrational frequency. However, the two-group and intertial terms in the Raman optical activity (7.3.33b,c) depend on ω/ω_p, the ratio of the exciting visible frequency to the vibrational frequency; whereas the corresponding terms in the infrared rotational strength (7.2.27b) have no such factor because now the exciting infrared frequency equals the vibrational frequency. Consequently, the Raman $I^R - I^L$ value tends to larger values with decreasing vibrational frequency, whereas the infrared $\epsilon^L - \epsilon^R$ value is comparable at high and low frequency. Thus the Raman Δ values are larger than the infrared g values by $\omega/\omega_p (= \lambda_p/\lambda)$, the ratio of the exciting frequency to the vibrational frequency. So the Raman approach to vibrational optical activity, because it uses visible exciting light, has a natural advantage over the infrared approach. For example, taking $\lambda_p = 500$ nm and $\lambda_p = 50\,000$ nm (corresponding to $\omega_p = 200\,\mathrm{cm}^{-1}$), the Raman experiment is 10^2 more favourable.

7.4.1 A simple two-group structure

We consider first the simple two-group structure of Fig. 5.4 where the principal axes of two neutral equivalent groups are in parallel planes. Since the structure has a twofold proper rotation axis, pairs of equivalent internal coordinates associated with the two groups, such as local bond stretchings or angle deformations, will always contribute with equal weight in symmetric and antisymmetric combinations to normal modes. The two idealized normal coordinates containing just symmetric and antisymmetric combinations of two equivalent internal coordinates localized on groups 1 and 2 are

$$Q_+ = N_+(s_1 + s_2), \tag{7.4.1a}$$

$$Q_- = N_-(s_1 - s_2), \tag{7.4.1b}$$

where $N_+ = N_- = N$ is a constant. The internal coordinates are

$$s_1 = \frac{1}{2N}(Q_+ + Q_-), \tag{7.4.1c}$$

$$s_2 = \frac{1}{2N}(Q_+ - Q_-), \tag{7.4.1d}$$

so that the L-matrix elements, defined in (7.2.15), are $L_{1+} = L_{1-} = L_{2+} = 1/2N, L_{2-} = -1/2N$.

We shall develop just the two-group terms in the infrared rotational strength (7.2.27b) and the Raman optical activity (7.3.33b,c). This means that the results apply when the two groups are axially symmetric (for then the intrinsic group optical activity terms are zero); and in addition the connecting bond is rigid (so that the internal coordinates are localized entirely on the two groups) in which case the intertial terms are zero provided we choose the local group origins to be the points where the connecting bond joins so that $(\partial r_{j\gamma}/\partial s_r)_0 = 0$.

Turning first to the infrared case, the dipole and rotational strengths (7.2.27) become

$$D(1_\pm \leftarrow 0) = \left(\frac{\hbar}{2\omega_\pm}\right)\left(\frac{1}{4N^2}\right)\left[\left(\frac{\partial\mu_{1\alpha}}{\partial s_1}\right)_0\left(\frac{\partial\mu_{1\alpha}}{\partial s_1}\right)_0 + \left(\frac{\partial\mu_{2\alpha}}{\partial s_2}\right)_0\left(\frac{\partial\mu_{2\alpha}}{\partial s_2}\right)_0\right.$$
$$\left.\pm 2\left(\frac{\partial\mu_{1\alpha}}{\partial s_1}\right)_0\left(\frac{\partial\mu_{2\alpha}}{\partial s_2}\right)_0\right], \qquad (7.4.2a)$$

$$R(1_\pm \leftarrow 0) = \pm\left(\frac{\hbar}{4}\right)\left(\frac{1}{4N^2}\right)\varepsilon_{\alpha\beta\gamma}R_{21_\beta}\left(\frac{\partial\mu_{1\alpha}}{\partial s_1}\right)_0\left(\frac{\partial\mu_{2\gamma}}{\partial s_2}\right)_0. \qquad (7.4.2b)$$

For totally symmetric local group internal coordinates (so that the relative orientation of the two groups does not change) we can write for the unit vectors along the bond axes at any instant during the vibrational excursion

$$u_{1\alpha} = I_\alpha, \qquad (7.4.3a)$$

$$u_{2\alpha} = I_\alpha\cos\theta + J_\alpha\sin\theta, \qquad (7.4.3b)$$

where **I, J, K** are unit vectors along the internal molecular axes *X, Y, Z* in Fig. 5.4. Since $\mu_{i\alpha} = \mu_i u_{i\alpha}$ where μ_i is the magnitude of the *i*th bond electric dipole moment, we have

$$\left(\frac{\partial\mu_{1\alpha}}{\partial s_1}\right)_0 = \left(\frac{\partial\mu_1}{\partial s_1}\right)_0 I_\alpha, \qquad (7.4.4a)$$

$$\left(\frac{\partial\mu_{2\alpha}}{\partial s_2}\right)_0 = \left(\frac{\partial\mu_2}{\partial s_2}\right)_0 (I_\alpha\cos\theta + J_\alpha\sin\theta). \qquad (7.4.4b)$$

After a little trigonometry, the infrared dissymmetry factors (5.2.27) for the two normal modes are found to be

$$g(1_+ \leftarrow 0) = -\frac{2\pi R_{21}\sin\theta}{\lambda_+(1 + \cos\theta)}, \qquad (7.4.5a)$$

$$g(1_- \leftarrow 0) = \frac{2\pi R_{21}\sin\theta}{\lambda_-(1 - \cos\theta)}. \qquad (7.4.5b)$$

These expressions are particularly pleasing since they involve only the geometry of the two-group structure. Although the dissymmetry factors have different magnitudes and opposite signs for the symmetric and antisymmetric bands, the numerators $\epsilon^L - \epsilon^R$ for the two bands have *equal* magnitudes and opposite signs: this has diagnostic value in the interpretation of infrared circular dichroism spectra.

Turning now to the corresponding Raman optical activity, the relevant products in (7.3.33) become

$$\langle 0|\alpha_{\alpha\beta}|1_\pm\rangle\langle 1_\pm|\alpha_{\alpha\beta}|0\rangle = \left(\frac{\hbar}{2\omega_\pm}\right)\left(\frac{1}{4N^2}\right)\left[\left(\frac{\partial\alpha_{1\alpha\beta}}{\partial s_1}\right)_0\left(\frac{\partial\alpha_{1\alpha\beta}}{\partial s_1}\right)_0\right.$$
$$\left.+\left(\frac{\partial\alpha_{2\alpha\beta}}{\partial s_2}\right)_0\left(\frac{\partial\alpha_{2\alpha\beta}}{\partial s_2}\right)_0 \pm 2\left(\frac{\partial\alpha_{1\alpha\beta}}{\partial s_1}\right)_0\left(\frac{\partial\alpha_{2\alpha\beta}}{\partial s_2}\right)_0\right],$$

$$(7.4.6a)$$

$$\langle 0|\alpha_{\alpha\beta}|1_\pm\rangle\langle 1_\pm|G'_{\alpha\beta}|0\rangle = \tfrac{1}{3}\omega\langle 0|\alpha_{\alpha\beta}|1_\pm\rangle\langle 1_\pm|\varepsilon_{\alpha\gamma\delta}A_{\gamma,\delta\beta}|0\rangle$$
$$= \mp\left(\frac{\hbar\omega}{4\omega_\pm}\right)\left(\frac{1}{4N^2}\right)\varepsilon_{\beta\gamma\delta}R_{21\gamma}\left(\frac{\partial\alpha_{1\alpha\beta}}{\partial s_1}\right)_0\left(\frac{\partial\alpha_{2\delta\alpha}}{\partial s_2}\right)_0,$$

$$(7.4.6b)$$

$$\langle 0|\alpha_{\alpha\alpha}|1_\pm\rangle\langle 1_\pm|G'_{\beta\beta}|0\rangle = 0. \qquad (7.4.6c)$$

Using the unit vectors (7.4.3) in the expression (7.3.14) for the polarizability tensor of an axially-symmetric group, we find from (3.5.36) the following polarized and depolarized Raman circular intensity difference components in 90° scattering:

$$\Delta_x(1_+ \leftarrow 0) = \frac{24\pi R_{21}[\partial(\alpha_i\kappa_i)/\partial s_i]_0^2\sin 2\theta}{\lambda\{40\,(\partial\alpha_i/\partial s_i)_0^2 + 7[\partial(\alpha_i\kappa_i)/\partial s_i]_0^2(5 + 3\cos 2\theta)\}}, \quad (7.4.7a)$$

$$\Delta_x(1_- \leftarrow 0) = -\frac{8\pi R_{21}\sin 2\theta}{7\lambda(1 - \cos 2\theta)}, \qquad (7.4.7b)$$

$$\Delta_z(1_+ \leftarrow 0) = \frac{2\pi R_{21}\sin 2\theta}{\lambda(5 + 3\cos 2\theta)}, \qquad (7.4.7c)$$

$$\Delta_z(1_- \leftarrow 0) = -\frac{2\pi R_{21}\sin 2\theta}{3\lambda(1 - \cos 2\theta)}. \qquad (7.4.7d)$$

Notice that only $\Delta_x(1_+ \leftarrow 0)$ depends on the derivatives of group polarizability tensor components with respect to group internal coordinates. These derivatives are usually difficult to evaluate, so empirical values, transferable from one molecule to another, are often used. Although the circular intensity differences have different magnitudes and opposite signs for the symmetric and antisymmetric bands, the numerators $I_z^R - I_z^L$ for the two bands have *equal* magnitudes and opposite signs.

Next consider an idealized normal coordinate containing just the internal coordinate of torsion s_t between groups 1 and 2 in the two-group structure:

$$Q_t = N_t s_t. \qquad (7.4.8)$$

Taking θ to be the equilibrium value of the torsion angle, as shown in Fig. 5.4, we may write the general torsion angle at some instant during the torsion vibration as

$\theta + \Delta\theta$, and so identify s_t with $\Delta\theta$. If **I, J, K** are unit vectors along the internal molecular axes X, Y, Z, the unit vectors along the principal axes of the two groups for some general torsion angle at a particular instant can be written

$$u_{1\alpha} = I_\alpha \cos\left(\tfrac{1}{2}\Delta\theta\right) - J_\alpha \sin\left(\tfrac{1}{2}\Delta\theta\right), \tag{7.4.9a}$$

$$u_{2\alpha} = I_\alpha \cos\left(\theta + \tfrac{1}{2}\Delta\theta\right) + J_\alpha \sin\left(\theta + \tfrac{1}{2}\Delta\theta\right). \tag{7.4.9b}$$

Again taking groups 1 and 2 to be axially symmetric, assuming the connecting bond to be rigid, and taking the local group origins to be the points where the connecting bond joins, we need only evaluate the two-group terms.

To calculate the Raman optical activity, we need to use the unit vectors (7.4.9) in group polarizability tensors of the form (7.3.14). This leads to the derivatives

$$\left(\frac{\partial \alpha_{1\alpha\beta}}{\partial\Delta\theta}\right)_0 = -\tfrac{3}{2}\alpha_1 \kappa_1 (I_\alpha J_\beta + J_\alpha I_\beta), \tag{7.4.10a}$$

$$\left(\frac{\partial \alpha_{2\alpha\beta}}{\partial\Delta\theta}\right)_0 = -\tfrac{3}{2}\alpha_2 \kappa_2 [(I_\alpha J_\beta - J_\alpha I_\beta)\sin 2\theta - (I_\alpha J_\beta + J_\alpha I_\beta)\cos 2\theta], \tag{7.4.10b}$$

which provide the following products:

$$\left(\frac{\partial \alpha_{i\alpha\beta}}{\partial\Delta\theta}\right)_0 \left(\frac{\partial \alpha_{i\alpha\beta}}{\partial\Delta\theta}\right)_0 = \tfrac{9}{2}\alpha_i^2 \kappa_i^2, \tag{7.4.11a}$$

$$\left(\frac{\partial \alpha_{1\alpha\beta}}{\partial\Delta\theta}\right)_0 \left(\frac{\partial \alpha_{2\alpha\beta}}{\partial\Delta\theta}\right)_0 = -\tfrac{9}{2}\alpha_1\alpha_2\kappa_1\kappa_2 \cos 2\theta, \tag{7.4.11b}$$

$$\left(\frac{\partial \alpha_{1\alpha\beta}}{\partial\Delta\theta}\right)_0 \varepsilon_{\beta\gamma\delta} R_{21\gamma} \left(\frac{\partial \alpha_{2\delta\alpha}}{\partial\Delta\theta}\right)_0 = \tfrac{9}{2}\alpha_1\alpha_2\kappa_1\kappa_2 R_{21} \sin 2\theta, \tag{7.4.11c}$$

$$\left(\frac{\partial \alpha_{1\alpha\alpha}}{\partial\Delta\theta}\right)_0 \left(\frac{\partial \alpha_{2\beta\beta}}{\partial\Delta\theta}\right)_0 = 0. \tag{7.4.11d}$$

When used in (7.3.33) and (3.5.36) these generate the following Raman circular intensity difference components in $90°$ scattering:

$$\Delta_x(1_t \leftarrow 0) = -\frac{8\pi R_{21} \sin 2\theta}{7\lambda(1 - \cos 2\theta)}, \tag{7.4.12a}$$

$$\Delta_z(1_t \leftarrow 0) = -\frac{2\pi R_{21} \sin 2\theta}{3\lambda(1 - \cos 2\theta)}. \tag{7.4.12b}$$

A similar procedure generates from the two-group term in (7.2.23) the following infrared dissymmetry factor:

$$g(1_t \leftarrow 0) = \frac{2\pi R_{21} \sin\theta}{\lambda_t(1 - \cos\theta)}. \tag{7.4.13}$$

An important point, not immediately apparent from g and Δ because the relevant factors have cancelled, is that infrared optical activity in the twisting mode requires the two groups to have permanent electric dipole moments, which is more restrictive than the corresponding Raman requirement that the two groups have a polarizability anisotropy.

There are two more idealized modes possible for our simple two-group structure: the symmetric and antisymmetric combinations of the equivalent internal coordinates corresponding to the deformations of the angles ϕ_1 and ϕ_2 between the group axes and the connecting bond. The general bond angles at some instant during these two vibrations are $\phi_1 + \Delta\phi_1$ and $\phi_2 + \Delta\phi_2$. Since in our special structure $\phi_1 = \phi_2 = 90°$, the unit vectors along the principal axes of the two groups at a particular instant are

$$u_{1\alpha} = I_\alpha \cos \Delta\phi_1 - K_\alpha \sin \Delta\phi_1, \tag{7.4.14a}$$

$$u_{2\alpha} = I_\alpha \cos \theta \cos \Delta\phi_2 + J_\alpha \sin \theta \cos \Delta\phi_2 + K_\alpha \sin \Delta\phi_2. \tag{7.4.14b}$$

Proceeding in a similar fashion to the torsion example above, it is easy to show that the Raman circular intensity differences generated in $90°$ scattering by these two normal modes are

$$\Delta_x(1_+ \leftarrow 0) = -\frac{4\pi R_{21} \sin \theta}{7\lambda(1 - \cos \theta)}, \tag{7.4.15a}$$

$$\Delta_x(1_- \leftarrow 0) = \frac{4\pi R_{21} \sin \theta}{7\lambda(1 + \cos \theta)}, \tag{7.4.15b}$$

$$\Delta_z(1_+ \leftarrow 0) = -\frac{\pi R_{21} \sin \theta}{3\lambda(1 - \cos \theta)}, \tag{7.4.15c}$$

$$\Delta_z(1_- \leftarrow 0) = \frac{\pi R_{21} \sin \theta}{3\lambda(1 + \cos \theta)}. \tag{7.4.15d}$$

The infrared optical activity associated with the two idealized deformation normal coordinates is found to be zero, at least within the approximation that $(\partial\mu_i/\partial\Delta\phi_i)_0 = 0$.

7.4.2 Methyl torsions in a hindered single-bladed propellor

We consider next the inertial contributions to the infrared and Raman optical activity; in particular, intrinsic group inertial terms ($i = j$). A good example is provided by methyl torsions: chiral organic molecules containing methyl groups often show large Raman optical activity at low wavenumber (between 100 and 300 cm^{-1}), some of which probably originates in methyl torsion vibrations (Barron, 1975c; Barron and Buckingham, 1979).

Since first- and second-rank tensorial properties of an object with a threefold or higher proper rotation axis are unaffected by rotations about that axis (see Section

4.2.6), the electric dipole moment, polarizability and optical activity of the methyl group do not change in the course of the torsion vibration. The origin of any infrared or Raman intensity and optical activity must therefore be sought in the rest of the molecule. Two mechanisms can be distinguished: a two-group mechanism involving coupling of the methyl torsion coordinate with other low frequency coordinates from the rest of the molecule so that the true normal coordinate embraces a chiral structural unit containing part of the frame together with the methyl group; and an inertial mechanism in which the interaction of the radiation field with the rest of the molecule, via the electric dipole moment vector, the polarizability tensor or the optical activity tensor intrinsic to the rest of the molecule, changes as the frame twists in space to compensate the twist of the methyl group so that the torsion vibration generates zero overall angular momentum (that is, the tail wags the dog!). Only the inertial mechanism is considered here.

The evaluation of the intrinsic group inertial term is simplified considerably if the methyl torsion axis is a principal inertial axis of the molecule. Thus our basic model consists of an anisotropic, intrinsically achiral, group i with a principal axis of polarizability along the unit vector $\mathbf{u}_i$ which is oriented relative to the threefold axis of the methyl group such that the anisotropic group and the threefold axis constitute a chiral structure. Group i is balanced dynamically by a spherical group so that, assuming the existence of a hindering potential, torsional oscillations are executed about the threefold axis of the methyl group (Fig. 7.4). If group i were

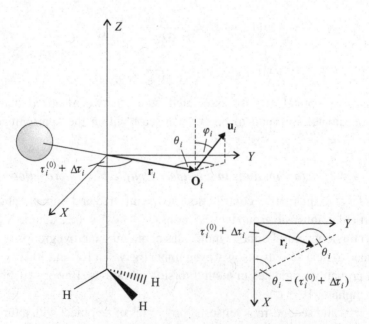

Fig. 7.4 The methyl torsion model based on a hindered single-bladed propellor.

an unsubstituted aromatic ring with $\mathbf{u}_i$ along the sixfold axis, the structure would have the appearance of a single-bladed propeller.

Two different molecule-fixed axes systems have been used for the internal rotation problem (Lister, Macdonald and Owen, 1978). The principal axis method uses the three principal inertial axes of the molecule, with the axis of internal rotation taken to be the symmetry axis of the top, which in general is not coincident with any one of the principal inertial axes. In the internal axis method, one of the molecule-fixed axes is taken to be parallel with the symmetry axis of the top. In our single-bladed propellor model, the symmetry axis of the top is contrived to be a principal inertial axis, so these two different axes systems coalesce in what we may refer to as a *principal internal* axis system.

We first separate the kinetic energy originating in rotation of the whole molecule about the torsion axis from that originating in the internal torsion mode of vibration. If χ_i and χ_{Me} are angles specifying the instantaneous orientations of the two parts of the molecule relative to some nonrotating axis perpendicular to the torsion axis, the total kinetic energy due to rotation about the torsion axis is

$$T = \tfrac{1}{2}I_i\dot{\chi}_i^2 + \tfrac{1}{2}I_{Me}\dot{\chi}_{Me}^2, \tag{7.4.16}$$

where I_i and I_{Me} are the moments of intertia of the two groups about the torsion axis (I_i refers to group i together with its balancing sphere). If new variables

$$\chi = (I_i\chi_i + I_{Me}\chi_{Me})/I \tag{7.4.17a}$$

and

$$\tau = \chi_i - \chi_{Me}, \tag{7.4.17b}$$

are defined, where $I = I_i + I_{Me}$, the kinetic energy (7.4.16) becomes (Townes and Schawlow, 1955)

$$T = \tfrac{1}{2}I\dot{\chi}^2 + \tfrac{1}{2}I_iI_{Me}\dot{\tau}^2/I. \tag{7.4.18}$$

The first term gives the kinetic energy originating in rotation of the complete molecule with the internal rotation frozen, and the second term gives that from the torsion vibration. Since $I\dot{\chi} = I_i\dot{\chi}_i + I_{Me}\dot{\chi}_{Me}$, all of the angular momentum about the torsion axis is associated with changes in the external coordinate χ; none is associated with changes in the internal torsion angle τ defining the relative orientation of the two groups. The complete Hamiltonian for rotation about the torsion axis is obtained by adding to (7.4.18) a potential energy term corresponding to the barrier hindering free rotation. For the single-bladed propellor, this would be a symmetric function dominated by

$$V = \tfrac{1}{2}V_3(1 - \cos 3\tau) \tag{7.4.19}$$

describing three potential minima with intervening barriers of height V_3.

The displacement of the internal torsion angle away from its equilibrium value $\tau^{(0)}$ during the course of the torsion vibration is denoted by the internal coordinate $\Delta\tau$ so that $\tau = \tau^{(0)} + \Delta\tau$. It is possible to write $\Delta\tau$ as the sum of displacements 'intrinsic' to each of the two groups,

$$\Delta\tau = \Delta\tau_i - \Delta\tau_{Me} \qquad (7.4.20a)$$

provided that the two displacements satisfy

$$I_i\Delta\tau_i = -I_{Me}\Delta\tau_{Me}. \qquad (7.4.20b)$$

This last condition follows from the requirement that the contribution to the angular momentum of the molecule about the torsion axis from the torsion vibration be zero (this is equivalent to the second Sayvetz condition: see Califano, 1976):

$$I_i\dot\tau_i = -I_{Me}\dot\tau_{Me}, \qquad (7.4.20c)$$

where $\tau_i = \tau_i^{(0)} + \Delta\tau_i$ and $\tau_{Me} = \tau_{Me}^{(0)} + \Delta\tau_{Me}$ specify the instantaneous orientations of the two groups relative to a principal internal axis, perpendicular to the torsion axis, that remains stationary during the torsion vibration. $\tau_i^{(0)}$ and $\tau_{Me}^{(0)}$ are the corresponding equilibrium orientations. In Fig. 7.4 this principal internal axis is the X axis.

For an idealized normal coordinate containing just the internal coordinate of torsion $\Delta\tau$,

$$Q_t = N_t\Delta\tau, \qquad (7.4.21)$$

and we can use (7.4.20) to write

$$Q_t = N_t\left(\frac{I}{I_{Me}}\right)\Delta\tau_i = -N_t\left(\frac{I}{I_i}\right)\Delta\tau_{Me}, \qquad (7.4.22)$$

from which $L_{it} = (I_{Me}/IN_t)$ and $L_{Met} = -(I_i/IN_t)$. Writing the electric dipole moment and polarizability of the molecule as sums of dipole moments and polarizabilities intrinsic to group i and the methyl group, these **L**-matrix elements multiply $(\partial\mu_{i_\alpha}/\partial\Delta\tau_i)_0$ and $(\partial\mu_{Me_\alpha}/\partial\Delta\tau_{Me})_0$ in the inertial terms of (7.2.27b), and $(\partial\alpha_{i_{\alpha\beta}}/\partial\Delta\tau_i)_0$ and $(\partial\alpha_{Me_{\alpha\beta}}/\partial\Delta\tau_{Me})_0$ in the inertial terms of (7.3.33). In fact

$$\left(\frac{\partial\mu_{Me_\alpha}}{\partial\Delta\tau_{Me}}\right)_0 = \left(\frac{\partial\alpha_{Me_{\alpha\beta}}}{\partial\Delta\tau_{Me}}\right)_0 = 0 \qquad (7.4.23)$$

because of the axial symmetry of the methyl group. This also means that there are no two-group contributions to the infrared and Raman methyl torsion optical activity in the hindered single-bladed propellor.

We shall show the details of just the Raman optical activity calculation. The corresponding infrared calculation is similar but less complicated. Referring to

Fig. 7.4, the unit vector along a symmetry axis of group i at some instant during the torsion vibration can be written in terms of the unit vectors $\mathbf{I}, \mathbf{J}, \mathbf{K}$ along the principal internal axes X, Y, Z:

$$u_{i\alpha} = -I_\alpha \sin\phi_i \cos\left[\theta_i - \left(\tau_i^{(0)} + \Delta\tau_i\right)\right]$$
$$+ J_\alpha \sin\phi_i \sin\left[\theta_i - \left(\tau_i^{(0)} + \Delta\tau_i\right)\right] + K_\alpha \cos\phi_i. \tag{7.4.24}$$

If group i is axially symmetric, we can write its polarizability tensor in the form (7.3.16), and using (7.4.24) we obtain

$$\alpha_{i\alpha\beta} = \alpha_i(1 - \kappa_i)\delta_{\alpha\beta}$$
$$+ 3\alpha_i\kappa_i\left\{ I_\alpha I_\beta \sin^2\phi_i \cos^2\left[\theta_i - \left(\tau_i^{(0)} + \Delta\tau_i\right)\right]\right.$$
$$+ J_\alpha J_\beta \sin^2\phi_i \sin^2\left[\theta_i - \left(\tau_i^{(0)} + \Delta\tau_i\right)\right] + K_\alpha K_\beta \cos^2\phi_i$$
$$- \tfrac{1}{2}(I_\alpha J_\beta + J_\alpha I_\beta) \sin^2\phi_i \sin 2\left[\theta_i - \left(\tau_i^{(0)} + \Delta\tau_i\right)\right]$$
$$- \tfrac{1}{2}(I_\alpha K_\beta + K_\alpha I_\beta) \sin 2\phi_i \cos\left[\theta_i - \left(\tau_i^{(0)} + \Delta\tau_i\right)\right]$$
$$\left. + \tfrac{1}{2}(J_\alpha K_\beta + K_\alpha J_\beta) \sin 2\phi_i \sin\left[\theta_i - \left(\tau_i^{(0)} + \Delta\tau_i\right)\right]\right\}. \tag{7.4.25}$$

We also need

$$r_{i\gamma} = r_i\left[I_\gamma \cos(\tau_i^{(0)} + \Delta\tau_i) + J_\gamma \sin(\tau_i^{(0)} + \Delta\tau_i)\right]. \tag{7.4.26}$$

The following partial derivatives are then obtained:

$$\left(\frac{\partial\alpha_{i\alpha\beta}}{\partial\Delta\tau_i}\right)_0 = 3\alpha_i\kappa_i\left[(I_\alpha I_\beta - J_\alpha J_\beta) \sin^2\phi_i \sin 2(\theta_i - \tau_i^{(0)})\right.$$
$$+ (I_\alpha J_\beta + J_\alpha I_\beta) \sin^2\phi_i \cos 2(\theta_i - \tau_i^{(0)})$$
$$- \tfrac{1}{2}(I_\alpha K_\beta + K_\alpha I_\beta) \sin 2\phi_i \sin\left(\theta_i - \tau_i^{(0)}\right)$$
$$\left. - \tfrac{1}{2}(J_\alpha K_\beta + K_\alpha J_\beta) \sin 2\phi_i \cos\left(\theta_i - \tau_i^{(0)}\right)\right], \tag{7.4.27a}$$

$$\left(\frac{\partial r_{i\gamma}}{\partial\Delta\tau_i}\right)_0 = R_i\left(- I_\gamma \sin\tau_i^{(0)} + J_\gamma \cos\tau_i^{(0)}\right). \tag{7.4.27b}$$

These provide the products

$$\left(\frac{\partial\alpha_{i\alpha\beta}}{\partial\Delta\tau_i}\right)_0 \left(\frac{\partial\alpha_{i\alpha\beta}}{\partial\Delta\tau_i}\right)_0 = 9\alpha_i^2\kappa_i^2(1 - \cos 2\phi_i), \tag{7.4.28a}$$

$$\left(\frac{\partial\alpha_{i\alpha\alpha}}{\partial\Delta\tau_i}\right)_0 \left(\frac{\partial\alpha_{i\beta\beta}}{\partial\Delta\tau_i}\right)_0 = 0, \tag{7.4.28b}$$

$$\left(\frac{\partial\alpha_{i\alpha\beta}}{\partial\Delta\tau_i}\right)_0 \varepsilon_{\beta\gamma\delta} (\alpha_{i\delta\alpha})_0 \left(\frac{\partial r_{i\gamma}}{\partial\Delta\tau_i}\right)_0 = -\tfrac{9}{2} R_i\alpha_i^2\kappa_i^2 \sin 2\phi_i \sin\theta_i. \tag{7.4.28c}$$

When used with the appropriate **L**-matrix elements in the inertial terms of (7.3.33), these generate from (3.5.36) the following Raman circular intensity difference components in 90° scattering:

$$\Delta_x(1_t \leftarrow 0) = \frac{8\pi R_i \sin 2\phi_i \sin \theta_i}{7\lambda_t(1 - \cos 2\phi_i)}, \qquad (7.4.29a)$$

$$\Delta_z(1_t \leftarrow 0) = \frac{2\pi R_i \sin 2\phi_i \sin \theta_i}{3\lambda_t(1 - \cos 2\phi_i)}. \qquad (7.4.29b)$$

The numerators reduce to zero if $\theta_i = 0°$ or $180°$ or if $\phi_i = 0°$ or $90°$.

Assuming group i is neutral, a similar procedure generates from the inertial dipole term in (7.2.27) the following infrared dissymmetry factor:

$$g(1_t \leftarrow 0) = -\frac{4\pi R_i \sin 2\phi_i \sin \theta_i}{\lambda_t(1 - \cos 2\phi_i)}. \qquad (7.4.30)$$

Apart from the opposite sign (which is purely a matter of convention), this infrared dissymmetry factor has the same dependence on the molecular geometry as the Raman circular intensity difference components (7.4.29). There are, however, important differences between the two methods of measuring methyl torsion optical activity. Methyl torsions occur in the far infrared, well beyond the range currently accessible to infrared circular dichroism instruments. Also group i needs to have a permanent electric dipole moment for the methyl torsion to be infrared optically active, which is more restrictive than the corresponding requirement for Raman optical activity, namely a polarizability anisotropy.

Notice that these optical activities are not exclusive to the methyl group: the same results would obtain whatever was driving the oscillations of the single-bladed propellor. But in practice such well-defined effects are only likely to be observed with methyl torsions because the corresponding frequencies occur in an accessible region of the Raman vibrational spectrum. Torsions of other groups with threefold symmetry, such as $-CF_3$, usually occur well below $100\,\mathrm{cm}^{-1}$ on account of the much greater mass. Groups such as $-OH$ and $-NH_2$ have torsion vibrations at accessible frequencies, but the above treatment would need to be extended to accommodate them because of their low symmetry.

Unfortunately, chiral molecules containing a single methyl group with its three-fold axis lying along a principal inertial axis are rare. But there is an intriguing extension to a more common situation. A molecule containing two adjacent methyl groups has normal vibrational coordinates containing symmetric and antisymmetric combinations of the two methyl torsions. These combinations can generate oscillations of the rest of the molecule about a principal inertial axis: in ortho-xylene (which is not chiral), for example, the symmetric combination generates a torsion about the twofold proper rotation axis resulting in a very intense Raman band at

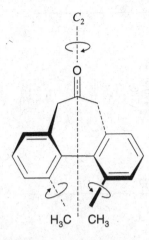

Fig. 7.5 (*R*)-(+)-dimethyldibenz-1,3-cycloheptadiene-6-one. The symmetric combination of the two methyl torsions induces oscillations of the double-bladed propellor about the molecular C_2 axis.

about $180\,\mathrm{cm}^{-1}$ on account of the large polarizability anisotropy of the aromatic group. The bridged biphenyl shown in Fig. 7.5 provides an interesting chiral example: the symmetric combination of the two methyl torsions generates oscillations in space of the rest of the molecule (which has the appearance of a double-bladed propellor) about the C_2 axis, and the associated optical activity is easily calculated (Barron and Buckingham, 1979). The Raman optical activity spectrum of this bridged biphenyl was measured in the early years of the subject (Barron, 1975*c*) and does indeed show large signals in the region appropriate for methyl torsions, but definitive assignments have not been made. A less complicated example is trans-2,3-dimethyloxirane in which the symmetric combination of the two methyl torsions generates oscillations of the rest of the molecule about the C_2 axis, for which a similar calculation provides good agreement with experimental Raman optical activity data (Barron and Vrbancich, 1983; Barron, Hecht and Polavarapu, 1992).

The extension of the methyl torsion theory to a completely asymmetric molecule is rather complicated since the threefold axis of the methyl group is no longer a principal inertial axis and it is necessary to resolve the methyl torsion angular momentum along all three principal axes (Barron and Buckingham, 1979). The methyl torsion is likely to mix considerably with other low-wavenumber modes in large completely asymmetric molecules, so assignments of bands to pure methyl torsions are not expected. Several possible examples of bands containing contributions from methyl torsions have been identified in the experimental Raman optical activity spectra of completely asymmetric chiral molecules (Barron and Buckingham, 1979), including the three bands below $300\,\mathrm{cm}^{-1}$ shown by (*R*)-(+)-3- methylclohexanone (see Fig. 7.6).

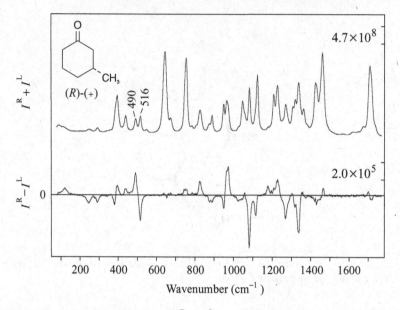

Fig. 7.6 The backscattered Raman ($I^R + I^L$) and Raman optical activity ($I^R - I^L$) spectra of a neat liquid sample of (R)-$(+)$-3-methylcyclohexanone. Recorded in the author's laboratory. The absolute intensities are not defined, but the relative Raman and Raman optical activity intensities are significant.

7.4.3 Intrinsic group optical activity tensors

Finally, we consider the terms in $\mu_i m_j$ in the infrared rotational strength (7.2.27b) and the terms in $\alpha_i G'_j$ and $\alpha_i A_j$ in the Raman optical activity (7.3.33b,c). The infrared terms in $\mu_i m_j$ are unlikely to be significant unless a group has a degenerate ground electronic state, otherwise all components of the intrinsic group magnetic moment $\mathbf{m}_j$ are zero, and we shall not consider these terms further. But achiral groups with symmetry lower than axial can have nonzero components of the optical activity tensors, and these can lead to significant contributions to the Raman optical activity.

Consider two idealized normal modes containing symmetric and antisymmetric combinations of two internal coordinates which are, in general, nonequivalent:

$$Q_+ = N_1 s_1 + N_2 s_2, \tag{7.4.31a}$$

$$Q_- = N_2 s_1 - N_1 s_2. \tag{7.4.31b}$$

The inverse expressions are

$$s_1 = \frac{1}{N_1^2 + N_2^2}(N_1 Q_+ + N_2 Q_-), \tag{7.4.32a}$$

$$s_2 = \frac{1}{N_1^2 + N_2^2}(N_2 Q_+ - N_1 Q_-), \tag{7.4.32b}$$

so that the **L**-matrix elements are $L_{1+} = N_1/(N_1^2 + N_2^2)$, $L_{1-} = N_2/(N_1^2 + N_2^2)$, $L_{2+} = N_2/(N_1^2 + N_2^2)$ and $L_{2-} = -N_1/(N_1^2 + N_2^2)$.

If s_1 and s_2 are localized on the same group i, the required contribution to (7.3.33b) is

$$\langle 0|\alpha_{\alpha\beta}|1_\pm\rangle\langle 1_\pm|G'_{\alpha\beta}|0\rangle = \pm\left(\frac{\hbar}{2\omega_\pm}\right)\frac{N_1 N_2}{(N_1^2 + N_2^2)^2}\left[\left(\frac{\partial\alpha_{i\alpha\beta}}{\partial s_1}\right)_0\left(\frac{\partial G'_{i\alpha\beta}}{\partial s_2}\right)_0\right.$$
$$\left. + \left(\frac{\partial\alpha_{i\alpha\beta}}{\partial s_2}\right)_0\left(\frac{\partial G'_{i\alpha\beta}}{\partial s_1}\right)_0\right] \qquad (7.4.33)$$

with an analogous contribution to (7.3.33c). Terms in

$$(\partial\alpha_{i\alpha\beta}/\partial s_q)_0(\partial G'_{i\alpha\beta}/\partial s_q)_0$$

are zero because group i is assumed to be intrinsically achiral. A possible example is the carbonyl group in molecules such as 3-methylcyclohexanone (Barron, Torrance and Vrbancich, 1982). The in-plane and out-of-plane deformation coordinates belong to symmetry species B_2 and B_1 in the local C_{2v} symmetry: B_2 is spanned by α_{YZ}, G'_{XZ} and G'_{ZX}; and B_1 is spanned by α_{XZ}, G'_{YZ} and G'_{ZY}. The skeletal chirality will lead to normal modes of vibration containing symmetric and antisymmetric combinations of the two locally orthogonal deformations, generating equal and opposite Raman optical activities. The absolute signs depend on N_1 and N_2, given by a normal coordinate analysis, and on

$$\left(\frac{\partial\alpha_{\alpha\beta}}{\partial s_{B_1}}\right)_0\left(\frac{\partial G'_{\alpha\beta}}{\partial s_{B_2}}\right)_0 + \left(\frac{\partial\alpha_{\alpha\beta}}{\partial s_{B_2}}\right)_0\left(\frac{\partial G'_{\alpha\beta}}{\partial s_{B_1}}\right)_0,$$

which is an intrinsic property of the carbonyl group. This term is now developed further by considering an idealized model of the carbonyl deformations.

Consider a carbonyl group with axes X, Y, Z oriented as in Fig. 5.6, but with the origin now at the carbon atom. We assume that the carbon atom remains fixed during the deformations and describe the deformations with the aid of axes X', Y', Z' that move with the carbonyl group relative to the axes X, Y, Z that remain fixed in the orientation corresponding to the equilibrium position of the carbonyl group. The internal coordinates s_{B_2} and s_{B_1} corresponding to the in-plane and out-of-plane deformations are identified with the displacement angles $\Delta\theta$ and $\Delta\phi$ illustrated in Fig. 7.7. Associating unit vectors $\mathbf{I}, \mathbf{J}, \mathbf{K}$ with X, Y, Z and $\mathbf{I}', \mathbf{J}', \mathbf{K}'$ with X', Y', Z' we have for the in-plane deformation

$$I'_\alpha = I_\alpha, \qquad (7.4.34a)$$
$$J'_\alpha = J_\alpha\cos\Delta\theta + K_\alpha\sin\Delta\theta, \qquad (7.4.34b)$$
$$K'_\alpha = K_\alpha\cos\Delta\theta - J_\alpha\sin\Delta\theta, \qquad (7.4.34c)$$

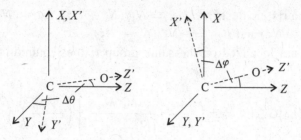

Fig. 7.7 Definition of the displacement angles $\Delta\theta$ and $\Delta\phi$ that characterize the in-plane (B_2) and out-of-plane (B_1) carbonyl deformations.

and for the out-of-plane deformation

$$I'_\alpha = I_\alpha \cos \Delta\phi - K_\alpha \sin \Delta\phi, \qquad (7.4.35a)$$

$$J'_\alpha = J_\alpha, \qquad (7.4.35b)$$

$$K'_\alpha = K_\alpha \cos \Delta\phi + I_\alpha \sin \Delta\phi. \qquad (7.4.35c)$$

Assuming that the carbonyl group maintains its intrinsic C_{2v} symmetry, it follows from Tables 4.2 that the only nonzero components of $G'_{\alpha\beta}$ are $G'_{XY} \neq G'_{YX}$. For some general orientation of the carbonyl group during a deformation, its intrinsic **G**′ tensor can therefore be written in the form

$$G'_{\alpha\beta} = G'_{XY} I'_\alpha J'_\beta + G'_{YX} J'_\alpha I'_\beta. \qquad (7.4.36)$$

Using (7.4.34) in this expression, we find for the in-plane deformation

$$\left(\frac{\partial G'_{\alpha\beta}}{\partial \Delta\theta} \right)_0 = G'_{XY} I_\alpha K_\beta + G'_{YX} K_\alpha I_\beta. \qquad (7.4.37a)$$

Similarly, using (7.4.35), we find for the out-of-plane deformation

$$\left(\frac{\partial G'_{\alpha\beta}}{\partial \Delta\phi} \right)_0 = -(G'_{XY} K_\alpha J_\beta + G'_{YX} J_\alpha K_\beta). \qquad (7.4.37b)$$

In C_{2v}, the only nonzero components of $\alpha_{\alpha\beta}$ are $\alpha_{XX} \neq \alpha_{YY} \neq \alpha_{ZZ}$, so the intrinsic polarizability tensor of the carbonyl group can be written in the form

$$\alpha_{\alpha\beta} = \alpha_{XX} I'_\alpha I'_\beta + \alpha_{YY} J'_\alpha J'_\beta + \alpha_{ZZ} K'_\alpha K'_\beta. \qquad (7.4.38)$$

Again using (7.4.34) and (7.4.35) the required derivatives are

$$\left(\frac{\partial \alpha_{\alpha\beta}}{\partial \Delta\theta} \right)_0 = (\alpha_{YY} - \alpha_{ZZ})(J_\alpha K_\beta + K_\alpha J_\beta), \qquad (7.4.39a)$$

$$\left(\frac{\partial \alpha_{\alpha\beta}}{\partial \Delta\phi} \right)_0 = (\alpha_{ZZ} - \alpha_{XX})(I_\alpha K_\beta + K_\alpha I_\beta). \qquad (7.4.39b)$$

We finally obtain for (7.4.33)

$$\langle 0|\alpha_{\alpha\beta}|1_{\pm}\rangle\langle 1_{\pm}|G'_{\alpha\beta}|0\rangle = \pm\left(\frac{\hbar}{2\omega_{\pm}}\right)\frac{N_1 N_2}{(N_1^2 + N_2^2)^2}(2\alpha_{ZZ}-\alpha_{XX}-\alpha_{YY})(G'_{XY}+G'_{YX}).$$

(7.4.40a)

Similarly,

$$\langle 0|\alpha_{\alpha\beta}|1_{\pm}\rangle\langle 1_{\pm}|\varepsilon_{\alpha\gamma\delta}A_{\gamma,\delta\beta}|0\rangle = \pm\left(\frac{\hbar}{2\omega_{\pm}}\right)\frac{N_1 N_2}{(N_1^2 + N_2^2)^2}(2\alpha_{ZZ} - \alpha_{XX} - \alpha_{YY})$$
$$\times (A_{Y,ZY} - A_{X,ZX} - A_{Z,YY} + A_{Z,XX}).$$

(7.4.40b)

The corresponding intensity is also required, and this is found to be

$$\langle 0|\alpha_{\alpha\beta}|1_{+}\rangle\langle 1_{+}|\alpha_{\alpha\beta}|0\rangle = \left(\frac{\hbar}{2\omega_{+}}\right)\frac{2}{(N_1^2 + N_2^2)^2}[N_2^2(\alpha_{ZZ} - \alpha_{XX})^2 + N_1^2(\alpha_{YY} - \alpha_{ZZ})^2],$$

(7.4.40c)

$$\langle 0|\alpha_{\alpha\beta}|1_{-}\rangle\langle 1_{-}|\alpha_{\alpha\beta}|0\rangle = \left(\frac{\hbar}{2\omega_{-}}\right)\frac{2}{(N_1^2 + N_2^2)^2}[N_1^2(\alpha_{ZZ} - \alpha_{XX})^2 + N_2^2(\alpha_{YY} - \alpha_{ZZ})^2].$$

(7.4.40d)

If the specified tensor components could be calculated, or extracted somehow from experimental data, and N_1 and N_2 were known from a normal coordinate analysis, the Raman circular intensity differences could be calculated. It follows directly from (7.4.40a) that similar deformations of an axially-symmetric group can generate no corresponding Raman optical activity because now $G'_{XY} = -G'_{YX}$.

Another significant contribution to the Raman optical activity can arise if s_1 and s_2 in (7.4.33) are localized on two different achiral groups that together constitute a chiral structure, but we will not develop this contribution explicitly since a detailed consideration of the relative disposition of the two groups is required.

It is tempting to try and explain the large conservative Raman optical activity couplet, positive at low wavenumber and negative at high, associated with Raman bands at 490 and 516 cm^{-1} in the Raman optical activity spectrum of (R)-$(+)$-3-methylcyclohexanone shown in Fig. 7.6 in terms of the mechanism developed above involving coupling of in-plane and out-of-plane carbonyl deformations. However, due to the complexity of the normal vibrational modes and the likely presence of more than one conformer, modern ab initio methods are mandatory for reliable assignments and quantitative analysis of Raman (and infrared) optical activity spectra of chiral molecules such as this (Devlin and Stephens, 1999).

Analogous mechanisms involving intrinsic group optical activity tensors can generate positive–negative optical activity couplets in locally degenerate modes

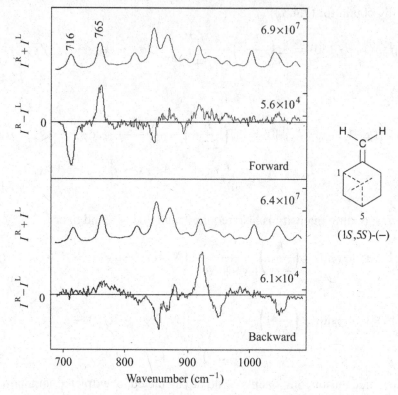

Fig. 7.8 The forward-scattered (upper pair) and backscattered (lower pair) of Raman ($I^R + I^L$) and Raman optical activity ($I^R - I^L$) spectra of a neat liquid sample of (1S,5S)-(−)-β-pinene. Adapted from Barron *et al.* (1990). The absolute intensities are not defined, but the relative Raman and Raman optical activity intensities are significant.

that are split by the chiral environment of the rest of the molecule. A good example is the methyl group: this has C_{3v} symmetry and can support three distinct sets of doubly-degenerate vibrations, namely the antisymmetric C−H stretches, the antisymmetric H−C−H deformations and the orthogonal H−C−C rockings. We refer to Nafie, Polavarapu and Diem (1980) for a detailed study of vibrational optical activity in perturbed degenerate modes.

Another example of Raman optical activity generated by the intrinsic optical activity tensors of a group has been observed in β-pinene (Barron *et al.*, 1990). As may be seen from Fig. 7.8, there is a large couplet, negative at low wavenumber and positive at high, in the forward-scattered but not in the backscattered Raman optical activity spectrum of the (1S,5S)-enantiomer that is associated with bands at 716 and 765 cm^{-1} in the parent Raman spectrum. This couplet also appears in the polarized but not the depolarized Raman optical activity spectrum measured in scattering at

90°. It follows from (7.3.9a–d) that these observations may only be reconciled if this couplet originates in pure isotropic scattering. It is possible to understand qualitatively how this isotropic Raman optical activity may be generated by considering, the symmetry aspects of the optical activity tensors intrinsic to the olefinic group $C = CH_2$. The olefinic methylene twist makes a significant contribution to the 716 cm^{-1} Raman band, whereas the 765 cm^{-1} Raman band originates in a pinane-type skeletal vibration, so this large Raman optical activity couplet appears to originate in coupling between these two modes. The methylene twist transforms as A_u in the D_{2h} point group of ethene itself, and as A_2 in a structure of C_{2v} symmetry, both of which irreducible representations are spanned by the tensor components G'_{XX}, G'_{YY} and G'_{ZZ}. Hence a fundamental vibrational Raman scattering transition associated with the methylene twist is allowed through G', the isotropic part of the axial electric dipole–magnetic dipole optical activity tensor, even in the parent structure of highest symmetry (D_{2h}) and so might be expected to show significant isotropic Raman optical activity if the effective symmetry of the olefinic group is reduced to that of a chiral point group as in β-pinene in which α, the isotropic part of the polar polarizability tensor, can also contribute to Raman scattering in the methylene twist. Alternatively the major contribution from α may arise through the pinane-type skeletal mode with which the methylene twist is coupled.

7.5 Coupling models

The infrared and Raman optical activity models discussed so far have not invoked coupling mechanisms of the type used in the theory of natural electronic optical activity in Chapter 5. The atom, bond or group electric and magnetic moments, and polarizability and optical activity tensors, were taken to pertain to the atom, bond or group unperturbed by nonbonded interactions with the rest of the molecule. We now discuss briefly how such interactions, electronic and vibrational, with the rest of the molecule can contribute to the vibrational optical activity. Such coupling can also make important contributions in some situations to the set of force constants that determine the normal modes of vibration.

Electronic coupling mechanisms are expected to be significant when a group frequency approximation is good, so that the normal mode is dominated by an internal coordinate localized on an intrinsically achiral group (so that there is very little intrinsic chirality in the normal mode), and when there are large highly polarizable groups nearby in a favourable relative orientation. One possible example of this mechanism has been identified by Barnett, Drake and Mason (1980) in the infrared circular dichroism associated with the symmetric N–H stretch in $-NH_2$ groups attached to aromatic rings in chiral binaphthyls. This situation is described by the first term of (5.3.24) with the $j_1 \leftarrow n_1$ transition now corresponding to a vibrational, rather than an electronic, absorption:

$$R(1_p \leftarrow 0) = - \left(\frac{1}{4\pi\epsilon_0}\right)\left(\frac{\omega_p}{2}\right)\varepsilon_{\alpha\beta\gamma}R_{12_\beta}\alpha_{2_{\gamma\delta}}T_{21_{\delta\epsilon}}\mathrm{Re}\left(\langle 0|\mu_{1\epsilon}|1_p\rangle\langle 1_p|\mu_{1\alpha}|0\rangle\right).$$

(7.5.1)

Groups 1 and 2 are the $-NH_2$ group and the perturbing naphthyl group, respectively. Notice that the polarizability tensor of the naphthyl group can be taken as the corresponding static polarizability since, at infrared frequencies, the frequency-dependent contribution is negligible.

Within the bond dipole model of infrared vibrational optical activity, electronic coupling can be incorporated formally by adding to the group or bond electric and magnetic dipole moments in each term contributions induced by static and dynamic fields from other groups within the molecule. These induced moments would be functions of group or bond internal coordinates and would change in the course of a normal mode excursion. Similarly, within the bond polarizability model of Raman optical activity, contributions to group or bond polarizability and optical activity tensors induced by static and dynamic coupling would be added to each term. The machinery for writing down explicit expressions for these induced bond moments and tensors has been given in Chapter 5. We shall not write out these generalized bond dipole and bond polarizability optical activity expressions because of their complexity.

An important example of *vibrational* coupling is found in the amide I vibrations of peptides and proteins. The amide I mode consists mainly of the $C = O$ stretch coordinate, with small contributions from the $C-N$ stretch and $N-H$ deformation coordinates. The relatively strong electric dipole vibrational transition moments, together with well-defined geometries in secondary structure motifs such as α-helix and β-sheet, leads to strong dipolar coupling interactions between the $C = O$ groups. Among other things, this coupling is manifest as a mixing of the degenerate (or near degenerate) excited state vibrational wavefunctions to form delocalized excited vibrational states analogous to the exciton states formed from excited electronic states and described in Section 5.3.4. For n interacting carbonyl transitions, n coupled vibrational excited states will be created with splittings determined by the dipole–dipole interaction potential (5.3.28). The incorporation of dipole–dipole interactions into computations of the normal modes of vibration of peptides was pioneered by Krimm (see Krimm and Bandekar, 1986, for a review), who treated these dipolar interactions as a set of additional force constants to modify the vibrational force field and hence the frequencies and intensities. Diem (1993) has developed a 'degenerate extended coupled oscillator' model for the vibrational circular dichroism generated by n interacting dipoles that is based on dipole–dipole coupling and has applied it to peptides and nucleic acids.

7.6 Raman optical activity of biomolecules

This chapter concludes with a brief account of applications of Raman optical activity in the realm of biomolecular science, which are highly promising. Raman optical activity is more incisive than conventional vibrational spectroscopy, infrared or Raman, in the study of biomolecules on account of its exquisite sensitivity to chirality. Using a backscattering geometry to maximize the signals (Section 7.3.6), Raman optical activity spectra may be measured routinely over a wide spectral range on the central molecules of life, namely proteins, carbohydrates, nucleic acids and viruses; all in aqueous solution to reflect their natural biological environment (Barron *et al.*, 2000, 2003). Even though the model theories and current ab initio computational methods described above are hopelessly inadequate for Raman optical activity calculations on structures the size and complexity of biomolecules, their experimental Raman optical activity spectra have nonetheless proved rich and transparent with regard to valuable information about structure and behaviour.

The normal modes of vibration of biomolecules can be highly complex, with contributions from vibrational coordinates within both the backbone and the side chains. Raman optical activity is able to cut through the complexity of the corresponding vibrational spectra, since the largest signals are often associated with vibrational coordinates which sample the most rigid and chiral parts of the structure. These are usually within the backbone and often give rise to Raman optical activity band patterns characteristic of the backbone conformation. Polypeptides in the standard conformations defined by characteristic Ramachandran ϕ, ψ angles found in secondary (α-helix and β-sheet), loop and turn structures within native proteins (Creighton, 1993) are particularly favourable in this respect since signals from the peptide backbone, illustrated in Fig. 7.9, usually dominate the Raman optical activity spectrum, unlike the parent conventional Raman spectrum in which bands from the amino acid side chains often obscure the peptide backbone bands. Carbohydrate Raman optical activity spectra are similarly dominated by signals from skeletal vibrations, in this case centred on the constituent sugar rings and the connecting glycosidic links. Although the parent Raman spectra of nucleic acids are dominated by bands from the intrinsic base vibrations, their Raman optical activity spectra tend to be dominated by bands characteristic of the stereochemical dispositions of the bases with respect to each other and to the sugar rings, together with signals from the sugar–phosphate backbone.

The determination of the structure and behaviour of proteins has been at the forefront of biomolecular science ever since the determination of the first protein structures from X-ray crystallography in the late 1950s by M. F. Perutz and J. C. Kendrew, a position reinforced in the post-genomic era. Although individual protein Raman optical activity bands may be assigned to elements of secondary

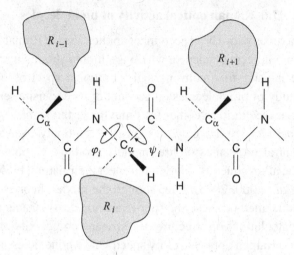

Fig. 7.9 A sketch of the polypeptide backbone of a protein, illustrating the Ramachandran ϕ, ψ angles and the amino acid side chains R.

structure such as helix and sheet, the presence of clear bands originating in the loops and turns connecting the secondary structure elements leads to overall Raman optical activity band patterns characteristic of the three-dimensional structure, or fold, of the protein. Unlike the parent Raman band patterns, the Raman optical activity band patterns for each fold type are therefore quite distinct. This enables structural information to be easily determined by comparing the Raman optical activity spectrum of a protein of unknown structure with a set of spectra of proteins of known structure. Indeed, the large number of structure-sensitive bands in the Raman optical activity spectra of proteins makes them ideal for the application of pattern recognition methods to automatically determine structural similarities (Barron *et al.*, 2003; McColl *et al.*, 2003). Hence despite the inadequacy of current theories for useful calculations of the Raman optical activity spectra of proteins and other large biomolecules, valuable structural information is nonetheless available from experimental Raman optical activity data.

The extended amide III spectral region of peptides and proteins, where coupled N–H and C_α–H deformations make large contributions to some of the normal modes of vibration (Diem, 1993), often shows large and informative Raman optical activity but only weak vibrational circular dichroism. A qualitative explanation for this may be provided by the results of Section 7.4.1 where it was demonstrated that the bond polarizability Raman optical activity generated by deformations of a simple two-group structure can be large, whereas the corresponding infrared optical activity is zero. From another perspective, the finding by Zuber and Hug (2004), mentioned in Section 7.3.1, that moderately diffuse *p*-type orbitals on hydrogen atoms make

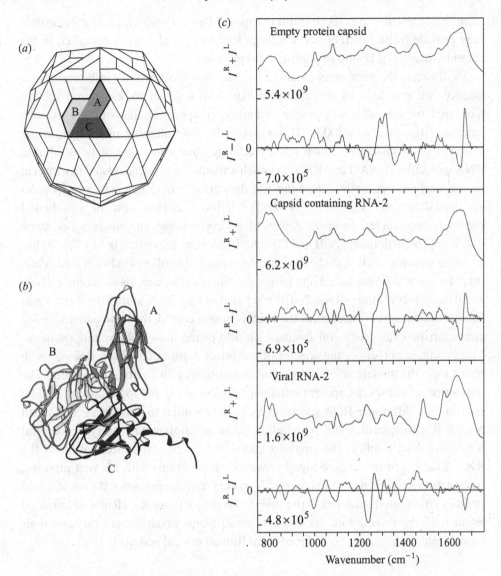

Fig. 7.10 (*a*) The icosahedral capsid of cowpea mosaic virus. (*b*) The asymmetric unit comprising three different protein domains, each having the same jelly-roll β-sandwich fold represented as MOLSCRIPT diagrams (Kraulis, 1991). (*c*) The backscattered Raman ($I^R + I^L$) and Raman optical activity ($I^R - I^L$) spectra measured in aqueous solution of the empty protein capsid (top pair), the intact capsid containing RNA-2 (middle pair), and the difference spectra obtained by subtracting the top from the middle spectra to reveal the spectra of the viral RNA-2 (bottom pair). Recorded in the author's laboratory. The absolute intensities are not defined, but the relative Raman and Raman optical activity intensities are significant.

significant contributions to ab initio computed Raman optical activity intensities may provide further insight into the large Raman optical activity observed in the extended amide III region of peptides and proteins.

To illustrate the enormous potential of Raman optical activity in biomolecular science, we conclude by displaying in Fig. 7.10 a set of backscattered spectra measured on a large biomolecular assembly, cowpea mosaic virus, in aqueous solution (Blanch *et al.*, 2002). This virus is the type member of the comovirus group of plant viruses. It has a nucleic acid genome consisting of two different RNA molecules (RNA-1 and RNA-2) which are separately encapsidated in identical icosahedral protein shells, called capsids, the structure of which is known from X-ray crystallography (Lin *et al.*, 1999). Fig. 7.10*a* illustrates how the icosahedral capsid is constructed from 60 copies of an asymmetric unit made up of three different protein domains A, B and C each of which, as illustrated in Fig. 7.10*b*, has a similar structure rich in β-sheet within the same jelly-roll β-sandwich fold. Virus preparations can be separated into empty protein capsids, capsids containing RNA-1 and capsids containing RNA-2. The top panel of Fig. 7.10*c* shows the Raman and Raman optical activity spectra of the empty protein capsid, the band patterns being characteristic of the jelly-roll β-sandwich fold of the individual protein domains. The middle panel shows the spectra of the intact capsid containing RNA-2, with bands from the nucleic acid now evident in addition to those from the protein. The bottom panel shows the spectra obtained by subtracting the top from the middle spectra. The difference ROA spectrum looks very similar to those of synthetic and natural RNA molecules and is therefore taken as originating mainly in the viral RNA: the details reflect the single-stranded A-type helical conformation of the RNA-2 packaged in the core together with its interactions with the coat proteins, information which is not available from X-ray crystallography since the nucleic acid in this virus is too disordered to provide useful diffraction data. Hence information about both the protein and nucleic acid constituents of an intact virus and their mutual interactions may be deduced from Raman optical activity!

8

Antisymmetric scattering and magnetic
Raman optical activity

If you think that I have got hold of something here please keep it to
yourself. I do not want some lousy Englishman to steal the idea. And it
will take a long time to get it into shape.

Friedrich Engels (in a letter to Marx)

8.1 Introduction

Antisymmetric matter tensors have been important for many years in connection with linear magneto–optical phenomena such as the Faraday effect. But until recently, Rayleigh and Raman scattering of light through antisymmetric tensors remained something of a minor curiosity, the only examples being in Rayleigh scattering from vapours of atoms such as sodium in spin degenerate ground states (Placzek, 1934; Penney, 1969; Tam and Au, 1976; Hamaguchi, Buckingham and Kakimoto, 1980). As discussed in Section 3.5.3, pure antisymmetric Rayleigh and Raman scattering is characterized by *inverse polarization* and its presence is usually detected via 'anomalies' in the depolarization ratio: estimates of the relative contributions of symmetric and antisymmetric scattering in a particular band require a set of complete polarization measurements that include the reversal coefficient in backscattering of circularly polarized light.

Antisymmetric scattering came to prominence with the observation by Spiro and Strekas (1972) of almost pure inverse polarization in many of the vibrational bands in the resonance Raman spectra of haem proteins, and antisymmetric scattering was found to dominate the resonance Raman spectra of many metalloporphyrins, at least using incident light at visible wavelengths. Antisymmetric scattering was subsequently observed in the resonance vibrational Raman spectra of iridium (IV) hexahalides (Hamaguchi, Harada and Shimanouchi, 1975; Hamaguchi and Shimanouchi, 1976) and has since been observed in a number of other odd-electron transition metal complexes. Prior to these observations of antisymmetric *vibrational* Raman scattering, Koningstein and Mortensen (1968) had observed

385

antisymmetric *electronic* Raman scattering in Eu^{3+} ions doped into yttrium aluminium garnet. Many other examples of antisymmetric scattering are now known.

As mentioned in Section 1.4, magnetic Rayleigh and Raman optical activity may be observed as a small difference in the scattered intensity in right- and left-circularly polarized incident light when an achiral molecular sample is placed in a magnetic field parallel to the incident light beam, or as a small circular component in the scattered light when the magnetic field is parallel to the scattered light beam. The signs of these observables reverse in the antiparallel arrangement. It was shown in Section 3.5.5 that magnetic Rayleigh and Raman optical activity originate in cross terms between the polarizability, or transition polarizability, and the same tensor perturbed to first order in the static magnetic field. Since antisymmetric scattering is important in magnetic Rayleigh and Raman optical activity, the two topics are discussed together in this chapter. We shall see that magnetic Rayleigh and Raman optical activity can provide a sensitive test for antisymmetric scattering, and also functions as 'Raman electron paramagnetic resonance' in that it may be used to determine electronic ground-state Zeeman splittings and how these change when a molecule is in an excited vibrational state. Following the first observations of magnetic resonance Raman optical activity in the 1970s (Barron, 1975a; Barron and Meehan, 1979), the phenomenon was explored in a number of molecular systems in dilute solution in the early 1980s (see Barron and Vrbancich, 1985, for a review of this work). Apart from reports of magnetic Raman optical activity in antiferromagnetic crystals of FeF_2 (Hoffman, Jia and Yen, 1990; Lockwood, Hoffman and Yen, 2002), the subject has languished ever since. Hopefully the account in this chapter will rekindle interest in magnetic Raman optical activity, for which many novel applications to studies of metal complexes, biological molecules and magnetic solids may be envisaged.

8.2 Symmetry considerations

It is well known that *Rayleigh* scattering from an atom or molecule in a nondegenerate state is pure symmetric. This follows from the symmetry of the Hamiltonian with respect to time reversal, as shown in Section 2.8.1. Thus antisymmetric Rayleigh scattering from systems in their ground states requires ground-state degeneracy. This can be electron spin degeneracy or orbital degeneracy, separately or together (although the Jahn–Teller effect removes orbital degeneracy in nonlinear molecules). And, as Baranova and Zel'dovich (1978) have pointed out, the degeneracy of the rotational states of a molecule in a nondegenerate electronic state can also generate antisymmetric Rayleigh scattering if the Coriolis force acting on the electrons is taken into account. Since ground vibrational states are always totally symmetric, vibrational degeneracy could only generate antisymmetric Rayleigh scattering from molecules in excited vibrational states.

On the other hand, degeneracy is not a prerequisite for antisymmetric vibrational *Raman* scattering. While, as shown below, electronic degeneracy is usually necessary in the initial state for the generation of antisymmetric contributions to vibrational Raman scattering in totally symmetric modes and in nontotally symmetric modes not spanning an antisymmetric irreducible representation, it is not necessary anywhere for modes that do span an antisymmetric irreducible representation. A clear example is the observation of antisymmetric resonance Raman scattering in B_{1g} modes of vibration in annulenes of D_{2h} symmetry for which no degeneracies exist (Fujimoto *et al.*, 1980). Although degeneracy is present in the intermediate excited electronic states of porphyrins of D_{4h} symmetry which, as discussed in Section 8.4.5 below, support antisymmetric scattering in A_{2g} modes of vibration, it is not essential.

As discussed in Section 4.4.3, the essence of the problem is that the effective complex operator determining the spatial symmetry aspects of any antisymmetric scattering tensor is antiHermitian and time odd. This realization enabled us to generate the following fundamental relation for the complex transition polarizability:

$$(\tilde{\alpha}_{\alpha\beta})_{mn} = (\tilde{\alpha}_{\beta\alpha})_{\Theta n\Theta m} = (\tilde{\alpha}_{\alpha\beta})^*_{\Theta m\Theta n}. \tag{4.4.2}$$

This was extended to the case of resonance scattering in (4.4.4). We make extensive use of these relations in the present chapter.

Time reversal and spatial symmetry arguments are combined in the generalized matrix element selection rule (4.3.37), which we now employ to elaborate the possibilities for antisymmetric scattering when the molecule is in a degenerate electronic state. We take V to be the effective polarizability operator $\hat{\alpha}_{\alpha\beta}$ (2.8.14a). Since the antisymmetric part $\hat{\alpha}^a_{\alpha\beta}$ is time odd, we obtain within the zeroth-order Herzberg–Teller approximation the following criteria for antisymmetric Rayleigh scattering and antisymmetric resonance Raman scattering in totally symmetric modes of vibration (the justification for this realm of applicability is given in the next section). If Γ_e is the irreducible representation spanned by the degenerate electronic states, and Γ_A is that spanned by an antisymmetric tensor (or axial vector) component, then for *even* electron systems $\{\Gamma^2_e\} \times \Gamma_A$, and for *odd* electron systems $[\Gamma^2_e] \times \Gamma_A$, must contain the totally symmetric irreducible representation. Remember that for odd-electron systems the irreducible representations refer to the appropriate double group.

If the zeroth-order Herzberg–Teller approximation is not invoked, there are additional possibilities for which explicit mechanisms can be developed (see the next section) by considering vibronic coupling between electronic states in different levels or between electronic states belonging to the same degenerate level, the latter being responsible for the Jahn–Teller effect. For these general situations, we obtain the following criteria for antisymmetric Raman scattering in a mode of vibration of symmetry species Γ_v. For *even* electron systems $\{\Gamma_e^2\} \times \Gamma_A$, and for *odd* electron systems $[\Gamma_e^2] \times \Gamma_A$, must contain Γ_v. Of course if Γ_v itself spans Γ_A, these generalized symmetry selection rules are superfluous because electronic degeneracy is no longer required. These generalized symmetry selection rules, which were first deduced by Child and Longuet-Higgins (1961) and Child (1962), are only valid in the absence of external magnetic fields.

Another relevant piece of information, deduced in Section 4.4.5, is that spatial symmetry arguments within the rotation group lead to the angular momentum selection rules $\Delta J = 0, \pm 1$ on antisymmetric scattering.

Although the main focus of the present chapter is *resonance* scattering, this being the only situation for which antisymmetric scattering and magnetic Raman optical activity has been observed to date, it should be mentioned that antisymmetric scattering is, in principle, possible for Raman scattering at *transparent* wavelengths in antisymmetric modes of vibration (Buckingham, 1988; Liu, 1991). This may be understood by invoking the Plazcek approximation (Section 2.8.3), which is valid for Raman scattering at transparent wavelengths. Within this approximation, the antisymmetric polarizability $\alpha'_{\alpha\beta}$ in the ground electronic state now acts as an effective operator bringing about vibrational transitions. However, being a time-odd property, $\alpha'_{\alpha\beta}$ must be expanded in the conjugate momentum $\dot{Q}$ of the normal vibrational coordinate, rather than the coordinate itself, similar to the development of the magnetic dipole moment operator in the theory of vibrational circular dichroism (Section 7.2.3). Such antisymmetric scattering is expected to be very weak.

The antisymmetric tensors discussed in this chapter cannot contribute to refringent scattering phenomena in the absence of an external time-odd influence. Although α'_{xy} contributes to optical rotation for light propagating along z in (3.4.16b), and is responsible for the Faraday effect in a static magnetic field along z, it cannot contribute in the absence of the field. Even though (4.4.6) show that an atom or molecule in a Kramers degenerate state can support such a tensor component, the sum over all the scattering transitions between the components of the Kramers degenerate set of states is zero. An external time-odd influence is required to lift this degeneracy and prevent complete cancellation. On the other hand, nonrefringent scattering is *incoherent* and each transition tensor contributes separately to the scattered intensity in the form $|(\tilde{\alpha}_{\alpha\beta})_{mn}|^2$, so antisymmetric Rayleigh and Raman scattering is possible without external fields.

8.3 A vibronic development of the vibrational Raman transition tensors

Both antisymmetric scattering and magnetic Rayleigh and Raman optical activity are only observed, at present, when the incident frequency is in the vicinity of an electronic absorption frequency of the atom or molecule. The scattered intensities can then show a tremendous resonance enhancement. We therefore need to cast the vibrational Raman transition tensors into a form suitable for application to resonance scattering.

Vibrational Raman scattering may be formulated in two distinct ways: Placzek's polarizability theory, given in Section 2.8.3, which considers the dependence of the ground state electronic polarizability on the normal coordinates of vibration, and vibronic theories which take detailed account of the coupling of electronic and vibrational motions. Although Placzek's theory provides a satisfactory treatment of the vibrational Raman transition tensors at transparent frequencies, like all ground-state theories it depends on a formal sum over all excited states and so is not applicable to the resonance situation. Here we develop the vibrational Raman transition tensors using the Herzberg–Teller approximation: this provides a convenient framework for a discussion of the symmetry aspects of resonance Raman scattering, but it is not a quantitative theory.

One approach is to extend the method given in Section 2.7 for the perturbed polarizabilities to the transition polarizabilities perturbed to first order in the vibrational–electronic interaction. This is satisfactory if the excited electronic states are orbitally nondegenerate. Although in principal degenerate states can be handled using the formalism of Section 2.7 by choosing components of a degenerate set to be diagonal in the perturbation, this procedure is not well defined in the crude adiabatic approximation: for example, in (2.8.43)

$$\sum_p (\partial H_e / \partial Q_p)_0 Q_p$$

is taken as a perturbation of just the electronic part of the crude adiabatic vibronic state. In the case that the electronic part of the resonant excited state is orbitally degenerate, we could go back to the unperturbed transition tensors and use proper Jahn–Teller states for $|j\rangle$ and Jahn–Teller energies in ω_{jn} at the outset: formally, this is valid when the Jahn–Teller splitting is greater than the absorption bandwidth so that the resonances with individual Jahn–Teller states are well resolved. However, we shall not develop the degenerate case explicitly here because we are concerned mainly with the symmetry aspects of the problem: as discussed later, it turns out that the correct relative values of the transition tensor components can be obtained from an essentially 'nondegenerate' development even for resonance with an orbitally-degenerate excited electronic state.

We develop the explicit symmetric and antisymmetric parts (2.8.8) of the transition polarizabilities. If the electronic parts of the states $|n\rangle$ and $|j\rangle$ are orbitally nondegenerate, we may use crude adiabatic vibronic states of the form (2.8.39) which can be represented by $|j\rangle = |e_j v_j\rangle = |e_j\rangle|v_j\rangle$. The electronic parts are taken to be perturbed to first order in

$$\sum_p (\partial H_e / \partial Q_p)_0 Q_p$$

and are therefore written as in (2.8.43). Although in principle the vibronic frequency separations should be written as the separations ω'_{jn} of the corresponding perturbed levels, similar to (2.7.4), we do not bother to do so because the contribution linear in

$$\sum_p (\partial H_e/\partial Q_p)_0 Q_p$$

is zero since there is no orbital electronic degeneracy. Taking account of the lifetime of the excited state, we find for the perturbed transition polarizabilities for resonance with the jth excited state

$$(\alpha_{\alpha\beta})^{\mathrm{s}}_{mn} = \frac{\omega_{jn}}{\hbar}(f+\mathrm{i}g)(X^{\mathrm{s}}_{\alpha\beta} + Z^{\mathrm{s}}_{\alpha\beta}), \tag{8.3.1a}$$

$$(\alpha_{\alpha\beta})^{\mathrm{a}}_{mn} = \frac{\omega}{\hbar}(f+\mathrm{i}g)(X^{\mathrm{a}}_{\alpha\beta} + Z^{\mathrm{a}}_{\alpha\beta}), \tag{8.3.1b}$$

$$(\alpha'_{\alpha\beta})^{\mathrm{s}}_{mn} = -\frac{\omega_{jn}}{\hbar}(f+\mathrm{i}g)(X'^{\mathrm{s}}_{\alpha\beta} + Z'^{\mathrm{s}}_{\alpha\beta}), \tag{8.3.1c}$$

$$(\alpha'_{\alpha\beta})^{\mathrm{a}}_{mn} = -\frac{\omega}{\hbar}(f+\mathrm{i}g)(X'^{\mathrm{a}}_{\alpha\beta} + Z'^{\mathrm{a}}_{\alpha\beta}), \tag{8.3.1d}$$

where the various parts of the **X**- and **Z**-tensors are

$$X^{\mathrm{s}}_{\alpha\beta} = \mathrm{Re}[(\langle e_m|\mu_\alpha|e_j\rangle\langle e_j|\mu_\beta|e_n\rangle$$
$$\pm \langle e_m|\mu_\beta|e_j\rangle\langle e_j|\mu_\alpha|e_n\rangle)\langle v_m|v_j\rangle\langle v_j|v_n\rangle], \tag{8.3.1e}$$

$$Z^{\mathrm{s}}_{\alpha\beta} = \mathrm{Re}\left\{\sum_{e_k\neq e_n}\frac{\langle e_k|\sum_p(\partial H_e/\partial Q_p)_0|e_n\rangle}{\hbar\omega_{e_n e_k}}\right.$$

$$\times(\langle e_m|\mu_\alpha|e_j\rangle\langle e_j|\mu_\beta|e_k\rangle \pm \langle e_m|\mu_\beta|e_j\rangle\langle e_j|\mu_\alpha|e_k\rangle)\langle v_m|v_j\rangle\langle v_j|Q_p|v_n\rangle$$

$$+ \sum_{e_k\neq e_m}\frac{\langle e_k|\sum_p(\partial H_e/\partial Q_p)_0|e_m\rangle^*}{\hbar\omega_{e_m e_k}}$$

$$\times(\langle e_k|\mu_\alpha|e_j\rangle\langle e_j|\mu_\beta|e_n\rangle \pm \langle e_k|\mu_\beta|e_j\rangle\langle e_j|\mu_\alpha|e_n\rangle)\langle v_m|Q_p^*|v_j\rangle\langle v_j|v_n\rangle$$

$$+ \sum_{e_k\neq e_j}\left[\frac{\langle e_k|\sum_p(\partial H_e/\partial Q_p)_0|e_j\rangle^*}{\hbar\omega_{e_j e_k}}\right.$$

$$\times(\langle e_m|\mu_\alpha|e_j\rangle\langle e_k|\mu_\beta|e_n\rangle \pm \langle e_m|\mu_\beta|e_j\rangle\langle e_k|\mu_\alpha|e_n\rangle)\langle v_m|v_j\rangle\langle v_j|Q_p^*|v_n\rangle$$

$$+ \frac{\langle e_k|\sum_p(\partial H_e/\partial Q_p)_0|e_j\rangle}{\hbar\omega_{e_j e_k}}$$

$$\left.\left.\times(\langle e_m|\mu_\alpha|e_k\rangle\langle e_j|\mu_\beta|e_n\rangle \pm \langle e_m|\mu_\beta|e_k\rangle\langle e_j|\mu_\alpha|e_n\rangle)\langle v_m|Q_p|v_j\rangle\langle v_j|v_n\rangle\right]\right\}. \tag{8.3.1f}$$

The upper and lower superscripts $_a^s$ belong to the upper and lower signs $\pm$ in the expressions. The corresponding primed tensors are given by (8.3.1e, f) with imaginary parts specified in place of real parts. These **X**- and **Z**-tensors are analogous to the A- and B-terms introduced by Albrecht (1961).

Before discussing resonance scattering, it is instructive to apply these results to scattering at *transparent* frequencies. In order to do so, we must first sum (8.3.1) over all the excited states $|j\rangle$. Introducing the approximation that the potential energy surfaces of the ground and excited electronic states are sufficiently similar that the vibrational states in the different electronic manifolds are orthonormal, that is

$$\langle v_j | v_n \rangle = \delta_{v_j v_n}, \tag{8.3.2}$$

the vibronic frequency factors ω_{jn} in (8.3.1) can be replaced by purely electronic factors $\omega_{e_j e_n}$ (Albrecht, 1961). The closure theorem in the space of the vibrational wavefunctions can then be invoked:

$$\sum_{v_j} \langle v_m | v_j \rangle \langle v_j | v_n \rangle = \langle v_m | v_n \rangle = \delta_{v_m v_m}. \tag{8.3.3}$$

The same arguments lead to the replacement of

$$\langle v_m | v_j \rangle \langle v_j | Q_p | v_n \rangle$$

and

$$\langle v_m | Q_p | v_j \rangle \langle v_j | v_n \rangle$$

by $\langle v_m | Q_p | v_n \rangle$ in (8.3.1 f). Thus at transparent frequencies the **X**-tensors contribute only to Rayleigh scattering. In this approximation, therefore, Raman scattering at transparent frequencies arises as a result of vibronic coupling, and is determined by appropriate parts of the **Z**-tensors. Even though we are excluding electronic Raman scattering, we have allowed the initial and final electronic states $|e_n\rangle$ and $|e_m\rangle$ to be different to allow for transitions between different components of degenerate ground states since, as discussed in Section 4.4.3, this has important implications for antisymmetric scattering. But if the initial and final electronic states are not degenerate, all the imaginary tensors vanish (in the absence of an external static magnetic field), as do the antisymmetric parts of the real tensors; so only the real symmetric parts survive at transparent frequencies in nondegenerate systems.

We now apply the transition tensors (8.3.1) to *resonance* scattering. It should first be realized that both the **X**- and **Z**-tensors can now contribute to vibrational Raman scattering, in contrast with the transparent case when **X** can only generate Rayleigh scattering. This is because it is no longer justifiable to replace the vibronic frequency factors by purely electronic factors since the precise values of the $\omega_{e_j v_j e_n v_n}$ are now critical and the small but finite values of the Franck–Condon overlap integrals (which were taken to be zero at transparent frequencies) can lead to significant effects. The factor $\langle v_m | v_j \rangle \langle v_j | v_n \rangle$ is determined by the magnitude of one or other of the Franck–Condon overlap integrals $\langle v_m | v_j \rangle$ and $\langle v_j | v_n \rangle$ (which are usually largest for totally symmetric vibrations), depending on whether $v_j = v_n$ or v_m. The factor

$$\langle v_m | v_j \rangle \langle v_j | Q_p | v_n \rangle$$

is largest when $v_j = v_m$ and v_j is the first excited vibrational state, associated with Q_p, in the first excited electronic state. The factor

$$\langle v_m | Q_p | v_j \rangle \langle v_j | v_n \rangle$$

is largest when $v_j = v_n$ and v_j is the ground vibrational state in the excited electronic state. In both cases the final vibrational state v_m is the same and corresponds to the first excited vibrational state, associated with Q_p, in the ground electronic state, and can be either totally symmetric or nontotally symmetric. Thus resonance Raman bands associated with totally symmetric vibrations are generated by appropriate parts of both the **X**- and **Z**-tensors, although the contributions from **X** are expected to be largest. Resonance Raman bands associated with nontotally symmetric vibrations are generated by appropriate parts of **Z**.

Notice that there are two conditions for the resonance enhancement of nontotally symmetric vibrations. (1) The incident frequency coincides with the transition frequency from the ground vibrational state in the ground electronic state to the ground vibrational state in the excited electronic state (the 0–0 transition), which boosts the contributions from terms depending on

$$\langle v_m | Q_p | v_j \rangle \langle v_j | v_n \rangle.$$

(2) The incident frequency coincides with the transition frequency from the ground vibrational state in the ground electronic state to one of the first excited vibrational states in the excited electronic state (the 0–1 transition), which boosts the contributions from terms depending on

$$\langle v_m | v_j \rangle \langle v_j | Q_p | v_n \rangle.$$

Although this is essentially a nondegenerate theory, the results remain applicable for contributions to resonance Raman scattering arising from coupling of a degenerate resonant excited electronic state with electronic states from other excited degenerate sets: in such a case the crude adiabatic state $|e_j\rangle|v_j\rangle$ is taken to be one of the components of a general vibronic state such as (2.8.44). For coupling of components within the same degenerate set of electronic states (the Jahn–Teller effect) a different formalism is required for quantitative considerations; but apart from the additional restrictions imposed by the generalized selection rule (4.3.37), the symmetry aspects are the same as for coupling between components of different degenerate sets of electronic states and so, even for a Jahn–Teller mechanism, we can deduce the correct relative values of the transition tensor components (Hamaguchi, 1977; Spiro and Stein, 1978).

8.4 Antisymmetric scattering

8.4.1 The antisymmetric transition tensors in the zeroth-order Herzberg–Teller approximation

In the zeroth-order Herzberg–Teller approximation, the transition polarizabilities (2.8.8) yield the **X**-tensors given in (8.3.1) and so may be used to describe Rayleigh scattering, both transparent and resonance, and resonance Raman scattering in totally symmetric modes of vibration. Bearing in mind the discussion in Section 4.4.3, we can write down the only allowed antisymmetric parts of the complex transition polarizability for the vibrational Raman transition $v_m \leftarrow v_n$ with the system returning either to the initial electronic state or, if the initial electronic state is part of a degenerate level, at most to some other degenerate electronic state within that level.

Thus for *diagonal* transitions within a degenerate electronic level of an odd-electron system,

$$(\alpha'_{\alpha\beta})^{\mathrm{a}}_{e_n v_m e_n v_n} = -\frac{2}{\hbar} \sum_{\substack{e_j \neq e_n \\ v_j}} \omega(f+\mathrm{i}g)$$

$$\times \mathrm{Im}\left(\langle e_n|\mu_\alpha|e_j\rangle\langle e_j|\mu_\beta|e_n\rangle\right)\langle v_m|v_j\rangle\langle v_j|v_n\rangle; \qquad (8.4.1a)$$

for *off-diagonal* transitions of both even- and odd-electron systems,

$$(\alpha_{\alpha\beta})^{\mathrm{a}}_{e'_n v_m e_n v_n} = \frac{1}{\hbar} \sum_{\substack{e_j \neq e_n \\ v_j}} \omega(f+\mathrm{i}g)$$

$$\times \mathrm{Re}(\langle e'_n|\mu_\alpha|e_j\rangle\langle e_j|\mu_\beta|e_n\rangle - \langle e'_n|\mu_\beta|e_j\rangle\langle e_j|\mu_\alpha|e_n\rangle)\langle v_m|v_j\rangle\langle v_j|v_n\rangle; \qquad (8.4.1b)$$

and for *off-diagonal* transitions of an odd-electron system,

$$(\alpha'_{\alpha\beta})^{\mathrm{a}}_{e'_n v_m e_n v_n} = -\frac{1}{\hbar} \sum_{\substack{e_j \neq e_n \\ v_j}} \omega(f+\mathrm{i}g)$$

$$\times \mathrm{Im}\left(\langle e'_n|\mu_\alpha|e_j\rangle\langle e_j|\mu_\beta|e_n\rangle - \langle e'_n|\mu_\beta|e_j\rangle\langle e_j|\mu_\alpha|e_n\rangle\right)\langle v_m|v_j\rangle\langle v_j|v_n\rangle. \qquad (8.4.1c)$$

These expressions are now applied to some simple examples in which the incident frequency coincides with a transition frequency to an excited spin–orbit state that is well resolved from other spin–orbit states. It should be mentioned that, if the spin–orbit states are not well resolved, a perturbation treatment analogous to that used in the Faraday effect can be applied, with the spin–orbit interaction replacing the interaction with the external magnetic field. Spin–orbit perturbed transition polarizabilities analogous to the Faraday A- and B-terms are then obtained. For details of this spin–orbit mechanism, we refer to Barron and Nørby Svendsen (1981).

Although antisymmetric Rayleigh scattering becomes very small away from resonance because of contributions with opposite signs from other electronic transitions, it remains finite and only tends to zero at very high and very low frequency (like natural and magnetic optical activity). On the other hand, antisymmetric Raman scattering in totally symmetric modes of vibration decreases much more rapidly away from an excitation band envelope because, in addition to cancellation from other electronic transitions, the closure theorem in the space of the vibrational wavefunctions associated with the excited electronic state can be invoked so that the transition polarizabilities (8.4.1) vanish on account of the orthogonality of the initial and final vibrational states.

8.4.2 Resonance Rayleigh scattering in atomic sodium

Perhaps the simplest case of the generation of an antisymmetric tensor is in resonance Rayleigh scattering in atomic sodium vapour. The essential features of this case were discussed by Placzek as early as 1934. The ground state of sodium has a twofold Kramers degeneracy, and we shall see that the antisymmetric scattering is generated at resonance with one or other of the components of the yellow doublet through both diagonal and off-diagonal transitions between the Kramers components. The yellow doublet originates in spin–orbit splitting of the excited states $^2P_{\frac{1}{2}}$ and $^2P_{\frac{3}{2}}$, which are generated by the $3p \leftarrow 3s$ electron promotion.

The relevant atomic states, specified as $|ls\,JM\rangle$ in the Russell–Saunders coupling scheme, are shown in Fig. 8.1. We use the results (4.4.26), derived from the Wigner–Eckart theorem, to work out the matrix elements of the cartesian components of the electric dipole moment operator between the atomic states. In this case the matrix elements of irreducible spherical tensor operators have the form

$$\langle l's'J'M'|T_q^k|ls\,JM\rangle = (-1)^{J'-M'}\langle l's'J'\|T^k\|ls\,J\rangle \begin{pmatrix} J' & k & J \\ -M' & q & M \end{pmatrix}, \quad (8.4.2a)$$

in which the reduced matrix element involving coupled spin and orbital angular momentum states can be broken down further into a reduced matrix element involving

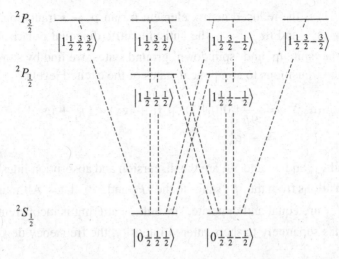

Fig. 8.1 The first few Na states $|ls\,JM\rangle$. The transitions shown are electric dipole allowed.

just the orbital parts:

$$\langle ls\,J||T^k||l's'J'\rangle = (-1)^{l+s+J'+k}[(2J'+1)(2J+1)]^{\frac{1}{2}}$$

$$\times \langle l||T^k||l'\rangle \begin{Bmatrix} l & J & s \\ J' & l' & k \end{Bmatrix}, \tag{8.4.2b}$$

where the object in curly brackets is the $6j$ symbol, associated with the coupling of three angular momenta, which has special symmetry properties. We refer to Silver (1976) for details of the background to (8.4.2b), and use the numerical values for $6j$ symbols tabulated by Rotenberg *et al.* (1959). It is now a simple matter to obtain all the allowed transition moments in the $P \leftarrow S$ manifold in terms of the same reduced matrix element $\langle l||\mu||l'\rangle$.

Consider first the xy components of the transition polarizability. Since, according to (4.4.26a,b), the operators μ_x, μ_y can only connect states with $\Delta M = \pm 1$, we anticipate that the xy components can only generate diagonal scattering transitions through (8.4.1a). Using (8.4.2) in (4.4.26), the following relationships are found:

$$\langle 0\tfrac{1}{2}\tfrac{1}{2}\tfrac{1}{2}|\mu_x|1\tfrac{1}{2}\tfrac{1}{2}-\tfrac{1}{2}\rangle\langle 1\tfrac{1}{2}\tfrac{1}{2}-\tfrac{1}{2}|\mu_y|0\tfrac{1}{2}\tfrac{1}{2}\tfrac{1}{2}\rangle$$

$$= -\langle 0\tfrac{1}{2}\tfrac{1}{2}-\tfrac{1}{2}|\mu_x|1\tfrac{1}{2}\tfrac{1}{2}\tfrac{1}{2}\rangle\langle 1\tfrac{1}{2}\tfrac{1}{2}\tfrac{1}{2}|\mu_y|0\tfrac{1}{2}\tfrac{1}{2}-\tfrac{1}{2}\rangle$$

$$= 2\langle 0\tfrac{1}{2}\tfrac{1}{2}\tfrac{1}{2}|\mu_x|1\tfrac{1}{2}\tfrac{3}{2}-\tfrac{1}{2}\rangle\langle 1\tfrac{1}{2}\tfrac{3}{2}-\tfrac{1}{2}|\mu_y|0\tfrac{1}{2}\tfrac{1}{2}\tfrac{1}{2}\rangle$$

$$= -2\langle 0\tfrac{1}{2}\tfrac{1}{2}-\tfrac{1}{2}|\mu_x|1\tfrac{1}{2}\tfrac{3}{2}\tfrac{1}{2}\rangle\langle 1\tfrac{1}{2}\tfrac{3}{2}\tfrac{1}{2}|\mu_y|0\tfrac{1}{2}\tfrac{1}{2}-\tfrac{1}{2}\rangle$$

$$= -\tfrac{2}{3}\langle 0\tfrac{1}{2}\tfrac{1}{2}\tfrac{1}{2}|\mu_x|1\tfrac{1}{2}\tfrac{3}{2}\tfrac{3}{2}\rangle\langle 1\tfrac{1}{2}\tfrac{3}{2}\tfrac{3}{2}|\mu_y|0\tfrac{1}{2}\tfrac{1}{2}\tfrac{1}{2}\rangle$$

$$= \tfrac{2}{3}\langle 0\tfrac{1}{2}\tfrac{1}{2}-\tfrac{1}{2}|\mu_x|1\tfrac{1}{2}\tfrac{3}{2}-\tfrac{3}{2}\rangle\langle 1\tfrac{1}{2}\tfrac{3}{2}-\tfrac{3}{2}|\mu_y|0\tfrac{1}{2}\tfrac{1}{2}-\tfrac{1}{2}\rangle$$

$$= -\tfrac{1}{9}i|\langle 0||\mu||1\rangle|^2, \tag{8.4.3}$$

where $\langle 0||\mu||1 \rangle$ is the reduced matrix element for an $p \leftarrow s$ transition. Thus, denoting by $(\alpha'_{\alpha\beta})^a_{\frac{1}{2}\frac{1}{2}}$ and $(\alpha'_{\alpha\beta})^a_{-\frac{1}{2}-\frac{1}{2}}$ the antisymmetric diagonal polarizabilities for the atom in the 'spin up' and 'spin down' ground states, we find by summing over all the allowed transitions to component states of the excited levels,

$$(\alpha'_{xy})^a_{\frac{1}{2}\frac{1}{2}} = \frac{2\omega}{9\hbar}|\langle 0||\mu||1 \rangle|^2[(f_{\frac{1}{2}} + ig_{\frac{1}{2}}) - (f_{\frac{3}{2}} + ig_{\frac{3}{2}})]$$
$$= -(\alpha'_{xy})^a_{-\frac{1}{2}-\frac{1}{2}}, \tag{8.4.4}$$

where $f_{\frac{1}{2}}$ and $g_{\frac{1}{2}}$, and $f_{\frac{3}{2}}$ and $g_{\frac{3}{2}}$, are the dispersion and absorption lineshape functions for transitions from the $^2S_{\frac{1}{2}}$ states to the $^2P_{\frac{1}{2}}$ and $^2P_{\frac{3}{2}}$ states. Although $(\alpha'_{\alpha\beta})^a_{\frac{1}{2}\frac{1}{2}}$ and $(\alpha'_{\alpha\beta})^a_{-\frac{1}{2}-\frac{1}{2}}$ are equal and opposite, since the scattering is incoherent each tensor contributes separately to the scattered intensity, the frequency dependence of which is

$$\omega^2[f_{\frac{1}{2}}^2 + g_{\frac{1}{2}}^2 + f_{\frac{3}{2}}^2 + g_{\frac{3}{2}}^2 - 2(f_{\frac{1}{2}}f_{\frac{3}{2}} + g_{\frac{1}{2}}g_{\frac{3}{2}})]. \tag{8.4.5}$$

If the transitions are well resolved, as in sodium vapour, this function shows two peaks, one at each transition frequency.

Consider next the xz and yz components of the transition polarizability. Since, according to (4.4.26c), the operator μ_z can only connect states with $\Delta M = 0$, we anticipate that the xz and yz components can only generate off-diagonal scattering transitions through (8.4.1b, c), it is found that

$$(\alpha_{xz})^a_{+\frac{1}{2}-\frac{1}{2}} = \frac{2\omega}{9\hbar}|\langle 0||\mu||1 \rangle|^2[(f_{\frac{1}{2}} + ig_{\frac{1}{2}}) - (f_{\frac{3}{2}} + ig_{\frac{3}{2}})]$$
$$= -(\alpha_{xz})^a_{-\frac{1}{2}+\frac{1}{2}}, \tag{8.4.6}$$

$$(\alpha'_{yz})^a_{+\frac{1}{2}-\frac{1}{2}} = \frac{2\omega}{9\hbar}|\langle 0||\mu||1 \rangle|^2[(f_{\frac{1}{2}} + ig_{\frac{1}{2}}) - (f_{\frac{3}{2}} + ig_{\frac{3}{2}})]$$
$$= (\alpha'_{yz})^a_{-\frac{1}{2}+\frac{1}{2}}. \tag{8.4.7}$$

These antisymmetric tensors also provide contributions to the scattered intensity with the frequency dependence (8.4.5).

The symmetric polarizability components may be calculated in a similar fashion, and it is found that

$$(\alpha_{xx})^s_{\pm\frac{1}{2}\pm\frac{1}{2}} = (\alpha_{yy})^s_{\pm\frac{1}{2}\pm\frac{1}{2}} = (\alpha_{zz})^s_{\pm\frac{1}{2}\pm\frac{1}{2}}$$

with

$$(\alpha_{xx})^s_{\pm\frac{1}{2}\pm\frac{1}{2}} = -\frac{2}{9\hbar}|\langle 0||\mu||1 \rangle|^2[\omega_{\frac{1}{2}}(f_{\frac{1}{2}} + ig_{\frac{1}{2}}) + 2\omega_{\frac{3}{2}}(f_{\frac{3}{2}} + ig_{\frac{3}{2}})], \tag{8.4.8}$$

where $\omega_{\frac{1}{2}}$ and $\omega_{\frac{3}{2}}$ are the $^2P_{\frac{1}{2}} \leftarrow ^2S_{\frac{1}{2}}$ and $^2P_{\frac{3}{2}} \leftarrow ^2S_{\frac{1}{2}}$ transition frequencies. All the other symmetric polarizability components are zero.

We can now calculate the depolarization ratio as a function of the incident frequency; but for simplicity we shall give just the values for exact resonance with one or other of the components of the doublet. The depolarization ratio when isotropic, anisotropic and antisymmetric scattering contribute to the same band was given in Chapter 3: for incident light linearly polarized perpendicular to the scattering plane we use (3.5.27) with $\beta(\alpha)^2$ and $\beta(\alpha')^2$ now interpreted as general anisotropic and antisymmetric invariants arising from either real or imaginary transition polarizabilities. Here, $\beta(\alpha)^2 = 0$. Since the different transition polarizabilities contribute incoherently to the intensity, we write the invariants $\beta(\alpha')^2$ and α^2 as sums of separate invariants, one for each distinct transition $+\frac{1}{2} \leftarrow +\frac{1}{2}, -\frac{1}{2} \leftarrow -\frac{1}{2}, +\frac{1}{2} \leftarrow -\frac{1}{2}$ and $-\frac{1}{2} \leftarrow +\frac{1}{2}$. Using (8.4.4) and (8.4.6–8), we find that $\rho(x)$ is 1 for the $^2P_{\frac{1}{2}} \leftarrow {}^2S_{\frac{1}{2}}$ resonance and $\frac{1}{4}$ for the $^2P_{\frac{3}{2}} \leftarrow {}^2S_{\frac{1}{2}}$ resonance. Notice that since the two coherent contributions to each antisymmetric transition polarizability from the two resonances are equal and opposite, they will tend to cancel if the incident frequency is far from the resonance region so that $\rho(x)$ tends to zero.

A useful visual representation of the relative values of the components of the complex transition polarizability $(\tilde{\alpha}_{\alpha\beta})_{e_m e_n} = (\alpha_{\alpha\beta})_{e_m e_n} - i(\alpha'_{\alpha\beta})_{e_m e_n}$ for resonance with each of the two components of the sodium yellow doublet is obtained by writing

(a) $^2P_{\frac{1}{2}}$ resonant term:

$$\begin{pmatrix} 1 & i & 0 \\ -i & 1 & 0 \\ 0 & 0 & 1 \end{pmatrix} \begin{pmatrix} 1 & -i & 0 \\ i & 1 & 0 \\ 0 & 0 & 1 \end{pmatrix} \begin{pmatrix} 0 & 0 & 1 \\ 0 & 0 & i \\ -1 & -i & 0 \end{pmatrix} \begin{pmatrix} 0 & 0 & -1 \\ 0 & 0 & i \\ 1 & -i & 0 \end{pmatrix}, \quad (8.4.9a)$$

$$\begin{array}{cccc} +\frac{1}{2} \leftarrow +\frac{1}{2} & -\frac{1}{2} \leftarrow -\frac{1}{2} & -\frac{1}{2} \leftarrow +\frac{1}{2} & +\frac{1}{2} \leftarrow -\frac{1}{2} \end{array}$$

(b) $^2P_{\frac{3}{2}}$ resonant term:

$$\begin{pmatrix} 2 & -i & 0 \\ i & 2 & 0 \\ 0 & 0 & 2 \end{pmatrix} \begin{pmatrix} 2 & i & 0 \\ -i & 2 & 0 \\ 0 & 0 & 2 \end{pmatrix} \begin{pmatrix} 0 & 0 & -1 \\ 0 & 0 & -i \\ 1 & i & 0 \end{pmatrix} \begin{pmatrix} 0 & 0 & 1 \\ 0 & 0 & -i \\ -1 & i & 0 \end{pmatrix}. \quad (8.4.9b)$$

$$\begin{array}{cccc} +\frac{1}{2} \leftarrow +\frac{1}{2} & -\frac{1}{2} \leftarrow -\frac{1}{2} & -\frac{1}{2} \leftarrow +\frac{1}{2} & +\frac{1}{2} \leftarrow -\frac{1}{2} \end{array}$$

The same factors multiply the components for the two resonances, so they can be compared directly.

8.4.3 Resonance Raman scattering in totally symmetric vibrations of iridium (IV) hexahalides

A similar mechanism to that in sodium can give rise to antisymmetric resonance Raman scattering in totally symmetric modes of vibration in molecules containing

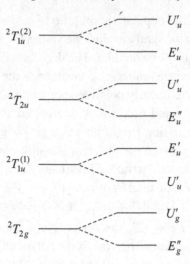

Fig. 8.2 The general pattern of spin–orbit states arising from the $\gamma_u^n t_{2g}^5$ and $\gamma_u^{n-1} t_{2g}^6$ configurations of iridium (IV) hexahalides.

an odd number of electrons. Good examples are the low-spin d^5 complexes $IrCl_6^{2-}$ and $IrBr_6^{2-}$.

The molecular orbital description of iridium (IV) hexahalides is similar to that of $Fe(CN)_6^{3-}$, outlined in Section 6.3.2. The orbital and state diagrams in Figs. 6.6 and 6.7 still apply, but now the spin–orbit splitting is large and must be included explicitly. Since the spin–orbit splitting here is less than the splitting of the atomic levels by the octahedral environment, we consider first the splitting of the spatial parts of the atomic states into states of symmetry species T_{2g}, T_{1u} and T_{2u} in the ordinary group O_h. The states generated by spin–orbit coupling must be classified with respect to the irreducible representations of the double group O_h^*, which has additional even-valued irreducible representations $E'_{g,u}$, $E''_{g,u}$ and $U'_{g,u}$ (using the nomenclature of Griffith, 1961). The representations T_{2g}, T_{1u} and T_{2u} in O_h become T'_{2g}, T'_{1u} and T'_{2u} in O_h^*. The symmetry species of the spin–orbit states are obtained from the direct product of the species of the space part of the wavefunction with E'_g, which is the species of the doublet spin part. Thus $T'_{2g} \times E'_g = E''_g + U'_g$, $T'_{1u} \times E'_g = E''_u + U'_u$ and $T'_{2u} \times E'_g = E''_u + U'_u$. The spin-orbit splitting pattern is shown in Fig. 8.2.

Thus the ground level is a Kramers doublet of symmetry species E''_g and the possible resonant excited levels are of species U'_u and E''_u arising from spin–orbit splitting of $^2T_{1u}$ and $^2T_{2u}$ levels (notice that resonant scattering via the excited E'_u level is electric dipole forbidden). We can apply the group theoretical criteria given in Section 8.2: antisymmetric Raman scattering is allowed in totally symmetric modes of vibration of IrX_6^{2-} because $[E''^2_g] = T_{1g}$, and T_{1g} is spanned by antisymmetric tensor components.

The calculation of the transition polarizability components proceeds along similar lines to the Na case, except that now we use the results (4.4.33), derived from Harnung's version of the Wigner–Eckart theorem for the finite molecular double groups, to work out the matrix elements of the cartesian components of the electric dipole moment operator. But, unlike the Na case, we will not get involved with the analogues of the $6j$ symbols and so will leave the results in terms of reduced matrix elements involving coupled spin–orbit states. As in the sodium case, we anticipate that the XY components can only generate diagonal scattering transitions through (8.4.1a), and the XZ and YZ components can only generate off-diagonal scattering transitions through (8.4.1b, c).

Consider, for example, resonance with an excited level of species U'_u. Using (4.4.33), the only nonzero products of matrix elements that generate the XY components, for example, are found to be

$$
\begin{aligned}
\langle E'' \tfrac{1}{2}| \, & \mu_X |U' \tfrac{3}{2}\rangle\langle U' \tfrac{3}{2}|\mu_Y|E'' \tfrac{1}{2}\rangle \\
&= -\langle E'' -\tfrac{1}{2}|\mu_X|U' -\tfrac{3}{2}\rangle\langle U' -\tfrac{3}{2}|\mu_Y|E'' -\tfrac{1}{2}\rangle \\
&= -\tfrac{1}{3}\langle E'' \tfrac{1}{2}|\mu_X|U' -\tfrac{1}{2}\rangle\langle U' -\tfrac{1}{2}|\mu_Y|E'' \tfrac{1}{2}\rangle \\
&= \tfrac{1}{3}\langle E'' -\tfrac{1}{2}|\mu_X|U' \tfrac{1}{2}\rangle\langle U' \tfrac{1}{2}|\mu_Y|E'' -\tfrac{1}{2}\rangle \\
&= \tfrac{1}{24}\mathrm{i}|\langle E''||\mu||U'\rangle|^2.
\end{aligned}
\tag{8.4.10}
$$

Then summing over all the allowed transitions to component states of the excited U'_u level gives

$$
\begin{aligned}
(\alpha'_{XY})^{\mathrm{a}}_{\tfrac{1}{2}\tfrac{1}{2}} &= \frac{\omega}{6\hbar}|\langle E''||\mu||U'\rangle|^2 \sum_{v_j}(f+\mathrm{i}g)\langle 1_n|v_j\rangle\langle v_j|0_n\rangle \\
&= -(\alpha'_{XY})^{\mathrm{a}}_{-\tfrac{1}{2}-\tfrac{1}{2}}.
\end{aligned}
\tag{8.4.11}
$$

Similarly for the other components:

$$
\begin{aligned}
(\alpha_{XZ})^{\mathrm{a}}_{+\tfrac{1}{2}-\tfrac{1}{2}} &= \frac{\omega}{6\hbar}|\langle E''||\mu||U'\rangle|^2 \sum_{v_j}(f+\mathrm{i}g)\langle 1_n|v_j\rangle\langle v_j|0_n\rangle \\
&= -(\alpha_{XZ})^{\mathrm{a}}_{-\tfrac{1}{2}+\tfrac{1}{2}},
\end{aligned}
\tag{8.4.12}
$$

$$
\begin{aligned}
(\alpha'_{YZ})^{\mathrm{a}}_{+\tfrac{1}{2}-\tfrac{1}{2}} &= \frac{\omega}{6\hbar}|\langle E''||\mu||U'\rangle|^2 \sum_{v_j}(f+\mathrm{i}g)\langle 1_n|v_j\rangle\langle v_j|0_n\rangle \\
&= (\alpha'_{YZ})^{\mathrm{a}}_{-\tfrac{1}{2}+\tfrac{1}{2}},
\end{aligned}
\tag{8.4.13}
$$

$$
\begin{aligned}
(\alpha_{XX})^{\mathrm{s}}_{\pm\tfrac{1}{2}\pm\tfrac{1}{2}} &= \frac{\omega}{3\hbar}|\langle E''||\mu||U'\rangle|^2 \sum_{v_j}(f+\mathrm{i}g)\langle 1_n|v_j\rangle\langle v_j|0_n\rangle \\
&= (\alpha_{YY})^{\mathrm{s}}_{\pm\tfrac{1}{2}\pm\tfrac{1}{2}} = (\alpha_{ZZ})^{\mathrm{s}}_{\pm\tfrac{1}{2}\pm\tfrac{1}{2}}.
\end{aligned}
\tag{8.4.14}
$$

For simplicity, we have not specified explicitly that the v_j are the vibrational states associated with the excited electronic level U'_u and that f and g are lineshape functions for the corresponding vibronic resonances. Similar calculations can be performed for the other possible resonance with an excited level of species E''_u.

The relative values of the components of the complex transition polarizability $(\tilde{\alpha}_{\alpha\beta})_{e_m e_n} = (\alpha_{\alpha\beta})_{e_m e_n} - i(\alpha'_{\alpha\beta})_{e_m e_n}$ for the two possible resonances are summarized in the form

(*a*) E''_u resonant level:

$$
\begin{pmatrix} 1 & i & 0 \\ -i & 1 & 0 \\ 0 & 0 & 1 \end{pmatrix}
\begin{pmatrix} 1 & -i & 0 \\ i & 1 & 0 \\ 0 & 0 & 1 \end{pmatrix}
\begin{pmatrix} 0 & 0 & 1 \\ 0 & 0 & i \\ -1 & -i & 0 \end{pmatrix}
\begin{pmatrix} 0 & 0 & -1 \\ 0 & 0 & i \\ 1 & -i & 0 \end{pmatrix}, \quad (8.4.15a)
$$
$$
+\tfrac{1}{2} \leftarrow +\tfrac{1}{2} \qquad -\tfrac{1}{2} \leftarrow -\tfrac{1}{2} \qquad -\tfrac{1}{2} \leftarrow +\tfrac{1}{2} \qquad +\tfrac{1}{2} \leftarrow -\tfrac{1}{2}
$$

(*b*) U'_u resonant level:

$$
\begin{pmatrix} 2 & -i & 0 \\ i & 2 & 0 \\ 0 & 0 & 2 \end{pmatrix}
\begin{pmatrix} 2 & i & 0 \\ -i & 2 & 0 \\ 0 & 0 & 2 \end{pmatrix}
\begin{pmatrix} 0 & 0 & -1 \\ 0 & 0 & -i \\ 1 & i & 0 \end{pmatrix}
\begin{pmatrix} 0 & 0 & 1 \\ 0 & 0 & -i \\ -1 & i & 0 \end{pmatrix}. \quad (8.4.15b)
$$
$$
+\tfrac{1}{2} \leftarrow +\tfrac{1}{2} \qquad -\tfrac{1}{2} \leftarrow -\tfrac{1}{2} \qquad -\tfrac{1}{2} \leftarrow +\tfrac{1}{2} \qquad +\tfrac{1}{2} \leftarrow -\tfrac{1}{2}
$$

The absolute values of the tensor components in (8.4.15*a,b*) depend on $|\langle E''||\mu||E''\rangle|^2$ and $|\langle E''||\mu||U'\rangle|^2$, respectively, which are functions of the detailed orbital configurations generating the states. Since in general the E''_u and U'_u excited levels could originate in different orbital configurations, the corresponding tensor components for the two resonances cannot be compared directly without detailed calculations of the appropriate reduced matrix elements.

Notice that the tensor patterns for the E''_u and U'_u resonances are identical with those for the $^2P_{\frac{1}{2}}$ and $^2P_{\frac{3}{2}}$ resonances in sodium. This means that the corresponding depolarization ratios are identical: thus $\rho(x)$ is 1 for the $E''_u \leftarrow E''_g$ resonance and $\tfrac{1}{4}$ for the $U'_u \leftarrow E''_g$ resonance. Indeed, Hamaguchi and Shimanouchi (1976) have observed the former value for resonance with an absorption band of $IrBr_6^{2-}$ assigned to an E''_u excited level, and Stein, Brown and Spiro (1977) have observed approximately the latter value for resonance with an absorption band of $IrCl_6^{2-}$ assigned to a U'_u excited level.

We refer to Hamaguchi (1977) and Stein, Brown and Spiro (1977) for alternative methods of calculation and further discussion.

8.4.4 Antisymmetric transition tensors generated through vibronic coupling

As discussed in Section 8.3, vibronic coupling is required for both resonance and nonresonance Raman scattering associated with non totally symmetric vibrations. The corresponding transition polarizabilities, which are perturbed to first order in

$$\sum_p (\partial H_e/\partial Q_p)_0 Q_p,$$

are given by the **Z**-tensors in (8.3.1).

Recalling that terms depending on

$$\langle v_m|Q_p|v_n\rangle\langle v_j|v_n\rangle$$

are largest for the 0–0 resonance, and that terms depending on

$$\langle v_m|v_j\rangle\langle v_j|Q_p|v_n\rangle$$

are largest for the 0–1 resonance, we may write the corresponding vibronically perturbed antisymmetric transition polarizabilities as the sum of contributions from the 0–0 and 0–1 vibronic resonances associated with a particular excited electronic state $|e_j\rangle$:

$$(\alpha_{\alpha\beta})^a_{e_m v_m e_n v_n} = \frac{\omega}{\hbar}\left\{(f_{0-0}+ig_{0-0})\,\mathrm{Re}\left[\sum_{e_k\neq e_j}\frac{\langle e_k|\sum_p(\partial H_e/\partial Q_p)_0|e_j\rangle}{\hbar\omega_{e_j e_k}}\right.\right.$$

$$\times((\langle e_m|\mu_\alpha|e_k\rangle\langle e_j|\mu_\beta|e_n\rangle - \langle e_m|\mu_\beta|e_k\rangle\langle e_j|\mu_\alpha|e_n\rangle)\langle 1_n|Q_p|0_j\rangle\langle 0_j|0_n\rangle\Bigg]$$

$$+(f_{0-1}+ig_{0-1})\,\mathrm{Re}\left[\sum_{e_k\neq e_j}\frac{\langle e_k|\sum_p(\partial H_e/\partial Q_p)_0|e_j\rangle^*}{\hbar\omega_{e_j e_k}}\right.$$

$$\times((\langle e_m|\mu_\alpha|e_j\rangle\langle e_k|\mu_\beta|e_n\rangle - \langle e_m|\mu_\beta|e_j\rangle\langle e_k|\mu_\alpha|e_n\rangle)\langle 1_n|1_j\rangle\langle 1_j|Q_p^*|0_n\rangle\Bigg]\Bigg\},$$

$$(8.4.16a)$$

$$(\alpha'_{\alpha\beta})^a_{e_m v_m e_n v_n} = -\frac{\omega}{\hbar}\left\{(f_{0-0}+ig_{0-0})\mathrm{Im}\left[\sum_{e_k\neq e_j}\frac{\langle e_k|\sum_p(\partial H_e/\partial Q_p)_0|e_j\rangle}{\hbar\omega_{e_j e_k}}\right.\right.$$

$$\times((\langle e_m|\mu_\alpha|e_k\rangle\langle e_j|\mu_\beta|e_n\rangle - \langle e_m|\mu_\beta|e_k\rangle\langle e_j|\mu_\alpha|e_n\rangle)\langle 1_n|Q_p|0_j\rangle\langle 0_j|0_n\rangle\Bigg]$$

$$+ (f_{0-1} + ig_{0-1})\text{Im} \left[\sum_{e_k \neq e_j} \frac{\langle e_k | \sum_p (\partial H_e / \partial Q_p)_0 | e_j \rangle^*}{\hbar \omega_{e_j e_k}} \right.$$

$$\left. \times (\langle e_m | \mu_\alpha | e_j \rangle \langle e_k | \mu_\beta | e_n \rangle - \langle e_m | \mu_\beta | e_j \rangle \langle e_k | \mu_\alpha | e_n \rangle) \langle 1_n | 1_j \rangle \langle 1_j | Q_p^* | 0_n \rangle \right] \Bigg\}.$$

$$(8.4.16b)$$

We have retained only vibronic mixing of $|e_j\rangle$ with other excited electronic states $|e_k\rangle$ since this will usually give much larger contributions than mixing of the ground state $|e_n\rangle$ with excited states.

The application of these antisymmetric tensors is rather different in the two cases of degenerate and nondegenerate initial states. If the initial electronic state is degenerate, the situation is analogous to that discussed in Section 8.4.1 for antisymmetric scattering without vibronic coupling: the same possibilities for diagonal and off-diagonal transitions exist, but now the inclusion of vibronic coupling allows antisymmetric scattering to be associated with resonance Raman scattering in nontotally symmetric modes of vibration, subject to the symmetry selection rule introduced in Section 8.2; namely that for even-electron systems $\{\Gamma_e^2\} \times \Gamma_A$, and for-odd electron systems $[\Gamma_e^2] \times \Gamma_A$, must contain Γ_v. But if the initial electronic state is nondegenerate, the symmetry species of the mode of vibration must be the same as that of an antisymmetric tensor component. These expressions are now applied to resonance Raman scattering associated with nontotally symmetric modes in porphyrins, which have nondegenerate ground electronic states (at least if they have an even number of electrons). The application to nontotally symmetric modes in molecules with degenerate ground electronic states is too complicated to illustrate here, and we refer to Hamaguchi (1977) for a detailed application to iridium (IV) hexahalides.

8.4.5 Resonance Raman scattering in porphyrins

Porphyrins provide good examples of the generation of antisymmetric resonance Raman scattering through vibronic coupling. The chromophore responsible is the conjugated ring system: the electronic states and transitions responsible for the visible and near ultraviolet absorptions are described in Section 6.3.1. Since this is an even-electron system without ground-state orbital degeneracy, antisymmetric scattering can only be associated with vibrational normal coordinates Q_a spanning antisymmetric irreducible representations; in this case A_{2g} and E_g in D_{4h} metal porphyrins, and B_{1g}, B_{2g} and B_{3g} in D_{2h} free-base porphyrins. In fact antisymmetric scattering has only been observed to date in A_{2g} modes of metal porphyrins, so we shall concentrate on this case.

The relevant expression obtains from (8.4.16a) by putting $e_m = e_n$, invoking the Hermiticity of the electric dipole moment operator and assuming that Q_{A_2} is real:

$$(\alpha_{\alpha\beta})^a_{e_n v_m e_n v_n} = \frac{\omega}{\hbar} \left[(f_{0-0} + ig_{0-0}) \langle 1_n | Q_{A_2} | 0_j \rangle \langle 0_j | 0_n \rangle \right.$$

$$\left. - (f_{0-1} + ig_{0-1}) \langle 1_n | 1_j \rangle \langle 1_j | Q_{A_2} | 0_n \rangle \right] Z^a_{\alpha\beta}, \quad (8.4.17a)$$

$$Z^a_{\alpha\beta} = \text{Re} \left[\sum_{e_k \neq e_j} \frac{\langle e_k | (\partial H_e / \partial Q_{A_2})_0 | e_j \rangle}{\hbar \omega_{e_j e_k}} \right.$$

$$\left. \times \left(\langle e_n | \mu_\alpha | e_k \rangle \langle e_j | \mu_\beta | e_n \rangle - \langle e_n | \mu_\beta | e_k \rangle \langle e_j | \mu_\alpha | e_n \rangle \right) \right]. \quad (8.4.17b)$$

It can now be seen that a fundamental criterion for antisymmetric scattering via vibronic coupling is that the 0–0 and 0–1 vibronic transitions be well resolved, otherwise their contributions tend to cancel. This is one of the reasons why strong antisymmetric scattering is observed in A_2 modes for excitation within the Q_0 and Q_1 bands of porphyrins, but not within the Soret band. Interference between the two contributions leads to characteristic variations of the resonance Raman intensity as the exciting laser frequency is swept through the region of the 0–0 and 0–1 absorptions (*excitation profiles*). We refer to Barron (1976) for the explicit form of these excitation profiles for both symmetric and antisymmetric scattering in the region of the porphyrin Q_0 and Q_1 absorption bands: the main features are that antisymmetric scattered intensity drops to zero outside the Q_0–Q_1 region much more rapidly than symmetric scattered intensity, but is much stronger between the Q_0 and Q_1 bands although both peak close to the 0–0 and 0–1 absorption peaks. Notice that (8.4.17) does not require degeneracy in any of the molecular states in order to be nonzero. Mortensen and Hassing (1979) have given a comprehensive general review of interference effects in resonance Raman scattering.

The electronic part of the excited resonant state, $|e_j\rangle$, is doubly degenerate for resonance within the Q_0 and Q_1 bands and belongs to E_{u_a} in the notation of Section 6.3.1. The nearest other excited state, corresponding to the Soret band, is also doubly degenerate, belonging to E_{u_b}. Even though the vibronic coupling is between components of degenerate sets of electronic states, as discussed at the end of Section 8.3 we can still deduce the correct symmetry aspects from the essentially nondegenerate result (8.4.17). Since

$$E_u^2 = [A_{1g}] + [B_{1g}] + [B_{2g}] + \{A_{2g}\},$$

it follows that vibrational coordinates of species A_{1g}, B_{1g}, B_{2g} and A_{2g} can effect vibronic coupling between components of the E_{u_a} and E_{u_b} sets; but only A_{1g}, B_{1g} and B_{2g} can effect coupling between components of the same set, either E_{u_a} or E_{u_b} (the Jahn–Teller effect). So we shall calculate relative values of the transition tensor components generated through vibronic coupling between components of the E_{u_a} and E_{u_b} sets of electronic states by the nontotally symmetric vibrational coordinates of species A_{2g}, B_{1g} and B_{2g}, and use the same results for the relative values of the transition tensors generated through Jahn–Teller coupling within the E_{u_a} set through the B_{1g} and B_{2g} vibrational coordinates. Incidentally, the fact that E_u^2 does not contain E_g or E_u appears to be sufficient to explain the absence of bands originating in E_g and E_u modes of vibration in the resonance Raman spectra of porphyrins.

In the absence of an external magnetic field, we may use the real version (4.4.28) of the Wigner–Eckart theorem for molecular symmetry groups together with Table D.3.2 (real) of Griffith (1962) for the V coefficients in D_4. The excited resonant state $|e_j\rangle$ can be either the X or Y component of the E_{u_a} state and, since

$$\langle EX|(\partial H_e/\partial Q_{A_2})_0|EY\rangle = \frac{1}{\sqrt{2}}\langle E\|(\partial H_e/\partial Q_{A_2})_0\|E\rangle, \qquad (8.4.18a)$$

$$\langle EY|(\partial H_e/\partial Q_{A_2})_0|EX\rangle = -\frac{1}{\sqrt{2}}\langle E\|(\partial H_e/\partial Q_{A_2})_0\|E\rangle, \qquad (8.4.18b)$$

$$\langle EX|(\partial H_e/\partial Q_{A_2})_0|EX\rangle = \langle EY|(\partial H_e/\partial Q_{A_2})_0|EY\rangle = 0, \qquad (8.4.18c)$$

it can couple via Q_{A_2} with the Y or X component, respectively, of the E_{u_b} state. Summing the contributions from the two degenerate X and Y components of the resonant level, we find from (8.4.17)

$$(\alpha_{XY})^a_{e_n v_m e_n v_n} = -\frac{\omega}{\sqrt{2}\hbar\Delta}[(f_{0-0} + ig_{0-0})\langle 1_n|Q_{A_2}|0_j\rangle\langle 0_j|0_n\rangle$$
$$- (f_{0-1} + ig_{0-1})\langle 1_n|1_j\rangle\langle 1_j|Q_{A_2}|0_n\rangle]$$
$$\times |\langle A_1\|\mu\|E\rangle|^2\langle E\|(\partial H_e/\partial Q_{A_2})_0\|E\rangle, \qquad (8.4.19)$$

where $\Delta = W_{E_b} - W_{E_a}$ is the energy separation of the Soret and Q levels.

The symmetric tensor components for scattering in B_1 and B_2 modes can be calculated in a similar fashion. The relative values of the components of the *complex* transition polarizability $(\tilde{\alpha}_{\alpha\beta})_{e_n v_m e_n e_n}$, calculated in a real basis, for B_1, B_2 and A_2 vibrations are summarized below:

(a) 0–0 vibronic resonance.

(i) $|E_{u_a} X\rangle$ electronic resonant state:

$$\begin{pmatrix} 2 & 0 & 0 \\ 0 & 0 & 0 \\ 0 & 0 & 0 \end{pmatrix} \begin{pmatrix} 0 & -1 & 0 \\ -1 & 0 & 0 \\ 0 & 0 & 0 \end{pmatrix} \begin{pmatrix} 0 & -1 & 0 \\ 1 & 0 & 0 \\ 0 & 0 & 0 \end{pmatrix}, \qquad (8.4.20a)$$
$$\quad Q_{B_1} \qquad\qquad Q_{B_2} \qquad\qquad Q_{A_2}$$

(ii) $|E_{u_a}Y\rangle$ electronic resonant state:

$$\underbrace{\begin{pmatrix} 0 & 0 & 0 \\ 0 & -2 & 0 \\ 0 & 0 & 0 \end{pmatrix}}_{Q_{B_1}} \underbrace{\begin{pmatrix} 0 & -1 & 0 \\ -1 & 0 & 0 \\ 0 & 0 & 0 \end{pmatrix}}_{Q_{B_2}} \underbrace{\begin{pmatrix} 0 & 1 & 0 \\ 1 & 0 & 0 \\ 0 & 0 & 0 \end{pmatrix}}_{Q_{A_2}}. \qquad (8.4.20b)$$

(*b*) 0–1 vibronic resonance.

(i) $|E_{u_a}X\rangle$ electronic resonant state:

$$\underbrace{\begin{pmatrix} 2 & 0 & 0 \\ 0 & 0 & 0 \\ 0 & 0 & 0 \end{pmatrix}}_{Q_{B_1}} \underbrace{\begin{pmatrix} 0 & -1 & 0 \\ -1 & 0 & 0 \\ 0 & 0 & 0 \end{pmatrix}}_{Q_{B_2}} \underbrace{\begin{pmatrix} 0 & 1 & 0 \\ -1 & 0 & 0 \\ 0 & 0 & 0 \end{pmatrix}}_{Q_{A_2}}, \qquad (8.4.20c)$$

(ii) $|E_{u_a}Y\rangle$ electronic resonant state:

$$\underbrace{\begin{pmatrix} 0 & 0 & 0 \\ 0 & -2 & 0 \\ 0 & 0 & 0 \end{pmatrix}}_{Q_{B_1}} \underbrace{\begin{pmatrix} 0 & -1 & 0 \\ -1 & 0 & 0 \\ 0 & 0 & 0 \end{pmatrix}}_{Q_{B_2}} \underbrace{\begin{pmatrix} 0 & 1 & 0 \\ -1 & 0 & 0 \\ 0 & 0 & 0 \end{pmatrix}}_{Q_{A_2}}. \qquad (8.4.20d)$$

We have distinguished the contributions from the X and Y components of the excited electronic resonant level because, if the fourfold symmetry axis of the porphyrin is destroyed by substituents, the degeneracy would be lifted and it might be possible to isolate the separate contributions. Notice that all the components are real: this is because the molecule has an even number of electrons and, since no external magnetic field is present, we have taken a real representation for the degenerate wavefunctions.

We shall also develop the porphyrin transition polarizabilities in a complex basis using the complex version (4.4.29) of the Wigner–Eckart theorem for molecular symmetry groups together with Table D.3.2 (complex) of Griffith (1962) for the V coefficients in D_4. This is to facilitate the subsequent calculations in Section 8.5.4 of magnetic resonance Raman optical activity in porphyrins. Since this section emphasizes antisymmetric scattering, the calculation is illustrated for Q_{A_2}. The excited resonant state $|e_j\rangle$ can now be either the $+1$ or -1 component of the E_{u_a} level and, since

$$\langle E \pm 1|(\partial H_e/\partial Q_{A_2})_0|E \pm 1\rangle = \pm \frac{i}{\sqrt{2}}\langle E\|(\partial H_e/\partial Q_{A_2})_0\|E\rangle, \qquad (8.4.21a)$$

$$\langle E1|(\partial H_e/\partial Q_{A_2})_0|E-1\rangle = \langle E-1|(\partial H_e/\partial Q_{A_2})_0|E1\rangle = 0, \qquad (8.4.21b)$$

it can couple via Q_{A_2} with the same component of the E_{u_b} state. In the absence of a magnetic field, we can again sum over the two degenerate components (this time $+1$ and -1) of the resonant level in (8.4.17) and recover (8.4.19), the real transition polarizability component in a real basis. But in the presence of a magnetic field, the degeneracy is lifted and each component transition must be considered separately. It transpires that, in addition to the real transition polarizability (8.4.16a), some components of the imaginary transition polarizability (8.4.16b) are nonzero, although these cancel when summed over the $+1$ and -1 components.

The relative values of the components of the *complex* transition polarizability $(\tilde{\alpha}_{\alpha\beta})_{e_n v_m e_n v_n}$, calculated in a complex basis, for B_1, B_2 and A_2 vibrations are summarized below.

(a) 0–0 vibronic resonance.

(i) $|E_{u_a}1\rangle$ electronic resonant state:

$$
\begin{pmatrix} 1 & -i & 0 \\ -i & -1 & 0 \\ 0 & 0 & 0 \end{pmatrix} \begin{pmatrix} -i & -1 & 0 \\ -1 & i & 0 \\ 0 & 0 & 0 \end{pmatrix} \begin{pmatrix} -i & -1 & 0 \\ 1 & -i & 0 \\ 0 & 0 & 0 \end{pmatrix},
\qquad (8.4.22a)
$$
$$
\quad Q_{B_1} \qquad\qquad Q_{B_2} \qquad\qquad Q_{A_2}
$$

(ii) $|E_{u_a}-1\rangle$ electronic resonant state:

$$
\begin{pmatrix} 1 & i & 0 \\ i & -1 & 0 \\ 0 & 0 & 0 \end{pmatrix} \begin{pmatrix} i & -1 & 0 \\ -1 & -i & 0 \\ 0 & 0 & 0 \end{pmatrix} \begin{pmatrix} i & -1 & 0 \\ 1 & i & 0 \\ 0 & 0 & 0 \end{pmatrix}.
\qquad (8.4.22b)
$$
$$
\quad Q_{B_1} \qquad\qquad Q_{B_2} \qquad\qquad Q_{A_2}
$$

(b) 0–1 vibronic resonance.

(i) $|E_{u_a}1\rangle$ electronic resonant state:

$$
\begin{pmatrix} 1 & i & 0 \\ i & -1 & 0 \\ 0 & 0 & 0 \end{pmatrix} \begin{pmatrix} i & -1 & 0 \\ -1 & -i & 0 \\ 0 & 0 & 0 \end{pmatrix} \begin{pmatrix} i & 1 & 0 \\ -1 & i & 0 \\ 0 & 0 & 0 \end{pmatrix},
\qquad (8.4.22c)
$$
$$
\quad Q_{B_1} \qquad\qquad Q_{B_2} \qquad\qquad Q_{A_2}
$$

(ii) $|E_{u_a}-1\rangle$ electronic resonant state:

$$
\begin{pmatrix} 1 & -i & 0 \\ -i & -1 & 0 \\ 0 & 0 & 0 \end{pmatrix} \begin{pmatrix} -i & -1 & 0 \\ -1 & i & 0 \\ 0 & 0 & 0 \end{pmatrix} \begin{pmatrix} -i & 1 & 0 \\ -1 & -i & 0 \\ 0 & 0 & 0 \end{pmatrix}.
\qquad (8.4.22d)
$$
$$
\quad Q_{B_1} \qquad\qquad Q_{B_2} \qquad\qquad Q_{A_2}
$$

Notice that, when summed over the $+1$ and -1 components of the electronic resonant state, the imaginary tensor components vanish and we recover the same results that are obtained by summing over the X and Y components in (8.4.20).

For completeness, we also give the transition polarizabilities for resonance Raman scattering in porphyrin A_1 modes: although these do not generate antisymmetric scattering (in the absence of a magnetic field), they are required for the magnetic resonance Raman optical activity calculations. Vibronic coupling is no longer necessary, so the **X**-tensors in (8.3.1) make the dominant contribution. Using both real and complex basis sets, the following relative values are found for both the 0–0 and 0–1 resonances.

(i) $|E_{u_a}X\rangle$ electronic resonant state:

$$\begin{pmatrix} 2 & 0 & 0 \\ 0 & 0 & 0 \\ 0 & 0 & 0 \end{pmatrix}_{Q_{A_1}}, \tag{8.4.23a}$$

(ii) $|E_{u_a}Y\rangle$ electronic resonant state:

$$\begin{pmatrix} 0 & 0 & 0 \\ 0 & 2 & 0 \\ 0 & 0 & 0 \end{pmatrix}_{Q_{A_1}}, \tag{8.4.23b}$$

(iii) $|E_{u_a}1\rangle$ electronic resonant state:

$$\begin{pmatrix} 1 & -i & 0 \\ i & 1 & 0 \\ 0 & 0 & 0 \end{pmatrix}_{Q_{A_1}}, \tag{8.4.23c}$$

(iv) $|E_{u_a}-1\rangle$ electronic resonant state:

$$\begin{pmatrix} 1 & i & 0 \\ -i & 1 & 0 \\ 0 & 0 & 0 \end{pmatrix}_{Q_{A_1}}. \tag{8.4.23d}$$

It should be mentioned that the mechanism elaborated here, involving intermanifold vibronic coupling between the E_{u_a} and E_{u_b} sets of electronic states, does not give a completely satisfactory description of all the characteristic features of resonance Raman scattering in metal porphyrins. It appears to be necessary to allow for interference between three distinct couplings: the intermanifold coupling between E_{u_a} and E_{u_b}; the intramanifold coupling within E_{u_a}; and the electronic configuration intraction between E_{u_a} and E_{u_b}. We refer to Zgierski, Shelnutt and Pawlikowski (1979) for further details.

8.5 Magnetic Rayleigh and Raman optical activity

8.5.1 The basic equations

General expressions for the magnetic optical activity observables in Rayleigh scattering were derived in Chapter 3 in terms of magnetically perturbed molecular property tensors. For example, the dimensionless circular intensity difference

$$\Delta = \frac{I^R - I^L}{I^R + I^L} \tag{1.4.1}$$

is given at transparent frequencies by (3.5.43–45). More general circular intensity difference components for resonance scattering can be deduced from the Stokes parameters (3.5.46–48). A more direct calculation can be found by referring to the treatment in Section 7.3.1 and retaining only terms in $\tilde{\alpha}^2$. Thus from (7.3.6) we may write, for scattering at 90°,

$$\Delta_x(90°) = \frac{2\mathrm{Im}(\tilde{\alpha}_{xy}\tilde{\alpha}_{xx}^*)}{(\tilde{\alpha}_{xx}\tilde{\alpha}_{xx}^* + \tilde{\alpha}_{xy}\tilde{\alpha}_{xy}^*)}, \tag{8.5.1a}$$

$$\Delta_z(90°) = \frac{2\mathrm{Im}(\tilde{\alpha}_{zy}\tilde{\alpha}_{zx}^*)}{(\tilde{\alpha}_{zx}\tilde{\alpha}_{zx}^* + \tilde{\alpha}_{zy}\tilde{\alpha}_{zy}^*)}, \tag{8.5.1b}$$

and these expressions can be developed by writing the polarizabilities perturbed to first order in a static magnetic field parallel to the incident light beam and taking a weighted Boltzmann average. The resulting averaged expressions must be used to describe magnetic Rayleigh and Raman optical activity associated with diagonal scattering transitions; but for off-diagonal scattering transitions it is advantageous to work directly from (8.5.1). We shall see that magnetic Rayleigh and Raman optical activity in off-diagonal scattering transitions is particularly interesting because it can probe directly the ground state Zeeman splitting.

Only magnetic optical activity in *resonance* Raman scattering is considered in the examples discussed below; magnetic optical activity has not yet been observed in transparent Raman scattering. Magnetic fields have little direct influence on molecular vibrational states (even if degenerate) and hence on vibrational spectra. The reason that magnetic optical activity is readily observable in vibrational resonance Raman scattering is that the Raman effect is essentially a scattering process mediated by excited *electronic* states, and electronic states can be influenced considerably by a magnetic field, particularly if they are degenerate. The much weaker direct effects of magnetic fields on vibrational states give rise to magnetic vibrational circular dichroism, mentioned in Section 1.5.

8.5.2 *Resonance Rayleigh scattering in atomic sodium*

Resonance Rayleigh scattering in atomic sodium vapour provides a simple introductory example; and we can apply immediately the tensor components calculated explicitly in Section 8.4.2. For the circular intensity differences associated with 90° scattering transitions between pure magnetic quantum states, the tensor components (8.4.9) are substituted directly into (8.5.1). Thus for resonance with each

Table 8.1 *Circular intensity difference components for resonance Rayleigh scattering transitions between pure magnetic quantum states of* Na

	$+\frac{1}{2} \leftarrow +\frac{1}{2}$	$-\frac{1}{2} \leftarrow -\frac{1}{2}$	$-\frac{1}{2} \leftarrow +\frac{1}{2}$	$+\frac{1}{2} \leftarrow -\frac{1}{2}$
$^2P_{\frac{1}{2}}$ resonant term				
$\Delta_x(90°)$	1	-1	0	0
$\Delta_z(90°)$	0	0	1	-1
$^2P_{\frac{3}{2}}$ resonant term				
$\Delta_x(90°)$	$-4/5$	$4/5$	0	0
$\Delta_z(90°)$	0	0	1	-1

of the two components of the sodium yellow doublet, we find the results listed in Table 8.1.

We see that a polarized circular intensity difference $\Delta_x(90°)$ is only generated by diagonal scattering transitions, whereas a depolarized circular intensity difference $\Delta_z(90°)$ is only generated by off-diagonal scattering transitions. Since $\Delta_x(90°)$ is equal and opposite for the two distinct diagonal scattering pathways, which give scattered lines at the central Rayleigh frequency, a nonzero result can only be obtained by using the developed version of (8.5.1) in which polarizabilities perturbed by the magnetic field are used and a weighted Boltzmann average taken: this gives residual effects due to the slightly different populations of the two components of the magnetically split $^2S_{\frac{1}{2}}$ initial term, and to mixing of the resonant states with other excited states. Residual diagonal effects might also be observed by tuning the incident frequency to one or other of the transitions $^2P_{\frac{1}{2},\pm\frac{1}{2}} \leftarrow {}^2S_{\frac{1}{2},\mp\frac{1}{2}}$, say, since these will have slightly different energies in the magnetic field. The off-diagonal scattering transitions are potentially much more interesting: even though $\Delta_z(90°)$ is equal and opposite for the two distinct off-diagonal scattering pathways, it is readily observable because the scattered lines are displaced on either side of the central Rayleigh frequency by the ground state Zeeman splitting.

These magnetic optical activity effects can be understood easily from simple considerations of angular momentum selection rules on atomic transitions. Since the electric vector of a right-circularly polarized light beam rotates in a clockwise sense when viewed towards the source of the beam, a right-circularly polarized photon has an angular momentum projection of $-\hbar$ along its direction of propagation if this is taken to be the magnetic quantization axis. Similarly, a left-circularly polarized photon has a projection of $+\hbar$. Consequently, absorption of a right-circularly polarized photon induces a $\Delta M = -1$ change in the atomic state, and absorption of a left-circularly polarized photon induces a $\Delta M = +1$ change. These conclusions also follow analytically from the Wigner–Eckart theorem (4.4.24). Also, it follows from (4.4.26) that absorption or emission of a photon linearly polarized along z induces a $\Delta M = 0$ change in the atom, and absorption or emission of an x- or y-polarized photon induces a $\Delta M = \pm 1$ change.

Fig. 8.3 illustrates the generation of magnetic optical activity in the off-diagonal Rayleigh scattering process for resonance with the $^2P_{\frac{1}{2}} \leftarrow {}^2S_{\frac{1}{2}}$ transition. Ignoring the hyperfine components, the magnetic field lifts the degeneracy of the $M_J = \pm\frac{1}{2}$ components of the two terms. If the magnetic field lies along the propagation direction of the incident light beam (that is $N \overset{z}{\rightarrow} S$), the right-circularly polarized incident photon connects the $M_J = +\frac{1}{2}$ component of the lower term with the $M_J = -\frac{1}{2}$ component of the upper term, and the z-polarized emitted photon originates in the subsequent emission down to the $M_J = -\frac{1}{2}$ component of the lower term (Fig. 8.3a). Thus if the incident photon has a frequency ω, the scattered photon has a frequency $\omega + \delta$, where δ is the Zeeman splitting of the lower term. Off-diagonal resonance Rayleigh scattering at 90° induced by right-circularly polarized incident light therefore generates a depolarized line displaced in frequency by $+\delta$ from the central Rayleigh line (Fig. 8.3b). Similarly, left-circularly polarized light generates a depolarized line displaced by $-\delta$. The spectrum of the corresponding depolarized circular intensity difference $\Delta_z(90°)$ is shown in Fig. 8.3c. In fact the lower frequency line will be slightly more intense on account of the higher Boltzmann population of the $M_J = -\frac{1}{2}$ state (although since Δ is dimensionless, the corresponding Δ-values for the two lines will be the same).

Notice that, if the two displaced off-diagonal Rayleigh lines were resolved (for a g-value of 2, a field of 1.07 T produces a splitting of 1 cm^{-1}), the ground state g-value could be measured as half the separation of the lines. Thus off-diagonal Rayleigh scattering provides the possibility of 'Rayleigh electron paramagnetic resonance'. The technique also gives immediately the sign of the g-value: if, as is usually the case, g is positive, the Δ-value on the higher frequency line is positive and that on the lower frequency line negative.

Overall, the atom suffers a $\Delta J = 0$, $\Delta M = \pm 1$ change. Such a change can only be induced in absorption through a magnetic dipole interaction, as in conventional electron paramagnetic resonance; but here the change is effected by an antisymmetric scattering tensor operator which, since it transforms as an axial vector, is associated with the same selection rules as a magnetic dipole operator. Another feature is that a transition between atomic spin states is effected by an operator (the antisymmetric scattering tensor) that contains no apparent spin operator, but simply two spatial electric dipole moment operators. However, it must be remembered that the intermediate resonant state is a resolved spin–orbit state in which spin and orbital components are intimately mixed by the spin–orbit coupling operator, thereby providing a scattering pathway connecting different initial and final spin states. We

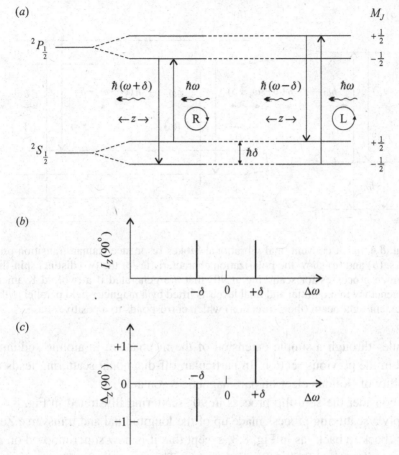

Fig. 8.3 (*a*) Off-diagonal spin-flip transitions for Rayleigh scattering at 90° for resonance with the $^2P_{\frac{1}{2}} \leftarrow {}^2S_{\frac{1}{2}}$ transition in a static magnetic field along z. $\hbar\delta$ is the ground-state Zeeman splitting. (*b*) The depolarized spectrum of the light scattered at 90°. The line displaced by $+\delta$ from the Rayleigh frequency originates in the $-\frac{1}{2} \leftarrow +\frac{1}{2}$ transition and is induced by right-circularly polarized incident light; that at $-\delta$ originates in the $+\frac{1}{2} \leftarrow -\frac{1}{2}$ transition and is induced by left-circularly polarized light. (*c*) The corresponding depolarized circular intensity difference spectrum for a positive magnetic field.

refer to this process, which more generally connects Kramers conjugate states, as a spin-flip scattering transition.

8.5.3 *Vibrational resonance Raman scattering in* $IrCl_6^{2-}$ *and* $CuBr_4^{2-}$: *spin-flip transitions and Raman electron paramagnetic resonance*

Since resonance Raman scattering in totally symmetric modes of vibration does not require vibronic coupling, it can generate magnetic optical activity in odd electron

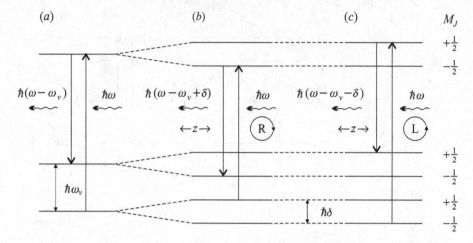

Fig. 8.4 (*a*) A conventional vibrational Stokes resonance Raman transition process. (*b*) and (*c*) show the polarization characteristics of the two distinct spin-flip Raman processes for scattering at 90° that are generated if a twofold Kramers degeneracy in the initial and final levels is lifted by a magnetic field parallel to the incident light beam (the z direction) which corresponds to a positive sense.

molecules through a simple extension of the mechanism in atomic sodium, discussed in the previous section. In particular, off-diagonal scattering leads to the possibility of 'Raman electron paramagnetic resonance'.

We consider the spin-flip process in 90° scattering illustrated in Fig. 8.4. This is simply a scattering process made up of the longitudinal and transverse Zeeman effects 'back to back' as in Fig. 8.3, except that it is now superimposed on a fundamental vibrational Stokes Raman process. The two spin-flip Raman transitions therefore lead to frequency shifts in the depolarized component, equal to the Zeeman splitting δ, on either side of the vibrational Raman frequency ω_v. In fact the frequency shifts will be the average of the ground state Zeeman splitting and that for the molecule in the final excited vibrational state, which is slightly different; but it suits our purpose here to assume that they are the same. Thus right-circularly polarized incident light generates a depolarized line displaced in frequency by $+\delta$ from the vibrational Stokes Raman line, and left-circularly polarized light generates a depolarized line displaced by $-\delta$. Depolarized magnetic circular intensity differences $I_z^R - I_z^L$ of equal magnitude and opposite sign are therefore associated with the two spin-flip transitions, so a couplet is observed.

Good examples of this spin-flip scattering mechanism are provided by the low-spin d^5 iridium(IV) complex $IrCl_6^{2-}$ and the d^9 copper(II) complex $CuBr_4^{2-}$ which have O_h and D_{2d} symmetry, respectively (Barron and Meehan, 1979). The depolarized resonance Raman and magnetic Raman optical activity spectra of dilute solutions of these two complexes are shown in Fig. 8.5. The large couplets which

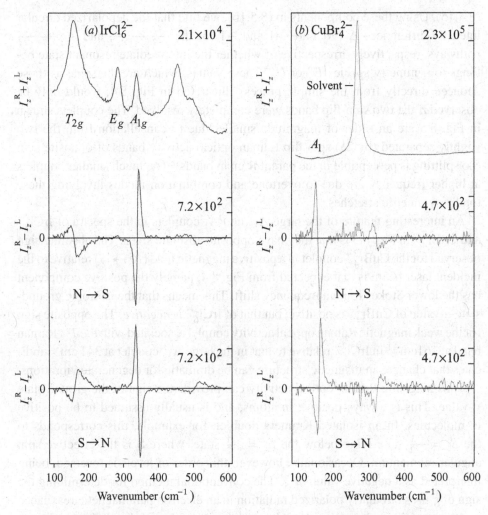

Fig. 8.5 The depolarized resonance Raman ($I_z^R + I_z^L$) and magnetic Raman optical activity ($I_z^R - I_z^L$) spectra for positive ($N \rightarrow S$) and negative ($S \rightarrow N$) magnetic fields, strength 1.2T, of (a) $IrCl_6^{2-}$ in dilute aqueous solution using 488.0 nm excitation, and (b) $CuBr_4^{2-}$ in dilute dichloromethane solution using 514.5 nm excitation. Recorded in the author's laboratory. The absolute intensities are not defined, but the relative Raman and Raman optical activity intensities are significant.

dominate the magnetic Raman optical activity spectra of both complexes are associated with resonance Raman bands assigned to totally symmetric stretching modes of vibration at $341 cm^{-1}$ in $IrCl_6^{2-} (A_{1g})$ and $174\ cm^{-1}$ in $CuBr_4^{2-} (A_1)$, the latter being so weak as to be barely perceptible. The explicit tensor components for the $\mp\frac{1}{2} \leftarrow \pm\frac{1}{2}$ Raman transitions in a totally symmetric vibrational mode of $IrCl_6^{2-}$ for resonance with excited electronic states of symmetry E_u'' and U_u' are given in

(8.4.15). Using these components in (8.5.1b), we find that the depolarized circular intensity differences $\Delta_z(90°)$ are $+1$ and -1 for the $-\frac{1}{2} \leftarrow +\frac{1}{2}$ and $+\frac{1}{2} \leftarrow -\frac{1}{2}$ pathways, respectively, irrespective of whether the intermediate resonant state belongs to symmetry species E_u'' or U_u'. These values, which are the same as those deduced directly from the simple process illustrated in Fig. 8.4, would only be observed if the two spin-flip bands were completely resolved. The couplets shown in Fig. 8.5 are an order of magnitude smaller due to cancellation from the two slightly separated (by 2δ) spin-flip Raman optical activity bands of opposite sign (no splitting is perceptible in the parent Raman bands). The much smaller couplets at higher frequency are due to overtone and combination modes involving these totally symmetric stretches.

An interesting feature of the large A_{1g} and A_1 couplets in the spectra of $IrCl_6^{2-}$ and $CuBr_4^{2-}$ in Fig. 8.5 is that they have opposite absolute signs. The absolute sign observed for the $CuBr_4^{2-}$ couplet in a positive magnetic field ($N \rightarrow S$) relative to the incident laser beam is that expected from Fig. 8.4, namely the positive component has the lower Stokes Raman frequency shift. This means that the isotropic ground-state g-value of $CuBr_4^{2-}$ is positive, but that of $IrCl_6^{2-}$ is *negative*. The opposite sign for the weak magnetic Raman optical activity couplet associated with the T_{2g} Raman band at $161 cm^{-1}$ in $IrCl_6^{2-}$ relative to that in the large A_{1g} couplet at $341 cm^{-1}$ indicates that changes in magnetic structure can be dramatic for degenerate vibrations.

Thus magnetic Raman optical activity can provide the sign of the ground state g-value. This is always positive in atoms, and is usually assumed to be positive in molecules. In an isolated Kramers doublet, for example, this corresponds to the $S_z = -\frac{1}{2}$ state lying below the $S_z = +\frac{1}{2}$ state, where S is the effective spin angular momentum. Occasionally, however, this order of levels is reversed, being interpreted as a negative value of g. The conventional method for determining the sign of g uses circularly polarized radiation in an electron paramagnetic resonance experiment: since absorption of right- or left-circularly polarized photons brings about $\Delta M = -1$ or $+1$ transitions, respectively, observation of which sense of circular polarization is effective in causing magnetic resonance transitions will determine experimentally the sign of g. However, such experiments are rarely performed, and the only example of a negative g-value determined in this way is for the ground state of NpF_6 (Hutchison and Weinstock, 1960). Negative g-values have been discussed in detail by Abragam and Bleaney (1970), who indicate that the isotropic g-value of $IrCl_6^{2-}$ should be negative on theoretical grounds.

8.5.4 Electronic resonance Raman scattering in uranocene

Spin-flip resonance Raman transitions connecting Kramers conjugate states, as described in the previous section, are not a prerequisite for magnetic Raman optical

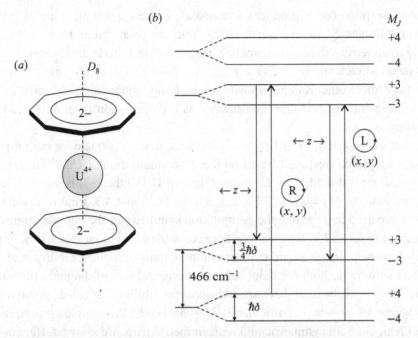

Fig. 8.6 (a) The uranocene molecule. (b) Electronic Raman scattering pathways for the $M_J = \pm 3 \leftarrow M_J = \pm 4$ transitions in uranocene induced by circularly polarized incident photons for z-polarized photons scattered at 90°, in a magnetic field parallel to the incident light beam (the z direction) which corresponds to a positive sense.

activity to function as Raman electron paramagnetic resonance. A number of examples of magnetic Raman optical activity have been observed in low-frequency electronic resonance Raman transitions in both even and odd electron molecules involving general $\Delta M = \pm 1$ transitions between Zeeman-split levels, of which uranocene is especially interesting (Barron and Vrbancich, 1983).

The even electron molecule uranocene, $U(C_8H_8)_2$, is the bis(cyclo-octatetraene) (COT) complex of uranium. This has a structure of D_{8h} symmetry with the central metal ion in the U(IV) oxidation state sandwiched between two COT^{2-} rings (Fig. 8.6a). In a ligand field treatment (Warren, 1977), the 20 π electrons of the two COT^{2-} rings are accommodated in predominantly ligand orbitals, with the two valence electrons of U(IV) occupying a metal $5f$ orbital. Overlap of the highest filled $e_u(\pi)$ orbitals of the COT^{2-} rings with the $l_z = \pm 2\,(f_{xyz},\ f_{z(x^2-y^2)})$ uranium $5f$ orbitals is favourable, and the occupation of the corresponding bonding orbital by the four $e_u\ COT^{2-}$ electrons then accounts for the remarkable stability of uranocene.

Uranocene has four moderately intense visible absorption bands between 600 and 700 nm attributed to charge transfer transitions from the COT^{2-} π orbitals to the U(IV) f orbitals. Laser excitation at wavelengths in the vicinity of these visible absorption bands produces a resonance Raman spectrum containing three bands (Dallinger, Stein and Spiro, 1978). Two are polarized and are assigned to totally symmetric vibrations, namely a ring breathing mode at 754 cm^{-1} and a ring–metal stretch mode at 211 cm^{-1}. The third band at 466 cm^{-1}, assigned to a pure electronic Raman transition involving nonbonding $5f$ orbitals, exhibits anomalous polarization and hence has a contribution from antisymmetric scattering.

Rather than use a full D_{8h} ligand field treatment, it is adequate for our purposes to use a simplified treatment based on the f^2 configuration of U(IV) in an effectively axial crystal field. The $^3H_4$ ground term of U(IV) then splits into five levels corresponding to $M_J = 0, \pm 1, \pm 2, \pm 3, \pm 4$. In Section 4.4.3, time reversal arguments were used to detemine the permutation symmetry of the complex transition polarizability for $\Delta J = 0$ transitions between a manifold of atomic states. For integral J, as here, it was found that the complex transition polarizability is always real and symmetric both for diagonal transitions and for off-diagonal transitions where $M' = -M$, but that there are additional possibilities for off-diagonal transitions where $M' \neq -M$. In particular, if $M + M'$ is odd, the complex polarizability is still real, but both symmetric and antisymmetric parts are allowed. Remembering the selection rules $\Delta M = 0, \pm 1$ on each of the two electric dipole transitions within the transition tensor, we may therefore associate symmetric scattering with $\Delta M_J = 0, \pm 1, \pm 2$ and antisymmetric scattering with $\Delta M_J = \pm 1$.

The antisymmetric scattering observed in the 466 cm^{-1} electronic resonance Raman band of uranocene may therefore be associated with $\Delta M = \pm 1$ transitions. From magnetic susceptibility data, Dallinger, Stein and Spiro (1978) suggested that the ground level is the $M_J = \pm 4$ component of the $^3H_4$ manifold and hence that the 466 cm^{-1} Raman band originates in the $M_J = \pm 3 \leftarrow M_J = \pm 4$ transition. Excitation within the 641.0 nm visible absorption band of a dilute solution of uranocene in tetrahydrofuran generates an enormous depolarized magnetic resonance Raman optical activity couplet in 90° scattering which, in a positive magnetic field, exhibits a positive lower-frequency component and a negative higher-frequency component (Barron and Vrbancich, 1983). This may be understood in terms of the magnetic components within this transition (Fig. 8.6b). Since the 641.0 nm absorption band is x,y-polarized, the transition to the excited resonant state must involve $\Delta M_J = \pm 1$. Assuming the g-values of the $M_J = \pm 4$ and ± 3 levels are the same and positive, the Zeeman splitting of the ± 3 level will be $\frac{3}{4}$ that of the ± 4 level, and it may be seen from Fig. 8.6b that, for a positive magnetic field, the lower-frequency component of the Stokes magnetic Raman optical activity couplet should be positive and the higher frequency component negative. The fact that this is what is observed provides good evidence that the $|M_J|$ value of the ground level is greater by unity than that of the first excited level. However, this conclusion that the $|M_J|$ value of the ground level is greater by unity than that of the first excited level is also consistent with an alternative possibility that the $M_J = \pm 3$ level lies lowest with the 466 cm^{-1} electronic Raman band arising from the $M_J = \pm 2 \leftarrow M_J = \pm 3$ transition, as suggested by Hager *et al.* (2004) in a resonance Raman study of crystalline uranocene. This reassignment was based on additional magnetic susceptibility data together with calculations utilizing a full D_{8h} ligand field treatment.

The resonance Raman spectrum of uranocene also shows a very weak band at 675 cm^{-1} that appears to be anomalously polarized. Excitation within the 641 nm visible absorption band generates a weak magnetic Raman optical activity couplet with the same sign as that of the 466 cm^{-1} band, which enabled the 675 cm^{-1} band to be confidently assigned to a combination of the 466 cm^{-1} electronic Raman transition with the totally symmetric vibrational transition at 211 cm^{-1} (Barron and Vrbancich, 1983).

8.5.5 Resonance Raman scattering in porphyrins

Since neutral porphyrins are even electron molecules with nondegenerate ground states, we must use the magnetically perturbed development of (8.5.1). The machinery for this development is given in Section 3.5.5. The effects in porphyrins originate in electronic degeneracy in the excited resonant state and so only Faraday A-tensors, which are generalizations of the Faraday A-term (6.2.2c), contribute. These follow from the magnetic analogues of the first terms of the perturbed polarizabilities (2.7.6a, b):

$$\alpha_{\alpha\beta,\gamma}^{(m)} = \frac{2}{\hbar} \sum_{j \neq n} \frac{\omega_{jn}^2 + \omega^2}{\hbar \left(\omega_{jn}^2 - \omega^2\right)^2} A_{\alpha\beta,\gamma}, \tag{8.5.2a}$$

$$\alpha_{\alpha\beta,\gamma}^{\prime(m)} = -\frac{2}{\hbar} \sum_{j \neq n} \frac{2\omega\omega_{jn}}{\hbar \left(\omega_{jn}^2 - \omega^2\right)^2} A_{\alpha\beta,\gamma}^{\prime}, \tag{8.5.2b}$$

where

$$A_{\alpha\beta,\gamma} = (m_{j\gamma} - m_{n\gamma}) \, \text{Re} \left(\langle n|\mu_\alpha|j\rangle\langle j|\mu_\beta|n\rangle\right), \tag{8.5.2c}$$

$$A_{\alpha\beta,\gamma}^{\prime} = (m_{j\gamma} - m_{n\gamma}) \, \text{Im} \left(\langle n|\mu_\alpha|j\rangle\langle j|\mu_\beta|n\rangle\right). \tag{8.5.2d}$$

Do not confuse these A-tensors with the electric dipole–electric quadrupole polarizability (2.6.27c, d).

These expressions are strictly valid only at transparent frequencies: the magnetic analogues of (2.7.8), which take account of the finite lifetimes of the excited states, should really be used in the region of an isolated absorption band, but the calculation becomes very complicated. However, we shall persist with the simpler expressions even in the region of an isolated absorption band because in calculating the dimensionless circular intensity difference (1.4.1) the same frequency dependence obtains (apart from a damping factor) since the additional complicating features cancel.

The perturbed polarizabilities (8.5.2) can be written as follows in terms of the corresponding unperturbed polarizabilities:

$$\alpha_{\alpha\beta,\gamma} = \frac{\omega_{jn}^2 + \omega^2}{\hbar\omega_{jn}(\omega_{jn}^2 - \omega^2)} m_{j\gamma}\alpha_{\alpha\beta}, \tag{8.5.3a}$$

$$\alpha'_{\alpha\beta,\gamma} = \frac{2\omega_{jn}}{\hbar(\omega_{jn}^2 - \omega^2)} m_{j\gamma}\alpha'_{\alpha\beta}, \tag{8.5.3b}$$

where we have retained only the contribution of one particular degenerate excited state and have dropped $m_{n\gamma}$ since the ground state is nondegenerate. For Raman scattering, the transition polarizability versions of these expressions are used, and at resonance both $(\alpha_{\alpha\beta,\gamma})_{mn}$ and $(\alpha'_{\alpha\beta,\gamma})_{mn}$ can contain both symmetric and antisymmetric parts.

All the required terms are included in the Stokes parameters written out in Chapter 3. The circular intensity difference components for 90° scattering can be extracted immediately from (3.5.6), (3.5.19) and (3.5.47) by recalling that $I_x \propto S_0 + S_1$ and $I_z \propto S_0 - S_1$.

Consider first magnetic resonance-Raman optical activity in porphyrin A_1 modes of vibration. Vibronic coupling is not required in this instance, the dominant contribution arising from the X-tensors in (8.3.1). The required transition polarizability components in a complex basis are given in (8.4.23c, d). Since

$$\langle E \pm 1|m_z|E \pm 1\rangle = \pm\frac{i}{\sqrt{2}}\langle E\|m\|E\rangle, \tag{8.5.4}$$

and just the Faraday A-tensor contributions (8.5.2) are retained, only $(\alpha_{\alpha\beta})_{mn}^s$ and $(\alpha'_{\alpha\beta,\gamma})_{mn}^a$ are nonzero when summed over the two components $|E1\rangle$ and $|E-1\rangle$ of the excited resonant state. Retaining only these tensors in the Stokes parameters, we find

$$\Delta_x(90°) = -\frac{8}{9}B_z\frac{(\alpha'_{XY,Z})_{mn}^a}{(\alpha_{XX})_{mn}^s} = -\frac{16i}{9\sqrt{2}}\frac{B_z\omega}{\hbar(\omega_{jn}^2 - \omega^2)}\langle E\|m\|E\rangle, \tag{8.5.5a}$$

$$\Delta_z(90°) = -2B_z\frac{(\alpha'_{XY,Z})_{mn}^a}{(\alpha_{XX})_{mn}^s} = -\frac{4i}{\sqrt{2}}\frac{B_z\omega}{\hbar(\omega_{jn}^2 - \omega^2)}\langle E\|m\|E\rangle, \tag{8.5.5b}$$

where we have used the fact that $(\alpha_{XX})_{mn} = (\alpha_{YY})_{mn}$ for an A_1 mode.

For B_1 modes vibronic coupling is required and the corresponding transition polarizability components are given in (8.4.22) for the 0–0 and 0–1 vibronic resonances. Now only $(\alpha_{\alpha\beta})_{mn}^s$ and $(\alpha'_{\alpha\beta,\gamma})_{mn}^s$ survive when summed over $|E1\rangle$ and

$|E-1\rangle$. Since $(\alpha_{XX})_{mn} = -(\alpha_{YY})_{mn}$ for a B_1 mode we find, for the 0–0 resonance,

$$\Delta_x(90°) = -\frac{8}{7}B_z\frac{(\alpha'_{XY,Z})^s_{mn}}{(\alpha_{XX})^s_{mn}} = -\frac{16i}{7\sqrt{2}}\frac{B_z\omega_{jn}}{\hbar(\omega^2_{jn}-\omega^2)}\langle E\|m\|E\rangle, \quad (8.5.6a)$$

$$\Delta_z(90°) = -\frac{2}{3}B_z\frac{(\alpha'_{XY,Z})^s_{mn}}{(\alpha_{XX})^s_{mn}} = -\frac{4i}{3\sqrt{2}}\frac{B_z\omega_{jn}}{\hbar(\omega^2_{jn}-\omega^2)}\langle E\|m\|E\rangle. \quad (8.5.6b)$$

The same magnitudes, but opposite signs, obtain for the 0–1 resonance.

For B_2 modes only $(\alpha_{\alpha\beta})^s_{mn}$ and $(\alpha'_{\alpha\beta,\gamma})^s_{mn}$ survive and since $(\alpha'_{XX})_{mn} = -(\alpha'_{YY})_{mn}$ we find, for the 0–0 resonance,

$$\Delta_x(90°) = \frac{8}{7}B_z\frac{(\alpha'_{XX,Z})^s_{mn}}{(\alpha_{XY})^s_{mn}} = -\frac{16i}{7\sqrt{2}}\frac{B_z\omega_{jn}}{\hbar(\omega^2_{jn}-\omega^2)}\langle E\|m\|E\rangle, \quad (8.5.7a)$$

$$\Delta_z(90°) = \frac{2}{3}B_z\frac{(\alpha'_{XX,Z})^s_{mn}}{(\alpha_{XY})^s_{mn}} = -\frac{4i}{3\sqrt{2}}\frac{B_z\omega_{jn}}{\hbar(\omega^2_{jn}-\omega^2)}\langle E\|m\|E\rangle, \quad (8.5.7b)$$

with the same magnitudes but opposite signs for the 0–1 resonance.

Finally, for A_2 modes only $(\alpha_{\alpha\beta})^a_{mn}$ and $(\alpha'_{\alpha\beta,\gamma})^s_{mn}$ survive and since $(\alpha'_{XX})^s_{mn} = (\alpha'_{YY})^s_{mn}$ we find, for the 0–0 resonance,

$$\Delta_x(90°) = \frac{8}{5}B_z\frac{(\alpha'_{XX,Z})^s_{mn}}{(\alpha_{XY})^a_{mn}} = -\frac{16i}{5\sqrt{2}}\frac{B_z\omega^2_{jn}}{\hbar\omega(\omega^2_{jn}-\omega^2)}\langle E\|m\|E\rangle \quad (8.5.8a)$$

$$\Delta_z(90°) = \frac{2}{5}B_z\frac{(\alpha'_{XX,Z})^s_{mn}}{(\alpha_{XY})^a_{mn}} = -\frac{4i}{5\sqrt{2}}\frac{B_z\omega^2_{jn}}{\hbar\omega(\omega^2_{jn}-\omega^2)}\langle E\|m\|E\rangle \quad (8.5.8b)$$

with the same magnitudes and the *same* signs for the 0–1 resonance.

An important feature that emerges from the calculation, although the explicit expressions have not been written down here, is that the *absolute magnitude* of $I^R_x - I^L_x$ is twice that for $I^R_z - I^L_z$ in all the modes.

The results (8.5.5–8) are summarized in Table 8.2. The predictions have been confirmed from measurements of the magnetic resonance-Raman optical activity spectrum of ferrocytochrome c using a range of excitation wavelengths spanning the visible absorption spectrum (Barron, Meehan and Vrbancich, 1982; Barron and Vrbancich 1985). Fig. 8.7 shows the visible absorption spectrum originating in transitions within the porphyrin ring of the haem group in this d^6 Fe(II) metalloprotein, together with the associated magnetic circular dichroism spectrum. The magnetic circular dichroism spectrum emphasizes the 0–1 vibronic structure since a Faraday A-curve like that for the 0–0 band is associated with each vibronic peak. Excitation within the 0–0 vibronic absorption band at 550 nm (the Q_0 band in the notation of Section 6.3.1) was observed to produce a magnetic resonance-Raman optical

Table 8.2 *Magnetic circular intensity difference components for vibrations of metal porphyrins. Each entry is to be multiplied by*

$$-(4i/\sqrt{2})B_z\omega\langle E||m||E\rangle/\hbar(\omega_{jn}^2 - \omega^2)$$

Normal mode	Polarizability components	0–0 resonance		0–1 resonance	
		Δ_x	Δ_z	Δ_x	Δ_z
A_1	$(\alpha_{\alpha\beta})^s_{mn}, (\alpha'_{\alpha\beta,\gamma})^a_{mn}$	4/9	1	4/9	1
A_2	$(\alpha_{\alpha\beta})^a_{mn}, (\alpha'_{\alpha\beta,\gamma})^s_{mn}$	4/5	1/5	4/5	1/5
B_1	$(\alpha_{\alpha\beta})^s_{mn}, (\alpha'_{\alpha\beta,\gamma})^s_{mn}$	4/7	1/3	−4/7	−1/3
B_2	$(\alpha_{\alpha\beta})^s_{mn}, (\alpha'_{\alpha\beta,\gamma})^s_{mn}$	4/7	1/3	−4/7	−1/3

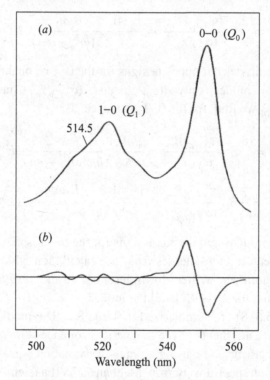

Fig. 8.7 The visible absorption (*a*) and magnetic circular dichroism (*b*) spectra of ferrocytochrome c in aqueous solution (arbitrary units). Adapted from Sutherland and Klein (1972).

activity spectrum with bands all of the same sign, as expected from the corresponding entries in Table 8.2, the signs being opposite for excitation on opposite sides of the 0–0 band centre due to the factors $(\omega_{jn}^2 - \omega^2)$ in the denominators. Since the magnetic circular intensity differences (8.5.5–8) depend on the same reduced

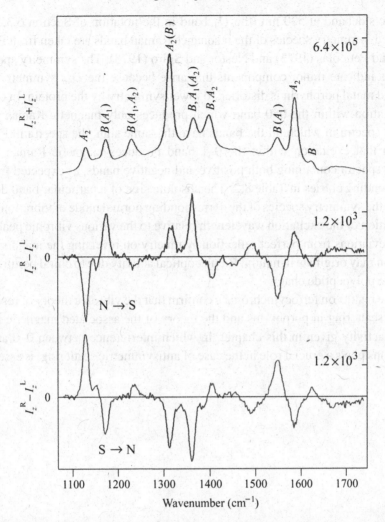

Fig. 8.8 The depolarized resonance Raman ($I_z^R + I_z^L$) and magnetic Raman optical activity ($I_z^R - I_z^L$) spectra for positive ($N \rightarrow S$) and negative ($S \rightarrow N$) magnetic fields, strength 1.2T, of ferrocytochrome c in aqueous solution using 514.5 nm excitation. Recorded in the author's laboratory. The absolute intensities are not defined, but the relative Raman and Raman optical activity intensities are significant.

magnetic dipole matrix element i$\langle E||m||E\rangle$ as the magnetic circular dichroism A/D value (6.3.4), the signs of the magnetic Raman optical activity bands may be directly related to those of the corresponding magnetic circular dichroism spectrum. Some of these characteristics are similar to those of natural resonance Raman optical activity in chiral molecules (Nafie, 1996; Vargek *et al.*, 1998). Fig. 8.8 shows the depolarized resonance Raman and magnetic resonance-Raman optical activity spectra of ferrocytochrome c for excitation at 514.5 nm, which falls within the 0–1 vibronic sideband at 520 nm (the Q_1 band in the notation of Section 6.3.1). The indicated symmetry species of the resonance Raman bands are taken from Pézolet, Nafie and Peticolas (1973) and Nestor and Spiro (1973). The symmetry species in brackets indicate minor components that arise because the D_{4h} symmetry of the unbound metal porphyrin is distorted to lower symmetry by the protein. In contrast to excitation within the 0–0 band which produces only magnetic Raman optical activity spectra in which all the bands have the same sign, the spectra in Fig. 8.8. confirm that excitation within the 0–1 band produces magnetic Raman optical activity spectra containing both positive and negative bands, as expected from the corresponding entries in Table 8.2. The absolute sign of a particular band depends on both the symmetry species of the corresponding normal mode of vibration and on the position of the excitation wavelength relative to the various vibronic peaks. The small deviations from perfect reflection symmetry on reversing the magnetic field direction may originate in natural Raman optical activity due to chiral perturbations from the polypeptide chains.

These results on ferrocytochrome c confirm that the vibronic theory of resonance Raman scattering in porphyrins and the theory of the associated magnetic Raman optical activity given in this chapter, in which interference between 0–0 and 0–1 transitions plays a crucial role in the case of antisymmetric scattering, is essentially correct.

References

Abragam, A. and Bleaney, B. (1970). *Electron Paramagnetic Resonance of Transition Ions*. Oxford: Clarendon Press.

Albrecht, A. C. (1961). *J. Chem. Phys.* **34**, 1476.

Alagna, L., Prosperi, T., Turchini, S., Goulon, J., Rogalev, A., Goulon-Ginet, C., Natoli, C. R., Peacock, R. D. and Stewart, B. (1998). *Phys. Rev. Lett.* **80**, 4799.

Altmann, S. L. (1992). *Icons and Symmetries*. Oxford: Clarendon Press.

Amos, R. D. (1982). *Chem. Phys. Lett.* **87**, 23.

—(1987). *Adv. Chem. Phys.* **67**, 99.

Anderson, P. W. (1972). *Science* **177**, 393.

—(1983). *Basic Notions of Condensed Matter Physics*. Menlo Park, California: Benjamin/Cummings.

Andrews, D. L. (1980). *J. Chem. Phys.* **72**, 4141.

—(1994). *Faraday Discuss.* **99**, 375.

Andrews, D. L. and Thirunamachandran, T. (1977a). *Proc. Roy. Soc.* **A358**, 297.

—(1977b). *Proc. Roy. Soc.* **A358**, 311.

—(1978). *J. Chem. Phys.* **68**, 2941.

Applequist, J. (1973). *J. Chem. Phys.* **58**, 4251.

Arago, D. F. J. (1811). *Mém. de l'Inst.* **12**, part 1, 93.

Arnaut, L. R. (1997). *J. Electromagnetic Waves and Applications* **11**, 1459.

Atkins, P. W. and Barron, L. D. (1969). *Mol. Phys.* **16**, 453.

Atkins, P. W. and Woolley, R. G. (1970). *Proc. Roy. Soc.* **A314**, 251.

Autschbach, J., Patchkovskii, S., Ziegler, T., van Gisbergen, S. J. A. and Baerends, E. J. (2002). *J. Chem. Phys.* **117**, 581.

Avalos, M., Babiano, R., Cintas, P., Jiménez, J. L., Palacios, J. C. and Barron, L. D. (1998). *Chem. Rev.* **98**, 2391.

Ballhausen, C. J. (1962). *Introduction to Ligand Field Theory*. New York: McGraw-Hill.

—(1979). *Molecular Electronic Structure of Transition Metal Complexes*. New York: McGraw-Hill.

Baranova, N. B., Bogdanov, Yu. V. and Zel'dovich, B. Ya. (1977). *Opt. Commun.* **22**, 243.

Baranova, N. B. and Zel'dovich, B. Ya (1978). *J. Raman Spectrosc.* **7**, 118.

—(1979a). *Mol. Phys.* **38**, 1085.

—(1979b). *Sov. Phys. Usp.* **22**, 143.

Barnett, C. J., Drake, A. F. and Mason, S. F. (1980). *J.C.S. Chem. Commun.*, 43.

Barron, L. D. (1971). *Mol. Phys.* **21**, 241.

—(1972). *Nature* **238**, 17.

—(1975*a*). *Nature* **257**, 372.

—(1975*b*). *J.C.S. Faraday Transactions II* **71**, 293.

—(1975*c*). *Nature* **255**, 458.

—(1976). *Mol. Phys.* **31**, 129.

—(1978). *Adv. Infrared Raman Spectrosc.* **4**, 271.

—(1979*a*). *J. Am. Chem. Soc.* **101**, 269.

—(1979*b*). In *Optical Activity and Chiral Discrimination*, ed. S. F. Mason, p. 219. Dordrecht: Reidel.

—(1981*a*). *Chem. Phys. Lett.* **79**, 392.

—(1981*b*). *Mol. Phys.* **43**, 1395.

—(1986*a*). *Chem. Phys. Lett.* **123**, 423.

—(1986*b*). *J. Am. Chem. Soc.* **108**, 5539.

—(1987). *Chem. Phys. Lett.* **135**, 1.

—(1993). *Physica* **B190**, 307.

—(1994). *Chem. Phys. Lett.* **221**, 311.

—(1996). *Chem. Eur. J.* **2**, 743.

Barron, L. D., Blanch, E. W., McColl, I. H., Syme, C. D., Hecht, L. and Nielsen, K. (2003). *Spectroscopy* **17**, 101.

Barron, L. D., Bogaard, M. P. and Buckingham, A. D. (1973). *J. Am. Chem. Soc.* **95**, 603.

Barron, L. D. and Buckingham, A. D. (1971). *Mol. Phys.* **20**, 1111.

—(1972). *Mol. Phys.* **23**, 145.

—(1973). *J. Phys.* **B6**, 1295.

—(1974). *J. Am. Chem. Soc.* **96**, 4769.

—(1979). *J. Am. Chem. Soc.* **101**, 1979.

—(2001). *Accs. Chem. Res.* **34**, 781.

Barron, L. D. and Clark, B. P. (1982). *Mol. Phys.* **46**, 839.

Barron, L. D. and Escribano, J. R. (1985). *Chem. Phys.* **98**, 437.

Barron, L. D., Escribano, J. R. and Torrance, J. F. (1986). *Mol. Phys.* **57**, 653.

Barron, L. D., Gargaro, A. R. and Wen, Z. Q. (1990). *J.C.S. Chem. Commun.* 1034.

Barron, L. D. and Gray, C. G. (1973). *J. Phys.* **A6**, 59.

Barron, L. D., Hecht, L., Gargaro, A. R. and Hug, W. (1990). *J. Raman Spectrosc.* **21**, 375.

Barron, L. D., Hecht, L. and Polavarapu, P. L. (1992) *Spectrochim. Acta* **48A**, 1193.

Barron, L. D., Hecht, L., Blanch, E. W. and Bell, A. F. (2000). *Prog. Biophys. Mol. Biol.* **73**, 1.

Barron, L. D. and Johnston, C. J. (1985). *J. Raman Spectrosc.* **16**, 208.

—(1987). *Mol. Phys.* **62**, 987.

Barron, L. D. and Meehan, C. (1979). *Chem. Phys. Lett.* **66**, 444.

Barron, L. D., Meehan, C. and Vrbancich, J. (1982). *J. Raman Spectrosc.* **12**, 251.

Barron, L. D. and Nørby Svendsen, E. (1981). *Adv. Infrared Raman Spectrosc.* **8**, 322.

Barron, L. D., Torrance, J. F. and Vrbancich, J. (1982). *J. Raman Spectrosc.* **13**, 171.

Barron, L. D. and Vrbancich, J. (1983). *J. Raman Spectrosc.* **14**, 118.

—(1984). *Mol. Phys.* **51**, 715.

—(1985). *Adv. Infrared Raman Spectrosc.* **12**, 215.

Berestetskii, V. B., Lifshitz, E. M. and Pitaevskii, L. P. (1982). *Quantum Electrodynamics*. Oxford: Pergamon Press.

Berova, N., Nakanishi, K. and Woody, R. W., eds. (2000). *Circular Dichroism. Principles and Applications*, 2nd edition. New York: Wiley-VCH.

Bhagavantam, S. (1942). *Scattering of Light and the Raman Effect*. New York: Chemical Publishing Company.

Bhagavantam, S. and Venkateswaran, S. (1930). *Nature* **125**, 237.

Bijvoet, J. M., Peerdeman, A. F. and van Bommel, A. J. (1951). *Nature* **168**, 271.

Biot, J. B. (1812). *Mém. de l'Inst.* **13**, part 1, 218.

—(1815). *Bull. Soc. Philomath.*, 190.

—(1818). *Ann. Chim.* **9**, 382; **10**, 63.

Biot, J. B. and Melloni, M. (1836). *Compt. Rend.* **2**, 194.

Binney, J. J., Dowrick, N. J., Fisher, A. J. and Newman, M. E. J. (1992). *The Theory of Critical Phenomena*. Oxford: Clarendon Press.

Birss, R. R. (1966). *Symmetry and Magnetism*, 2nd edition. Amsterdam: North-Holland.

Birss, R. R. and Shrubsall, R. G. (1967). *Phil. Mag.* **15**, 687.

Blaizot, J.-P. and Ripka, G. (1986). *Quantum Theory of Finite Systems*. Cambridge, MA: MIT Press.

Blanch, E. W., Hecht, L., Syme, C. D., Volpetti, V., Lomonossoff, G. P., Nielsen, K. and Barron, L. D. (2002). *J. Gen. Virol.* **83**, 2593.

Bloembergen, N. (1996). *Nonlinear Optics*, 4th edition. Singapore: World Scientific.

Blum, K. and Thompson, D. G. (1997). *Adv. At. Mol. Opt. Phys.* **38**, 39.

Bohm, D. (1951). *Quantum Theory*. Englewood Cliffs, NJ: Prentice-Hall.

Born, M. (1915). *Phys. Zeit.* **16**, 251.

—(1933). *Optik*. Berlin: Springer.

Born, M. and Huang, K. (1954). *Dynamical Theory of Crystal Lattices*. Oxford: Clarendon Press.

Born, M. and Jordan, P. (1930). *Elementare Quantenmechanik*. Berlin: Springer.

Born, M. and Oppenheimer, J. R. (1927). *Ann. Phys.* **84**, 457.

Born, M. and Wolf, E. (1980). *Principles of Optics*, 6th edition. Oxford: Pergamon Press.

Bose, P. K., Barron, L. D. and Polavarapu, P. L. (1989). *Chem. Phys. Lett.* **155**, 423.

Bosnich, B., Moskovits, M. and Ozin, G. A. (1972). *J. Am. Chem. Soc.* **94**, 4750.

Bouchiat, M. A. and Bouchiat, C. (1974). *J. de Physique* **35**, 899.

—(1997). *Rep. Prog. Phys.* **60**, 1351.

Bour, P., Tam, C. N. and Keiderling, T. A. (1996). *J. Phys. Chem.* **100**, 2062.

Bourne, D. E. and Kendall, P. C. (1977). *Vector Analysis and Cartesian Tensors*, 2nd edition. Sunbury-on-Thames, Middlesex: Nelson.

Boyle, L. L. and Mathews, P. S. C. (1971). *Int. J. Quant. Chem.* **5**, 381.

Boys, S. F. (1934). *Proc. Roy. Soc.* **A144**, 655 and 675.

Branco, G. C., Lavoura, L. and Silva, J. P. (1999). *CP Violation*. Oxford: Clarendon Press.

Bridge, N. J. and Buckingham, A. D. (1966). *Proc. Roy. Soc.* **A295**, 334.

Brown, W. F., Shtrikman, S. and Treves, D. (1963). *J. Appl. Phys.* **34**, 1233.

Brożek, Z., Stadnicka, K., Lingard, R. J. and Glazer, A. M. (1995). *J. Appl. Cryst.* **28**, 78.

Buckingham, A. D. (1958). *J. Chem. Phys.* **30**, 1580.

—(1962). *Proc. Roy. Soc.* **A267**, 271.

—(1967). *Adv. Chem. Phys.* **12**, 107.

—(1970). In *Physical Chemistry*, ed. D. Henderson, Vol. 4, p. 349. New York: Academic Press.

—(1972). In *MTP International Review of Science, Physical Chemistry, Series One, Spectroscopy*, ed. D. A. Ramsay, Vol. 3, p. 73. London: Butterworths.

—(1978). In *Intermolecular Interactions: From Biopolymers to Diatomics*, ed. B. Pullman, p. 1. New York: Wiley.

—(1988). In *Proceedings of the Eleventh International Conference on Raman Spectroscopy*, eds. R. J. H. Clark and D. A. Long, p. 3. Chichester: Wiley.

—(1994). *Faraday Discuss.* **99**, 1.

Buckingham, A. D. and Dunn, M. B. (1971). *J. Chem. Soc. A*, 1988.

Buckingham, A. D. and Fischer, P. (2000). *Phys. Rev.* **A61**, art. no. 035801.

Buckingham, A. D., Fowler, P. W. and Galwas, P. A. (1987). *Chem. Phys.* **112**, 1.

Buckingham, A. D., Graham, C. and Raab, R. E. (1971). *Chem. Phys. Lett.* **8**, 622.

Buckingham, A. D. and Joslin, C. G. (1981). *Chem. Phys. Lett.* **80**, 615.

Buckingham, A. D. and Longuet-Higgins, H. C. (1968). *Mol. Phys.* **14**, 63.

Buckingham, A. D. and Pople, J. A. (1955). *Proc. Phys. Soc.* **A68**, 905.

Buckingham, A. D. and Raab, R. E. (1975). *Proc. Roy. Soc.* **A345**, 365.

Buckingham, A. D. and Shatwell, R. A. (1978). *Chem. Phys.* **35**, 353.

—(1980). *Phys. Rev. Lett.* **45**, 21.

Buckingham, A. D. and Stephens, P. J. (1966). *Annu. Rev. Phys. Chem.* **17**, 399.

Buckingham, A. D. and Stiles, P. J. (1972). *Mol. Phys.* **24**, 99.

—(1974). *Accs. Chem. Res.* **7**, 258.

Bungay, A. R., Svirko, Yu. P. and Zheludev, N. I. (1993). *Phys. Rev. Lett.* **70**, 3039.

Caldwell, D. J. and Eyring, H. (1971). *The Theory of Optical Activity.* New York: John Wiley & Sons.

Califano, S. (1976). *Vibrational States.* New York: John Wiley & Sons.

Charney, E. (1979). *The Molecular Basis of Optical Activity.* New York: John Wiley & Sons.

Child, M. S. (1962). *Phil. Trans. Roy. Soc.* **A255**, 31.

Child, M. S. and Longuet-Higgins, H. C. (1961). *Phil. Trans. Roy. Soc.* **A254**, 259.

Chisholm, C. D. H. (1976). *Group Theoretical Techniques in Quantum Chemistry.* London: Academic Press.

Christenson, J. H., Cronin, J. W., Fitch, V. L. and Turlay, R. (1964). *Phys. Rev. Lett.* **13**, 138.

Clark, R., Jeyes, S. R., McCaffery, A. J. and Shatwell, R. A. (1974). *J. Am. Chem. Soc.* **96**, 5586.

Condon, E. U. (1937). *Rev. Mod. Phys.* **9**, 432.

Condon, E. U., Altar, W. and Eyring, H. (1937). *J. Chem. Phys.* **5**, 753.

Condon, E. U. and Shortley, G. H. (1935). *The Theory of Atomic Spectra.* Cambridge: Cambridge University Press.

Coriani, S., Pecul, M., Rizzo, A., Jørgensen, P. and Jaszunski, M. (2002). *J. Chem. Phys.* **117**, 6417.

Costa de Beauregard, O. (1987). *Time, the Physical Magnitude.* Dordrecht: Reidel.

Costante, J., Hecht, L., Polavarapu, P. L., Collet, A. and Barron, L. D. (1997). *Ang. Chem. Int. Ed. Engl.* **36**, 885.

Cotton, A. (1895). *Compt. Rend.* **120**, 989 and 1044.

Cotton, A. and Mouton, H. (1907). *Annls. Chim. Phys.* **11**, 145 and 289.

Craig, D. P. and Thirunamachandran, T. (1984). *Molecular Quantum Electrodynamics.* London: Academic Press. Reprinted 1998. New York: Dover.

Creighton, T. E. (1993). *Proteins,* 2nd edition. New York: W. H. Freeman.

Cronin, J. W. (1981). *Rev. Mod. Phys.* **53**, 373.

Crum Brown, A. C. (1890). *Proc. Roy. Soc. Edin.* **17**, 181.

Curie, P. (1894). *J. Phys. Paris* (3) **3**, 393.

—(1908). *Oeuvres de Pierre Curie.* Paris: Société Francaise de Physique.

Dallinger, R. F., Stein, P. and Spiro, T. G. (1978). *J. Am. Chem. Soc.* **100**, 7865.

Davydov, A. S. (1976). *Quantum Mechanics,* 2nd edition. Oxford: Pergamon Press.

de Figueiredo, I. M. B. and Raab, R. E. (1980). *Proc. Roy. Soc.* **A369**, 501.

—(1981). *Proc. Roy. Soc.* **A375**, 425.

de Gennes, P. G. and Prost, J. (1993).*The Physics of Liquid Crystals,* 2nd edition. Oxford: Clarendon Press.

Dekkers, H. P. J. M. (2000). In *Circular Dichroism: Principles and Applications*, 2nd edition, eds. N. Berova, K. Nakanishi and R. W. Woody, p. 185. New York: Wiley-VCH.

de Lange, O. L. and Raab, R. E. (2004). *Mol. Phys.* **102**, 125.

de Mallemann, R. (1925). *Compt. Rend.* **181**, 371.

Deutsche, C. W. and Moscowitz, A. (1968). *J. Chem. Phys.* **49**, 3257.

—(1970). *J. Chem. Phys.* **53**, 2630.

Devlin, F. J. and Stephens, P. J. (1999). *J. Am. Chem. Soc.* **121**, 7413.

Diem, M. (1993). *Introduction to Modern Vibrational Spectroscopy*. New York: John Wiley & Sons.

Diem, M., Fry, J. L. and Burow, D. F. (1973). *J. Am. Chem. Soc.* **95**, 253.

Dirac, P. A. M. (1958). *The Principles of Quantum Mechanics*, 4th edition. Oxford: Clarendon Press.

Djerassi, C. (1960). *Optical Rotatory Dispersion*. New York: McGraw-Hill.

Dobosh, P. A. (1972). *Phys. Rev.* **A5**, 2376.

—(1974). *Mol. Phys.* **27**, 689.

Drude, P. (1902). *The Theory of Optics*. London, Longmans, Green. Reprinted 1959. New York: Dover.

Dudley, R. J., Mason, S. F. and Peacock, R. D. (1972). *J. C. S. Chem. Commun.*, 1084.

Eades, J., ed. (1993). *Antihydrogen: Proceedings of the Antihydrogen Workshop*. Basel: J. C. Baltzer. Reprinted from *Hyperfine Interactions*, Vol. 76, No. 1–4.

Edmonds, A. R. (1960). *Angular Momentum in Quantum Mechanics*, 2nd edition. Princeton, NJ: Princeton University Press.

Eliel, E. L. and Wilen, S. H. (1994). *Stereochemistry of Organic Compounds*. New York: John Wiley & Sons.

Englman, R. (1972). *The Jahn-Teller Effect in Molecules and Crystals*. New York: John Wiley & Sons.

Escribano, J. R. and Barron, L. D. (1988). *Mol. Phys.* **65**, 327.

Evans, M. W. (1993). *Adv. Chem. Phys.* **85**, 97.

Eyring, H., Walter, J. and Kimball, G. E. (1944). *Quantum Chemistry*. New York: John Wiley & Sons.

Fabelinskii, I. L. (1968). *Molecular Scattering of Light*. New York: Plenum Press.

Fano, U. (1957). *Rev. Mod. Phys.* **29**, 74.

Fano, U. and Racah, G. (1959). *Irreducible Tensorial Sets*. New York: Academic Press.

Faraday, M. (1846). *Phil. Mag.* **28**, 294; *Phil. Trans. Roy. Soc.* **136**, 1.

Fasman, G. D., ed. (1996). *Circular Dichroism and the Conformational Analysis of Biomolecules*. New York: Plenum Press.

Faulkner, T. R., Marcott, C., Moscowitz, A. and Overend, J. (1977). *J. Am. Chem. Soc.* **99**, 8160.

Fieschi, R. (1957). *Physica* **24**, 972.

Fitts, D. D. and Kirkwood, J. G. (1955). *J. Am. Chem. Soc.* **77**, 4940.

Fortson, E. N. and Wilets, L. (1980). *Adv. At. Mol. Phys.* **16**, 319.

Fredericq, E. and Houssier, C. (1973). *Electric Dichroism and Electric Birefringence*. Oxford: Clarendon Press.

Fresnel, A. (1824). *Bull. Soc. Philomath.*, 147.

—(1825). *Ann. Chim.* **28**, 147.

Fumi, F. G. (1952). *Nuovo Cim.* **9**, 739.

Fujimoto, E., Yoshimizu, N., Maeda, S., Iyoda, M. and Nakagawa, M. (1980). *J. Raman Spectrosc.* **9**, 14.

Gans, R. (1923). *Z. Phys.* **17**, 353.

Gibson, W. M. and Pollard, B. R. (1976). *Symmetry Principles in Elementary Particle Physics.* Cambridge: Cambridge University Press.

Giesel, F. (1910). *Phys. Zeit.* **11**, 192.

Gottfried, K. and Weisskopf, V. F. (1984). *Concepts of Particle Physics*, Vol. 1. Oxford: Clarendon Press.

Goulon, J., Rogalev, A., Wilhelm, F., Jaouen, N., Goulon-Ginet, C. and Brouder, C. (2003). *J. Phys.: Condens. Matter* **15**, S633.

Gouterman, M. (1961). *J. Mol. Spectrosc.* **6**, 138.

Graham, C. (1980). *Proc. Roy. Soc.* **A369**, 517.

Graham, E. B. and Raab, R. E. (1983). *Proc. Roy. Soc.* **A390**, 73.

—(1984). *Mol. Phys.* **52**, 1241.

Griffith, J. S. (1961). *The Theory of Transition Metal Ions.* Cambridge: Cambridge University Press.

—(1962). *The Irreducible Tensor Method for Molecular Symmetry Groups.* Englewood Cliffs, NJ: Prentice-Hall.

Grimme, S., Furche, F. and Ahlrichs, R. (2002). *Chem. Phys. Lett.* **361**, 321.

Gunning, M. J. and Raab, R. E. (1997). *J. Opt. Soc. Am.* **B14**, 1692.

Gutowsky, H. S. (1951). *J. Chem. Phys.* **19**, 438.

Guye, P. A. (1890). *Compt. Rend.* **110**, 714.

Hager, J. S., Zahradnís, J., Pagni, R. H. and Compton, R. N. (2004). *J. Chem. Phys.* **120**, 2708.

Haidinger, W. (1847). *Ann. Phys.* **70**, 531.

Halperin, B. I., March-Russell, J. and Wilczek, F. (1989). *Phys. Rev.* **B40**, 8726.

Hamaguchi, H. (1977). *J. Chem. Phys.* **66**, 5757.

—(1985). *Adv. Infrared Raman Spectrosc.* **12**, 273.

Hamaguchi, H., Buckingham, A. D. and Kakimoto, M. (1980). *Opt. Lett.* **5**, 114.

Hamaguchi, H., Harada, I. and Shimanouchi, T. (1975). *Chem. Phys. Lett.* **32**, 103.

Hamaguchi, H. and Shimanouchi, T. (1976). *Chem. Phys. Lett.* **38**, 370.

Hamermesh, M. (1962). *Group Theory.* Reading, MA: Addison-Wesley.

Harnung, S. E. (1973). *Mol. Phys.* **26**, 473.

Harris, R. A. (1966). *J. Chem. Phys.* **43**, 959.

—(1980). In *Quantum Dynamics of Molecules*, ed. R. G. Woolley, p. 357. New York: Plenum Press.

—(2001). *J. Chem. Phys.* **115**, 10577.

—(2002). *Chem. Phys. Lett.* **365**, 343.

Harris, R. A. and Stodolsky, L. (1978). *Phys. Lett.* **78B**, 313.

—(1981). *J. Chem. Phys.* **74**, 2145.

Hassing, S. and Nørby Svendsen, E. (2004). *J. Raman Spectrosc.* **35**, 87.

Hecht, L. and Barron, L. D. (1989). *Spectrochim. Acta* **45A**, 671.

—(1990). *Appl. Spectrosc.* **44**, 483.

—(1993a). *Ber. Bunsenges. Phys. Chem.* **97**, 1453.

—(1993b). *Mol. Phys.* **79**, 887.

—(1993c). *Mol. Phys.* **80**, 601.

—(1994). *Chem. Phys. Lett.* **225**, 525.

—(1996). *Mol. Phys.* **89**, 61.

Hecht, L., Barron, L. D. and Hug, W. (1989). *Chem. Phys. Lett.* **158**, 341.

Hecht, L. and Nafie, L. A. (1990). *Chem. Phys. Lett.* **174**, 575.

Hediger, H. J. and Günthard, Hs.H. (1954). *Helv. Chim. Acta* **37**, 1125.

Hegstrom, R. A., Chamberlain, J. P., Seto, K. and Watson, R. G. (1988). *Am. J. Phys.* **56**, 1086.

Hegstrom, R. A., Rein, D. W. and Sandars, P. G. H. (1980). *J. Chem. Phys.* **73**, 2329.

Heine, V. (1960). *Group Theory in Quantum Mechanics.* Oxford: Pergamon Press.

Heisenberg, W. (1966). *Introduction to the Unified Field Theory of Elementary Particles.* New York: John Wiley & Sons.

Henning, G. N., McCaffery, A. J., Schatz, P. N. and Stephens, P. J. (1968). *J. Chem. Phys.* **48**, 5656.

Herschel, J. F. W. (1822). *Trans. Camb. Phil. Soc.* **1**, 43.

Herzberg, G. and Teller, E. (1933). *Z. Phys. Chem.* **B21**, 410.

Hobden, M. V. (1967). *Nature* **216**, 678.

Hicks, J. M., Petralli-Mallow, T. and Byers, J. D. (1994). *Faraday Discuss.* **99**, 341.

Hoffman, K. R., Jia, W. and Yen, W. M. (1990). *Opt. Lett.* **15**, 332.

Höhn, E. G. and Weigang, O. E. (1968). *J. Chem. Phys.* **48**, 1127.

Hollister, J. H., Apperson, G. R., Lewis, L. L., Emmons, T. P., Vold, T. G. and Fortson, E. N. (1981). *Phys. Rev. Lett.* **46**, 643.

Holzwarth, G., Hsu, E. C., Mosher, H. S., Faulkner, T. R. and Moscowitz, A. (1974). *J. Am. Chem. Soc.* **96**, 251.

Hornreich, R. M. and Shtrikman, S. (1967). *Phys. Rev.* **161**, 506.

—(1968). *Phys. Rev.* **171**, 1065..

Hsu, E. C. and Holzwarth, G. (1973). *J. Chem. Phys.* **59**, 4678.

Hug, W. (2003). *Appl. Spectrosc.* **57**, 1.

Hug, W., Kint, S., Bailey, G. F. and Scherer, J. R. (1975). *J. Am. Chem. Soc.* **97**, 5589.

Hug, W. (2001). *Chem. Phys.* **264**, 53.

Hug, W. (2002). In *Handbook of Vibrational Spectroscopy*, eds. J. M. Chalmers and P. R. Griffiths, p. 745. Chichester: John Wiley & Sons.

Hug, W., Zuber, G., de Meijere, A., Khlebnikov, A. F. and Hansen, H.-J. (2001). *Helv. Chim. Acta* **84**, 1.

Hund, W. (1927). *Z. Phys.* **43**, 805.

Hutchison, C. A. and Weinstock, B. (1960). *J. Chem. Phys.* **32**, 56.

Jeffreys, H. (1931). *Cartesian Tensors.* Cambridge: Cambridge University Press.

Jeffreys, H. and Jeffreys, B. S. (1950). *Methods of Mathematical Physics.* Cambridge: Cambridge University Press.

Jenkins, F. A. and White, H. E. (1976). *Fundamentals of Optics*, 4th edition. New York: McGraw-Hill.

Jones, R. C. (1941). *J. Opt. Soc. Am.* **31**, 488.

—(1948). *J. Opt. Soc. Am.* **38**, 671.

Jones, R. V. (1976). *Proc. Roy. Soc.* **A349**, 423.

Joshua, S. J. (1991). *Symmetry Principles and Magnetic Symmetry in Solid State Physics.* Bristol: IOP Publishing.

Judd, B. R. (1975). *Angular Momentum Theory for Diatomic Molecules.* New York: Academic Press.

Judd, B. R. and Runciman, W. A. (1976). *Proc. Roy. Soc.* **A352**, 91.

Jungwirth, P., Skála, L. and Zahradník, R. (1989). *Chem. Phys. Lett.* **161**, 502.

Kaempffer, F. A. (1965). *Concepts in Quantum Mechanics.* New York: Academic Press.

Kaminsky, W. (2000). *Rep. Prog. Phys.* **63**, 1575.

Kastler, A. (1930). *Compt. Rend.* **191**, 565.

Katzin, L. I. (1964). *J. Phys. Chem.* **68**, 2367.

Kauzmann, W. (1957). *Quantum Chemistry.* New York: Academic Press.

Keiderling, T. A. (1981). *J. Chem. Phys.* **75**, 3639.

—(1986). *Nature* **322**, 851.

Keiderling, T. A. and Stephens, P. J. (1979). *J. Am. Chem. Soc.* **101**, 1396

Lord Kelvin (1904). *Baltimore Lectures.* London: C. J. Clay & Sons.

Kerker, M. (1969). *The Scattering of Light.* New York: Academic Press.

Kerr, J. (1875). *Phil. Mag.* **50**, 337.

—(1877). *Phil. Mag.* **3**, 321.

Khriplovich, I. B. (1991). *Parity Nonconservation in Atomic Phenomena.* Philadelphia: Gordon and Breach.

Kielich, S. (1961). *Acta Phys. Polon.* **20**, 433.

—(1968/69). *Bulletin de la Société des Amis des Sciences et des Lettres de Poznan'* **B21**, 47.

King, G. W. (1964). *Spectroscopy and Molecular Structure.* New York: Holt, Rinehart & Winston.

King, R. B. (1991). In *New Developments in Molecular Chirality*, ed. P. G. Mezey, p. 131. Dordrecht: Kluwer Academic Publishers.

Kirkwood, J. G. (1937). *J. Chem. Phys.* **5**, 479.

Kleindienst, P. and Wagnière, G. H. (1998). *Chem. Phys. Lett.* **288**, 89.

Kliger, D. S., Lewis, J. W. and Randall, C. E. (1990). *Polarized Light in Optics and Spectroscopy.* Boston: Academic Press.

Klingbiel, R. T. and Eyring, H. (1970). *J. Phys. Chem.* **74**, 4543.

Kobayashi, J and Uesu, Y. (1983). *J. Appl. Crystallogr.* **16**, 204.

Kondru, R. K., Wipf, P. and Beratan, D. N. (1998). *Science* **282**, 2247.

Koningstein, J. A. and Mortensen, O. S. (1968). *Nature* **217**, 445.

Koslowski, A., Sreerama, N. and Woody, R. W. (2000). In *Circular Dichroism. Principles and Applications*, 2nd edition, eds. N. Berova, K. Nakanishi and R. W. Woody, p. 55. New York: Wiley-VCH.

Kraulis, P. J. (1991). *J. Appl. Cryst.* **24**, 946.

Krimm, S. and Bandekar, J. (1986). *Adv. Protein Chem.* **38**, 181.

Krishnan, R. S. (1938). *Proc. Indian Acad. Sci.* **A7**, 91.

Kruchek, M. P. (1973). *Opt. Spectrosc.* **34**, 340.

Kuball, H. G. and Singer, D. (1969). *Z. Electrochem.* **73**, 403.

Kuhn, W. (1930). *Trans. Faraday Soc.* **26**, 293.

Landsberg, G. and Mandelstam, L. (1928). *Naturwiss.* **16**, 557.

Landau, L. D. and Lifshitz, E. M. (1960). *Electrodynamics of Continuous Media.* Oxford: Pergamon Press.

—(1975). *The Classical Theory of Fields*, 4th edition. Oxford: Pergamon Press.

—(1977). *Quantum Mechanics*, 3rd edition. Oxford: Pergamon Press.

Lee, T. D. and Yang, C. N. (1956). *Phys. Rev.* **104**, 254.

Lifshitz, E. M. and Pitaevskii, L. P. (1980). *Statistical Physics*, part 1. Oxford: Pergamon Press.

Lightner, D. A. (2000). In *Circular Dichroism. Principles and Applications*, 2nd edition, eds. N. Berova, K. Nakanishi and R. W. Woody, p. 261. New York: Wiley-VCH.

Lightner, D. A., Hefelfinger, D. T., Powers, T. W., Frank, G. W. and Trueblood, K. N. (1972). *J. Amer. Chem. Soc.* **94**, 3492.

Lin, T., Chen, Z., Usha, R., Stauffacher, C. V., Dai, J.-B., Schmidt, T. and Johnson, J. E. (1999). *Virology* **265**, 20.

Lindell, I. V., Sihvola, A. H., Tretyakov, S. A. and Viitanen (1994). *Electromagnetic Waves in Chiral and Bi-Isotropic Media.* Boston: Artech House.

Linder, R. E., Morrill, K., Dixon, J. S., Barth, G., Bunnenberg, E., Djerassi, C., Seamans, L. and Moscowitz, A. (1977). *J. Am. Chem. Soc.* **99**, 727.

Lister, D. G., MacDonald, J. N. and Owen, N. L. (1978). *Internal Rotation and Inversion.* London: Academic Press.

Liu, F.-C. (1991). *J. Phys. Chem.* **95**, 7180.

Lockwood, D. J., Hoffman, K. R. and Yen, W. M. (2002). *J. Luminesc*, **100**, 147.

Long, D. A. (2002). *The Raman Effect*. Chichester: John Wiley & Sons.

Longuet-Higgins, H. C. (1961). *Adv. Spectrosc.* **2**, 429.

Loudon, R. (1983). *The Quantum Theory of Light*, 2nd edition. Oxford: Clarendon Press.

Lowry, T. M. (1935). *Optical Rotatory Power.* London: Longmans, Green. Reprinted 1964. New York: Dover.

Lowry, T. M. and Snow, C. P. (1930). *Proc. Roy. Soc.* **A127**, 271.

MacDermott, A. J. (2002). In *Chirality in Natural and Applied Science*, ed. W. J. Lough and I. W. Wainer, p. 23. Oxford: Blackwell Science.

Maestre, M. F., Bustamante, C., Hayes, T. L., Subirana, J. A. and Tinoco, I., Jr. (1982). *Nature* **298**, 773.

Maker, P. D., Terhune, R. W. and Savage, C. M. (1964). *Phys. Rev. Lett.* **12**, 507.

Mason, S. F. (1973). In *Optical Rotatory Dispersion and Circular Dichroism*, eds. F. Ciardelli and P. Salvadori, p. 196. London: Heyden & Son.

—(1979). *Accs. Chem. Res.* **12**, 55.

—(1982). *Molecular Optical Activity and the Chiral Discriminations*. Cambridge: Cambridge University Press.

McCaffery, A. J. and Mason, S. F. (1963). *Mol. Phys.* **6**, 359.

McCaffery, A. J., Mason, S. F. and Ballard, R. E. (1965). *J. Chem. Soc.*, 2883.

McClain, W. M. (1971). *J. Chem. Phys.* **55**, 2789.

McColl, I. H., Blanch, E. W., Gill, A. C., Rhie, A. G. O., Ritchie, M. A., Hecht, L., Nielsen, K. and Barron, L. D. (2003). *J. Am. Chem. Soc.* **125**, 10019.

McHugh, A. J., Gouterman, M. and Weiss, C. (1972). *Theor. Chim. Acta* **24**, 346.

Mead, C. A. (1974). *Topics Curr. Chem.* **49**, 1.

Mead, C. A. and Moscowitz, A. (1967). *Int. J. Quant. Chem.* **1**, 243.

Michel, L. (1980). *Rev. Mod. Phys.* **52**, 617.

Michl, J. and Thulstrup, E. W. (1986). *Spectroscopy with Polarized Light*. Deerfield Beach, FL: VCH Publishers.

Milne, E. A. (1948). *Vectorial Mechanics*. London: Methuen.

Mislow, K. (1999). *Topics Stereochem.* **22**, 1.

Moffit, W., Woodward, R. B., Moscowitz, A., Klyne, W. and Djerassi, C. (1961). *J. Am. Chem. Soc.* **83**, 4013.

Mortensen, O. S. and Hassing, S. (1979). *Adv. Infrared Raman Spectrosc.* **6**, 1.

Moscowitz, A. (1962). *Adv. Chem. Phys.* **4**, 67.

Mueller, H. (1948). *J. Opt. Soc. Am.* **38**, 671.

Nafie, L. A. (1983). *J. Chem. Phys.* **79**, 4950.

—(1996). *Chem. Phys.* **205**, 309.

Nafie, L. A. and Che, D. (1994). *Adv. Chem. Phys.* **85** (Part 3), 105.

Nafie, L. A. and Freedman, T. B. (1989). *Chem. Phys. Lett.* **154**, 260.

—(2000). In *Circular Dichroism: Principles and Applications*, 2nd edition, eds. N. Berova, K. Nakanishi and R. W. Woody, p. 97. New York: Wiley-VCH.

Nafie, L. A., Keiderling, T. A. and Stephens, P. J. (1976). *J. Am. Chem. Soc.* **98**, 2715.

Nafie, L. A., Polavarapu, P. L. and Diem, M. (1980). *J. Chem. Phys.* **73**, 3530.

Nestor, J. and Spiro, T. G. (1973). *J. Raman Spectrosc.* **1**, 539.

Neumann, F. E. (1885). *Vorlesungen über die Theorie Elastizität der festen Körper und des Lichtäthers*. Leipzig: Teubner.

Newman, M. S. and Lednicer, D. (1956). *J. Am. Chem. Soc.* **78**, 4765.

Newton, R. G. (1966). *Scattering Theory of Waves and Particles*. New York: McGraw-Hill.

Nye, J. F. (1985). *Physical Properties of Crystals*, 2nd edition. Oxford: Clarendon Press.

Okun, L. B. (1985). *Particle Physics: The Quest for the Substance of Substance*. Chur: Harwood Academic Publishers.

Oseen, C. W. (1915). *Ann. Phys.* **48**, 1.

Özkan, I. and Goodman, L. (1979). *Chem. Revs.* **79**, 275.

Papakostas, A., Potts, A., Bagnall, D. M., Prosvirnin, S. L., Coles, H. J. and Zheludev, N. I. (2003). *Phys. Rev. Lett.* **90**, art. no. 107404.

Partington, J. R. (1953). *An Advanced Treatise on Physical Chemistry*, vol. 4. London: Longmans, Green.

Pasteur, L. (1848). *Compt. Rend.* **26**, 535.

Peacock, R. D. and Stewart, B. (2001). *J. Phys. Chem.* **105**, 351.

Penney, C. M. (1969). *J. Opt. Soc. Am.* **59**, 34.

Perrin, F. (1942). *J. Chem. Phys.* **10**, 415.

Pèzolet, M., Nafie, L. A. and Peticolas, W. L. (1973). *J. Raman Spectrosc.* **1**, 455.

Pfeifer, P. (1980). Chiral Molecules- a Superselection Rule Induced by the Radiation Field. Doctoral Thesis, Swiss Federal Institute of Technology, Zürich. (Diss. ETH No. 6551).

Piepho, S. B. and Schatz, P. N. (1983). *Group Theory in Spectroscopy with Applications to Magnetic Circular Dichroism*. New York: John Wiley & Sons.

Placzek, G. (1934). In *Handbuch der Radiologie*, ed. E. Marx, vol. 6, part 2, p. 205. Leipzig: Akademische Verlagsgesellschaft. English translation UCRL Trans. 526 (L) from the US Dept. of Commerce clearing house for Federal Scientific and Technical Information.

Polavarapu, P. L. (1987). *J. Chem. Phys.* **86**, 1136.

—(1990). *J. Phys. Chem.* **94**, 8106.

—(1997). *Mol. Phys.* **91**, 551.

—(1998). *Vibrational Spectra: Principles and Applications with Emphasis on Optical Activity*. Amsterdam: Elsevier.

—(2002a). *Ang. Chem. Int. Ed. Engl.* **41**, 4544.

—(2002b). *Chirality* **14**, 768.

Pomeau, Y. (1973). *J. Chem. Phys.* **58**, 293.

Portigal, D. L. and Burstein, E. (1971). *J. Phys. Chem. Solids* **32**, 603.

Post, E. J. (1962). *Formal Structure of Electromagnetics*. Amsterdam: Elsevier. Reprinted 1997. New York: Dover.

Prasad, P. L. and Nafie, L. A. (1979). *J. Chem. Phys.* **70**, 5582.

Quack, M. (1989). *Ang. Chem. Int. Ed. Engl.* **28**, 571.

—(2002). *Ang. Chem. Int. Ed. Engl.* **41**, 4618.

Raab, R. E. (1975). *Mol. Phys.* **29**, 1323.

Raab, R. E. and Cloete, J. H. (1994). *J. Electromagnetic Waves and Applications* **8**, 1073.

Raab, R. E. and de Lange, O. L. (2003). *Mol. Phys.* **101**, 3467.

Raab, R. E. and Sihvola, A. H. (1997). *J. Phys.* **A30**, 1335.

Ramachandran, G. N. and Ramaseshan, S. (1961). In *Handbuch der Physik*, ed. S. Flügge, **25**, (1) 1.

Raman, C. V. and Krishnan, K. S. (1928). *Nature* **121**, 501.

Lord Rayleigh (then the Hon. J. W. Strutt) (1871). *Phil. Mag.* **41**, 107.

Lord Rayleigh (1900). *Phil. Mag.* **49**, 324.

Rein, D. W. (1974). *J. Mol. Evol.* **4**, 15.

Richardson, F. S. (1979). *Chem. Revs.* **79**, 17.

Richardson, F. S. and Riehl, J. P. (1977). *Chem. Revs.* **77**, 773.

Richardson, F. S. and Metcalf, D. H. (2000). In *Circular Dichroism: Principles and Applications*, 2nd edition, eds. N. Berova, K. Nakanishi and R. W. Woody, p. 217. New York: Wiley-VCH.

Rikken, G. L. J. A., Fölling, J. and Wyder, P. (2001). *Phys. Rev. Lett.* **87**, art. no. 236602

Rikken, G. L. J. A. and Raupach, E. (1997). *Nature* **390**, 493.

—(1998). *Phys. Rev.* **E58**, 5081.

—(2000). *Nature* **405**, 932.

Rikken, G. L. J. A., Strohm, C. and Wyder, P. (2002). *Phys. Rev. Lett.* **89**, art. no. 133005.

Rinard, P. M. and Calvert, J. W. (1971). *Am. J. Phys.* **39**, 753.

Rizzo, A. and Coriani, S. (2003). *J. Chem. Phys.* **119**, 11064.

Rodger, A. and Norden, B. (1997). *Circular Dichroism and Linear Dichroism.* Oxford: Oxford University Press.

Rosenfeld, L. (1928). *Z. Phys.* **52**, 161.

—(1951). *Theory of Electrons.* Amsterdam: North-Holland. Reprinted 1965. New York: Dover.

Ross, H. J., Sherbourne, B. S. and Stedman, G. E. (1989). *J. Phys.* **B22**, 459.

Rotenberg, M., Bivens, R., Metropolis, N. and Wooten, J. K. (1959). *The 3-j and 6-j Symbols.* Cambridge, MA: Technology Press, MIT.

Roth, T. and Rikken, G. L. J. A. (2000). *Phys. Rev. Lett.* **85**, 4478.

—(2002). *Phys. Rev. Lett.* **88**, art. no. 063001.

Ruch, E. (1972). *Accs. Chem. Res.* **5**, 49.

Ruch, E. and Schönhofer, A. (1970). *Theor. Chim. Acta* **19**, 225.

Ruch, E., Schönhofer, A. and Ugi, I. (1967). *Theor. Chim. Acta* **7**, 420.

Sachs, R. G. (1987). *The Physics of Time Reversal.* Chicago: University of Chicago Press.

Salzman, W. R. (1977). *J. Chem. Phys.* **67**, 291.

Sandars, P. G. H. (1968). *J. Phys.* **B1**, 499.

—(2001). *Contemp. Phys.* **42**, 97.

Saxe, J. D., Faulkner, T. R. and Richardson, F. S. (1979). *J. Appl. Phys.* **50**, 8204.

Schatz, P. N., McCaffery, A. J., Suëtaka, W., Henning, G. N., Ritchie, A. B. and Stephens, P. J. (1966). *J. Chem. Phys.* **45**, 722.

Schellman, J. A. (1968). *Accs. Chem. Res.* **1**, 144.

—(1973). *J. Chem. Phys.* **58**, 2882.

Schrader, B. and Korte, E. H. (1972). *Ang. Chem. Int. Ed. Eng.* **11**, 226.

Schütz, G., Wagner, W., Wilhelm, W., Kienle, P., Zeller, R., Frahm, R. and Materlik, G. (1987). *Phys. Rev. Lett.* **58**, 737.

Schwanecke, A. S., Krasavin, A., Bagnall, D. M., Potts, A., Zayats, A. V. and Zheludev, N. I. (2003). *Phys. Rev. Lett.* **91**, art. no. 247404.

Seamans, L. and Moscowitz, A. (1972). *J. Chem. Phys.* **56**, 1099.

Seamans, L., Moscowitz, A., Barth, G., Bunnenberg, E. and Djerassi, C. (1972). *J. Am. Chem. Soc.* **94**, 6464.

Seamans, L., Moscowitz, A., Linder, R. E., Morrill, K., Dixon, J. S., Barth, G., Bunnenberg, E. and Djerassi, C. (1977). *J. Am. Chem. Soc.* **99**, 724.

Sellmeier, W. (1872). *Ann. Phys.* **147**, 386.

Serber, R. (1932). *Phys. Rev.* **41**, 489.

Shubnikov, A. V. and Koptsik, V. A. (1974). *Symmetry in Science and Art.* New York: Plenum Press.

Silver, B. L. (1976). *Irreducible Tensor Methods.* New York: Academic Press.

Silverman, M. P., Badoz, J. and Briat, B. (1992). *Opt. Lett.* **17**, 886.

Spiro, T. G. and Stein, P. (1978). *Indian J. Pure and Appl. Phys.* **16**, 213.

Spiro, T. G. and Strekas, T. C. (1972). *Proc. Natl. Acad. Sci. USA* **69**, 2622.

Stedman, G. E. and Butler, P. H. (1980). *J. Phys.* **A13**, 3125.

Stein, P., Brown, J. M. and Spiro, T. G. (1977). *Chem. Phys.* **25**, 237.

Stephens, P. J. (1964). *Theoretical Studies of Magneto-Optical Phenomena.* Doctoral Thesis, University of Oxford.

—(1970). *J. Chem. Phys.* **52**, 3489.

—(1985). *J. Phys. Chem.* **89**, 748.

Stephens, P. J. and Devlin, F. J. (2000). *Chirality* **12**, 172.

Stephens, P. J., Devlin, F. J., Cheeseman, J. R., Frisch, M. J. and Rosini, C. (2002). *Org. Lett.* **4**, 4595.

Stephens, P. J., Suëtaka, W. and Schatz, P. N. (1966). *J. Chem. Phys.* **44**, 4592.

Stewart, B., Peacock, R. D., Alagna, L., Prosperi, T., Turchini, S., Goulon, J., Rogalev, A. and Goulon-Ginet, C. (1999). *J. Am. Chem. Soc* **121**, 10233.

Stokes, G. G. (1852). *Trans. Camb. Phil. Soc.* **9**, 399.

Stone, A. J. (1975). *Mol. Phys.* **29**, 1461.

—(1977). *Mol. Phys.* **33**, 293.

Sullivan, R., Pyda, M., Pak, J., Wunderlich, B., Thompson, J. R., Pagni, R., Pan, H., Barnes, C., Schwerdtfeger, P. and Compton, R. N. (2003). *J. Phys. Chem.* **A107**, 6674.

Sutherland, J. G. and Klein, M. P. (1972). *J. Chem. Phys.* **57**, 76.

Sverdlov, L. M., Kovner, M. A., and Krainov, E. P. (1974). *Vibrational Spectra of Polyatomic Molecules.* Jerusalem: Israel Program for Scientific Translations.

Svirko, Yu. P. and Zheludev, N. I. (1994). *Faraday Discuss.* **99**, 359.

—(1998). *Polarization of Light in Nonlinear Optics.* Chichester: John Wiley & Sons.

Tam, A. C. and Au, C. K. (1976). *Opt. Commun.* **19**, 265.

Tellegen, B. D. H. (1948). *Philips Res. Repts.* **3**, 81.

Temple, G. (1960). *Cartesian Tensors.* London: Methuen.

Theron, I. P. and Cloete, J. H. (1996). *J. Electromagnetic Waves and Applications* **10**, 539.

Tinoco, I. (1957). *J. Am. Chem. Soc.* **79**, 4248.

—(1962). *Adv. Chem. Phys.* **4**, 113.

Tinoco, I. and Williams, A. L. (1984). *Ann. Rev. Phys. Chem.* **35**, 329.

Townes, C. H. and Schawlow, A. L. (1955). *Microwave Spectroscopy.* New York: McGraw-Hill. Reprinted 1975. New York: Dover.

Turner, D. H., Tinoco, I. and Maestre, M. (1974). *J. Am. Chem. Soc.* **96**, 4340.

Tyndall, J. (1869). *Phil. Mag.* **37**, 384; **38**; 156.

Ulbricht, T. L. V. (1959). *Quart. Rev. Chem. Soc.* **13**, 48.

Vager, Z. (1997). *Chem. Phys. Lett.* **273**, 407.

Vallet, M., Ghosh, R., Le Floch, A., Ruchon, T., Bretenaker, F. and Thépot, J.-Y. (2001). *Phys. Rev. Lett.* **87**, art. no. 183003.

Van Bladel, J. (1991). *IEEE Antennas and Propagation Magazine* **33**, 69.

Van de Hulst, H. C. (1957). *Light Scattering by Small Particles.* New York: John Wiley & Sons. Reprinted 1981. New York: Dover.

Van Vleck, J. H. (1932). *The Theory of Electric and Magnetic Susceptibilities.* Oxford: Oxford University Press.

Vargek, M., Freedman, T. B., Lee, E. and Nafie, L. A. (1998). *Chem. Phys. Lett.* **287**, 359.

Velluz, L., Legrand, M. and Grosjean, M. (1965). *Optical Circular Dichroism.* Weinheim: Verlag Chemie, and New York: Academic Press.

Verdet, E. (1854). *Compt. Rend.* **39**, 548.

Wagnière, G. H. and Meier, A. (1982). *Chem. Phys. Lett.* **93**, 78.

Walz, J., Fendel, P., Herrmann, M., König, M., Pahl, A., Pittner, H., Schatz, B. and Hänsch, T. W. (2003). *J. Phys.* **B36**, 649.

Warren, K. D. (1977). *Structure and Bonding* **33**, 97.

Weigang, O. E. (1965). *J. Chem. Phys.* **43**, 3609.

Weiglhofer, W. S. and Lakhtakia, A. (1998). *AEU Int. J. Electronics and Communications* **52**, 276.

Weinberg, S. (1995). *The Quantum Theory of Fields*, Vol. 1. Cambridge: Cambridge University Press.

—(1996). *The Quantum Theory of Fields*, Vol. 2. Cambridge: Cambridge University Press.

Weisskopf, V. and Wigner, E. (1930). *Z. Phys.* **63**, 54; **65**, 18.

West, C. D. (1954). *J. Chem. Phys.* **22**, 749.

Wesendrup, R., Laerdahl, J. K., Compton, R. N. and Schwerdtfeger, P. (2003). *J. Phys. Chem.* **A107**, 6668.

Whittet, D. C. B. (1992). *Dust in the Galactic Environment*. Bristol: Institute of Physics Publishing.

Wigner, E. P. (1927). *Z. Phys.* **43**, 624.

—(1959). *Group Theory*. New York: Academic Press.

—(1965). *Scientific American* **213**, No. 6, 28.

Williams, R. (1968). *Phys. Rev. Lett.* **21**, 342.

Wilson, E. B., Decius, J. C. and Cross, P. C. (1955). *Molecular Vibrations*. New York: McGraw-Hill. Reprinted 1980. Dover, New York.

Wood, C. S., Bennett, S. C., Cho, D., Masterton, B. P., Roberts, J. L., Tanner, C. E. and Wieman, C. E. (1997). *Science* **275**, 1759.

Woolley, R. G. (1975a). *Adv. Chem. Phys.* **33**, 153.

—(1975b). *Adv. Phys.* **25**, 27.

—(1981). *Chem. Phys. Lett.* **79**, 395.

—(1982). *Structure and Bonding* **52**, 1.

Wu, C. S., Ambler, E., Hayward, R. W., Hoppes, D. D. and Hudson, R. P. (1957). *Phys. Rev.* **105**, 1413.

Wyss, H. R. and Günthard, Hs.H. (1966). *Helv. Chim. Acta* **49**, 660.

Zeeman, P. (1896). *Phil. Mag.* **43**, 226.

Zgierski, M. Z., Shelnutt, J. A. and Pawlikowski, M. (1979). *Chem. Phys. Lett.* **68**, 262.

Zocher, H. and Török, C. (1953). *Proc. Natl. Acad. Sci. USA* **39**, 681.

Zuber, G. and Hug, W. (2004). *J. Phys. Chem.* **A108**, 2108.

Index